T0214065

Lecture Notes in Computer Science　12012

Founding Editors

Gerhard Goos
Karlsruhe Institute of Technology, Karlsruhe, Germany
Juris Hartmanis
Cornell University, Ithaca, NY, USA

Editorial Board Members

Elisa Bertino
Purdue University, West Lafayette, IN, USA
Wen Gao
Peking University, Beijing, China
Bernhard Steffen
TU Dortmund University, Dortmund, Germany
Gerhard Woeginger
RWTH Aachen, Aachen, Germany
Moti Yung
Columbia University, New York, NY, USA

More information about this series at http://www.springer.com/series/7409

Andreas Herzig · Juha Kontinen (Eds.)

Foundations of Information and Knowledge Systems

11th International Symposium, FoIKS 2020
Dortmund, Germany, February 17–21, 2020
Proceedings

 Springer

Editors
Andreas Herzig ⓘ
CNRS, University of Toulouse
Toulouse, France

Juha Kontinen ⓘ
University of Helsinki
Helsinki, Finland

ISSN 0302-9743   ISSN 1611-3349   (electronic)
Lecture Notes in Computer Science
ISBN 978-3-030-39950-4   ISBN 978-3-030-39951-1   (eBook)
https://doi.org/10.1007/978-3-030-39951-1

LNCS Sublibrary: SL3 – Information Systems and Applications, incl. Internet/Web, and HCI

© Springer Nature Switzerland AG 2020
This work is subject to copyright. All rights are reserved by the Publisher, whether the whole or part of the material is concerned, specifically the rights of translation, reprinting, reuse of illustrations, recitation, broadcasting, reproduction on microfilms or in any other physical way, and transmission or information storage and retrieval, electronic adaptation, computer software, or by similar or dissimilar methodology now known or hereafter developed.
The use of general descriptive names, registered names, trademarks, service marks, etc. in this publication does not imply, even in the absence of a specific statement, that such names are exempt from the relevant protective laws and regulations and therefore free for general use.
The publisher, the authors and the editors are safe to assume that the advice and information in this book are believed to be true and accurate at the date of publication. Neither the publisher nor the authors or the editors give a warranty, expressed or implied, with respect to the material contained herein or for any errors or omissions that may have been made. The publisher remains neutral with regard to jurisdictional claims in published maps and institutional affiliations.

This Springer imprint is published by the registered company Springer Nature Switzerland AG
The registered company address is: Gewerbestrasse 11, 6330 Cham, Switzerland

Preface

These proceedings contain the papers selected for presentation at the 11th International Symposium on Foundations of Information and Knowledge Systems (FoIKS 2020). The symposium was held during February 17–21, 2020, at the TU Dortmund University, Germany.

The FoIKS symposia provide a biennial forum for presenting and discussing theoretical and applied research on information and knowledge systems. The goal is to bring together researchers with an interest in this subject, share research experiences, promote collaboration, and identify new issues and directions for future research. Previous FoIKS meetings were organized in Schloss Salzau (Germany, 2002), Vienna (Austria, 2004), Budapest (Hungary, 2006), Pisa (Italy, 2008), Sofia (Bulgaria, 2010), Kiel (Germany, 2012), Bordeaux (France, 2014), Linz (Austria, 2016), and Budapest (Hungary, 2018).

The call for papers solicited original contributions dealing with any foundational aspect of information and knowledge systems, including submissions that apply ideas, theories, or methods from specific disciplines to information and knowledge systems. Examples of such disciplines are discrete mathematics, logic and algebra, model theory, information theory, complexity theory, algorithmics and computation, statistics, and optimization.

The FoIKS symposia are traditionally a forum for intense discussions where speakers are given sufficient time to present their ideas and results within the larger context of their research; furthermore, participants are asked to prepare a first response to another contribution in order to initiate discussion.

FoIKS 2020 received 33 paper submissions, which were evaluated by the Program Committee on the basis of their significance, novelty, technical soundness, and appropriateness for the FoIKS audience. Each paper was subjected to three reviews (only two in one case). In the end, 19 papers were selected for presentation at the symposium and publication in the proceedings.

We were delighted to have four outstanding keynote speakers. The abstracts of their talks were included in this volume:

- Jan Van den Bussche: "First-Order Logic of Information Flows: Expressibility, Primitivity, Input-Output Properties, and Limited Access Patterns"
- Bernhard Nebel: "Implicit Coordination of Mobile Agents"
- Klaus-Dieter Schewe: "Computation on Structures: Behavioural Theory, Logic and Complexity"
- Anni-Yasmin Turhan: "Description Logics for Typicality and Nonmonotonic Reasoning"

We would like to thank all the people who contributed to make FoIKS 2020 a success. In particular, we thank the invited speakers for their inspirational talks, the authors for providing their high-quality submissions, revising and presenting their work, and all the attendees for contributing to the symposium discussion. We thank the Program Committee and the external reviewers for their prompt, careful reviewing and discussion of the submissions on a quite tight schedule.

We extend our thanks to the Local Organizing Committee chaired by Gabriele Kern-Isberner and Christoph Beierle. We gratefully acknowledge the support of FoIKS 2020 by the TU Dortmund University, the FernUniversität in Hagen (University of Hagen), and the European Association for Theoretical Computer Science (EATCS). Finally, we are grateful to EasyChair for allowing us to use their excellent conference management system.

February 2020   Andreas Herzig
                                                        Juha Kontinen

Organization

Program Chairs

Andreas Herzig   CNRS, IRIT, University of Toulouse, France
Juha Kontinen   University of Helsinki, Finland

Program Committee

Yamine Ait Ameur   IRIT/INPT-ENSEEIHT, France
Alessandro Artale   Free University of Bolzano-Bozen, Italy
Kim Bauters   University of Bristol, UK
Christoph Beierle   FernUniversität in Hagen, Germany
Leopoldo Bertossi   Relational AI Inc. and Carleton University, Canada
Philippe Besnard   CNRS, IRIT, University of Toulouse, France
Nicole Bidoit   LRI, Universitè de Paris, France
Meghyn Bienvenu   CNRS, LaBRI, University of Bordeaux, France
Joachim Biskup   Technische Universität Dortmund, Germany
Elena Botoeva   Imperial College London, UK
Arina Britz   CAIR, Stellenbosch University, South Africa
Dragan Doder   University of Belgrade, Serbia
Thomas Eiter   Vienna University of Technology, Austria
Christian Fermüller   Vienna University of Technology, Austria
Flavio Ferrarotti   Software Competence Centre Hagenberg, Austria
Nina Gierasimczuk   Technical University of Denmark, Denmark
Dirk Van Gucht   Indiana University Bloomington, USA
Marc Gyssens   Universiteit Hasselt, Belgium
Andreas Herzig   CNRS, IRIT, University of Toulouse, France
Tomi Janhunen   Tampere University, Finland
Matti Järvisalo   University of Helsinki, Finland
Gabriele Kern-Isberner   Technische Universität Dortmund, Germany
Attila Kiss   Eötvös Loránd University, Hungary
Juha Kontinen   University of Helsinki, Finland
Markus Krötzsch   TU Dresden, Germany
Antti Kuusisto   Tampere University, Finland
Sebastian Link   The University of Auckland, New Zealand
Thomas Lukasiewicz   University of Oxford, UK
Alessandra Palmigiano   Technical University of Delft, The Netherlands
Sebastian Rudolph   TU Dresden, Germany
Attila Sali   Alfréd Rényi Institute of Mathematics, Hungary
Klaus-Dieter Schewe   Zhejiang University, China
Steven Schockaert   Cardiff University, UK
Kostyantyn Shchekotykhin   Universität Klagenfurt, Austria

Guillermo R. Simari	Universidad del Sur, Argentina
Bernhard Thalheim	Christian-Albrechts-Universität zu Kiel, Germany
Alex Thomo	University of Victoria, Canada
Mirek Truszczynski	University of Kentucky, USA
José Turull-Torres	Universidad Nacional de La Matanza, Argentina
Ivan Varzinczak	CNRS, University of Artois, France
Jonni Virtema	Universiteit Hasselt, Belgium
Qing Wang	The Australian National University, Australia
Stefan Woltran	Vienna University of Technology, Austria

Additional Reviewers

Sabine Frittella
Aurona Gerber
Adrian Haret
Rafael Kiesel
Timo Lang
Andrea Mazzullo
Vladislav Ryzhikov

Local Organizing Chairs

| Gabriele Kern-Isberner | Technische Universität Dortmund, Germany |
| Christoph Beierle | University of Hagen, Germany |

Local Organizing Team

Kai Sauerwald
Marco Wilhelm

Abstracts of Invited Talks

First-Order Logic of Information Flows: Expressibility, Primitivity, Input-Output Properties, and Limited Access Patterns

Jan Van den Bussche

Universiteit Hasselt, Belgium
jan.vandenbussche@uhasselt.be

Abstract. The logic of information flows has been proposed by Ternovska as a general framework in the field of knowledge representation. The general aim of LIF is to model how information propagates in complex systems. In this work, we focus on the first-order (FO) version of LIF. Unique to LIF is the dynamic semantics which relates input assignments to output assignments. We formulate semantic definitions of input and output variables of a LIF expression. These semantic definitions are then approximated by syntactic definitions. The expressive power of LIF is compared to that of FO logic, and bounded-variable fragments of FO. In this setting, we investigate the primitivity of the composition operator in LIF. We also introduce the "forward" fragment of LIF and show how it can offer a fresh, navigational perspective on querying information sources with limited access patterns.

This is joint work with Evgenia Ternovska (Simon Fraser University), Bart Bogaerts (Vrije Universiteit Brussel), and Heba Aamer and Dimitri Surinx (Universiteit Hasselt).

Short Biography. Jan Van den Bussche is professor of databases and theoretical computer science at Hasselt University in Belgium. He received his PhD from the University of Antwerp in 1993, under the supervision of Jan Paredaens. He served as PC chair, and chair of the council, for the International Conference on Database Theory, and also as PC chair, and chair of the Executive Committee, for the ACM Symposium on Principles of Database Systems. His main research interest is in data models and query languages for a wide variety of data applications, ranging from spatial data to data stored in DNA. Most recently he is leading a work package on distributed data intelligence within the context of the Artificial Intelligence Research Flanders initiative.

Implicit Coordination of Mobile Agents

Bernhard Nebel

Institut für Informatik, Albert-Ludwigs-Universität Freiburg, Germany
nebel@informatik.uni-freiburg.de

Abstract. In multi-agent path finding (MAPF), it is usually assumed that planning is performed centrally and that the destinations of the agents are common knowledge. We will drop both assumptions and analyze under which conditions it can be guaranteed that the agents reach their respective destinations using implicitly coordinated plans without communication. Furthermore, we will analyze what the computational costs associated with such a coordination regime are. As it turns out, guarantees can be given assuming that the agents are of a certain type. However, the implied computational costs are quite severe. The plan existence problem becomes PSPACE-complete, while the original MAPF plan existence problem is in P. This clearly demonstrates the value of communicating about plans before execution starts.

Short Biography. Bernhard Nebel received his first degree in Computer Science (Dipl.-Inform.) from the University of Hamburg and his PhD (Dr. rer. nat.) from the University of Saarland in 1989. Between 1982 and 1993 he worked on different AI projects at the University of Hamburg, the Technical University of Berlin, ISI/USC, IBM Germany, and the German Research Center for AI (DFKI). From 1993 to 1996 he held an associate professor position at the University of Ulm. Since 1996, he has been a professor at Albert-Ludwigs-Universität Freiburg and head of the research group on Foundations of Artificial Intelligence. Bernhard Nebel is an EurAI and AAAI fellow and member of the Leopoldina. His research interests are knowledge representation, planning, and the application of methods from these areas in robotic contexts.

Computation on Structures: Behavioural Theory, Logic and Complexity

Klaus-Dieter Schewe

UIUC Institute, Zhejiang University, China
kd.schewe@intl.zju.edu.cn
kdschewe@acm.org

Abstract. Database theory is closely linked to finite model theory. Each relational database defines a finite relational structure, and queries as well as database transformations are computations on these structures. Furthermore, descriptive complexity theory provides many links between the expressiveness of logics concerning queries and common space/time complexity classes. Examples are the celebrated results by Fagin concerning the capture of NP by existential second-order logic and by Immerman/Vardi concerning the capture of PTIME by first-order logic plus inflationary fixed-point over ordered structures.

Another related development is provided by behavioural theories of classes of algorithms and proofs that these are captured by variants of abstract state machines, which operate on isomorphism classes of structures. Such theories with purely logical definitions of classes of computations of interest have been developed for sequential, parallel, concurrent, and reflective algorithms.

Therefore, we will argue for a shift of paradigm making computations on structures as provided by abstract state machines the standard model of computation with the advantage that computations on arbitrary levels of abstraction are enabled. Behavioural theories provide the means for logical characterisation of classes of computations, and also the logical characterisation of complexity in connection with the capture of complexity classes by specific classes of abstract state machines seems possible and adequate. We will provide evidence for the rationale of this shift.

Short Biography. Klaus-Dieter Schewe studied mathematics and computer science at the University of Bonn, Germany. Originally he worked in group representation theory, which was also the field of his PhD. After some years in industry he returned to academia working in software engineering, knowledge representation, database theory, and rigorous methods, bringing these diverse fields together. His particular interest is on mathematical and logical foundations, semantics, and expressiveness. He investigated methods for consistency enforcement, contributed to many results in dependency theory for complex-value databases (together with Sali, Link, and Hartmann), developed a thorough methodology for the design and development of web information systems (together with Thalheim), created a client-centric middleware for cloud computing (together with Bosa and others), and developed a theory of knowledge patterns for entity resolution (together with Qing Wang). He developed behavioural theories for unbounded parallel and reflective algorithms (together with Ferrarotti,

Wang, and Tec) and for concurrent systems (together with Börger), with which he contributed to the foundations of rigorous methods. Recently, he linked this research to descriptive complexity theory. He graduated in Pure Mathematics at University of Bonn, received a PhD in 1985 from University of Bonn in Mathematics, and later in 1995 received his DSc from Brandenburg Technical University in Theoretical Computer Science. He was Chair of Information Systems at Massey University, Director of the Information Science Research Centre in New Zealand, and Scientific Director of the Software Competence Center Hagenberg in Austria. In September 2019, he joined the University of Illinois at Urbana Champaign and the Institute of Zhejiang University at the International Campus in Haining, China.

Description Logics for Typicality and Nonmonotonic Reasoning

Anni-Yasmin Turhan

Institute for Theoretical Computer Science,
Technische Universität Dresden, Germany
anni-yasmin.turhan@tu-dresden.de

Abstract. Description Logics (DLs) are an intensively studied class of logics tailored towards building ontologies. Most DLs are decidable fragments of first-order logic. A DL knowledge base consists of two parts: the terminological part that captures the concepts of the application at hand by means of concept axioms and of the assertional part that captures the data from the application. In practical applications where exceptions occur frequently, such as biology or medicine, classical, monotone reasoning is often undesirable. Likewise, for applications where data may be erroneous and causes inconsistencies with respect to the DL terminology, monotonic reasoning is unsuitable. An approach to achieve nonmonotonic behaviour of DLs is to employ *defeasible* description logics, which admit the use of defeasible concept axioms. During reasoning these axioms can be omitted for those instances that otherwise would cause an inconsistency. Defeasible DLs can also be employed to model typicality. Instances that fulfill more defeasible axioms are simply regarded more typical than others.

Over the last decade a whole range of defeasible DLs has been proposed in the literature. The proposals have very different strengths and weaknesses and can be rather hard to compare with each other—in particular, as the semantics are often given in a procedural way.

In this talk I introduce defeasible DLs and demonstrate different approaches to reasoning in these DLs. I discuss our recent approach for the defeasible Description Logic $\mathcal{EL}_\perp$ that has model-theoretic semantics and remedies some of the severe shortcomings of earlier approaches. We also investigate the complexity of reasoning in this defeasible DL in different settings.

Short Biography. Anni-Yasmin Turhan is an assistant professor at the Institute of Theoretical Computer Science of Dresden University of Technology. Having obtained her Diploma in Computer Science from the University of Hamburg, she started her scientific career in a research project dedicated to non-standard inferences in Description Logics (DLs). She received her doctoral (2008) and habilitation degree (2014) from TU Dresden. Her research interests are rooted in the fields of artificial intelligence (AI) and theoretical computer science and are dedicated to knowledge representation and reasoning. She has been a Program Committee member for many conferences in the fields of knowledge representation and reasoning, Semantic Web, and AI. In her research she studied inferences that can be employed to build and maintain DL knowledge bases. Her recent research projects are dedicated to situation recognition by DL reasoning, reasoning using similarity, and reasoning under non-monotonic semantics.

Contents

Functional Dependencies in Incomplete Databases with Limited Domains ...   1
  Munqath Alattar and Attila Sali

Normal Forms of Conditional Knowledge Bases Respecting Entailments
and Renamings...   22
  Christoph Beierle and Jonas Haldimann

On Matrices and K-Relations   42
  Robert Brijder, Marc Gyssens, and Jan Van den Bussche

Social Consolidations: Rational Belief in a Many-Valued Logic
of Evidence and Peerhood   58
  Yuri David Santos

ASPARTIX-V19 - An Answer-Set Programming Based System
for Abstract Argumentation   79
  *Wolfgang Dvořák, Anna Rapberger, Johannes P. Wallner,
  and Stefan Woltran*

Proper Hierarchies in Polylogarithmic Time and Absence
of Complete Problems......................................   90
  *Flavio Ferrarotti, Senén González, Klaus-Dieter Schewe,
  and José María Turull-Torres*

Diversity, Dependence and Independence.........................   106
  Pietro Galliani and Jouko Väänänen

Towards Probabilistic Reasoning in Type Theory - The Intersection
Type Case   122
  *Silvia Ghilezan, Jelena Ivetić, Simona Kašterović, Zoran Ognjanović,
  and Nenad Savić*

Measuring Inconsistency in a General Information Space   140
  John Grant and Francesco Parisi

Parameterised Complexity of Model Checking and Satisfiability
in Propositional Dependence Logic   157
  Yasir Mahmood and Arne Meier

Utilizing Deep Learning and RDF to Predict Heart
Transplantation Survival.....................................   175
  Dennis Medved, Johan Nilsson, and Pierre Nugues

Game Description Logic with Integers: A GDL Numerical Extension   191
  Munyque Mittelmann and Laurent Perrussel

Craig Interpolation of Epistemic Logics with Distributed Knowledge.   211
  Ryo Murai and Katsuhiko Sano

On the Dynamics of Structured Argumentation: Modeling Changes
in Default Justification Logic   222
  Stipe Pandžić

Logic-Based Approach to Incremental Monitoring and Optimization
on Strongly Distributed Data Streams   242
  Elena V. Ravve

Realisability of Choreographies.   263
  Klaus-Dieter Schewe, Yamine Aït-Ameur, and Sarah Benyagoub

Schema Optimisation Instead of (Local) Normalisation   281
  Bernhard Thalheim

Strongly Minimal MapReduce Algorithms: A TeraSort Case Study   301
  Daniel Xia, Michael Simpson, Venkatesh Srinivasan, and Alex Thomo

Event Sequence Interpretation of Structural Geomodels:
A Knowledge-Based Approach for Extracting Tectonic Sequences   318
  Xianglin Zhan, Cai Lu, and Guangmin Hu

Author Index   335

Functional Dependencies in Incomplete Databases with Limited Domains

Munqath Alattar[1] and Attila Sali[1,2(✉)]

[1] Department of Computer Science and Information Theory,
Budapest University of Technology and Economics, Budapest, Hungary
`m.attar@cs.bme.hu`
[2] Alfréd Rényi Institute of Mathematics, Budapest, Hungary
`sali.attila@renyi.hu`

Abstract. Missing data value is an extensive problem in both research and industrial developers. Two general approaches are there to deal with the problem of missing values in databases, they either could be ignored (removed) or imputed (filled in) with new values [9]. In the present paper, we use the second method. Possible worlds were introduced in [14,16] and possible and certain keys, as well as weak and strong functional dependencies were studied. We introduced the intermediate concept of strongly possible worlds that are obtained by imputing values already existing in the table in a preceding paper. This results in strongly possible keys and strongly possible functional dependencies. We give a polynomial algorithm to verify a single spKey, and show that in general, it is NP-complete to verify an arbitrary collection of spKeys. We give a graph theoretical characterization of the validity of a given spFD $X \rightarrow_{sp} Y$. We analyze which weak/strong functional dependency axioms remain sound for strongly possible functional dependencies and give appropriate modifications of the not sound ones.

Keywords: Strongly possible functional dependencies · `NULL` values · Armstrong tables · Data imputation · Matchings in bipartite graphs · List coloring

1 Introduction

Many systems today allow entering incomplete tuples into a database. For example, in case of data warehousing, if different sources of raw data are merged, some attributes may exist in some of the sources while not available in some of the others. This makes it necessary to treat keys over incomplete tables. It is

Research of the second author was partially supported by the National Research, Development and Innovation Office (NKFIH) grant K–116769. This work is also connected to the scientific program of the "Development of quality-oriented and harmonized R+D+I strategy and functional model at BME" project, supported by the New Hungary Development Plan (Project ID: TMOP-4.2.1/B-09/1/KMR-2010-0002).

© Springer Nature Switzerland AG 2020
A. Herzig and J. Kontinen (Eds.): FoIKS 2020, LNCS 12012, pp. 1–21, 2020.
https://doi.org/10.1007/978-3-030-39951-1_1

common to encounter databases having up to half of the entries missing, making it very difficult to mine them using data analysis methods that can work only with complete data [10].

There are different reasons why incompleteness occurs in database tables. Date [7] determined more than one kind of missing data and identified seven distinct types of null as follows: value not applicable, value unknown, value does not exist, value undefined, value not valid, value not supplied, and value is the empty set. The present paper deals with data consumption with missing values in a database table, we take the second, third, and seventh types. For the other types of missing data we assume that symbol N/A belongs to each domain, and we treat it as regular domain element in comparisons.

The issue of missing values complicates data analysis for the analysts. Other problems are usually associated with missing values such as loss of data efficiency and effectiveness [9]. Although some methods of data analysis may overcome the missing value problem, many others require complete databases. Two general approaches are there to deal with the problem of missing values in databases, incomplete tuples either could be ignored (removed) or imputed (filled in) with new values [9].

Köhler et al. [14] used possible worlds by replacing each occurrence of a null with a value from the corresponding attribute's (possibly infinite) domain. Each possible world is considered as a table of total tuples that may contain duplicated tuples. They defined a possible key as a key that is satisfied by some possible world of a non total database table and a certain key as a key that is satisfied by every possible world of the table. For example, Fig. 1a has some possible world that satisfies the possible key {Course Name} while there is no possible world of the table that satisfies key {Lecturer} and, furthermore, every possible world of the table satisfies the certain key {Course Name, Year, Semester}. Also, weak functional dependencies were defined as FD's that are satisfied by some possible world and strong functional dependencies are satisfied by each possible world.

In many cases we have no proper reason to assume the existence of any other attribute value than the ones already existing in the table. Such examples could be types of cars, diagnoses of patients, applied medications, dates of exams, course descriptions, etc. We define a strongly possible world as a possible world that is obtained by replacing each occurrence of null value from the corresponding attribute's existing values. This is a data mining type approach, our idea is that we are given a raw table with nulls and we would like to identify possible key sets and functional dependencies based on the data only. A strongly possible key is some key satisfied by a strongly possible world and similarly, a strongly possible functional dependency is an FD satisfied by some strongly possible world. Set of the values that are already shown in each attribute of a table represent the part of that attribute's domain which is certain. For the cases of incomplete databases with unknown domain values, choosing values that are not in the real attribute domain may distort the data. For example, the clients' information table bellow, it could be inappropriate to use any other value than the ones appearing in the *marriage status* attribute to fill the ⊥ in the second

Name	gender	mrg status	age
⊥	female	married	32
Sarah	female	⊥	⊥
David	⊥	divorced	38
James	male	single	⊥
⊥	male	widower	47

(a) Incomplete Table

Name	gender	mrg status	age
30	female	married	32
Sarah	female	lawyer	high
David	Apple	divorced	38
James	male	single	-12
operation	male	widower	47

(b) Possible World

Name	gender	mrg status	age
David	female	married	32
Sarah	female	single	32
David	male	divorced	38
James	male	single	38
James	male	widower	47

(c) Strongly Possible World

Fig. 1. Possible and strongly possible worlds

row. Otherwise, some other values like numbers, symbols, or any other strings with distant meanings cause the distortion. So, using one of the already shown values in the attribute provides more meaningful and semantically acceptable possible (strongly possible) world, as the one in Fig. 1c is preferred to the one in Fig. 1b.

The present paper continues the work started in [2]. There properties of strongly possible keys were analyzed and a (worst case exponential time) algorithm was given for verifying a single spKey. Here we give a polynomial time solution for the same problem and show that verifying an arbitrary system of strongly possible keys is NP-complete. Then we turn to spFD's and give a graph theoretical characterization when spFD $X \rightarrow_{sp} Y$ holds in an SQL table T containing NULL's. Then we analyze several rules/axioms of weak and strong functional dependencies whether they are sound for spFD's, as well. For those which are not, we give several possible weakenings or restrictions that keep soundness.

The paper is organized as follows. Section 2 contains the necessary definitions, Sect. 3 discusses related work. Section 4 deals with the complexity of verifying strongly possible keys. Section 5 studies strongly possible functional dependencies. Finally, Sect. 6 contains concluding remarks and future research directions.

2 Basic Definitions

We start with summarizing some basic definitions and terminologies. Let $R = \{A_1, A_2, \ldots A_n\}$ be a relation schema. The set of all the possible values for each attribute $A_i \in R$ is called the domain of A_i and denoted as $D_i = dom(A_i)$ for $i = 1, 2, \ldots n$. For $X \subseteq R$ then $D_X = \prod_{\forall A_i \in K} D_i$.

An instance $T = (t_1, t_2, \ldots t_s)$ over R is a list of tuples that each tuple is a function $t : R \rightarrow \bigcup_{A_i \in R} dom(A_i)$ and $t[A_i] \in dom(A_i)$ for all A_i in R. By taking a list of tuples we use the <u>bag semantics</u> that allows several occurrences of the same tuple. For a tuple $t_r \in T$ and $X \subset R$, let $t_r[X]$ be the restriction of t_r to X.

It is assumed that $\perp$ is an element of each attribute's domain that denotes missing information. t_r is called V-total for a set V of attributes if $t_r[A] \neq \perp$, $\forall A \in V$. Also a tuple t_r is a total tuple if it is a R-total. t_1 and t_2 are <u>weakly similar</u> on $X \subseteq R$ denoted as $t_1[X] \sim_w t_2[X]$ defined by Köhler et al. [14] if:

$$\forall A \in X \quad (t_1[A] = t_2[A] \text{ or } t_1[A] = \perp \text{ or } t_2[A] = \perp).$$

Furthermore, t_1 and t_2 are <u>strongly similar</u> on $X \subseteq T$ denoted by $t_1[X] \sim_s t_2[X]$ if:

$$\forall A \in X \quad (t_1[A] = t_2[A] \neq \perp).$$

For the sake of convenience we write $t_1 \sim_w t_2$ if t_1 and t_2 are weakly similar on R and use the same convenience for strong similarity. Let $T = (t_1, t_2, \ldots t_s)$ a table instance over R. $T' = (t'_1, t'_2, \ldots t'_s)$ is a <u>possible world</u> of T, if $t_i \sim_w t'_i$ for all $i = 1, 2, \ldots s$ and T' is completely NULL-free. That is, we replace the occurrences of $\perp = t[A_i]$ with a value from the domain D_i different from $\perp$ for all tuples and all attributes.

<u>Weak functional dependency</u> $X \rightarrow_w Y$ holds in T if there exists a possible world T' such that $T' \models X \rightarrow Y$ in the classical sense, that is functional dependency $X \rightarrow Y$ holds in T' meaning that if $t'_i[X] = t'_j[X]$ then $t'_i[Y] = t'_j[Y]$ is satisfied, for all pairs of tuples $t'_i, t'_j \in T'$.

<u>Strong functional dependency</u> $X \rightarrow_s Y$ holds in T if functional dependency $X \rightarrow Y$ holds in all possible worlds T' of T. If $X \rightarrow_w R$ holds in T then X is a <u>possible key</u> and if $X \rightarrow_s R$ holds in T, then X is a <u>certain key</u>. The following was proven in [14].

Theorem 2.1. *$X \subseteq R$ is a certain (possible) key iff $\forall t_1, t_2 \in T : t_1[X] \not\sim_w t_2[X] (t_1[X] \not\sim_s t_2[X])$.*

2.1 Strongly Possible Worlds

The concept of <u>strongly possible world</u> was introduced in [2].

Definition 2.1. *The <u>visible domain</u> of an attribute A_i (VD_i) is the set of all distinct values except $\perp$ that are already used by tuples in T:*

$$VD_i = \{t[A_i] : t \in T\} \setminus \{\perp\} \text{ for } A_i \in R$$

Then the VD_1 in Fig. 2a is {Mathematics, Datamining}. The term visible domain refers to the data that already exist in a given dataset. For example, if we have a dataset with no information about the attributes' domains definitions, then we use the data itself to define their own structure and domains. This may provide

Course Name	Year	Lecturer	Credits	Semester
Mathematics	2019	$\perp$	5	1
Datamining	2018	Sarah	7	$\perp$
$\perp$	2019	Sarah	$\perp$	2

(a) Incomplete Dataset

Course Name	Year	Lecturer	Credits	Semester
Mathematics	2019	Sarah	5	1
Datamining	2018	Sarah	7	2
Datamining	2019	Sarah	7	2

(b) Complete Dataset

Fig. 2. Complete and incomplete datasets

more realistic results when extracting the relationship between data so it is more reliable to consider only what information we have in a given dataset.

While a possible world is obtained by using the domain values instead of the occurrence of NULL, a strongly possible world is obtained by using the visible domain values.

Definition 2.2. *A possible world T' of T is called* <u>strongly possible world</u> *if $t'[A_i] \in VD_i$ for all $t' \in T'$ and $A_i \in R$.*

Strongly possible worlds allow us to define <u>strongly possible keys</u> and <u>strongly possible functional dependencies (spFD's)</u>.

Definition 2.3. *Strongly possible functional dependency $X \rightarrow_{sp} Y$ holds in table T over schema R if there exists a strongly possible world T' of T such that $T' \models X \rightarrow Y$. X is a strongly possible key, if there exists a strongly possible world T' of T such that X is a key in T', in notation $sp\langle X \rangle$. Note that this is not equivalent with spFD $X \rightarrow_{sp} R$, since we use the bag semantics.*

For example, {Course Name, Year} is a strongly possible key of the table in Fig. 2a as the strongly possible world in Fig. 2b shows it.

2.2 NULL-Free Subschema

The <u>NULL-free subschema (NFS)</u> of a schema R is a subset $R_S \subseteq R$. This corresponds to SQL's NOT NULL constraint. A table T over R satisfies NFS R_S, if it is R_S-total, that is every tuple $t \in T$ is R_S-total, i.e. $\forall A \in R_S \colon t[A] \neq \perp$. If T satisfies NFS R_S, then we say T is over (R, R_s). Also, if Σ is a set of integrity constraints (for example spFD's) then a table T over (R, R_s) is an <u>Armstrong instance</u> of Σ if

(1) $T \models \sigma$ iff $\Sigma \models \sigma$ for any constraint σ, and
(2) For any attribute $A \in R \setminus R_S$ there is a tuple $t \in T$ such that $t[A] = \bot$.

This is the classical definition of Armstrong instance extended with the requirement that if an attribute is not in the NULL-free subschema, then it indeed contains NULL's.

3   Related Work

Keys and functional dependencies are important constraints that enforce the semantics of relational database systems. Database tables of real database systems usually contain occurrences of null values and for some cases this includes candidate key columns. Various studies have been done for the purpose of handling missing values.

Sree Dhevi [4] shows that it is necessary to impute the missing values based on other information in the dataset to overcome the biased results that affect the accuracy of classification generated by missing values. Similarly, we use the attribute's existing values for each null in that attribute. Cheng et al. [6] utilize clustering algorithms to cluster data, and calculate coefficient values between different attributes by generating minimum average error.

Farhangfar et al. introduced a framework of imputation methods in [9] and evaluates how the choice of different imputation methods affects the performance in [10]. Experimental analyses of several algorithms for imputation of missing values were performed by [1,5,8,12] An approach introduced by Zhang et al. [21] discusses and compares several strategies that utilize only known values.

Interactions of functional dependencies and other integrity constraints with null values have long been investigated. Early studies concentrated on "fixing" the database using the Chase procedure, such as Grahne did in [11]. Imlienski and Lipski [13] also studied the properties of Chase with respect to NULL's.

The two main interpretations of NULL's are "value unknown at present" and "no information". The first one leads to possible world semantics that is NULL's are replaced by domain values to obtain total tables. Vassilou [19] gave a three-valued model of FD satisfaction, namely all possible words of a table T are considered and a functional dependency either holds, does not hold or may hold on T. This latter means that in some possible worlds it holds, and in some other ones it does not hold. Levene and Loizou defined weak and strong functional dependencies based on possible world semantics. A weak FD holds in some of the possible worlds and a strong FD holds in all possible worlds. They gave a sound and complete axiom system for them in [16].

The "no information" approach unifies the treatment of unknown as well as non-existing data. Lien [17] defined functional dependencies to hold if strong similarity on the LHS implies equality on RHS. Here equality means that if an attribute value is NULL in a tuple then the same attribute must also be NULL in the other tuple. This corresponds to p-FD's of Köhler and Link [15]. The main novelty of the latter paper is the concept of c-FD's, that is certain functional dependencies. A c-FD holds if weak similarity of LHS implies equality of the RHS, where equality is in the same sense as for p-FD's.

Atzeni and Morfuni gave some axiom system for functional dependencies with NULL's [3]. The drawback of their approach that they allowed no NULL's on the left hand sides of functional dependencies.

Our strongly possible functional dependency is also based on possible world semantics and fits between weak and strong FD's of Levene and Loizou. In case that a table instance has at least one non-NULL value in each attribute, then satisfaction of a c-FD $X_w \rightarrow Y$ implies that $X \rightarrow_{sp} Y$ holds, as well. On the other hand, satisfaction of $X \rightarrow_{sp} Y$ does not imply that p-FD $X_s \rightarrow Y$ holds. A brief comparison of the different notions is given in the following example, which is a modification of Example 3 of [15].

Example 3.1. Let T be the following SQL table.

Employee	Dept	Manager	Salary
Knuth	NULL	Chomsky	100,000
Turing	CS	von Neumann	NULL
Turing	NULL	Gödel	NULL

We compare 3-valued [19], weak and strong [16], possible [17], certain [15] and strongly possible.

	3-valued	Weak	Strong	Possible	Certain	Strongly possible
$e \rightarrow d$	unk	T	F	F	F	T
$e \rightarrow m$	F	F	F	F	F	F
$e \rightarrow s$	unk	T	F	T	T	T
$d \rightarrow d$	T	T	T	T	F	T
$d \rightarrow m$	unk	T	F	T	F	F
$m \rightarrow e$	T	T	T	T	T	T
$m \rightarrow d$	unk	T	F	T	T	T

Köhler et al. [14] introduced possible and certain keys. A set K of attributes is possible key if there is a possible world where K is a key. On the other hand, K is a certain key if it is a key in every possible world. The main concept of the present paper is between these two, since a strongly possible world is a possible world, as well. Possible worlds may use any value from an attribute domain to replace a null. This effectively allows an infinite pool of values. Strongly possible worlds are created from finite attribute domains. Some of the results in [14] essentially use that some attribute domains are infinite. In the present paper we investigate what can be stated without that assumption.

Finally let us mention the very recent concept of embedded functional dependency introduced by Wei and Link [20]. Their aim is to define a robust interpretation of functional dependencies that does not depend on the interpretation of NULL's. Both papers [15,20] contain work on normalization of databases based on the appropriate functional dependencies. Normalization is important to eliminate redundancy that may cause inconsistency at updates. It is a topic of future research how our spFD's could be applied for the same goal.

4   Complexity of Strongly Possible Keys

The algorithmic question we study here is the following. Given SQL table T and collection of strongly possible key constraints Σ, does $T \models \Sigma$ hold? We may assume without loss of generality that all tables treated have at least one strongly possible world. Indeed, the degenerate case of non-existence of strongly possible world occurs only if a table contains only NULLs in an attribute.

An algorithm using bipartite matchings was given in [2] for the case of $\Sigma = \{sp\langle K\rangle\}$, that is when a single strongly possible key needs to be checked. Let $K = \{A_1, A_2 \ldots A_b\}$. The running time of that algorithm is $O(|R|(|T| + |T^*|)) + O((|T| + |T^*|)|E|)$, where T^* is the set of total tuples $T^* = \{t^* \in \Pi_{i=1}^b VD_i : \exists t \in T$ such that $t[K] \sim_w t^*[K]\}$. However, the size of T^* can easily be exponential function of the size of T. Here we give a polynomial time refinement of that algorithm.

If a single strongly possible key $sp\langle K\rangle$ is to be checked, then it is enough to consider $T|_K$, since K is a key iff the tuples are pairwise distinct on K. Let $T^* \subseteq VD_1 \times VD_2 \times \ldots \times VD_b$ be defined by $T^* = \{t^* \in VD_1 \times VD_2 \times \ldots \times VD_b : \exists t \in T : t[K] \sim_w t^*\}$ and define bipartite graph $G = (T, T^*; E)$ with $\{t, t^*\} \in E \iff t[K] \sim_w t^*[K]$.

Proposition 4.1. $T \models sp\langle K\rangle$ iff there exists a matching in $G = (T, T^*; E)$ covering T.

Proof. Let $T = \{t_1, t_2, \ldots t_m\}$. Suppose first that $T \models sp\langle K\rangle$. This means that there exists a strongly possible world T' such that $\forall i \neq j$, $t_i', t_j' \in T' : t_i'[K] \neq t_j'[K]$. Then clearly $M = \{\{t_i, t_i'[K]\} : t_i'$ is the extension of t_i $i = 1, 2, \ldots, m\}$ is a matching in G covering T.

On the other hand, if a matching $M = \{\{t_i, t_i^*\} : i = 1, 2, \ldots m$ exists in G, then let t_i' be an arbitrary extension of t_i^* with visible domain values on attributes in $R \setminus K$. Then $T' = \{t_i' : i = 1, 2, \ldots, m\}$ is a strongly possible world that has no two distinct tuples that agree on K, that is K is a key in T'.   □

In order to make our algorithm run in polynomial time we only generate part of T^*. Let $T = \{t_1, t_2 \ldots t_m\}$ and $\ell(t_i) = |\{t^* \in VD_1 \times VD_2 \times \ldots \times VD_b : t^* \sim_w t_i[K]\}|$. Note that $\ell(t_i) = \prod_{j : t_i[A_j] = \perp} |VD_j|$, hence these values can be calculated by scanning T once and using appropriate search tree data structures to hold values of visible domains of each attribute. Sort tuples of T in non-decreasing $\ell(t_i)$ order, i.e. assume that $\ell(t_1) \leq \ell(t_2) \leq \ldots \leq \ell(t_p)$.

Let $j = \max\{i : \ell(t_i) < i\}$ and $T_j = \{t_1, t_2, \dots t_j\}$, furthermore $T_j^\star = \{t^\star : \exists t \in T_j : t^\star \sim_w t[K]\} \subseteq VD_1 \times VD_2 \times \dots \times VD_b$. Note that $|T_j^\star| \leq \frac{1}{2}j(j-1)$. If $\forall i = 1, 2, \dots, m : \ell(t_i) \geq i$, then define $j = 0$ and $T_j^\star = \emptyset$.

Proposition 4.2. $T \models sp\langle K \rangle$ iff $j = 0$ or there exists a matching in $G' = (T_j, T_j^\star; E|_{T_j \times T_j^\star})$ covering T_j.

Proof. $\Rightarrow$: According to Proposition 4.1 $T \models sp\langle K \rangle$ implies a matching M in G covering T. The edges of M incident with tuples from T_j can only go to tuples in $T_j^\star$, hence they form a matching covering T_j in bipartite graph G'.
$\Leftarrow$: Assume that a matching M' in $G' = (T_j, T_j^\star; E|_{T_j \times T_j^\star})$ covering T_j exists. For $j < k \leq m$ pick a tuple $t_k^\star \in T^\star$ such that $t_k[K] \sim_w t_k^\star$ and $t_k^\star$ has not been used for $t_i : i = 1, 2, \dots k - 1$ yet. Since $\ell(t_k) \geq k$, such $t_k^\star$ exists. Then $M = M' \cup \{\{t_k, t_k^\star\} : j < k \leq p\}$ is a matching verifying that $T \models sp\langle K \rangle$ holds by Proposition 4.1. $\qquad\square$

Proposition 4.2 gives a polynomial time algorithm for deciding whether $T \models sp\langle K \rangle$ holds.

Algorithm 1. Verifying $T \models sp\langle K \rangle$

Input: Table T over schema R, $K \subseteq R$
Output: Strongly possible world $T^\star$ showing $T \models sp\langle K \rangle$ if exists

1: **procedure spKey**(item T, item R, item K)
2:     Calculate $\ell(t) : t \in T$
3:     Sort T_i in non-decreasing $\ell(t_i)$ order
4:     $j \leftarrow \max\{i : \ell(t_i) < i\}$
5:     Construct bipartite graph $G' = (T_j, T_j^\star; E|_{T_j \times T_j^\star})$
6:     $M = \textbf{MaxMatching}(G')$
7:     **if** $|M| < j$ **then return** $T \not\models sp\langle K \rangle$
8:         $T^\star \leftarrow M \cap T_j^\star$
9:         **for** $k = j + 1$ **to** $|T|$ **do**
10:             Generate $t_k^\star \notin T^\star$ such that $t_k \sim_w t_k^\star$
11:             $T^\star \leftarrow T^\star \cup \{t_k^\star\}$
12:         **end for**
13:         **return** $T^\star$
14:     **end if**
15: **end procedure**

The running time of Algorithm 1 can roughly be estimated as follows. Line 2 can be done using binary search trees in time $O(|K| \cdot |T| \log |T|)$. Sorting in line 3 adds another $O(|T| \log |T|)$. Finding the maximum in line 4 takes $O(|T|)$ time. Constructing bipartite graph G' takes $O(|T_j| \cdot |T_j^\star|) = O(|T|^3)$ time. Finding maximum matching using Augmenting Path method [18] is done in $O((|T_j| + |T_j^\star|)|E|_{T_j \times T_j^\star}|) = O(|T|^5)$ time. In the for loop of lines 9–11 we have to generate at most $|T|$ tuples in each round that have to be checked against at most $|T|$ tuples each, and we have at most $|T|$ rounds, so in total the loop takes $O(|T|^3)$ time. Summing up we get that the running time of Algorithm 1 is $O(|K| \cdot |T| \log |T| + |T|^5)$.

The question whether $T \models \Sigma$ for a collection Σ of strongly possible key constraints was reduced to the problem of finding maximal common independent set of three or more matroids in [2]. That problem is known to be NP-complete, but this does not yet prove that our problem is NP-complete. However, modifying an argument of [14] we can prove that the latter is the case.

Theorem 4.1. *The strongly possible key satisfaction problem is NP-complete.*

Proof. Let $\Sigma = \{sp\langle K_1 \rangle, sp\langle K_2 \rangle, \ldots, sp\langle K_m \rangle\}$ be a collection of strongly possible key constraints and T be relational table instance containing NULLs. The question whether $T \models \Sigma$ belongs to NP, since we can guess a strongly possible world T' of T and then check in polynomial time that $K_i \colon i = 1, 2, \ldots, m$ are keys in T'.

In order to prove that the question is NP-hard, the known NP-complete problem 3-SAT is reduced to it. Let $\mathcal{I}$ be an instance of 3-SAT with set of variables $V = \{v_1, v_2, \ldots v_n\}$ and set of clauses $C = \{c_1, c_2, \ldots, c_m\}$. We may assume without loss of generality that C does not contain tautologies. Construct a table T over scheme R and collection strongly possible key constraints Σ as follows. $R = \{A_1, A_2, \ldots A_m, B_1, B_2, \ldots B_n\}$. $\Sigma = \{sp\langle K_1 \rangle, sp\langle K_2 \rangle, \ldots, sp\langle K_m \rangle\}$ where $K_i = \{A_i, B_{i_1}, B_{i_2}, B_{i_3}\}$ for $c_i = [(\neg)v_{i_1} \vee (\neg)v_{i_2} \vee (\neg)v_{i_3}]$. T consists of tuples $T = \{t_0, t_1, \ldots t_m, t_{m+1}, t_{m+2}\}$ such that

$$t_0[A_i] = 0 \ i = 1, 2, \ldots m, \text{ and } t_0[B_j] = \bot \ j = 1, 2, \ldots, n.$$

and for all $i = 1, 2, \ldots m$

$$t_i[A_i] = i, t_i[A_k] = 0 \ k \neq i, \text{ and } t_i[B_j] = F \text{ if } v_j \in c_i, \ t_i[B_j] = T \text{ if } v_j \notin c_i \ j = 1, 2, \ldots, n.$$

Finally, for $s = m + 1, m + 2$

$$t_s[A_i] = s \ i = 1, 2, \ldots m, t_{m+1}[B_j] = T \text{ and } t_{m+2}[B_j] = F \ j = 1, 2, \ldots, n.$$

A strongly possible world of T' of T corresponds to a truth assignment of the variables by setting $v_j = t_0'[B_j]$ for $j = 1, 2, \ldots, n$. Each tuple $t_i \ i > 0$ of T is total and so $t_j'[A_i] \neq t_k'[A_i]$ for $0 < j < k \leq m + 2$, hence no two tuples of positive indices of T' agree on any of the attribute sets $K_i \colon i = 1, 2, \ldots m$. On the other hand, it is easy to see that $t_i'[K_i] = t_0'[K_i]$ iff c_i evaluates *false* by the truth assignment given above. $\qquad \square$

In some special cases, more than one strongly possible keys can be verified in polynomial time.

Proposition 4.3. *Let T be a table over schema R, $\Sigma = \{sp\langle K_1 \rangle, sp\langle K_2 \rangle, \ldots, sp\langle K_m \rangle\}$ be a collection of strongly possible key constraints. If $K_i \cap K_j = \emptyset$ for $i \neq j$, then $T \models \Sigma$ can be decided in polynomial time.*

Proof. Since the K_i's are pairwise distinct, Algorithm 1 can be applied successively and then the tuples of the strongly possible worlds of $T|_{K_i}$ for $i = 1, 2, \ldots m$ can be concatenated. □

5   Strongly Possible Functional Dependencies

Having investigated strongly possible keys, it is natural to extend investigations to strongly possible FD's. In this section we first give a graph theoretical characterization of when does a table T satisfy $X \to_{sp} Y$. Then we characterize what set systems could be left hand sides and right hand sides of spFD's if the other side is a fixed attribute set. In the remaining part of the section we give the first step towards a possible axiomatisation of spFD's by comparing them with the axioms given for weak and strong FD's by Levene and Loizou [16].

Recall that a table T over Schema R satisfies $X \to_{sp} Y$ iff there exists a strongly possible world T' of T such that $T' \models X \to Y$. If $T = \{t_1, t_2, \ldots, t_p\}$ and $T' = \{t'_1, t'_2, \ldots, t'_p\}$ with $t_i \sim_w t'_i$, then t'_i is called an sp-extension or in short an extension of t_i. Let $X \subseteq R$ be a set of attributes and let $t_i \sim_w t'_i$ such that for each $A \in X \colon t'_i[A] \in VD(A)$, then $t'_i[X]$ is an strongly possible extension of t_i on X. A useful tool in investigations of strongly possible functional dependencies is (spFD's in short) is the concept of weak similarity graph.

Definition 5.1. *Let $T = \{t_1, t_2, \ldots t_p\}$ be a table (instance) over schema R. The weak similarity graph G_Y with respect to Y is defined as $G_Y = (T, E)$, where $\{t_i, t_j\} \in E \iff t_i[Y] \sim_w t_j[Y]$.*

We can characterize when $T \models X \to_{sp} Y$ holds using weak similarity graphs.

Theorem 5.1. *Let $T = \{t_1, t_2 \ldots t_m\}$ be a table over schema R. For $X, Y \subseteq R$, $T \models X \to_{sp} Y$ holds iff $\overline{G_Y}$ can be list colored[1] using lists $\{t^1_i, t^2_i \ldots t^{r_i}_i\}$ for $t_i \in T$, where $\overline{G_Y}$ is the complement of the weak similarity graph on Y and t^j_i's are the strongly possible extensions of t_i on X.*

Proof. $X \to_{sp} Y$ holds iff there exists strongly possible world T' such that $T' \models X \to Y$.

$\Rightarrow$: Assume $T \models X \to_{sp} Y$ holds. Take a strongly possible world in which $X \to Y$. This gives a coloring from the lists by giving color $t'_i[X]$ to t_i. That coloring is proper because if $t_1[Y] \not\sim_w t_2[Y]$, then $t'_1[Y] \neq t'_2[Y]$ and also $t'_1[X] \neq t'_2[X]$ where $t'_1[X]$ and $t'_2[X]$ are colors from lists of t_1 and t_2 respectively.

$\Leftarrow$: Assume there exist a proper list coloring, then $t'_1[X], t'_2[X] \ldots t'_m[X]$ are partial extensions of $t_i[X]$. Suppose $t'_{i_1}[X] = t'_{i_2}[X] = \ldots = t'_{i_b}[X]$, then $t_{i_h}[Y] \sim_w t_{i_k}[Y]$ for $1 \leq h < k \leq b$ that forms a weak similarity clique, see Table 1. Thus in each attribute of Y there is at most one non-NULL value in tuples $t_{i_1}, t_{i_2}, \ldots t_{i_b}$, hence they all can be made equal in Y. □

---

[1] Let $G(V, E)$ be a graph and $L \colon V \to 2^{\mathbb{N}}$ be a mapping that assigns each vertex a list of colors $L(v)$. A list coloring of G using lists $\{L(v) \colon v \in V\}$ is a mapping $c \colon V \to \bigcup_{v \in V} L(v)$ such that $c(v) \in L(v)$ and $c(u) \neq c(v)$ if $\{u, v\} \in E$.

Table 1. Color classes and weak similarity cliques.

$$X \qquad\qquad\qquad\qquad\qquad Y$$

$$\begin{bmatrix} \text{Same color} \end{bmatrix} \qquad \begin{bmatrix} \text{Weak Similarity Clique} \end{bmatrix}$$

$$\begin{bmatrix} \text{Same color} \end{bmatrix} \qquad \begin{bmatrix} \text{Weak Similarity Clique} \end{bmatrix}$$

$$\begin{bmatrix} \text{Same color} \end{bmatrix} \qquad \begin{bmatrix} \text{Weak Similarity Clique} \end{bmatrix}$$

If we fix the left hand side of an spFD, then the possible right hand sides clearly form a down-set, that is if $T \models X \to_{sp} Y$ holds and $Y' \subset Y$, then $T \models X \to_{sp} Y'$ also holds. The following Proposition tells, that there is no other restriction for the right hand sides of spFD's with fixed left hand sides.

Proposition 5.1. *Let (R, R_S) be a schema and $X \subset R$. Let $\mathcal{Y}$ be a down-set of subsets of $R\backslash X$. Then there exists a table T over (R, R_S) such that $T \models X \to_{sp} Y$ holds iff $Y \cap (R \setminus X) \in \mathcal{Y}$.*

Proof. Let the maximal elements of $\mathcal{Y}$ be $Y_1, Y_2, \ldots, Y_s$, that is $\mathcal{Y} = \{A : \exists i$ such that $A \subseteq Y_i\}$ and $Y_i \not\subseteq Y_j$ for $i \neq j$. Let $A_0 \in X \setminus R_S$ be a fixed attribute, $A_1, A_2, \ldots A_n$ be the other attributes of R. Table T contains tuples $t_0, t_1, \ldots t_s$ such that

$$t_0[A_i] = \begin{cases} \bot & \text{if } i = 0 \\ 1 & \text{if } i > 0 \end{cases}$$

and

$$t_i[A_j] = \begin{cases} 1 & \text{if } A_j \in Y_i \\ i & \text{if } A_j \notin Y_i \end{cases} \quad \text{for } i = 1, 2, \ldots, s \text{ and } j = 1, 2, \ldots, n.$$

If $X \subseteq R_S$, then tuple t_0 is not included in T.

$$\begin{array}{cccc} A_0 & A_1 & \cdots & A_n \\ \hline \bot & 1 & \cdots & 1 \\ i & 1 & \cdots 1 \quad i \cdots & i \end{array}$$

$$\underbrace{\qquad\qquad}_{Y_i}$$

$Y \in \mathcal{Y} \iff \exists 1 \leq i_Y \leq s: Y \subseteq Y_{i_Y}$, so FD $X \to Y$ holds in the strongly possible world obtained by replacing $\bot$ in $t_0[A_0]$ by i_Y. On the other hand, if $Y \notin \mathcal{Y}$, then for every $1 \leq i \leq s$ there exists an attribute $A_{j_i} \in Y \setminus Y_i$, so whichever element $i \in VD_{A_0}$ is put in place of $\bot$ in $t_0[A_0]$, we get that $t_0[X] = t_i[X]$, but $t_0[Y] \neq t_i[Y]$, hence $T \not\models X \to_{sp} Y$. $\qquad\square$

The case when the right hand side of an spFD is fixed can also be characterized. It is clear that for any given table T over a schema R and a fixed set $Y \subset R$ the collection of attribute sets $\mathcal{X} = \{X : T \models X \rightarrow_{sp} Y\}$ forms an up-set. But this is the only condition we have, as the following Proposition shows.

Proposition 5.2. *Let (R, R_S) be a schema, $Y \subset R$ be a fixed set of attributes, furthermore $\mathcal{X}$ be an upset of subsets of $R \setminus Y$. Then there exists a table T over (R, R_S) such that $T \models X \rightarrow_{sp} Y \iff X \in \mathcal{X}$.*

For the proof we recall the Armstrong instance construction for strongly possible keys from [2].

Theorem 5.2. *Suppose that $\Sigma = \{sp \langle K \rangle : K \in \mathcal{K}\}$ is a collection of strongly possible key constraints such that if $|K| = 1$, then $K \subseteq R_S$. Then there exists an Armstrong table for (R, R_S, Σ).*

Proof of Proposition 5.2. If $\emptyset \in \mathcal{X}$, then any table T over (R, R_S) that has exactly one non-NULL value in each attribute of Y works. So assume that $\forall X \in \mathcal{X} : |X| > 0$. Let $U = \{X_1, X_2, \ldots, X_z, X_{z+1}, \ldots, X_u\}$ be the collection of singleton sets in $\mathcal{X}$ such that $X_1, X_2, \ldots, X_z \in R \setminus R_S$ and $X_{z+1}, \ldots, X_u \in R_S$. Construct table T as follows. Let t_0, t_1 be two tuples defined by

$$t_0[A] = \begin{cases} \perp \text{ if } A \in \{X_1, X_2, \ldots, X_z\} \\ 1 \text{ otherwise} \end{cases}$$

$$t_1[A] = \begin{cases} 1 \text{ if } A \in U \cup Y \\ 2 \text{ otherwise} \end{cases}$$

Tuples $t_2, t_3, \ldots t_g$ are given by the construction of the proof of Theorem 5.2 as follows. $t_j[A] = j$ if $A \in U \cup Y$. On columns (attributes) of $R \setminus (U \cup Y)$ we put the Armstrong table constructed for strongly possible key system $\Sigma = \{sp \langle X \rangle : X \cap U = \emptyset$ and $X \in \mathcal{X}\}$ with the minor modification that attribute values are shifted up, so values 1 and 2 do not occur on these tuples.

$U \setminus R_s$		$U \cap R_s$		$R_s \setminus (U \cup Y)$	Y	
X_1	......		X_u			
$\perp$	...	$\perp$ 1	...	1 1	...	1 1 ... 1
1	...	1 1	...	1 2	...	2 1 ... 1
2	...	2 2	...	2	spKey	2 ... 2
:		: :		:	construction	: :
i	...	i i	...	i	(3 ... i)	i ... i

Distinct

We need to show that the table T defined above satisfies $T \models X \rightarrow_{sp} Y \iff X \in \mathcal{X}$. If $X \cap U \neq \emptyset$, then substituting 1 in place of $\perp$ in t_0 we get a strongly possible world that satisfies $X \rightarrow Y$. If $X \cap U = \emptyset$, then the strongly possible world for key system Σ that shows $sp\langle X \rangle$ holds also shows $X \rightarrow_{sp} Y$. Now, for $Z \subseteq R \setminus Y$, if $Z \notin \mathcal{X}$, then $Z \cap U = \emptyset$. Also, since $\mathcal{X}$ is an up-set $\Sigma \not\models sp\langle Z \rangle$ by the characterization of implication between strongly possible keys given in Theorem 4.1 of [2]. Thus, in any strongly possible world there are two tuples from $t_2, t_3, \ldots t_g$ that are equal on Z, hence $T \not\models Z \rightarrow_{sp} Y$.   □

5.1   Basic Properties of spFD's

Levene and Loizou [16] gave an axiomatization of weak and strong functional dependencies together. As a first step towards axiomatisation of strongly possible functional dependencies, here we check these axioms and see which are sound for spFD's as well. As we will see, there are significant differences caused by the limited domains. Let T be any SQL table over (R, R_S).

(1) **Reflexivity** If $Y \subseteq X \subseteq R$ then $T \models X \rightarrow_{sp} Y$ holds. Indeed, in any possible world $X \rightarrow Y$ holds.
(2) **Augmentation** if $T \models X \rightarrow_{sp} Y$ and $W \subseteq R$, then $T \models XW \rightarrow_{sp} YW$. Again, this holds in any possible world. Augmentation rule is a Disjoint Augmentation if $W \cap X = \emptyset$.
(3) **Union** If $T \models X \rightarrow_{sp} Y$ and $T \models X \rightarrow_{sp} Z$ then $T \models X \rightarrow_{sp} YZ$. This is not sound, see the counterexample (Table 2).

Table 2. Union counterexample

X	Y	Z
1	1	1
⊥	1	2
2	2	2

The visible domain of X is $\{1, 2\}$. If $\perp$ is replaced by 1, then $X \rightarrow Y$ is satisfied but $X \rightarrow Z$ does not hold, if $\perp$ is replaced by 2, then just the opposite is the case.

Possible weakenings are
(a) **Mixed-union** If $T \models X \rightarrow_s Y$ and $T \models X \rightarrow_{sp} Z$ then $T \models X \rightarrow_{sp} YZ$ or if $T \models X \rightarrow_{sp} Y$ and $T \models X \rightarrow_s Z$ then $T \models X \rightarrow_{sp} YZ$. This is sound, since strong functional dependencies hold in each possible world, in particular in the strongly possible one giving the other dependency.
(b) **NULL-free union** $T \models X \rightarrow_{sp} Y$ and $T \models X \rightarrow_{sp} Z$ and $X \subseteq R_S$, then $T \models X \rightarrow_{sp} YZ$. To prove the soundness, we need that $\overline{G_{YZ}}$, the complement of the weak similarity graph with respect to YZ can be list colored using extensions on X. It is not hard to see that $\{t_i, t_j\} \in E(\overline{G_{YZ}})$ iff $\{t_i, t_j\} \in E(\overline{G_Y})$ or $\{t_i, t_j\} \in E(\overline{G_Z})$. Note that $X \subseteq R_S$ implies that every tuple is X-total, so by $T \models X \rightarrow_{sp} Y$ if $\{t_i, t_j\} \in E(\overline{G_Y})$, then $t_i[X] \neq t_j[X]$. Similarly, $t_i[X] \neq t_j[X]$ if $\{t_i, t_j\} \in E(\overline{G_Z})$.

(4) **Decomposition** if $T \models X \rightarrow_{sp} YZ$, then $T \models X \rightarrow_{sp} Y$ and $T \models X \rightarrow_{sp} Z$. The strongly possible world T' that satisfies $X \rightarrow YZ$ also satisfies $X \rightarrow Y$ and $X \rightarrow Z$.

(5) **Transitivity** If $T \models X \rightarrow_{sp} Y$ and $T \models Y \rightarrow_{sp} Z$ hold, then $T \models X \rightarrow_{sp} Z$. This is not sound (Table 3).

Table 3. Transitivity counterexample

X	Y	Z
1	1	1
2	$\perp$	1
2	2	2

Again, there is no strongly possible world that satisfies both $T \models X \rightarrow_{sp} Y$ and $T \models Y \rightarrow_{sp} Z$.

(a) **AM-transitivity** Atzeni and Morfuni [3] used a weaker version. If $T \models X \rightarrow_{sp} Y$ and $T \models Y \rightarrow_{sp} Z$ hold, and $Y - X \subseteq R_s$, then $T \models X \rightarrow_{sp} Z$. This is not sound either, as shown by the following counterexample.

Y			
X		**Z**	
1	1	1	1
1	$\perp$	1	2
1	2	2	1
2	2	2	$\perp$

However, some weaker versions are sound.

(b) **Sp-transitivity** $T \models X \rightarrow_{sp} Y$ and $T \models Y \rightarrow_{sp} Z$ hold, and $Y \subseteq R_s$, then $T \models X \rightarrow_{sp} Z$. Indeed, $T \models Y \rightarrow_{sp} Z$ is equivalent list coloring of $\overline{G_Z}$ by extensions of tuples on Y. However, as Y is NULL-free, we obtain that $\{t_i, t_j\} \in E(\overline{G_Z})$ implies $t_i[Y] \neq t_j[Y]$, so $\{t_i, t_j\} \in E(\overline{G_Y})$. That is $\overline{G_Z}$ is a subgraph of $\overline{G_Y}$, hence the list coloring of $\overline{G_Y}$ by extensions of tuples on X is proper also for $\overline{G_Z}$, thus $T \models X \rightarrow_{sp} Z$.

(c) **Mixed-transitivity** We may mix different dependencies like in the Union case. If $T \models X \rightarrow_s Y$ and $T \models Y \rightarrow_{sp} Z$ or if $T \models X \rightarrow_{sp} Y$ and $T \models Y \rightarrow_s Z$ hold, then $T \models X \rightarrow_{sp} Z$. This is sound in the same reason as Mixed-union is sound.

(6) **Pseudo-transitivity** If $T \models X \rightarrow_{sp} Y$ and $T \models YZ \rightarrow_{sp} V$ hold, then $T \models XZ \rightarrow_{sp} V$. Or if $T \models X \rightarrow_{sp} YZ$ and $T \models Y \rightarrow_{sp} V$ hold, then $T \models X \rightarrow_{sp} ZV$. These are not sound as the following counterexamples show (Fig. 3). Sound weaker versions are as follows.

X	Y	Z	V
1	1	1	1
1	⊥	2	1
1	1	2	2
2	2	1	2

X	Y	Z	V
1	1	1	1
1	⊥	1	2
2	2	2	1
3	3	3	2

Fig. 3. Pseudo transitivity counterexamples

(a) **Mixed-pseudo transitivity** Of course, if one of the given spFD's is strong, then the property is sound, since strong functional dependencies hold in every possible world, in particular in the strongly possible world that satisfies the other spFD.

(b) **NULL-free pseudo transitivity** Assume that $Y \subseteq R_S$. Then $T \models X \to_{sp} YZ$ and $T \models Y \to_{sp} V$ imply $T \models X \to_{sp} ZV$. Indeed, $T \models X \to_{sp} YZ$ means that there is a proper coloring of $\overline{G_{YZ}}$ by extensions of tuples on X. $E(\overline{G_{YZ}}) = E(\overline{G_Y}) \cup E(\overline{G_Z})$, thus the coloring satisfies $t_i[X] \neq t_j[X]$ if $t_i[Y] \not\sim_w t_j[Y]$ or $t_i[Z] \not\sim_w t_j[Z]$. On the other hand, $T \models Y \to_{sp} V$ means that there is a proper coloring by extensions of tuples on Y of $\overline{G_V}$. Since $Y \subseteq R_S$, there is one extension of each tuple on Y, so we get that if $t_i[V] \not\sim_w t_j[V]$, then $t_i[Y] \not\sim_w t_j[Y]$. Thus the coloring of tuples by extensions on X given by $T \models X \to_{sp} YZ$ also has the property that $t_i[X] \neq t_j[X]$ if $t_i[V] \not\sim_w t_j[V]$ or $t_i[Z] \not\sim_w t_j[Z]$ yielding $T \models X \to_{sp} ZV$.

(c) **Disjoint NULL-free pseudo transitivity** Unfortunately, the assumption $Y \subseteq R_S$ is not enough to make $(T \models X \to_{sp} Y$ and $T \models YZ \to_{sp} V) \Rightarrow T \models XZ \to_{sp} V$ sound. For example, if $X = Z$, then the following is a counterexample (Table 4).

Table 4. NULL-free pseudo transitivity counterexample

X	Y	V
1	1	1
⊥	2	2
2	2	3

Replacing ⊥ by 1 we have $T \models XY \to_{sp} V$, replacing by 2 we get $T \models X \to_{sp} Y$. However, as $|VD_X| = 2$, $T \not\models XZ \to_{sp} V$. This can be mended by requiring disjointness. Assume that $Y \subseteq R_S$. If $X \cap Z = \emptyset$ then $T \models X \to_{sp} Y$ and $T \models YZ \to_{sp} V$ imply $T \models XZ \to_{sp} V$. Indeed, $T \models YZ \to_{sp} V$ means that $\overline{G_V}$ can be properly colored by extensions of tuples on YZ, that is if $t_i[V] \not\sim_w t_j[V]$, then $t_i[YZ] \neq t_j[YZ]$. To have

$T \models XZ \rightarrow_{sp} V$ we need a coloring such that if $t_i[V] \not\sim_w t_j[V]$, then $t_i[XZ] \neq t_j[XZ]$. On Z take the extensions given by $t_i[YZ] \neq t_j[YZ]$. Now, if $t_i[V] \not\sim_w t_j[V]$ but $t_i[Z] = t_j[Z]$, then $t_i[Y] \neq t_j[Y]$ which implies $t_i[Y] \not\sim_w t_j[Y]$ since $Y \subseteq R_s$. For X then take the extensions obtained from $T \models X \rightarrow_{sp} Y$. Since $X \cap Z = \emptyset$, these are independent of the extensions on Z, we get that $t_i[XZ] \neq t_j[XZ]$ if $t_i[V] \not\sim_w t_j[V]$.

(7) **Chaining** If $T \models X \rightarrow_{sp} Y$, $T \models XY \rightarrow_s W$, and $T \models XW \rightarrow_{sp} Z$, with $W \cap Y = \emptyset$, then $T \models X \rightarrow_{sp} Z$. This not sound, see the counterexample (Table 5).

Table 5. Chaining counterexample

X	Y	W	Z
3	3	2	2
1	$\perp$	1	$\perp$
2	$\perp$	1	2
$\perp$	2	1	3
1	1	1	1

We get sound rule if one of the spFD of the premise is changed into strong FD.

(a) **Strongly possible Chaining:** If $T \models X \rightarrow_s Y$, $T \models XY \rightarrow_s W$, and $T \models XW \rightarrow_{sp} Z$, or $T \models X \rightarrow_{sp} Y$, $T \models XY \rightarrow_s W$, and $T \models XW \rightarrow_s Z$, with $W \cap Y = \emptyset$, then $T \models X \rightarrow_{sp} Z$. Indeed, the strongly possible world that satisfies the spFD in the premise also satisfies the strong FD's, so by the soundness follows from the fact that the rule is sound for ordinary FD's.

(8) **Weakening** If $T \models X \rightarrow_s Y$ then $T \models X \rightarrow_{sp} Y$. Also, if $T \models X \rightarrow_{sp} Y$ then $T \models X \rightarrow_w Y$. This follows from the definitions.

(9) **Strengthening** If $T \models X \rightarrow_s Y$, $T \models Y \rightarrow_{sp} W$, and $T \models XW \rightarrow_s Z$ with $W \cap Z = \emptyset$, then $T \models X \rightarrow_s Z$ provided the following condition is satisfied:

$$\exists \text{ a sequence } \{A_i\} \, (i \in I), \text{ with } \bigcup_{i \in I} A_i = X \cap Y, \text{ such that } \forall i \in I \, \exists V_i \subseteq R$$

such that $S_i = (Y - X) \cup \{A_j | A_j \in X \cap Y \text{ and } j < i\}, T \models S_i \rightarrow_{sp} V_i$ and $\exists V \subseteq XYV_i$ such that $V - A_i \rightarrow_s A_i$.

It was proven in [16] that strengthening is sound if the spFD's are replaced by wFD's. However that means that the conclusion follows from weaker conditions then as stated above, so they remain true if we assume spFD's in place of wFD's.

(10) **Composition** If $T \models X \rightarrow_{sp} Y$ and $T \models A \rightarrow_{sp} B$ hold, then $T \models XA \rightarrow_{sp} YB$. This is not sound as the following counterexample shows (Table 6).

Table 6. Composition counterexample

X/A	X ∩ A	A/X	Y	B
1	1	1	1	1
1	⊥	1	2	1
1	2	1	2	2

We obtain sound rules as follows.

(a) **Mixed composition** Replace one of the spFD's of the premise by strongFD, so if $T \models X \rightarrow_s Y$ and $T \models A \rightarrow_{sp} B$ hold, then $T \models XA \rightarrow_{sp} YB$, or if $T \models X \rightarrow_{sp} Y$ and $T \models A \rightarrow_s B$ hold, then $T \models XA \rightarrow_{sp} YB$. Both of these versions are sound, because the strongly possible world that verifies the spFD in the premise satisfies the strongFD, as well, hence the conclusion is valid in the same strongly possible world, by properties of ordinary FD's.

(b) **NULL-free composition** If $T \models X \rightarrow_{sp} Y$ and $T \models A \rightarrow_{sp} B$ hold with $YA \subseteq R_S$, then $T \models XA \rightarrow_{sp} YB$. This is sound, since $T \models X \rightarrow_{sp} Y$ implies $T \models XA \rightarrow_{sp} YA$ and $T \models A \rightarrow_{sp} B$ implies $T \models YA \rightarrow_{sp} YB$ by Augmentation rule. These latter two imply $T \models XA \rightarrow_{sp} YB$ by Sp-transitivity.

(c) **Disjoint composition** If $T \models X \rightarrow_{sp} Y$ and $T \models A \rightarrow_{sp} B$ hold with $X \cap A = \emptyset$, then $T \models XA \rightarrow_{sp} YB$. This is sound, since $XA \rightarrow_{sp} YB$ holds if and only if $\overline{G_{YB}}$ can be list colored by extensions of tuples on XA. $\{t_1, t_2\} \in E(\overline{G_{YB}})$ means $t_1[YB] \sim_w t_2[YB]$ if and only if $\exists$ an attribute $C \in Y \cup B$ such that $\bot \neq t_1[C] \neq t_2[C] \neq \bot$. If $C \in Y$ then $\{t_1, t_2\} \in E(\overline{G_Y})$. If $C \in B$ then $\{t_1, t_2\} \in E(\overline{G_B})$, hence $E(\overline{G_{YB}}) = E(\overline{G_Y}) \cup E(\overline{G_B})$. An extension of a tuple $t \in T$ on XA can be obtained by taking an extension on X and matching it up with an extension on A.

   i. $X \rightarrow_{sp} Y$: $\exists$ a coloring tuples of T by extensions on X that is a proper coloring of $\overline{G_Y}$.

   ii. $A \rightarrow_{sp} B$: $\exists$ a coloring tuples of T by extensions on A that is a proper coloring of $\overline{G_B}$.

   Now for each $t \in T$, for $t[X]$ use (a) and for $t[A]$ use (b). So that let $\{t_1, t_2\} \in E(\overline{G_{YB}})$, if $\{t_1, t_2\} \in E(\overline{G_Y})$ then the coloring differs on the X part and if $\{t_1, t_2\} \in E(\overline{G_B})$ then the coloring differs on the A part.

From this list of rules we may deduce that if a rule contains only one spFD in its premise then it remains sound. However, more than one spFD's in the premise usually cause problems, since by the limitations of visible domains, the different spFD's do not hold in the same strongly possible world. For a complete axiomatization this problem must be handled. In particular, the fact that composition does not hold in general makes usual proof methods of completeness virtually unusable.

An interesting special case of transitivity is the following proposition.

Proposition 5.3. *Let T be a table over (R, R_S) and $X_1, X_2, \ldots X_m \in R$ be attributes such that $T \models X_i \to_{sp} X_{i+1}$ for $i = 1, 2, \ldots, m$ where indices are understood modulo m. Then $T \models X_{i+1} \to_{sp} X_i$ for $i = 1, 2, \ldots, m$ also holds.*

Proof. $T \models X_i \to_{sp} X_{i+1}$ implies by pigeon hole principle that $|VD_{i+1}| \leq |VD_i|$, hence by the cyclicity we get that $|VD_{i+1}| = |VD_i|$ for $i = 1, 2, \ldots, m$. Let i be an arbitrarily fixed index and $VD_i = \{a_1, a_2, \ldots a_k\}$. Also assume without loss of generality that $t_j[X_i] = a_\ell$ for $p_{\ell-1} < j \leq p_\ell$, where $0 = p_0 < p_1 < \ldots < p_k \leq n$ and $t_j[X_i] = \perp$ for $p_k < j \leq n$ where the number of tuples of the table is n. Clearly, there cannot be two tuples t_g, t_h such that $p_d < g < h \leq p_{d+1}$ with $t_g[X_{i+1}] \neq t_h[X_{i+1}]$ and both not NULL. If there exist $j \leq p_b < j' \leq p_\ell$ such that $t_j[X_{i+1}] = t_{j'}[X_{i+1}] \neq \perp$, then there must be visible domain value β of X_{i+1} that occurs only in tuples where X_i has the value $\perp$. However, in that case any strongly possible world would have two tuples that agree on X_i but differ on X_{i+1} (one value in X_{i+1} is β). From this it follows, that there is a substitution of the NULL's with visible domain values such that $t_a[X_i] = t_b[X_i] \iff t_a[X_{i+1}] = t_b[X_{i+1}]$, so $T \models X_{i+1} \to_{sp} X_i$. $\qquad\qquad\square$

It would be tempting to guess that more could be proven, that is $T \models X_j \to_{sp} X_i$ holds for any i, j. However, that is not the case, as the following example shows (Table 7).

Table 7. $X_1 \not\to_{sp} X_3$

X_1	X_2	X_3	X_4
1	1	1	1
1	$\perp$	2	$\perp$
2	2	$\perp$	2
2	$\perp$	$\perp$	$\perp$

6   Conclusions

In the present paper we continued the investigations started in [2] of strongly possible worlds of SQL tables with NULL's. We gave a polynomial time algorithm to verify whether a given set K of attributes is a strongly possible key. We also showed that to do the same for an arbitrary system of $\Sigma = \{sp\langle K_i\rangle\colon i = 1, 2, \ldots n\}$ of strongly possible key constraints is NP-complete. Then we investigated rules established for strong and weak functional dependencies and determined which ones remain sound for strongly possible functional dependencies and for those that failed we gave appropriate weakenings. This is a large step toward a possible axiomatization of spFD's. The main obstacle is that different spFD's in the premises of rules may hold in different strongly possible worlds that are incompatible with each other. Further research is needed to find how to incorporate this into the axiom system.

Another direction of future investigations is finding a robust definition of closures with respect to strongly possible functional dependencies. As the composition rule is not sound, the usual way of $X^+ = \{A \colon T \models X \rightarrow_{sp} A\}$ may result in $T \not\models X \rightarrow_{sp} X^+$, which is an undesired event.

Finally, the main application of functional dependencies is lossless decompositions of database tables to eliminate redundancy and the possibilities of inconsistent updates. Recent works of Köhler and Link [15], furthermore Wei and Link [20] show how to use c-FD's or embedded functional dependencies for that. Our future research will include similar analysis for spFD's.

References

1. Acuña, E., Rodriguez, C.: The treatment of missing values and its effect on classifier accuracy. In: Banks, D., McMorris, F.R., Arabie, P., Gaul, W. (eds.) Classification, Clustering, and Data Mining Applications. STUDIES CLASS. Springer, Berlin, Heidelberg (2004). https://doi.org/10.1007/978-3-642-17103-1_60

2. Alattar, M., Sali, A.: Keys in relational databases with nulls and bounded domains. In: Welzer, T., Eder, J., Podgorelec, V., Kamišalić Latifić, A. (eds.) ADBIS 2019. LNCS, vol. 11695, pp. 33–50. Springer, Cham (2019). https://doi.org/10.1007/978-3-030-28730-6_3

3. Atzeni, P., Morfuni, N.M.: Functional dependencies and constraints on null values in database relations. Inf. Process. Lett. **18**(4), 233–238 (1984)

4. Sree Dhevi, A.T.: Imputing missing values using Inverse Distance Weighted Interpolation for time series data. In: Sixth International Conference on Advanced Computing (ICoAC), Chennai, pp. 255–259 (2014). https://doi.org/10.1109/ICoAC.2014.7229721

5. Chang, G., Ge, T.: Comparison of missing data imputation methods for traffic flow. In: Proceedings 2011 International Conference on Transportation, Mechanical, and Electrical Engineering (TMEE), Changchun, pp. 639–642 (2011). https://doi.org/10.1109/TMEE.2011.6199284

6. Cheng, C., Wei L., Lin, T.: Improving relational database quality based on adaptive learning method for estimating null value. In: Second International Conference on Innovative Computing, Information and Control (ICICIC 2007), Kumamoto, p. 81 (2007). https://doi.org/10.1109/ICICIC.2007.350

7. Date, C.J.: *NOT Is Not "Not"!* (Notes on Three-Valued Logic and Related Matters) in Relational Database Writings 1985–1989. Addison-Wesley, Reading (1990)

8. Farhangfar, A., Kurgan, L.A., Pedrycz, W.: Experimental analysis of methods for imputation of missing values in databases. In: Proceedings of the SPIE, Intelligent Computing: Theory and Applications II, 12 April 2004, vol. 5421 (2004). https://doi.org/10.1117/12.542509

9. Farhangfar, A., Kurgan, L.A., Pedrycz, W.: A novel framework for imputation of missing values in databases. IEEE Trans. Syst. Man Cybern. Part A Syst. Hum. **37**(5), 692–709 (2007)

10. Farhangfar, A., Kurgan, L., Dy, J.: Impact of imputation of missing values on classification error for discrete data. Pattern Recogn. **41**(12), 3692–3705 (2008)

11. Grahne, G.: Dependency satisfaction in databases with incomplete information. In: Proceedings of the Tenth International Conference of Very Large Data Base, Singapore, August 1984. vldb.org

12. Grzymala-Busse, J.W., Hu, M.: A comparison of several approaches to missing attribute values in data mining. In: Ziarko, W., Yao, Y. (eds.) RSCTC 2000. LNCS (LNAI), vol. 2005, pp. 378–385. Springer, Heidelberg (2001). https://doi.org/10.1007/3-540-45554-X_46
13. Imielinski, T., Lipski, W.: Incomplete information in relational databases. JACM **31**(4), 761–791 (1984)
14. Köhler, H., Leck, U., Link, S., Zhou, X.: Possible and certain keys for SQL. VLDB J. **25**(4), 571–596 (2016)
15. Köhler, H., Link, S.: SQL schema design: foundations, normal forms, and normalization. Inf. Syst. **76**, 88–113 (2018)
16. Levene, M., Loizou, G.: Axiomatisation of functional dependencies in incomplete relations. J. Theor. Comput. Sci. **206**(1–2), 283–300 (1998)
17. Lien, Y.E.: On the equivalence of database models. J. ACM **29**(2), 333–362 (1982)
18. Lovász, L., Plummer, M.D.: Matching Theory, vol. 367. American Mathematical Society, Providence (2009)
19. Vassilou, Y.: Functional dependencies and incomplete information. In: Proceedings of International Conference on Very Large Databases, VLDB 1980, pp. 260–269 (1980)
20. Wei, Z., Link, S.: Embedded functional dependencies and data-completeness tailored database design. PVLDB **12**(11), 1458–1470 (2019)
21. Zhang, S., Qin, Z., Ling, C.X., Sheng, S.: "Missing is Useful": missing values in cost-sensitive decision trees. IEEE Trans. Knowl. Data Eng. **17**(12), 1689–1693 (2005)

Normal Forms of Conditional Knowledge Bases Respecting Entailments and Renamings

Christoph Beierle and Jonas Haldimann$^{(\boxtimes)}$ [iD]

Faculty of Mathematics and Computer Science,
FernUniversität in Hagen, 58084 Hagen, Germany
`jonas.haldimann@fernuni-hagen.de`

Abstract. Normal forms of conditional knowledge bases are useful to create, process and compare the knowledge represented by them. In this paper, we propose the reduced antecedent normal form (RANF) for conditional knowledge bases. Compared to the antecedent normal form, it represents conditional knowledge with significantly fewer conditionals. A set of transformation rules maps every knowledge base to a model equivalent knowledge base in RANF. The new notion of renaming normal form (ρNF) of a conditional knowledge base takes signature renamings into account. We develop an algorithm for systematically generating conditional knowledge bases over a given signature that are both in RANF and in ρNF. The generated knowledge bases are consistent, pairwise not antecedentwise equivalent and pairwise not equivalent under signature renaming. Furthermore, the algorithm is complete in the sense that, taking signature renamings and model equivalence into account, every consistent knowledge base is generated.

Keywords: Conditional knowledge base · Antecedent normal form · ANF · Reduced antecedent normal form · RANF · Signature renaming · Renaming normal form · ρNF · Knowledge base generation

1 Introduction

For describing objects, situations, relationships, problems, solutions, etc. in a formal representation, adequacy, expressivity, and manageability of the used representation language are of crucial importance. This typically leads to many possibilities to express something in syntactically different, albeit semantically equivalent ways. On the other hand, the idea of normal forms is to reduce this redundancy by focussing on some standardized way of representation. Besides supporting standardized representations, benefits of normal forms include easier comparisons, avoidance of notorious borderline cases, fewer exceptions and thus clearer algorithms, and easier organization of proofs. Regarding the cost of normal forms, it has to be noted, though, that some machine-oriented normal forms

© Springer Nature Switzerland AG 2020
A. Herzig and J. Kontinen (Eds.): FoIKS 2020, LNCS 12012, pp. 22–41, 2020.
https://doi.org/10.1007/978-3-030-39951-1_2

may reduce human readability. Normal forms have been studied and employed extensively e.g. in logic formalisms or in automated theorem proving [18].

In this paper, we deal with normal forms of conditional knowledge bases. A conditional knowledge base is a set of conditionals representing a defeasible rule *If A then usually B*. Such conditionals play a central role in nonmonotonic reasoning, and different semantic approaches have been defined for them (cf. [1,9–15,17]). Normal forms of conditional knowledge bases have been investigated in e.g. [3,4,7]. Here, we will propose two new normal forms for conditional knowledge bases. The *reduced antecedent normal form (RANF)* extends the notion of antecedent normal form introduced in [7] by taking non-monotonic entailments sanctioned by system P [14] into account. The *renaming normal form (ρNF)* which we define for knowledge bases consisting only of so-called *normal form conditionals*, respects renamings of the underlying signature. While the RANF takes the semantic notion of model equivalence into account, the ρNF refers to renaming equivalence on the syntactic level. For both normal forms, we study their key properties. After providing a transformation system converting a conditional knowledge base into a model-equivalent knowledge base in RANF, we address the question of systematically generating knowledge bases that are both in RANF and in ρNF. We present an algorithm $KB^{\rho ra}$ generating knowledge bases over a given signature that are consistent, pairwise not antecedent-wise equivalent and pairwise not equivalent under signature renaming. We also show that, when taking signature renamings and model equivalence into account, every consistent knowledge base is generated, thus providing an excellent base for empirical evaluations and comparisons of conditional knowledge bases.

The rest of this paper is organized in the following way. After recalling the required basics in Sect. 2, the reduced antecedentwise normal form is introduced in Sect. 3, and the transformation system yielding a knowledge base in RANF is presented in Sect. 4. The renaming normal form ρNF is developed in Sect. 5, and the systematic generation of knowledge bases is presented in Sect. 6. In Sect. 7 we conclude and point out future work.

2   Background: Conditional Logic

Let $\mathcal{L}$ be a propositional language over a finite signature Σ of atoms $a, b, c, \dots$. The formulas of $\mathcal{L}$ will be denoted by letters $A, B, C, \dots$. We write AB for $A \wedge B$ and $\overline{A}$ for $\neg A$. We identify the set of all complete conjunctions over Σ with the set Ω of possible worlds over $\mathcal{L}$. For $\omega \in \Omega$, $\omega \models A$ means that $A \in \mathcal{L}$ holds in ω, and the set of worlds satisfying A is $\Omega_A = \{\omega \mid \omega \models A\}$. By introducing a new binary operator $\mid$, we obtain the set $(\mathcal{L} \mid \mathcal{L}) = \{(B|A) \mid A, B \in \mathcal{L}\}$ of *conditionals* over $\mathcal{L}$. For a conditional $r = (B|A)$, $ant(r) = A$ is the *antecedent* of r, and $cons(r) = B$ is its *consequent*. The *counter conditional* of $r = (B|A)$ is $\overline{r} = (\overline{B}|A)$. As semantics for conditionals, we use *ordinal conditional functions (OCF)* [19]. An OCF is a function $\kappa : \Omega \to \mathbb{N}$ expressing degrees of plausibility of possible worlds where a lower degree denotes "less surprising". At least one world must be regarded as being normal; therefore, $\kappa(\omega) = 0$ for at least one

$\omega \in \Omega$. Each κ uniquely extends to a function mapping sentences to $\mathbb{N} \cup \{\infty\}$ given by $\kappa(A) = \min\{\kappa(\omega) \mid \omega \models A\}$ where $\min \emptyset = \infty$. An OCF κ *accepts* a conditional $(B|A)$, written $\kappa \models (B|A)$, if the verification of the conditional is less surprising than its falsification, i.e., if $\kappa(AB) < \kappa(A\overline{B})$; equivalently, $\kappa \models (B|A)$ iff for every $\omega' \in \Omega_{A\overline{B}}$ there is $\omega \in \Omega_{AB}$ with $\kappa(\omega) < \kappa(\omega')$. A conditional $(B|A)$ is trivial if it is *self-fulfilling* $(A \models B)$ or *contradictory* $(A \models \overline{B})$; a set of conditionals is self-fulfilling if every conditional in it is self-fulfilling. A finite set $\mathcal{R} \subseteq (\mathcal{L}|\mathcal{L})$ of conditionals is called a *knowledge base*. An OCF κ accepts $\mathcal{R}$ if κ accepts all conditionals in $\mathcal{R}$, and $\mathcal{R}$ is *consistent* if an OCF accepting $\mathcal{R}$ exists [11]. We extend the set of knowledge bases by the special symbol $\diamond$, denoting an inconsistent knowledge base. $Mod(\mathcal{R})$ denotes the set of all OCFs κ accepting $\mathcal{R}$. Two knowledge bases $\mathcal{R}, \mathcal{R}'$ are *model equivalent*, denoted by $\mathcal{R} \equiv_{mod} \mathcal{R}'$, if $Mod(\mathcal{R}) = Mod(\mathcal{R}')$. We say $(B|A) \equiv (B'|A')$ if $A \equiv A'$ and $AB \equiv A'B'$, where $\equiv$ denotes classical propositional equivalence. Example 1 presents a knowledge base we will use for illustration.

Example 1 ($\mathcal{R}_{book}$). Let $\Sigma_{book} = \{b, p, t\}$ be a propositional signature where b indicates whether something is a book, p indicates whether something is made out of paper, and t indicates whether something contains text. The knowledge base $\mathcal{R}_{book}$ contains five conditionals:

q_1: $(p|pb)$ "Books printed on paper are usually made out of paper."
q_2: $(p|b)$ "Books are usually made out of paper."
q_3: $(t|b)$ "Books usually contain text."
q_4: $(t|bp)$ "Books made of paper usually contain text"
q_5: $(\overline{t}|b\overline{p})$ "Books that are not made out of paper usually do not contain text."

System P [14] allows reasoning about conditional knowledge bases. It consists of the six axioms that are displayed in Fig. 1. If a conditional $(B|A)$ can be derived from a conditional knowledge base $\mathcal{R}$ by applying the rules in system P, we denote this by $A \mathrel{\vfill\vartriangleright}_{\mathcal{R}} B$. It has been shown (see [1,10,14,16]) that system P inference coincides with p-entailment [11] where A p-entails B in the context of $\mathcal{R}$ iff all models of $\mathcal{R}$ accept the conditional $(B|A)$. Moreover, $A \mathrel{\vfill\vartriangleright}_{\mathcal{R}} B$ holds iff $\mathcal{R} \cup \{(\overline{B}|A)\}$ is inconsistent [11].

3   Reduced Antecedentwise Normal Form

Instead of the semantic equivalence notion $\equiv_{mod}$ we will now consider syntactic notions of equivalences of knowledge bases and corresponding normal forms. For comparing or generating knowledge bases, it is useful to abstract from merely syntactic variants. In particular, it is desirable to have minimal versions and normal forms of knowledge bases at hand. The idea of elementwise equivalence [4] is that each piece of knowledge in one knowledge base directly corresponds to a piece of knowledge in the other knowledge base.

(RE) *Reflexivity*: $\qquad\qquad\qquad\qquad A \mathrel{\vdash\mkern-7mu\sim} A$

(LLE) *Left Logical Equivalence*: $\dfrac{\models A \leftrightarrow B,\quad A \mathrel{\vdash\mkern-7mu\sim} C}{B \mathrel{\vdash\mkern-7mu\sim} C}$

(RW) *Right Weakening*: $\dfrac{\models B \rightarrow C,\quad A \mathrel{\vdash\mkern-7mu\sim} B}{A \mathrel{\vdash\mkern-7mu\sim} C}$

(AND) *And*: $\dfrac{A \mathrel{\vdash\mkern-7mu\sim} B,\quad A \mathrel{\vdash\mkern-7mu\sim} C}{A \mathrel{\vdash\mkern-7mu\sim} B \wedge C}$

(OR) *Or*: $\dfrac{A \mathrel{\vdash\mkern-7mu\sim} C,\quad B \mathrel{\vdash\mkern-7mu\sim} C}{A \vee B \mathrel{\vdash\mkern-7mu\sim} C}$

(CM) *Cautious Monotonicity*: $\dfrac{A \mathrel{\vdash\mkern-7mu\sim} B,\quad A \mathrel{\vdash\mkern-7mu\sim} C}{A \wedge B \mathrel{\vdash\mkern-7mu\sim} C}$

Fig. 1. Axioms of system P (cf. [14]) with $A, B, C \in \mathcal{L}$.

Definition 1 (equivalence $\equiv_{ee}$ [4]). *Let $\mathcal{R}$, $\mathcal{R}'$ be knowledge bases.*

- *$\mathcal{R}$ is an elementwise equivalent sub-knowledge base of $\mathcal{R}'$, denoted by $\mathcal{R} \ll_{ee}$ $\mathcal{R}'$, if for every conditional $(B|A) \in \mathcal{R}$ that is not self-fulfilling there is a conditional $(B'|A') \in \mathcal{R}'$ such that $(B|A) \equiv (B'|A')$.*
- *$\mathcal{R}$ and $\mathcal{R}'$ are strictly elementwise equivalent if $\mathcal{R} \ll_{ee} \mathcal{R}'$ and $\mathcal{R}' \ll_{ee} \mathcal{R}$.*
- *$\mathcal{R}$ and $\mathcal{R}'$ are elementwise equivalent, denoted by $\mathcal{R} \equiv_{ee} \mathcal{R}'$, if either both are inconsistent, or both are consistent and strictly elementwise equivalent.*

The knowledge bases $\mathcal{R}_1 = \{(a|\top), (b|\top), (ab|\top)\}$ and $\mathcal{R}_2 = \{(a|\top), (b|\top)\}$ are model equivalent, but not elementwise equivalent since for $(ab|\top) \in \mathcal{R}_1$ there is no corresponding conditional in $\mathcal{R}_2$. The idea of antecedentwise equivalence [7] is to take into account the *set* of conditionals having the same (or propositionally equivalent) antecedent when comparing to knowledge bases.

Definition 2 ($Ant(\mathcal{R})$, $\mathcal{R}_{|A}$, ANF [7]). *Let $\mathcal{R}$ be a knowledge base.*

- *$Ant(\mathcal{R}) = \{A \mid (B|A) \in \mathcal{R}\}$ is the set of antecedents of $\mathcal{R}$.*
- *For $A \in Ant(\mathcal{R})$, the set $\mathcal{R}_{|A} = \{(B'|A') \mid (B'|A') \in \mathcal{R} \text{ and } A \equiv A'\}$ is the set of A-conditionals in $\mathcal{R}$.*
- *$\mathcal{R}$ is in antecedent normal form (ANF) if either $\mathcal{R}$ is inconsistent and $\mathcal{R} = \diamond$, or $\mathcal{R}$ is consistent, does not contain any self-fulfilling conditional, contains only conditionals of the form $(AB|A)$, and $|\mathcal{R}_{|A}| = 1$ for all $A \in Ant(\mathcal{R})$.*

Definition 3 ($\ll_{ae}$, equivalence $\equiv_{ae}$ [7]). *Let $\mathcal{R}$, $\mathcal{R}'$ be knowledge bases.*

- *$\mathcal{R}$ is an antecedentwise equivalent sub-knowledge base of $\mathcal{R}'$, denoted by $\mathcal{R} \ll_{ae} \mathcal{R}'$, if for every $A \in Ant(\mathcal{R})$ such that $\mathcal{R}_{|A}$ is not self-fulfilling there is an $A' \in Ant(\mathcal{R}')$ with $\mathcal{R}_{|A} \equiv_{mod} \mathcal{R}'_{|A'}$.*

- $\mathcal{R}$ and $\mathcal{R}'$ are strictly antecedentwise equivalent if $\mathcal{R} \ll_{ae} \mathcal{R}'$ and $\mathcal{R}' \ll_{ae} \mathcal{R}$.
- $\mathcal{R}$ and $\mathcal{R}'$ are antecedentwise equivalent, denoted by $\mathcal{R} \equiv_{ae} \mathcal{R}'$, if either both are inconsistent, or both are consistent and strictly antecedentwise equivalent.

Note that any two inconsistent knowledge bases are also antecedentwise equivalent e.g., $\{(b|a), (\bar{b}|a)\} \equiv_{ae} \{(b|b), (a\bar{a}|\top)\}$, thus avoiding cumbersome case distinctions when dealing with consistent and inconsistent knowledge bases. In general, we have:

Proposition 1 ($\equiv_{ae}$ [7]). *Let $\mathcal{R}, \mathcal{R}'$ be consistent knowledge bases.*

1. *If $\mathcal{R} \ll_{ae} \mathcal{R}'$ then $Mod(\mathcal{R}') \subseteq Mod(\mathcal{R})$.*
2. *If $\mathcal{R} \equiv_{ae} \mathcal{R}'$ then $\mathcal{R} \equiv_{mod} \mathcal{R}'$.*
3. *If $\mathcal{R} \ll_{ee} \mathcal{R}'$ then $\mathcal{R} \ll_{ae} \mathcal{R}'$.*
4. *If $\mathcal{R} \equiv_{ee} \mathcal{R}'$ then $\mathcal{R} \equiv_{ae} \mathcal{R}'$.*
5. *None of the implications (1.)–(4.) holds in general in the reverse direction.*

Thus, under antecedentwise equivalence more knowledge bases are equivalent than under elementwise equivalence. For instance, for $\mathcal{R}_5 = \{(bc|a), (cd|a)\}$ and $\mathcal{R}_6 = \{(bd|a), (bcd|a)\}$ we have $\mathcal{R}_5 \equiv_{ae} \mathcal{R}_6$, but $\mathcal{R}_5 \not\equiv_{ee} \mathcal{R}_6$.

While $\equiv_{ae}$ is a better approximation of $\equiv_{mod}$ than $\equiv_{ee}$, the following example demonstrates a characteristic feature of model equivalence that is not covered by antecedentwise equivalence.

Example 2 ($\equiv_{ae}$). Let $\mathcal{R}_7 = \{(bc|a), (c|ab)\}$ and $\mathcal{R}_8 = \{(bc|a), (b|ac)\}$. Then $\mathcal{R}_7 \not\equiv_{ae} \mathcal{R}_8$, but $\mathcal{R}_7 \equiv_{mod} \mathcal{R}_8$. Furthermore, also for $\mathcal{R}'_7 = \{(abc|a), (abc|ab)\}$ and $\mathcal{R}'_8 = \{(abc|a), (abc|ac)\}$ which are both in ANF, we have $\mathcal{R}'_7 \not\equiv_{ae} \mathcal{R}'_8$, but $\mathcal{R}'_7 \equiv_{mod} \mathcal{R}'_8$.

The knowledge bases $\mathcal{R}_7$ and $\mathcal{R}_8$ from Example 2 both contain a conditional that is entailed from the other conditionals in that knowledge base using the axioms of system P, i.e., $(c|ab)$ in $\mathcal{R}_7$ and $(b|ac)$ in $\mathcal{R}_8$, respectively. Therefore, we now introduce a new normal form prohibiting this kind of redundancy.

Definition 4 (reduced form, RANF). *Let $\mathcal{R}$ be a knowledge base.*

- $\mathcal{R}$ *is in* reduced form *(with respect to system P) if there is no conditional $(B|A) \in \mathcal{R}$ such that $A \mathrel{\vdash\!\!\!\!\sim}_{\mathcal{R}\setminus(B|A)} B$.*
- $\mathcal{R}$ *is in* reduced antecedent normal form (RANF) *if $\mathcal{R}$ is in ANF and in reduced form.*

Example 3 ($\equiv_{ae}$). None of the four knowledge bases from Example 2 is in RANF, while, e.g., $\mathcal{R}_9 = \{(abc|a)\}$ is in RANF. Both $\mathcal{R}_{10} = \{(ac|a), (bc|b), (ac \vee bc|a \vee b)\}$ and $\mathcal{R}_{11} = \{(ac|a), (bc|b)\}$ are in ANF, but only the obviously simpler knowledge base $\mathcal{R}_{11}$ is in RANF.

4   Transforming Knowledge Bases into Reduced ANF

In order to be able to deal with normal forms of formulas in $\mathcal{L}$ without having to select a specific representation, we assume a function ν mapping a propositional formula A to a unique normal form $\nu(A)$ such that $A \equiv A'$ iff $\nu(A) = \nu(A')$. We also use a function Π with $\Pi(\mathcal{R}) = \diamond$ iff $\mathcal{R}$ is inconsistent; Π can easily be implemented by the tolerance test for conditional knowledge bases [11]. Using Π and the propositional normalization function ν, the system Θ^{ra} given in Fig. 2 contains five transformation rules:

(RE) *reduction*: $\qquad \dfrac{\mathcal{R} \cup \{(B|A)\}}{\mathcal{R}} \qquad \Pi((\mathcal{R} \setminus \{(B|A)\}) \cup \{(\overline{B}|A)\}) = \diamond,\ A \not\equiv \bot$

(SF) *self-fulfilling*: $\qquad \dfrac{\mathcal{R} \cup \{(B|A)\}}{\mathcal{R}} \qquad A \models B,\ A \not\equiv \bot$

(AE) *antecedence*: $\qquad \dfrac{\mathcal{R} \cup \{(B|A),(B'|A')\}}{\mathcal{R} \cup \{(BB'|A)\}} \quad A \equiv A'$

(NO) *normalization*: $\qquad \dfrac{\mathcal{R} \cup \{(B|A)\}}{\mathcal{R} \cup \{(\nu(AB)|\nu(A))\}} \quad A \neq \nu(A) \text{ or } B \neq \nu(AB)$

(IC) *inconsistency*: $\qquad \dfrac{\mathcal{R}}{\diamond} \qquad \mathcal{R} \neq \diamond, \Pi(\mathcal{R}) = \diamond$

Fig. 2. Transformation rules Θ^{ra} and their applicability conditions for the transformation of knowledge bases into RANF; Π is a consistency test, e.g. the tolerance criterion [11], and ν a normalization function for propositional formulas.

(RE) removes a conditional $(B|A)$ with $A \not\equiv \bot$ from $\mathcal{R}$ if A p-entails B in the context of $\mathcal{R} \setminus \{(B|A)\}$.

(SF) removes a self-fulling conditional $(B|A)$ with $A \not\equiv \bot$.

(AE) merges two conditionals $(B|A)$ and $(B'|A')$ with propositionally equivalent antecedents to a conditional having this antecedent and the conjunction of the consequents.

(NO) transforms a conditional $(B|A)$ by sharpening its consequent to the conjunction with its antecedent and propositionally normalizes the antecedent and the resulting consequent.

(IC) transforms an inconsistent knowledge base into $\diamond$.

Example 4. Consider the knowledge base $\mathcal{R}_{book}$ from Example 1.

(SF) The conditional q_1 is self-fulfilling. Application of (SF) removes q_1 from the knowledge base.

(AE) The conditionals q_2 and q_3 share the same antecedence. The application of (AE) replaces them with $q_6 : (pt|b)$.

(RE) The knowledge base $\{q_6, q_5\}$ p-entails q_4. Therefore q_4 is removed by the application of (RE).

(NO) Normalizing the conditionals q_5 and q_6 in the knowledge base yields q'_5 : $(\nu(tbp)|\nu(bp))$ and $q'_6 : (\nu(\overline{tb}\overline{p})|\nu(b\overline{p}))$.

By applying Θ^{ra} exhaustively to $\mathcal{R}_{book}$, we get $\mathcal{R}'_{book} = \{(\nu(tbp)|\nu(bp)), (\nu(\overline{tb}\overline{p})|\nu(b\overline{p}))\}$ which is in RANF and model equivalent to $\mathcal{R}_{book}$.

In contrast to the transformation system Θ given in [7] yielding the unique ANF for every conditional knowledge base, the system Θ^{ra} is not confluent.

Example 5 (Non-confluence of Θ^{ra}). Consider the knowledge base $\mathcal{R} = \{(\nu(ab)|\nu(a)), (\nu(ab)|\nu(b)), (\nu((a \lor c)d)|\nu(a \lor c)), (\nu((b \lor c)d)|\nu(b \lor c)), (\nu(aef)|\nu(ae))\}$. It is already in ANF, but it is not reduced.

The third or the forth conditional can be removed with (RE), yielding either $\mathcal{R}' = \{(\nu(ab)|\nu(a)), (\nu(ab)|\nu(b)), (\nu((a \lor c)d)|\nu(a \lor c)), (\nu(aef)|\nu(ae))\}$ or $\mathcal{R}'' = \{(\nu(ab)|\nu(a)), (\nu(ab)|\nu(b)), (\nu((b \lor c)d)|\nu(b \lor c)), (\nu(aef)|\nu(ae))\}$ after exhaustive application of (NO). Neither $\mathcal{R}'$ nor $\mathcal{R}''$ can be reduced further by Θ^{ra}. Both $\mathcal{R}'$ and $\mathcal{R}''$ are in RANF, and they are two different normal forms of $\mathcal{R}$ with respect to Θ^{ra}.

Because Θ^{ra} is not confluent, the result of applying the rules Θ^{ra} to a knowledge base $\mathcal{R}$ is not uniquely determined in general. In the following, $\Theta^{ra}(\mathcal{R})$ will denote the set of all knowledge bases that can be obtained from $\mathcal{R}$ by exhaustively applying the transformation rules of Θ^{ra} to $\mathcal{R}$ in some arbitrary order.

Proposition 2 (properties of Θ^{ra}). *Let $\mathcal{R}$ be a knowledge base, and let $\mathcal{R}' \in \Theta^{ra}(\mathcal{R})$.*

1. **(termination)** Θ^{ra} *is terminating.*
2. **(consistency)** $\mathcal{R}$ *is consistent iff $\Theta^{ra}(\mathcal{R}) \neq \{\diamond\}$.*
3. **($\equiv_{mod}$ soundness)** $\mathcal{R} \equiv_{mod} \mathcal{R}'$.
4. **(RANF)** $\mathcal{R}'$ *is in reduced antecedent normal form.*

Proof. (1.) The transformation rules (RE), (SF) and (AE) each remove one conditional. (NO) can be applied at most once per conditional, and (IC) can be applied at most once in the whole transformation. Therefore Θ^{ra} is terminating.

(3.) We will check that $\equiv_{mod}$ is preserved by every rule in Θ^{ra}. (RE) is applied if $(\mathcal{R} \setminus \{(B|A)\}) \cup \{(\overline{B}|A)\}$ is inconsistent. In this case $A \mathrel{\vdash\!\!\!\sim}_{\mathcal{R} \setminus \{(B|A)\}} B$, i.e. every model of $\mathcal{R} \setminus \{(B|A)\}$ accepts $(B|A)$. Therefore removing the conditional $(B|A)$ from $\mathcal{R}$ does not affect the set of models. It was shown that (SF), (AE), (NO), and (IC) preserve model equivalence in [7, Proof of Prop. 2, ($\equiv_{mod}$ correctness)]. Therefore, $\mathcal{R} \equiv_{mod} \mathcal{R}'$.

(2.) Let $\mathcal{R}$ be an inconsistent knowledge base. (3.) implies that the application of the rules in Θ^{ra} to $\mathcal{R}$ cannot lead to a consistent knowledge base. Thus every exhaustive application of rules in Θ^{ra} will apply (IC) at one point. Therefore $\Theta^{ra}(\mathcal{R}) = \{\diamond\}$.

(4.) $\mathcal{R}'$ is in ANF, otherwise (SF), (AE), (NO), or (IC) could be applied, and $\mathcal{R}'$ is in reduced form, otherwise (RE) could be applied.

□

5   Renamings and Renaming Normal Form (ρNF)

There are knowledge bases that are identical except for the names of their variables. E.g., the knowledge bases $\mathcal{R}_1 = \{(a|b), (a|c)\}$ and $\mathcal{R}_2 = \{(c|b), (c|a)\}$ become equal if we swap the names for the variables a and c in one of them. When generating knowledge bases, we are only interested to store one of such knowledge bases that are identical except for a signature renaming.

Definition 5 (renaming, $\simeq$). *Let Σ be a signature. We call a bijective function $\rho : \Sigma \to \Sigma$ a (signature) renaming. A renaming is lifted canonically to formulas, worlds, conditionals, knowledge bases, and sets thereof as usual.*

Two formulas, worlds, conditionals, knowledge bases, or sets thereof X, X' are called equivalent under signature renaming, *denoted as $X \simeq X'$, if there exists a renaming ρ such that $X' = \rho(X)$.*

For a set M, $m \in M$, and an equivalence relation $\equiv$ on M, the set of equivalence classes induced by $\equiv$ is denoted by $[M]_{/\equiv}$, and the unique equivalence class containing m is denoted by $[m]_{\equiv}$. It is easy to see that equivalence under signature renaming is an equivalence relation. Thus, for instance, for $\Sigma_{ab} = \{a, b\}$ the only non-identity renaming is the function ρ_{ab} with $\rho_{ab}(a) = b$ and $\rho_{ab}(b) = a$, $[\Omega_{\Sigma_{ab}}]_{/\simeq} = \{[ab], [a\bar{b}, \bar{a}b], [\bar{a}\bar{b}]\}$ are the three equivalence classes of worlds over Σ_{ab}, and we have $[(ab|ab \vee a\bar{b})]_{\simeq} = [(ab|ab \vee \bar{a}b)]_{\simeq}$.

An important observation is that renaming has no influence on whether a knowledge base is in ANF or in RANF, respectively, or not:

Proposition 3. *Let R, R' be knowledge bases such that $\mathcal{R} \simeq \mathcal{R}'$. It holds that:*

- *$\mathcal{R}$ is in ANF iff $\mathcal{R}'$ is in ANF.*
- *$\mathcal{R}$ is in RANF iff $\mathcal{R}'$ is in RANF.*

For developing a method for the systematic generation of knowledge bases in RANF while also taking renamings into account, we will represent each formula $A \in \mathcal{L}$ uniquely by its set Ω_A of satisfying worlds. The two conditions $B \subsetneq A$ and $B \neq \emptyset$ then ensure the falsifiability and the verifiability of a conditional $(B|A)$, thereby excluding any trivial conditional [6]. This yields a propositional normalization function ν as employed in Θ^{ra} (Fig. 2), giving us:

Proposition 4 (NFC(Σ) [8]). *For $NFC(\Sigma) = \{(B|A) \mid A \subseteq \Omega_\Sigma, B \subsetneq A, B \neq \emptyset\}$, the set of normal form conditionals over a signature Σ, the following holds:*

(nontrivial) *$NFC(\Sigma)$ does not contain any trivial conditional.*
(complete) *For every nontrivial conditional over Σ there is an equivalent conditional in $NFC(\Sigma)$.*
(minimal) *All conditionals in $NFC(\Sigma)$ are pairwise non-equivalent.*

For instance, for Σ_{ab} we have $(\{ab, a\bar{b}\}|\{ab, \bar{a}b\}) \equiv (\{ab\}|\{ab, \bar{a}b\})$ where the latter is in $NFC(\Sigma_{ab})$. Out of the different 256 conditionals over Σ_{ab} obtained when using sets of worlds as formulas, only 50 are in $NFC(\Sigma_{ab})$ [8].

For defining a linear order on $NFC(\Sigma)$, we use the following notation. For an ordering relation $\leqslant$ on a set M, its lexicographic extension to strings over M is denoted by $\leqslant_{lex}$. For ordered sets $S, S' \subseteq M$ with $S = \{e_1, \ldots, e_n\}$ and $S' = \{e'_1, \ldots, e'_{n'}\}$ where $e_i \leqslant e_{i+1}$ and $e'_j \leqslant e'_{j+1}$ its extension $\leqslant_{set}$ to sets is:

$$S \leqslant_{set} S' \text{ iff } n < n', \text{ or } n = n' \text{ and } e_1 \ldots e_n \leqslant_{lex} e'_1 \ldots e'_{n'} \tag{1}$$

For Σ with ordering $<$, $[\![\omega]\!]_<$ is the usual interpretation of a world ω as a binary number; e.g., for Σ_{ab} with $a < b$, $[\![ab]\!]_< = 3$, $[\![a\bar{b}]\!]_< = 2$, $[\![\bar{a}b]\!]_< = 1$, and $[\![\bar{a}\bar{b}]\!]_< = 0$.

Definition 6 (induced ordering on formulas and conditionals). *Let Σ be a signature with linear ordering $<$. The orderings induced by $<$ on worlds ω, ω' and conditionals $(B|A), (B'|A')$ over Σ are given by:*

$$\omega \overset{w}{\leqslant} \omega' \text{ iff } [\![\omega]\!]_< \geqslant [\![\omega']\!]_< \tag{2}$$

$$(B|A) \overset{c}{\leqslant} (B'|A') \text{ iff } \Omega_A \overset{w}{<}_{set} \Omega_{A'}, \text{ or } \Omega_A = \Omega_{A'} \text{ and } \Omega_B \overset{w}{\leqslant}_{set} \Omega_{B'} \tag{3}$$

In order to ease our notation, we will omit the upper symbol in $\overset{w}{<}$ and $\overset{c}{<}$, and write just $<$ instead, and analogously $\leqslant$ for the non-strict variants. For instance, for Σ_{ab} with $a < b$ we have $ab < a\bar{b} < \bar{a}b < \bar{a}\bar{b}$ for worlds, and $(ab|ab \vee a\bar{b}) < (ab|ab \vee \bar{a}b)$ and $(ab \vee \bar{a}b|ab \vee a\bar{b} \vee \bar{a}b) < (\bar{a}b|ab \vee a\bar{b} \vee \bar{a}b \vee \bar{a}\bar{b})$ for conditionals.

Proposition 5 ($NFC(\Sigma)$, $<$ [8]). *For a linear ordering $<$ on a signature Σ, the induced ordering $<$ according to Definition 6 is a linear ordering on $NFC(\Sigma)$.*

Obviously, a signature renaming can map a conditional only to a conditional in the same equivalence class induced by $\simeq$, leading to the following observations.

Proposition 6. *Let Σ be a signature, $\mathcal{R}, \mathcal{R}' \subseteq NFC(\Sigma)$ knowledge bases, and ρ a renaming on Σ. Then it holds that:*

$$\rho(\mathcal{R}) = \mathcal{R}' \quad \Leftrightarrow \quad \text{for every } [r]_\simeq \in [NFC(\Sigma)]_{/\simeq} : \rho(\mathcal{R} \cap [r]_\simeq) = \mathcal{R}' \cap [r]_\simeq$$

$$\mathcal{R} \simeq \mathcal{R}' \quad \Rightarrow \quad \text{for every } [r]_\simeq \in [NFC(\Sigma)]_{/\simeq} : |\mathcal{R} \cap [r]_\simeq| = |\mathcal{R}' \cap [r]_\simeq|$$

The first observation states that a renaming ρ maps $\mathcal{R}$ to $\mathcal{R}'$ if and only if for every equivalence class $[r]_\simeq$ wrt. $\simeq$ it maps the conditionals in $\mathcal{R}$ from $[r]_\simeq$ to the conditionals in $\mathcal{R}'$ from $[r]_\simeq$. Later on, we will use this observation to check whether two knowledge bases $\mathcal{R}, \mathcal{R}'$ are equivalent under renaming by checking if $\rho(\mathcal{R} \cap [r]_\simeq) = \mathcal{R}' \cap [r]_\simeq$ for every equivalence class $[r]_\simeq \in [NFC(\Sigma)]_{/\simeq}$. The second observation in Proposition 6 is an immediate consequence of the first observation.

Given the ordering $<$ on $NFC(\Sigma)$ from Proposition 5, we will now define a new ordering $\prec$ on these conditionals that takes signature renamings into account and prioritizes the conditionals according to the $<$-minimal elements in each $\simeq$-induced equivalence class.

Definition 7 ($NFC(\Sigma)$, $\prec$). *Given a signature Σ with linear ordering $<$, let $[NFC(\Sigma)]_{/\simeq} = \{[r_1]_\simeq, \ldots, [r_m]_\simeq\}$ be the equivalence classes of $NFC(\Sigma)$ induced by renamings such that for each $i \in \{1, \ldots, m\}$, the conditional r_i is the minimal element in $[r_i]_\simeq$ with respect to $<$, and $r_1 < \ldots < r_m$. The conditionals $\{r_1, \ldots, r_m\}$ are the* canonical normal form conditionals *over Σ. Let $M_i = [r_i]_\simeq \setminus \{r_i\}$ denote the equivalence class of the canonical normal form conditional r_i without r_i itself. The* canonical ordering *on $NFC(\Sigma)$ induced by $<$, denoted by $\prec$, is given by the schema*

$$r_1 \prec M_1 \prec r_2 \prec M_2 \prec \ldots \prec r_m \prec M_m$$

where $r \prec r'$ iff $r < r'$ for all $r, r' \in M_i$ with $i \in \{1, \ldots, m\}$.

For instance, while $NFC(\Sigma_{ab})$ contains 50 conditionals, there are 31 equivalence classes in $[NFC(\Sigma_{ab})]_{/\simeq}$; hence $NFC(\Sigma_{ab})$ has 31 canonical normal form conditionals. Note that the ordering defined in Definition 7 differs form the corresponding ordering $\prec$ defined in [8] because in [8] all canonical conditionals are $\prec$-smaller than all non-canonical conditionals. The complete ordering $\prec$ on $NFC(\Sigma_{ab})$ is given in Table 1.

Proposition 7 ($NFC(\Sigma)$, $\prec$). *For a linear ordering $<$ on a signature Σ, the induced ordering $\prec$ according to Definition 7 is a linear ordering on $NFC(\Sigma)$. Furthermore, its extension to sets of conditionals $\prec_{set}$ as given by Eq. (1) is a linear ordering on the set of knowledge bases using $NFC(\Sigma)$.*

In the following, we will abbreviate $\mathcal{R} \prec_{set} \mathcal{R}'$ simply by $\mathcal{R} \prec \mathcal{R}'$ for knowledge bases $\mathcal{R}, \mathcal{R}'$, and analogously for the non-strict version $\preccurlyeq_{set}$. Furthermore, for every set $R \subseteq NFC(\Sigma)$, the maximal element of $\mathcal{R}$ with respect to $\prec$ will be denoted by $\max_\prec(\mathcal{R})$ or simply by $\max(\mathcal{R})$. Using these notations, we can now introduce a new normal form for conditional knowledge bases that takes renamings into account.

Definition 8 (ρNF). *A knowledge base $\mathcal{R} \subseteq NFC(\Sigma)$ is in* renaming normal form *(ρNF) if for every knowledge base $\mathcal{R}'$ with $\mathcal{R} \simeq \mathcal{R}'$ it holds that $\mathcal{R} \preccurlyeq \mathcal{R}'$.*

Note that we defined the ρNF just for knowledge bases containing only normal form conditionals. However, for every knowledge base $\mathcal{R}$, a corresponding knowledge base in ρNF exists that modulo renaming has the same models as $\mathcal{R}$, or that is even renaming equivalent to $\mathcal{R}$, depending on whether $\mathcal{R}$ contains arbitrary or only normal form conditionals. More precisely, we have:

Proposition 8 (ρ**NF**). *Let $\mathcal{R}$ be a consistent knowledge base over Σ.*

(i) There is a knowledge base $\mathcal{R}'$ over Σ in ρNF and a renaming ρ such that $R \equiv_{mod} \rho(R')$.

(ii) Furthermore, if $\mathcal{R}$ contains only normal form conditionals from $NFC(\Sigma)$, then there is a unique knowledge base $\mathcal{R}'$ in ρNF such that $R \simeq R'$.

Table 1. The 50 conditionals $r_{01.1} \prec r_{01.2} \prec \ldots \prec r_{30.2} \prec r_{31.1}$ in $NFC(\Sigma_{ab})$ for $\Sigma_{ab} = \{a, b\}$, and their 31 equivalence classes $[01], \ldots, [31]$. Formulas in conditionals are given by sets of worlds, and worlds are represented by their binary number interpretation; e.g., $r_{03.1}$: $(\{3\}|\{3, 0\})$ stands for $(\{ab\}|\{ab, \overline{a}\overline{b}\})$.

Class	First conditional	Second conditional
[01]	$r_{01.1}$: $(\{3\}|\{3, 2\})$	$r_{01.2}$: $(\{3\}|\{3, 1\})$
[02]	$r_{02.1}$: $(\{2\}|\{3, 2\})$	$r_{02.2}$: $(\{1\}|\{3, 1\})$
[03]	$r_{03.1}$: $(\{3\}|\{3, 0\})$	
[04]	$r_{04.1}$: $(\{0\}|\{3, 0\})$	
[05]	$r_{05.1}$: $(\{2\}|\{2, 1\})$	$r_{05.2}$: $(\{1\}|\{2, 1\})$
[06]	$r_{06.1}$: $(\{2\}|\{2, 0\})$	$r_{06.2}$: $(\{1\}|\{1, 0\})$
[07]	$r_{07.1}$: $(\{0\}|\{2, 0\})$	$r_{07.2}$: $(\{0\}|\{1, 0\})$
[08]	$r_{08.1}$: $(\{3\}|\{3, 2, 1\})$	
[09]	$r_{09.1}$: $(\{2\}|\{3, 2, 1\})$	$r_{09.2}$: $(\{1\}|\{3, 2, 1\})$
[10]	$r_{10.1}$: $(\{3, 2\}|\{3, 2, 1\})$	$r_{10.2}$: $(\{3, 1\}|\{3, 2, 1\})$
[11]	$r_{11.1}$: $(\{2, 1\}|\{3, 2, 1\})$	
[12]	$r_{12.1}$: $(\{3\}|\{3, 2, 0\})$	$r_{12.2}$: $(\{3\}|\{3, 1, 0\})$
[13]	$r_{13.1}$: $(\{2\}|\{3, 2, 0\})$	$r_{13.1}$: $(\{1\}|\{3, 1, 0\})$
[14]	$r_{14.1}$: $(\{0\}|\{3, 2, 0\})$	$r_{14.1}$: $(\{0\}|\{3, 1, 0\})$
[15]	$r_{15.1}$: $(\{3, 2\}|\{3, 2, 0\})$	$r_{15.2}$: $(\{3, 1\}|\{3, 1, 0\})$
[16]	$r_{16.1}$: $(\{3, 0\}|\{3, 2, 0\})$	$r_{16.2}$: $(\{3, 0\}|\{3, 1, 0\})$
[17]	$r_{17.1}$: $(\{2, 0\}|\{3, 2, 0\})$	$r_{17.2}$: $(\{1, 0\}|\{3, 1, 0\})$
[18]	$r_{18.1}$: $(\{2\}|\{2, 1, 0\})$	$r_{18.2}$: $(\{1\}|\{2, 1, 0\})$
[19]	$r_{19.1}$: $(\{0\}|\{2, 1, 0\})$	
[20]	$r_{20.1}$: $(\{2, 1\}|\{2, 1, 0\})$	
[21]	$r_{21.1}$: $(\{2, 0\}|\{2, 1, 0\})$	$r_{21.2}$: $(\{1, 0\}|\{2, 1, 0\})$
[22]	$r_{22.1}$: $(\{3\}|\{3, 2, 1, 0\})$	
[23]	$r_{23.1}$: $(\{2\}|\{3, 2, 1, 0\})$	$r_{23.2}$: $(\{1\}|\{3, 2, 1, 0\})$
[24]	$r_{24.1}$: $(\{0\}|\{3, 2, 1, 0\})$	
[25]	$r_{25.1}$: $(\{3, 2\}|\{3, 2, 1, 0\})$	$r_{25.2}$: $(\{3, 1\}|\{3, 2, 1, 0\})$
[26]	$r_{26.1}$: $(\{3, 0\}|\{3, 2, 1, 0\})$	
[27]	$r_{27.1}$: $(\{2, 1\}|\{3, 2, 1, 0\})$	
[28]	$r_{28.1}$: $(\{2, 0\}|\{3, 2, 1, 0\})$	$r_{28.2}$: $(\{1, 0\}|\{3, 2, 1, 0\})$
[29]	$r_{29.1}$: $(\{3, 2, 1\}|\{3, 2, 1, 0\})$	
[30]	$r_{30.1}$: $(\{3, 2, 0\}|\{3, 2, 1, 0\})$	$r_{30.2}$: $(\{3, 1, 0\}|\{3, 2, 1, 0\})$
[31]	$r_{31.1}$: $(\{2, 1, 0\}|\{3, 2, 1, 0\})$	

While in case (ii) in Proposition 8 the knowledge base R' is uniquely determined, there may be several different knowledge bases $\mathcal{R}'$ satisfying case (i) of Proposition 8. Naturally, requiring a renaming ρ such that $R = \rho(R')$ as in case

(ii) is a syntax oriented restriction that is a much stricter condition than the model oriented restriction requiring $R \equiv_{mod} \rho(R')$ as in case (i). In the remaining parts of this paper, we will focus on automatically enumerating knowledge bases in ρNF. The task of designing an algorithm for transforming a knowledge base $\mathcal{R}$ into a knowledge base $\mathcal{R}'$ that is in ρNF and that satisfies Proposition 8 will be addressed to future work.

Removing the $\prec$-maximal element form a knowledge base that is in ρNF yields a knowledge base that is also in ρNF.

Proposition 9. *Let $\mathcal{R}$ be a knowledge base in ρNF and $r = \max(\mathcal{R})$. Then $\mathcal{R} \setminus \{r\}$ is in ρNF as well.*

Proof. Let $\mathcal{Q} = \mathcal{R} \setminus \{r\}$. Assume there is a renaming ρ, such that $\rho(\mathcal{Q}) \prec \mathcal{Q}$. Then is $\rho(\mathcal{R}) \prec \mathcal{R}$ because r is the maximum of $\mathcal{R}$. This is a contradiction to the assumption that $\mathcal{R}$ is in ρNF. $\square$

This property will be exploited in the following for systematically generating knowledge bases in ρNF.

6   Generating Knowledge Bases in RANF and ρNF

In this section, we will introduce two algorithms to generate consistent knowledge bases that are both in RANF and ρNF. The first algorithm, $KB_{basic}^{\rho ra}$, clarifies the basic approach used here to generate knowledge bases. The second algorithm, $KB^{\rho ra}$, obtains the same result as $KB_{basic}^{\rho ra}$ but is computationally more efficient.

6.1   Basic Algorithm $KB_{basic}^{\rho ra}$

$KB_{basic}^{\rho ra}$ (Algorithm 1) is a basic algorithm to generate consistent knowledge bases in RANF, that are not pairwise equivalent under renaming. To do so, it starts with an empty set of conditionals and tries to extend it by one conditional at a time. Proposition 2 ensures that it is sufficient to generate only knowledge bases over $NFC(\Sigma)$ (cf. transformation rule (NO) in Fig. 2). To avoid generating the same knowledge base multiple times, the algorithm $KB_{basic}^{\rho ra}$ will add conditionals to a knowledge base in the order of $\prec$ as induced by the order $\lessdot$ on Σ (cf. Definition 7).

Proposition 10 $(KB_{basic}^{\rho ra})$. *Let Σ be a signature with a linear ordering $\lessdot$. Then applying $KB_{basic}^{\rho ra}$ terminates and returns a set $\mathcal{KB}$ for which the following holds:*

1. (correctness)	*If $\mathcal{R} \in \mathcal{KB}$ then $\mathcal{R}$ is a knowledge base over Σ.*
2. (consistency)	*If $\mathcal{R} \in \mathcal{KB}$ then $\mathcal{R}$ is consistent.*
3. (RANF)	*If $\mathcal{R} \in \mathcal{KB}$ then $\mathcal{R}$ is in RANF.*
4. (ρNF)	*If $\mathcal{R} \in \mathcal{KB}$ then $\mathcal{R}$ is in ρNF.*
5. ($\simeq$ minimality)	*If $\mathcal{R}, \mathcal{R}' \in \mathcal{KB}$ and $\mathcal{R} \neq \mathcal{R}'$ then $\mathcal{R} \not\simeq \mathcal{R}'$.*

Algorithm 1. KB_{basic}^{pra} – Basic algorithm to generate knowledge bases over Σ with order $\prec$ that are in RANF and ρNF

Require: signature Σ with order $\prec$
Ensure: set $\mathcal{KB}$ of knowledge bases over Σ in *reduced antecedentwise normal form* (RANF) and in *renaming normal form* (ρNF) that are consistent, pairwise not antecedentwise equivalent and pairwise not equivalent under signature renaming
1: $k \leftarrow 0$
2: $L_0 \leftarrow \{\emptyset\}$
3: **while** $L_k \neq \emptyset$ **do** $\triangleright L_k$ contains all generated KBs with k conditionals
4:    $L_{k+1} \leftarrow \emptyset$
5:    **for** $\mathcal{R} \in L_k$ **do**
6:       **for** $r \in NFC(\Sigma)$ **do** $\triangleright$ For every possible extension...
7:          **if** $\max(\mathcal{R}) \prec r$
8:             **and** $\mathcal{R} \cup \{r\}$ is consistent $\triangleright$ ...that is consistent...
9:             **and** $\mathcal{R} \cup \{r\}$ is in RANF $\triangleright$ ...in RANF ...
10:            **and** $\mathcal{R} \cup \{r\}$ is in ρNF $\triangleright$ ...and in ρNF ...
             then
11:               $L_{k+1} \leftarrow L_{k+1} \cup \{\mathcal{R} \cup \{r\}\}$ $\triangleright$ ...add it to L_{k+1}.
12:    $k \leftarrow k + 1$
13: **return** $\mathcal{KB} \leftarrow L_1 \cup \cdots \cup L_k$

---

6. ($\equiv_{ae}$ **minimality**)   If $\mathcal{R}, \mathcal{R}' \in \mathcal{KB}$ and $\mathcal{R} \neq \mathcal{R}'$ then $\mathcal{R} \not\equiv_{ae} \mathcal{R}'$.
7. ($\simeq$ **completeness**)   If $\mathcal{R}$ is a consistent knowledge base in RANF over Σ then there is $\mathcal{R}' \in \mathcal{KB}$ and a signature renaming ϱ such that $\mathcal{R} = \varrho(\mathcal{R}')$.
8. ($\equiv_{mod}$ **completeness**)   If $\mathcal{R}$ is a consistent knowledge base over Σ then there is $\mathcal{R}' \in \mathcal{KB}$ and a signature renaming ϱ such that $\mathcal{R} \equiv_{mod} \varrho(\mathcal{R}')$.

Proof. We prove the stated properties one by one.

(1.) is clear, as the algorithm returns a set of sets of conditionals by design.

(2.) is ensured by the condition in Line 8.

(3.) is ensured by the condition in Line 9.

(4.) is ensured by the condition in Line 10.

(5.) Assume that there are two knowledge bases $\mathcal{R}, \mathcal{R}' \in \mathcal{KB}$ such that $\mathcal{R} \neq \mathcal{R}'$ but $\mathcal{R} \simeq \mathcal{R}'$. Then (4.) implies that both knowledge bases are in ρNF. With Proposition 8 it follows that $\mathcal{R} = \mathcal{R}'$ because $\mathcal{R} \simeq \mathcal{R}'$ and both are in ρNF. This is a contradiction to the assumption.

(6.) Assume that there are two knowledge bases $\mathcal{R}, \mathcal{R}' \in \mathcal{KB}$ such that $\mathcal{R} \neq \mathcal{R}'$ but $\mathcal{R} \equiv_{ae} \mathcal{R}'$. Because $\mathcal{R}$ and $\mathcal{R}'$ are in ANF, the sets $\mathcal{R}_{|A}$ and $\mathcal{R}'_{|A}$ contain at most one element for every antecedent A. Therefore, $\mathcal{R} \equiv_{ae} \mathcal{R}'$ implies that every conditional in $\mathcal{R}$ is equivalent to a conditional in $\mathcal{R}'$ and vice versa, i.e. $\mathcal{R} \equiv_{ee} \mathcal{R}'$. Because all conditionals in $\mathcal{R}$ and $\mathcal{R}'$ are in normal form, this implies $\mathcal{R} = \mathcal{R}'$, a contradiction to the assumption.

(7.) For proving (7.), we first prove the following by induction over k:

For $k \geqslant 0$ it holds that every consistent knowledge base $\mathcal{R}$ in *RANF* and in ρNF with $|\mathcal{R}| = k$ is contained in L_k.   (4)

Base case: The empty set is the only knowledge base with 0 conditionals. Therefore the statement holds for $k = 0$, as $L_0 = \{\emptyset\}$.

Induction step: Let $\mathcal{R}$ be a consistent knowledge base in RANF and ρNF with $|\mathcal{R}| = k$. Let $q := \max(\mathcal{R})$. Then $\mathcal{Q} := \mathcal{R} \setminus \{q\}$ is in ρNF as well (cf. Proposition 9). Moreover, $\mathcal{Q}$ is consistent and in RANF. Hence, the induction hypothesis implies $\mathcal{Q} \in L_{k-1}$. Therefore, $\mathcal{R} = \mathcal{Q} \cup \{q\}$ is added to L_k by $KB_{basic}^{\rho ra}$ because $\max(\mathcal{Q}) \prec q$ and $\mathcal{R}$ is consistent, in RANF, and in ρNF.

Now let $\mathcal{R}$ be a consistent knowledge base in RANF. With Proposition 8 (ii) it follows that there is a renaming ρ and a knowledge base $\mathcal{R}'$ in ρNF with $\mathcal{R}' = \rho(\mathcal{R})$. Because $\mathcal{R}$ is consistent and in RANF, $\mathcal{R}'$ is consistent and in RANF. With (4) it follows that $\mathcal{R}' \in L_{|\mathcal{R}'|} \subseteq \mathcal{KB}$.

(8.) Let $\mathcal{R}$ be a consistent knowledge base over Σ. Then there is a knowledge base $\mathcal{R}''$ in RANF with $\mathcal{R}'' \equiv_{mod} \mathcal{R}$ (cf. Proposition 2). From (7.) it follows that there is a renaming ρ and a knowledge base $\mathcal{R}' \in \mathcal{KB}$ such that $\mathcal{R}'' = \rho(\mathcal{R}')$. Therefore, $\mathcal{R} \equiv_{mod} \rho(\mathcal{R}')$. □

Example 6. Let $\Sigma = \{a, b\}$ and, using the notation as in Table 1,

$$\mathcal{R}_1 = \{r_{01.1}\colon (\{3\}|\{3, 2\}), r_{01.2}\colon (\{3\}|\{3, 1\}), r_{05.1}\colon (\{2\}|\{2, 1\})\},$$
$$\mathcal{R}_2 = \{r_{01.1}\colon (\{3\}|\{3, 2\}), r_{05.2}\colon (\{1\}|\{2, 1\})\},$$
$$\mathcal{R}_3 = \{r_{01.1}\colon (\{3\}|\{3, 2\}), r_{01.2}\colon (\{3\}|\{3, 1\}), r_{05.2}\colon (\{1\}|\{2, 1\})\}.$$

Then $\mathcal{R}_1, \mathcal{R}_2 \in KB_{basic}^{\rho ra}(\Sigma)$, but $\mathcal{R}_3 \notin KB_{basic}^{\rho ra}(\Sigma)$ because $\mathcal{R}_3 \simeq \mathcal{R}_1$ and $\mathcal{R}_1 \prec \mathcal{R}_3$.

6.2   Improved Algorithm $KB^{\rho ra}$

In order to improve Algorithm $KB_{basic}^{\rho ra}$ by decreasing its search space, we take a closer look on how to evaluate the conditions in lines 7 to 10 in $KB_{basic}^{\rho ra}$. Similar to the well-known apriori algorithm determining candidates for extending frequent item sets [2] and also similar to the algorithms introduced in [7,8] that generate systematically knowledge bases, but in less elaborated normal forms, we can keep track of a set C of conditionals that might be extensions of $\mathcal{R}$. In this section, we will thus refine $KB_{basic}^{\rho ra}$, thereby refining and largely extending the concept of candidates and candidate elimination for knowledge base generation. Furthermore, we employ significant search space improvements for ensuring that all generated knowledge bases are in ρNF.

Definition 9 (excluded conditionals, $EXC(r)$). *Given a signature Σ, let $r \in NFC(\Sigma)$. The set $EXC(r)$, called the set of conditionals excluded by r, is:*

$$EXC(r) = \{r' \in NFC(\Sigma) \mid r' \preccurlyeq r \text{ or } ant(r) = ant(r') \text{ or } \{r, r'\} \text{ is inconsistent}\}$$

Whenever Algorithm $KB_{basic}^{\rho ra}$ adds a conditional r to a knowledge base, all conditionals r' in $EXC(r)$ can be excluded from the set of candidates for further extensions. If $r' \preccurlyeq r$ (or $\{r, r'\}$ is inconsistent or $ant(r) = ant(r')$, respectively) the condition in Line 7 (or in 8 or in 9, respectively) of KB_{basic}^{pra} will prohibit the extension of $\mathcal{R} \cup \{r\}$ by r'. Note that $EXC(r)$ is independent of $\mathcal{R}$ and can be pre-computed before running KB^{pra}. For instance, for $r = r_{04.1}$: $(\{0\}|\{3,0\}) \in NFC(\Sigma_{ab})$ (cf. Table 1) we have:

$$EXC(r) = \{r_{01.1}: (\{3\}|\{3,2\}), r_{01.2}: (\{3\}|\{3,1\}), \dots, r_{03.1}: (\{3\}|\{3,0\}),$$
$$r_{12.1}: (\{3\}|\{3,2,0\}), r_{12.2}: (\{3\}|\{3,1,0\}), r_{22.1}: (\{3\}|\{3,2,1,0\}))\}$$

Another improvement regards the check for ρNF in Line 10 of Algorithm KB_{basic}^{pra}. Checking for ρNF in Line 10 naively requires to consider all signature renamings over Σ. For convenience, we will identify the renamings of Σ with the set Perm_Σ of permutations over Σ, with id being the identity. Thus, there are $|\mathrm{Perm}_\Sigma| = |\Sigma|!$ many renamings over Σ. The following proposition provides a criterion for reducing the number of renamings that have to be taken into account if the conditional to be added to $\mathcal{R}$ is $\prec$-larger than the maximal conditional in $\mathcal{R}$.

Proposition 11. *Let $\mathcal{R}$ be a knowledge base and $r \in NFC(\Sigma)$ such that $\max(\mathcal{R}) \prec r$ and $\mathcal{R} \setminus [r]_\simeq$ is in ρNF. Then $\mathcal{R} \cup \{r\}$ is in ρNF iff for all renamings ρ it holds that:*

$$\rho(\mathcal{R} \setminus [r]_\simeq) = \mathcal{R} \setminus [r]_\simeq \quad \Rightarrow \quad (\mathcal{R} \cup \{r\}) \cap [r]_\simeq \preccurlyeq \rho((\mathcal{R} \cup \{r\}) \cap [r]_\simeq)$$

Proof. Due to Proposition 8, from $\mathcal{R}$ being in ρNF, we conclude that $\mathcal{R} \setminus [r]_\simeq$ is in ρNF. Let $\mathcal{S} = \mathcal{R} \cup \{r\}$. Then we have:

$$\mathcal{S} \text{ in } \rho\text{NF} \tag{5}$$

$$\Leftrightarrow \text{ for every renaming } \rho \text{ it holds that } \mathcal{S} \preccurlyeq \rho(\mathcal{S}) \tag{6}$$

$$\Leftrightarrow \text{ for every renaming } \rho \text{ it holds that } \mathcal{S} \setminus [r]_\simeq \prec \rho(\mathcal{S} \setminus [r]_\simeq)$$
$$\text{or } \mathcal{S} \setminus [r]_\simeq = \rho(\mathcal{S} \setminus [r]_\simeq) \text{ and } \mathcal{S} \cap [r]_\simeq \preccurlyeq \rho(\mathcal{S} \cap [r]_\simeq) \tag{7}$$

$$\Leftrightarrow \text{ for every renaming } \rho \text{ it holds that}$$
$$\mathcal{S} \setminus [r]_\simeq = \rho(\mathcal{S} \setminus [r]_\simeq) \quad \Rightarrow \quad \mathcal{S} \cap [r]_\simeq \preccurlyeq \rho(\mathcal{S} \cap [r]_\simeq) \tag{8}$$

The equivalence between (6) and (7) holds, because a renaming can only map a conditional to another conditional in the same equivalence class (cf. Proposition 6) and the conditionals in $[r]_\simeq$ are greater than all conditionals in $\mathcal{S} \setminus [r]_\simeq$. The equivalence between (7) and (8) holds because $\mathcal{S} \setminus [r]_\simeq$ is in ρNF and therefore, $\mathcal{S} \setminus [r]_\simeq \neq \rho(\mathcal{S} \setminus [r]_\simeq)$ implies $\mathcal{S} \setminus [r]_\simeq \prec \rho(\mathcal{S} \setminus [r]_\simeq)$. $\square$

Proposition 11 implies that for verifying that $\mathcal{R} \cup \{r\}$ is in ρNF if $\max(\mathcal{R}) \prec r$ and $\mathcal{R} \setminus [r]_\sim$ is in ρNF, it is sufficient to check

$$(\mathcal{R} \cup \{r\}) \cap [r]_\sim \preccurlyeq \rho((\mathcal{R} \cup \{r\}) \cap [r]_\sim)$$

for all $\rho \in P := \{\rho \mid (\mathcal{R} \cup \{r\}) \setminus [r]_\sim = \rho((\mathcal{R} \cup \{r\}) \setminus [r]_\sim)\} \setminus \{\mathrm{id}\}$ instead of checking

$$\mathcal{R} \cup \{r\} \preccurlyeq \rho'(\mathcal{R} \cup \{r\})$$

for all $\rho' \in \mathrm{Perm}_\Sigma$. As conditionals from the same equivalence class are added directly after each other, it is possible to keep track of the set P during the algorithm.

Algorithm $KB^{\rho ra}$ (Algorithm 2) is an refined version of Algorithm $KB^{\rho ra}_{basic}$ that makes use of these concepts and observations. The set L_k contains *states* instead of knowledge bases. A state is a triple consisting of a knowledge base $\mathcal{R}$, the set P of all signature renamings ρ such that $\rho(\mathcal{R} \setminus E) = \mathcal{R} \setminus E$ where $E := [\max(\mathcal{R})]_\sim$, and the set C of conditionals that are candidates for the extension of $\mathcal{R}$. In the initial state, we have $\mathcal{R} = \emptyset$, $P = \mathrm{Perm}_\Sigma \setminus \{\mathrm{id}\}$, and $C = NFC(\Sigma)$. If a conditional from a new equivalence class is selected, the set P has to be updated. The set C is updated with every extension.

Proposition 12 ($KB^{\rho ra}$). *Let Σ be a signature with a linear ordering $<$. Then applying $KB^{\rho ra}$ terminates and $KB^{\rho ra}_{basic}(\Sigma) = KB^{\rho ra}(\Sigma)$.*

Proof. There are three differences between the algorithm $KB^{\rho ra}_{basic}$ and $KB^{\rho ra}$: The usage of candidate sets, the modified check for ρNF, and omission of the checks in Lines 7 to 9 in $KB^{\rho ra}_{basic}$. In order to prove that $KB^{\rho ra}$ is a correct refinement of $KB^{\rho ra}_{basic}$, we have to show that

- it is sufficient to consider the conditionals in the candidate set of a knowledge base maintained in $KB^{\rho ra}$ as extensions,
- the check for ρNF in $KB^{\rho ra}$ is correct, and
- the checks for $\max(\mathcal{R}) \prec r$, consistency, and RANF done in $KB^{\rho ra}_{basic}$ are not necessary due to the filtering of the candidate sets done in $KB^{\rho ra}$.

Let us first consider the candidate sets. Initially a candidate set contains all normal form conditionals. So far, no possible extension is excluded. Now let us consider the conditionals that are removed from candidate sets in the Lines 17 to 20. Assume the knowledge base $\mathcal{R}$ was just extended by the conditional r. The conditionals in $EXC(r)$ cannot extend $\mathcal{R} \cup \{r\}$ any more by the construction of $EXC(r)$. If $\mathcal{R} \cup \{r, d\}$ is inconsistent, d cannot extend $\mathcal{R} \cup \{r\}$ or any superset thereof to a consistent knowledge base. Every candidate d, such that $\mathcal{R} \cup \{r, \overline{d}\}$ is inconsistent, is an entailment of $\mathcal{R} \cup \{r\}$. Therefore, an extension of $\mathcal{R} \cup \{r\}$ or a any superset thereof by d cannot be in reduced form. Analogously, if $(\mathcal{R} \cup \{r, d, \overline{r'}\}) \setminus \{r'\}$ is inconsistent for any $r' \in \mathcal{R} \cup \{r\}$, then $(\mathcal{R} \cup \{r, d\}) \setminus \{r'\}$ entails r'. Therefore, $\mathcal{R} \cup \{r, d\}$ or any superset thereof is not in reduced form.

Algorithm 2. $KB^{\rho ra}$ – Improved algorithm to generate knowledge bases over Σ with order $\lessdot$ that are in RANF and in ρNF

Require: signature Σ with order $\lessdot$
Ensure: set $\mathcal{KB}$ of knowledge bases over Σ in *reduced antecedentwise normal form* (RANF) and in *renaming normal form* (ρNF) that are consistent, pairwise not antecedentwise equivalent and pairwise not equivalent under signature renaming

1: $k \leftarrow 0$
2: $L_0 \leftarrow \{\langle \emptyset, \mathrm{Perm}_\Sigma \setminus \{\mathrm{id}\}, NFC(\Sigma)\rangle\}$ ▷ Initialize the set of states
3: **while** $L_k \neq \emptyset$ **do**
4:      $L_{k+1} \leftarrow \emptyset$
5:      **for** $\langle \mathcal{R}, P, C \rangle \in L_k$ **do**
6:          $E \leftarrow$ **if** $k = 0$ **then** $[\max(\mathcal{R})]_\simeq$ **else** $\emptyset$ ▷ equiv. class of last conditional
7:          **for** $r \in C$ **do**
8:              **if** $r \notin E$ **then**   ▷ If r is from a new equivalence class. . .
9:                  $P \leftarrow \{\varrho \in P \mid \varrho(\mathcal{R}) = \mathcal{R}\}$ ▷ . . . update P and . . .
10:                  $E \leftarrow [r]_\simeq$ ▷ . . . update E.
11:              *is_minimal* $\leftarrow$ *true*
12:              **for** $\varrho \in P$ **do**   ▷ Check if $\mathcal{R} \cup \{r\}$ is in ρNF
13:                  **if** $\varrho((\mathcal{R} \cup \{r\}) \cap E) \prec (\mathcal{R} \cup \{r\}) \cap E$ **then**
14:                      *is_minimal* $\leftarrow$ *false*
15:                      **break**
16:              **if** *is_minimal* **then**   ▷ Keep the extension if it is in ρNF
17:                  $C_{new} \leftarrow C \setminus EXC(r)$ ▷ Reduce candidate set
18:                  $C_{new} \leftarrow C_{new} \setminus \{d \in C_{new} \mid \mathcal{R} \cup \{r, d\}$ inconsistent$\}$
19:                  $C_{new} \leftarrow C_{new} \setminus \{d \in C_{new} \mid \mathcal{R} \cup \{r, \overline{d}\}$ inconsistent$\}$
20:                  $C_{new} \leftarrow C_{new} \setminus \{d \in C_{new} \mid r' \in \mathcal{R} \cup \{r\},$
                                               $(\mathcal{R} \cup \{r, d, \overline{r'}\}) \setminus \{r'\}$ inconsistent$\}$
21:              $L_{k+1} \leftarrow L_{k+1} \cup \{\langle \mathcal{R} \cup \{r\}, P, C_{new}\rangle\}$
22:      $k \leftarrow k + 1$
23: **return** $\mathcal{KB} \leftarrow R_1 \cup \cdots \cup R_k$ with $R_i = \{\mathcal{R} \mid \langle \mathcal{R}, P, C\rangle \in L_i\}$

The check for ρNF in Lines 9 to 15 of $KB^{\rho ra}$ is correct because of Proposition 11.

The checks in Lines 7 to 9 of $KB^{\rho ra}_{basic}$ were moved to the reduction of the candidate sets. The exclude set $EXC(r)$ for a conditional r contains all smaller conditionals $r' \preccurlyeq r$. All extensions by a single conditional d such that r and d are pairwise inconsistent are already removed by using $EXC(r)$, and all other extensions of $R \cup \{r\}$ that would lead to an inconsistency are removed in Line 18. The exclude set $EXC(r)$ also contains all conditionals with the same antecedence as r, and all conditionals considered here are already normal form conditionals, implying that the extension of a knowledge base with a candidate from the corresponding candidate set is always in ANF. The Lines 19 and 20 ensure that the extension is in reduced form. Therefore the omitted checks from $KB^{\rho ra}_{basic}$ are not necessary in $KB^{\rho ra}$. In summary this implies the claim $KB^{\rho ra}_{basic}(\Sigma) = KB^{\rho ra}(\Sigma)$.   □

Thus, all properties (1.)–(8.) as stated in Proposition 10 also hold for the set of knowledge base returned by $KB^{\rho ra}$. At the same time, $KB^{\rho ra}$ requires less computational effort than $KB^{\rho ra}_{basic}$, because it uses a significantly reduced search space. For the check for ρNF, only the intersection of the equivalence class of the conditional to be added with the knowledge base to be extended has to be checked, and only with respect to the permutations maintained by $KB^{\rho ra}$. For many knowledge bases, this set of permutations still to be taken into account will shrink very fast after a few conditionals have been selected. Furthermore, the improved algorithm checks unsuccessful extensions less often. If $KB^{\rho ra}_{basic}$ does not use a conditional in one iteration of the **while** loop, it will check the same conditional again in the next iteration of the loop. If $KB^{\rho ra}$ does not use a conditional for another reason than the extension not being in ρNF, it is removed from the candidate set permanently.

7 Conclusions and Further Work

In this paper, we introduced two new normal forms for conditional knowledge bases, the semantically motivated reduced antecedent normal form (RANF) for arbitrary knowledge bases and the syntax oriented renaming normal form (ρNF) for knowledge bases over normal form conditionals, and studied their properties. The RANF is an improved version of the antecedent normal form (ANF) and further reduces the number of conditionals in the knowledge base. The transformation system Θ^{ra} transforms any conditional knowledge base into one in RANF. To systematically generate knowledge bases over a given signature that are both in RANF and in ρNF, we developed the algorithm $KB^{\rho ra}$ and proved its correctness and key characteristics.

We are currently working on extending the results of this paper in several directions. Here, the notion of ρNF has been defined only for knowledge bases containing no other conditionals than normal form conditionals. While for such knowledge bases there is a unique knowledge base that is equivalent to a given knowledge base under signature renaming, there is not yet an algorithm doing this transformation into ρNF; designing such an algorithm as well as the extension of ρNF to knowledge bases containing arbitrary conditionals is addressed in our current work. Furthermore, the ordering $\prec$ induced by the underlying $<$ on the signature is only one of many orders that could be used in our approach; it could be replaced by other orderings that respect the equivalence classes induced by renamings. Moreover, instead of renamings over the same signature, also mappings between different signatures could be taken into account. Both, the normalizing system Θ^{ra} and the generating algorithm $KB^{\rho ra}$ will be studied with respect to efficiency and complexity, and using implementations of these methods and the InfOCF system [5], we will empirically evaluate properties of given and of generated conditional knowledge bases.

References

1. Adams, E.W.: The Logic of Conditionals: An Application of Probability to Deductive Logic. Synthese Library. Springer, Dordrecht (1975). https://doi.org/10.1007/978-94-015-7622-2
2. Agrawal, R., Mannila, H., Srikant, R., Toivonen, H., Verkamo, A.: Fast discovery of association rules. In: Fayyad, U., Piatetsky-Shapiro, G., Smyth, P., Uthurusamy, R. (eds.) Advances in Knowledge Discovery and Data Mining, pp. 307–328. MIT Press, Cambridge (1996)
3. Beierle, C.: Inferential equivalence, normal forms, and isomorphisms of knowledge bases in institutions of conditional logics. In: Hung, C., Papadopoulos, G.A. (eds.) The 34th ACM/SIGAPP Symposium on Applied Computing (SAC 2019), pp. 1131–1138. ACM, New York (2019)
4. Beierle, C., Eichhorn, C., Kern-Isberner, G.: A transformation system for unique minimal normal forms of conditional knowledge bases. In: Antonucci, A., Cholvy, L., Papini, O. (eds.) ECSQARU 2017. LNCS (LNAI), vol. 10369, pp. 236–245. Springer, Cham (2017). https://doi.org/10.1007/978-3-319-61581-3_22
5. Beierle, C., Eichhorn, C., Kutsch, S.: A practical comparison of qualitative inferences with preferred ranking models. KI - Künstliche Intelligenz **31**(1), 41–52 (2017)
6. Beierle, C., Kutsch, S.: Computation and comparison of nonmonotonic skeptical inference relations induced by sets of ranking models for the realization of intelligent agents. Appl. Intell. **49**(1), 28–43 (2019)
7. Beierle, C., Kutsch, S.: On the antecedent normal form of conditional knowledge bases. In: Kern-Isberner, G., Ognjanović, Z. (eds.) ECSQARU 2019. LNCS (LNAI), vol. 11726, pp. 175–186. Springer, Cham (2019). https://doi.org/10.1007/978-3-030-29765-7_15
8. Beierle, C., Kutsch, S.: Systematic generation of conditional knowledge bases up to renaming and equivalence. In: Calimeri, F., Leone, N., Manna, M. (eds.) JELIA 2019. LNCS (LNAI), vol. 11468, pp. 279–286. Springer, Cham (2019). https://doi.org/10.1007/978-3-030-19570-0_18
9. Benferhat, S., Dubois, D., Prade, H.: Possibilistic and standard probabilistic semantics of conditional knowledge bases. J. Logic Comput. **9**(6), 873–895 (1999)
10. Dubois, D., Prade, H.: Conditional objects as nonmonotonic consequence relationships. IEEE Trans. Syst. Man Cybern. **24**(12), 1724–1740 (1994). Special Issue on Conditional Event Algebra
11. Goldszmidt, M., Pearl, J.: Qualitative probabilities for default reasoning, belief revision, and causal modeling. Artif. Intell. **84**, 57–112 (1996)
12. Kern-Isberner, G.: Conditionals in Nonmonotonic Reasoning and Belief Revision. LNCS (LNAI), vol. 2087. Springer, Heidelberg (2001). https://doi.org/10.1007/3-540-44600-1
13. Kraus, S., Lehmann, D., Magidor, M.: Nonmonotonic reasoning, preferential models and cumulative logics. Artif. Intell. **44**, 167–207 (1990)
14. Lehmann, D., Magidor, M.: What does a conditional knowledge base entail? Artif. Intell. **55**, 1–60 (1992)
15. Lewis, D.: Counterfactuals. Harvard University Press, Cambridge (1973)
16. Pearl, J.: Probabilistic Reasoning in Intelligent Systems. Morgan Kaufmann, San Mateo (1988)

17. Pearl, J.: System Z: a natural ordering of defaults with tractable applications to nonmonotonic reasoning. In: Parikh, R. (ed.) Proceedings of the 3rd Conference on Theoretical Aspects of Reasoning About Knowledge (TARK 1990), pp. 121–135. Morgan Kaufmann Publishers Inc., San Francisco (1990)
18. Robinson, J.A., Voronkov, A. (eds.): Handbook of Automated Reasoning (in 2 Volumes). Elsevier and MIT Press, Cambridge (2001)
19. Spohn, W.: The Laws of Belief: Ranking Theory and Its Philosophical Applications. Oxford University Press, Oxford (2012)

On Matrices and K-Relations

Robert Brijder, Marc Gyssens$^{(\boxtimes)}$, and Jan Van den Bussche

Data Science Institute, Hasselt University, Martelarenlaan 42, 3500 Hasselt, Belgium
marc.gyssens@uhasselt.be

Abstract. We show that the matrix query language MATLANG corresponds to a natural fragment of the positive relational algebra on K-relations. The fragment is defined by introducing a composition operator and restricting K-relation arities to two. We then proceed to show that MATLANG can express all matrix queries expressible in the positive relational algebra on K-relations, when intermediate arities are restricted to three. Thus we offer an analogue, in a model with numerical data, to the situation in classical logic, where the algebra of binary relations is equivalent to first-order logic with three variables.

Keywords: Expressive power · Provenance semirings · Annotated relations · Data science

1 Introduction

Motivated by large-scale data science, there is recent interest in supporting linear algebra operations, such as matrix multiplication, in database systems. This has prompted investigations comparing the expressive power of common matrix operations with the operations on relations provided by the relational algebra and SQL [2,6,7,10].

For *boolean* matrices, the connection between matrices and relations is very natural and well known. An $m \times n$ boolean matrix A can be viewed as a binary relation $R \subseteq \{1, \ldots, m\} \times \{1, \ldots, n\}$, where R consists of those pairs (i, j) for which $A_{i,j} = 1$. Boolean matrix multiplication then amounts to composition of binary relations. Composition is the central operation in the *algebra of binary relations* [13,15,16]. Besides composition, this algebra has operations such as converse, which corresponds to transposition of a boolean matrix; union and complement, which correspond to disjunction and negation of boolean matrices; and the empty and identity relations, which correspond to the zero and identity matrices.

A common theme in research in the foundations of databases is the expressive power of query languages [1]. When we employ a query language, we would like to understand as well as possible what we can do with it. Of this kind is the classical Codd theorem, stating the equivalence between the standard relational algebra and first-order logic. Likewise, for the algebra of binary relations, a classical result [17] is that it has the same expressive power as the formulas with two free

© Springer Nature Switzerland AG 2020
A. Herzig and J. Kontinen (Eds.): FoIKS 2020, LNCS 12012, pp. 42–57, 2020.
https://doi.org/10.1007/978-3-030-39951-1_3

variables in FO(3), the three-variable fragment of first-order logic. In this sense, we understand quite well the expressive power of a natural set of operations on boolean matrices.

What can now be said in this regard about more general matrices, with entries that are not just boolean values? An $m \times n$ matrix with entries in some semiring K is essentially a mapping from $\{1, \ldots, m\} \times \{1, \ldots, n\}$ to K. This perfectly fits the data model of K-*relations* introduced by Green, Garvounarakis and Tannen [5]. In general, consider an infinite domain **dom** and a supply of attributes. In a database instance, we assign to each attribute a range of values, in the form of a finite subset of **dom**. Attributes can be declared to be compatible; compatible attributes have the same range. A relation schema S is a finite set of attributes. Tuples over S are mappings that assign to each attribute a value of the appropriate range. Now a K-relation over S is a mapping that assigns to each tuple over S an element of K.

So, an $m \times n$ matrix X can be seen as a K-relation over two attributes A and B where the range of A is $\{1, \ldots, m\}$ and the range of B is $\{1, \ldots, n\}$. We can assume an order on all attributes and choose $A < B$ so that we know which values are row indices and which are column indices. Then an $n \times k$ matrix Y is modeled using attributes $C < D$ where we choose C and B compatible, to reflect that the number of columns of matrix X equals the number of rows of matrix Y. We can view vectors as K-relations over a single attribute, and scalars as K-relations over the empty schema. In general, a K-relation of arity r is essentially an r-dimensional tensor (multidimensional array). (Because we need not necessarily assume an order on **dom**, the tensor is unordered.)

Green et al. defined a generalization of the positive relation algebra working on K-relations, which we denote here by ARA.[1] When we restrict ARA to arities of at most three, which we denote by ARA(3), we obtain an analogue to FO(3) mentioned above. So, ARA provides a suitable scenario to reinvestigate, in a data model with numerical values, the equivalence between the algebra of binary relations and FO(3). In this paper we make the following contributions.

1. We define a suitable generalization, to K-relations, of the composition operation of classical binary relations. When we add this composition operator to ARA, but restrict arities to at most two, we obtain a natural query language for matrices. We refer to this language here as "ARA(2) plus composition".
2. We show that ARA(2) plus composition actually coincides with the matrix query language MATLANG, introduced by two of the present authors with Geerts and Weerwag [2] in an attempt to formalize the set of common matrix operations found in numerical software packages.
3. We show that a matrix query is expressible in ARA(3) if and only if it is expressible in MATLANG, thus providing an analogue to the classical result about FO(3) and the algebra of binary relations. More generally, for any arity r, we show that an r-ary query over r-ary K-relations is expressible in

---

[1] ARA stands for Annotated-Relation Algebra, as the elements from K that a K-relation assigns to its tuples are usually viewed as annotations.

ARA$(r+1)$ if and only if it is expressible in ARA(r) plus composition. For this result, we need the assumption that K is commutative. We stress that the proof is not a trivial adaptation of the proof of the classical result, because we can no longer rely on familiar classical properties like idempotence of union and join.

ARA has been a very influential vehicle for data provenance.[2] The elements from K are typically viewed as annotations, or as identifiers, and the semantics of ARA operations was originally designed to show how these annotations are propagated in the results of data manipulations. Other applications, apart from provenance, have been identified from the outset, such as security levels, or probabilities [5]. By doing the present work, we have understood that ARA can moreover serve as a fully-fledged query language for tensors (multidimensional arrays), and matrices in particular. This viewpoint is backed by the recent interest in processing Functional Aggregate Queries (FAQ [11,12], also known as AJAR [8]). Indeed, FAQ and AJAR correspond to the project-join fragment of ARA, without self-joins.

This paper is further organized as follows. Section 2 recalls the data model of K-relations and the associated query language ARA. Section 3 presents the result on ARA$(r + 1)$ and ARA(r) plus composition. Section 4 relates ARA(2) plus composition to MATLANG. Section 5 draws conclusions, discusses related work, and proposes directions for further research.

2   Annotated-Relation Algebra

By *function* we will always mean a total function. For a function $f : X \to Y$ and $Z \subseteq X$, the *restriction* of f to Z, denoted by $f|_Z$, is the function $Z \to Y$ where $f|_Z(x) = f(x)$ for all $x \in Z$.

Recall that a *semiring* K is a set equipped with two binary operations, addition $(+)$ and multiplication $(*)$, such that (1) addition is associative, commutative, and has an identity element 0; (2) multiplication is associative, has an identity element 1, and has 0 as an annihilating element; and (3) multiplication distributes over addition. A semiring is called *commutative* when multiplication is commutative. We fix a semiring K.

Remark 1. We will explicitly indicate where we assume commutativity of K.

From the outset, we also fix countable infinite sets **rel**, **att**, and **dom**, the elements of which are called *relation names*, *attributes*, and *domain elements*, respectively. We assume an equivalence relation $\sim$ on **att** that partitions **att** into an infinite number of equivalence classes that are each infinite. When $A \sim B$, we say that A and B are *compatible*. Intuitively, $A \sim B$ will mean that A and B have the same set of domain values. A function $f : X \to Y$ with X and Y sets of attributes is called *compatible* if $f(A) \sim A$ for all $A \in X$.

---

[2] The paper [5] received the PODS 2017 test-of-time award.

A *relation schema* is a finite subset of **att**. A *database schema* is a function S on a finite set N of relation names, assigning a relation schema $S(R)$ to each $R \in N$. The *arity* of a relation name R is the cardinality $|S(R)|$ of its schema. The *arity* of S is the largest arity among relation names $R \in N$.

We now recursively define the expressions of the *Annotated-Relation Algebra*, abbreviated by ARA. At the same time we assign a relation schema to each ARA expression by extending S from relation names to arbitrary ARA expressions. An ARA *expression* e over a database schema S is one of the following:

Relation name a relation name R of S;
One the one operation $\mathbf{1}(e')$, where e' is an ARA expression, and $S(e) := S(e')$;
Union the union $e_1 \cup e_2$, where e_1 and e_2 are ARA expressions with $S(e_1) = S(e_2)$, and $S(e) := S(e_1)$;
Projection the projection $\pi_Y(e')$, where e' is an ARA expression and $Y \subseteq S(e')$, and $S(e) := Y$;
Selection the selection $\sigma_Y(e')$, where e' is an ARA expression, $Y \subseteq S(e')$, the elements of Y are mutually compatible, and $S(e) := S(e')$;
Renaming the renaming $\rho_\varphi(e')$, where e' is an ARA expression and $\varphi : S(e') \to Y$ a compatible one-to-one correspondence with $Y \subseteq \mathbf{att}$, and $S(e) := Y$; or
Join the join $e_1 \bowtie e_2$, where e_1 and e_2 are ARA expressions, and $S(e) := S(e_1) \cup S(e_2)$.

A *domain assignment* is a function $D : \mathbf{att} \to \mathcal{D}$, where $\mathcal{D}$ is a set of nonempty finite subsets of **dom**, such that $A \sim B$ implies $D(A) = D(B)$. Let X be a relation schema. A *tuple* over X with respect to D is a function $t : X \to \mathbf{dom}$ such that $t(A) \in D(A)$ for all $A \in X$. We denote by $\mathcal{T}_D(X)$ the set of tuples over X with respect to D. Note that $\mathcal{T}_D(X)$ is finite. A *relation* r over X with respect to D is a function $r : \mathcal{T}_D(X) \to K$. So a relation annotates every tuple over X with respect to D with a value from K. If S is a database schema, then an *instance* $\mathcal{I}$ of S *with respect to* D is a function that assigns to every relation name R of S a relation $\mathcal{I}(R) : \mathcal{T}_D(S(R)) \to K$.

Remark 2. In practice, a domain assignment need only be defined on the attributes that are used in the database schema (and on attributes compatible to these attributes). Thus, it can be finitely specified. While here we have chosen to keep the notion of domain assignment and instance separate, it may

$\mathcal{I}(\text{no\_courses}) =$

student	dptm	K
Alice	CS	5
Alice	Math	2
Alice	Bio	0
Bob	CS	2
Bob	Math	1
Bob	Bio	3

$\mathcal{I}(\text{course\_fee}) =$

dptm	K
CS	300
Math	250
Bio	330

Fig. 1. Example of a database instance.

perhaps be more natural to think of the domain assignment as being part of the instance.

Example 1. Let us record for a university both the number of courses each student takes in each department and the course fee for each department. Let K be the set of integers and let S be a database schema on $\{\text{no\_courses}, \text{course\_fee}\}$ with $S(\text{no\_courses}) = \{\text{student}, \text{dptm}\}$ and $S(\text{course\_fee}) = \{\text{dptm}\}$. Let D be a domain assignment with $D(\text{student}) = \{\text{Alice}, \text{Bob}\}$ and $D(\text{dptm}) = \{\text{CS}, \text{Math}, \text{Bio}\}$. A database instance $\mathcal{I}$ of S with respect to D is shown in Fig. 1.

We now define the relation $\mathbf{1}_X$, as well as the generalizations of the classical operations from the positive relational algebra to work on K-relations.

One. For every relation schema X, we define $\mathbf{1}_X : \mathcal{T}_D(X) \to K$ as $\mathbf{1}_X(t) = 1$.

Union. Let $r_1, r_2 : \mathcal{T}_D(X) \to K$. Define $r_1 \cup r_2 : \mathcal{T}_D(X) \to K$ as $(r_1 \cup r_2)(t) = r_1(t) + r_2(t)$.

Projection. Let $r : \mathcal{T}_D(X) \to K$ and $Y \subseteq X$. Define $\pi_Y(r) : \mathcal{T}_D(Y) \to K$ as

$$(\pi_Y(r))(t) = \sum_{\substack{t' \in \mathcal{T}_D(X), \\ t'|_Y = t}} r(t').$$

Selection. Let $r : \mathcal{T}_D(X) \to K$ and $Y \subseteq X$ where the elements of Y are mutually compatible. Define $\sigma_Y(r) : \mathcal{T}_D(X) \to K$ such that

$$(\sigma_Y(r))(t) = \begin{cases} r(t) & \text{if } t(A) = t(B) \text{ for all } A, B \in Y; \\ 0 & \text{otherwise.} \end{cases}$$

Renaming. Let $r : \mathcal{T}_D(X) \to K$ and $\varphi : X \to Y$ a compatible one-to-one correspondence. We define $\rho_\varphi(r) : \mathcal{T}_D(Y) \to K$ as $\rho_\varphi(r)(t) = r(t \circ \varphi)$.

Join. Let $r_1 : \mathcal{T}_D(X_1) \to K$ and $r_2 : \mathcal{T}_D(X_2) \to K$. Define $r_1 \bowtie r_2 : \mathcal{T}_D(X_1 \cup X_2) \to K$ as $(r_1 \bowtie r_2)(t) = r_1(t|_{X_1}) * r_2(t|_{X_2})$.

The above operations provide semantics for ARA in a natural manner. Formally, let S be a database schema, let e be an ARA expression over S, and let $\mathcal{I}$ be an instance of S. The *output* relation $e(\mathcal{I})$ of e under $\mathcal{I}$ is defined as follows. If $e = R$ with R a relation name of S, then $e(\mathcal{I}) := \mathcal{I}(R)$. If $e = 1(e')$, then $e(\mathcal{I}) := \mathbf{1}_{S(e')}$. If $e = e_1 \cup e_2$, then $e(\mathcal{I}) := e_1(\mathcal{I}) \cup e_2(\mathcal{I})$. If $e = \pi_X(e')$, then $e(\mathcal{I}) := \pi_X(e'(\mathcal{I}))$. If $e = \sigma_Y(e')$, then $e(\mathcal{I}) := \sigma_Y(e'(\mathcal{I}))$. If $e = \rho_\varphi(e')$, then $e(\mathcal{I}) := \rho_\varphi(e'(\mathcal{I}))$. Finally, if $e = e_1 \bowtie e_2$, then $e(\mathcal{I}) := e_1(\mathcal{I}) \bowtie e_2(\mathcal{I})$.

Remark 3. The language ARA is a slight variation of the K-annotated relational algebra as originally defined by Green et al. [5] to better suit our purposes.

First of all, the original definition does not have a domain assignment $D : \text{att} \to \mathcal{D}$ but instead a single domain common to all attributes (and it therefore also does not have a compatibility relation $\sim$). As such, the original definition corresponds to the case where database schemas and ARA expressions use only mutually compatible attributes. We need our more general setting when we compare ARA to MATLANG in Sect. 4.

Also, here, we focus on equality selections, while the original paper does not fix the allowed selection predicates. Finally, the original definition assumes zero-relations $\mathbf{0}_X$, while we instead use one-relations $\mathbf{1}_X$.

The following observations, to the effect that some (but not all) classical relational-algebra equivalences carry over to the K-annotated setting, were originally made by Green et al.

Proposition 1 ([5, **Proposition 3.4**]). *The following properties and equivalences hold, where, for each given equivalence, we assume that the left-hand side is well defined.*

- *Union is associative and commutative.*
- *Join is associative and distributive over union, i.e., $(r_1 \cup r_2) \bowtie r_3 = (r_1 \bowtie r_3) \cup (r_2 \bowtie r_3)$.*
- *Any two selections commute.*
- *Projection and selection commute when projection retains the attributes on which selection takes place.*
- *Projection distributes over union, i.e., $\pi_Y(r_1 \cup r_2) = \pi_Y(r_1) \cup \pi_Y(r_2)$.*
- *Selection distributes over union, i.e., $\sigma_Y(r_1 \cup r_2) = \sigma_Y(r_1) \cup \sigma_Y(r_2)$.*
- *We have $\sigma_Y(r_1) \bowtie r_2 = \sigma_Y(r_1 \bowtie r_2)$ and $r_1 \bowtie \sigma_Y(r_2) = \sigma_Y(r_1 \bowtie r_2)$.*
- *If K is commutative, then join is commutative.*

Note that idempotence of union and of join, i.e., $r \bowtie r = r \cup r = r$, which holds for the classical relational algebra, does *not* in general hold for ARA.

We supplement Proposition 1 with the following easy-to-verify properties.

Lemma 1. *Let $r_1 : \mathcal{T}_D(X_1) \to K$ and $r_2 : \mathcal{T}_D(X_2) \to K$.*

- *If $X_1 \cap X_2 \subseteq X \subseteq X_1 \cup X_2$, then $\pi_X(r_1 \bowtie r_2) = \pi_{X \cap X_1}(r_1) \bowtie \pi_{X \cap X_2}(r_2)$.*
- *If $Y_1, Y_2 \subseteq X_1$ where $Y_1 \cap Y_2 \neq \emptyset$ and the attributes of Y_1 and of Y_2 are mutually compatible, then $\sigma_{Y_2}(\sigma_{Y_1}(r_1)) = \sigma_{Y_1 \cup Y_2}(r_1)$.*
- *If $\varphi : X_1 \cup X_2 \to X$ is a compatible one-to-one correspondence, then $\rho_\varphi(r_1 \bowtie r_2) = \rho_{\varphi|_{X_1}}(r_1) \bowtie \rho_{\varphi|_{X_2}}(r_2)$. If moreover $X_1 = X_2$, then $\rho_\varphi(r_1 \cup r_2) = \rho_\varphi(r_1) \cup \rho_\varphi(r_2)$.*
- *If $Y \subseteq X_1$ and $\varphi : X_1 \to X$ is a compatible one-to-one correspondence, then $\rho_\varphi(\sigma_Y(r_1)) = \sigma_{\varphi(Y)}(\rho_\varphi(r_1))$, where $\varphi(Y) = \{\varphi(y) \mid y \in Y\}$.*

We also use the operation of projecting away an attribute, i.e., $\hat{\pi}_A(e) := \pi_{\mathcal{S}(e) \setminus \{A\}}(e)$ if $A \in \mathcal{S}(e)$. Note that conversely, $\pi_X(e) = (\hat{\pi}_{A_m} \cdots \hat{\pi}_{A_1})(e)$ where $X = \mathcal{S}(e) \setminus \{A_1, \ldots, A_m\}$ and the A_i's are mutually distinct. Projecting away, allowing one to deal with one attribute at a time, is sometimes notationally more convenient.

3   Composition and Equivalence

In this section we define an operation called k-composition and show that augmenting ARA by composition allows one to reduce the required arity of the relations that are computed in subexpressions. The intuition is to provide a generalization of classical composition of two binary relations to annotated relations, so that we can compose up to k relations of arity up to k. Specifically, the classical composition of a binary relation r with a binary relation s amounts to viewing these relations as relations over schemas $\{B, A\}$ and $\{A, C\}$, respectively, and performing $\hat{\pi}_A(r \bowtie s)$. Thus we arrive at the following generalization.

Definition 1. *Let k be a nonnegative integer and let $l \in \{1, \ldots, k\}$. Let $r_i :$ $T_D(X_i) \to K$ for $i \in \{1, \ldots, l\}$, let $X = X_1 \cup \cdots \cup X_l$, and let $A \in X_1 \cap \cdots \cap X_l$.*
Define the k-composition $\zeta_{A,k}(r_1, \ldots, r_l) : T_D(X \setminus \{A\}) \to K$ as

$$(\zeta_{A,k}(r_1, \ldots, r_l))(t) = (\hat{\pi}_A(r_1 \bowtie \cdots \bowtie r_l))(t)$$

for all $t \in T_D(X \setminus \{A\})$.

Note that $\zeta_{A,k}$ takes at most k arguments. We emphasize that $\zeta_{A,k}$ is defined as a new operator (albeit one that can be defined by an ARA expression) and not as a shorthand for an ARA expression.

We denote by $\mathsf{ARA} + \zeta_k$ the language obtained by extending ARA with k-composition. Consequently, if $e_1, \ldots, e_l$ are $\mathsf{ARA} + \zeta_k$ expressions with $l \leq k$ and $A \in \mathcal{S}(e_1) \cap \cdots \cap \mathcal{S}(e_l)$, then $e = \zeta_{A,k}(e_1, \ldots, e_l)$ is an $\mathsf{ARA} + \zeta_k$ expression. Also, we let $\mathcal{S}(e) := (\mathcal{S}(e_1) \cup \cdots \cup \mathcal{S}(e_l)) \setminus \{A\}$.

Let k be a nonnegative integer. We denote by $\mathsf{ARA}(k)$ the fragment of ARA in which the database schemas are restricted to arity at most k and the relation schema of each subexpression is of cardinality at most k. In particular, join $e_1 \bowtie e_2$ is only allowed if $|\mathcal{S}(e_1 \bowtie e_2)| \leq k$. The fragment $(\mathsf{ARA} + \zeta_k)(k)$ is defined similarly.

From Definition 1 it is apparent that $(\mathsf{ARA} + \zeta_k)(k)$ is subsumed by $\mathsf{ARA}(k+1)$. One of our main results (Corollary 1) provides the converse inclusion, when the database schemas and outputs are restricted to arity at most k. To this end, we establish a normal form for ARA expressions.

We use the following terminology. Let $\mathcal{F}$ be any family of expressions. A *selection of $\mathcal{F}$-expressions* is an expression of the form $\sigma_{Y_n} \cdots \sigma_{Y_1}(f)$, where f is an $\mathcal{F}$-expression and $n \geq 0$. Note the slight abuse of terminology as we allow multiple selection operations. Also, when we say that e is a *union of $\mathcal{F}$-expressions* or a *join of $\mathcal{F}$-expressions*, we allow e to be just a single expression in $\mathcal{F}$ (so union and join may be skipped).

We are now ready to formulate a main result of this paper. This result is inspired by the classic equivalence of FO(3) and the algebra of binary relations [17]. A compact proof of this classical result is given by Marx and Venema [14, Theorem 3.4.5, Claim 2], and a self-contained exposition is also available [3].

Two ARA expressions e_1 and e_2 over the same database schema are called *equivalent*, naturally, if they yield the same output relation for every domain assignment and every database instance respecting that domain assignment.

Theorem 1. *Let S be a database schema of arity at most k and assume that K is commutative. Every $\mathsf{ARA}(k+1)$ expression over S is equivalent to a union of selections of joins of $(\mathsf{ARA} + \zeta_k)(k)$ expressions over S.*

The proof of Theorem 1 uses Proposition 1, Lemma 1, and the following technical lemma to effectively construct the expression in the form given by Theorem 1. This effective construction is illustrated in Example 2.

Lemma 2. *Let $r_1, \ldots, r_n$ be relations with relation schemas $X_1, \ldots, X_n$, respectively, and with respect to a domain assignment D. Assume that $A, B \in X_1 \cup \cdots \cup X_n$ are distinct and compatible. Define, for $i \in \{1, \ldots, n\}$,*

$$
r_i' := \begin{cases}
r_i & \text{if } A \notin X_i; \\
\rho_{A \to B}(r_i) & \text{if } A \in X_i, B \notin X_i; \\
\hat{\pi}_A(\sigma_{\{A,B\}}(r_i)) & \text{if } A, B \in X_i,
\end{cases}
$$

where $A \to B$ denotes the one-to-one correspondence from X_i to $(X_i \setminus \{A\}) \cup \{B\}$ that assigns A to B and keeps the remaining attributes fixed. Then

$$
\hat{\pi}_A(\sigma_{\{A,B\}}(r_1 \bowtie \cdots \bowtie r_n)) = r_1' \bowtie \cdots \bowtie r_n'.
$$

Example 2. Assume that K is commutative and consider the $\mathsf{ARA}(3)$ expression $e = \pi_{\{B,C\}}(\sigma_{\{B,C\}}(R \bowtie R \bowtie S \bowtie T \bowtie \rho_\varphi(T)) \cup \sigma_{\{A,B\}}(R \bowtie S \bowtie T))$, where $S(R) = \{A, B\}$, $S(S) = \{B, C\}$, $S(T) = \{A, C\}$ (A, B, C are mutually distinct), and φ sends A to B and C to itself. The proof of Theorem 1 obtains an equivalent expression in normal form by using Proposition 1, Lemmas 1 and 2 as follows.

$$
\begin{aligned}
e &= \hat{\pi}_A(\sigma_{\{B,C\}}(R \bowtie R \bowtie S \bowtie T \bowtie \rho_\varphi(T)) \cup \sigma_{\{A,B\}}(R \bowtie S \bowtie T)) \\
&\equiv \hat{\pi}_A(\sigma_{\{B,C\}}(R \bowtie R \bowtie S \bowtie T \bowtie \rho_\varphi(T))) \cup \hat{\pi}_A(\sigma_{\{A,B\}}(R \bowtie S \bowtie T)) \\
&\equiv \sigma_{\{B,C\}}(\hat{\pi}_A(R \bowtie R \bowtie S \bowtie T \bowtie \rho_\varphi(T))) \cup \hat{\pi}_A(\sigma_{\{A,B\}}(R \bowtie S \bowtie T)) \\
&\equiv \sigma_{\{B,C\}}(S \bowtie \rho_\varphi(T) \bowtie \hat{\pi}_A(R \bowtie R \bowtie T)) \cup \hat{\pi}_A(\sigma_{\{A,B\}}(R \bowtie S \bowtie T)) \\
&\equiv \sigma_{\{B,C\}}(S \bowtie \rho_\varphi(T) \bowtie \zeta_{A,2}(R \bowtie R, T)) \cup \hat{\pi}_A(\sigma_{\{A,B\}}(R \bowtie S \bowtie T)) \\
&\equiv \sigma_{\{B,C\}}(S \bowtie \rho_\varphi(T) \bowtie \zeta_{A,2}(R \bowtie R, T)) \cup \left(\hat{\pi}_A(\sigma_{\{A,B\}}(R)) \bowtie S \bowtie \rho_{A \to B}(T)\right).
\end{aligned}
$$

The first two equivalences follow from Proposition 1, the third equivalence follows from Lemma 1, the fourth equivalence is by the definition of $\zeta_{A,2}$, and the last equivalence is by Lemma 2. The last expression is in the normal form since the subexpressions S, $\rho_\varphi(T)$, $\zeta_{A,2}(R \bowtie R, T)$, $\hat{\pi}_A(\sigma_{\{A,B\}}(R))$, and $\rho_{A \to B}(T)$ are all $(\mathsf{ARA} + \zeta_2)(2)$ expressions.

Note that we likely cannot omit the "selections of" in the above theorem. For example, for $k = 2$ consider $\sigma_{\{A,C\}}(R \bowtie S)$ where R and S are relation names with $S(R) = \{A, B\}$ and $S(S) = \{B, C\}$.

Remark 4. Theorem 1 still holds if the $\mathbf{1}$ operator is omitted from the definition of ARA.

Since union, selection, and join do not decrease the number of attributes of relations, we have the following corollary to Theorem 1, which establishes the main result announced in the Introduction.

Corollary 1. *Let S be a database schema of arity at most k and assume that K is commutative. Every $\mathsf{ARA}(k+1)$ expression e over S with $|S(e)| \leq k$ is equivalent to an $(\mathsf{ARA} + \zeta_k)(k)$ expression over S.*

Remark 5. We remark that transforming an expression into the normal form of Theorem 1 may lead to an exponential increase in expression length. The reason is that the proof uses distributivity of join over union. Indeed, each time we replace an expression of the form $(e_1 \cup e_2) \bowtie e_3$ by $(e_1 \bowtie e_3) \cup (e_2 \bowtie e_3)$ there is a duplication of e_3. The proof of the classic translation of FO(3) to the algebra of binary relations also induces an exponential increase of expression length for similar reasons. A proof that this blowup is unavoidable remains open, both for our result and for the classical result (to the best of our knowledge).

3.1   Connection with FO(k)

The connection between $\mathsf{ARA}(k)$ and $\mathsf{FO}(k)$, to which we have hinted several times already, can be made explicit as follows.

Let K be the Boolean semiring. Consider a database schema S of arity at most k. Let D be a domain assignment such that $D(A) = D(B)$ for all attributes A and B. In other words, D just fixes a single nonempty finite subset of **dom**. A tuple (D, I), with I an instance over S, is a classical relational structure over S.

We can consider $\mathsf{FO}(k)$ formulas as first-order logic formulas (also known as relational-calculus formulas [1]) over S that use only k distinct variables. It is then an easy exercise to see that every $\mathsf{FO}(k)$ formula that does not use negation or universal quantification, and has $k' \leq k$ free variables can be translated to an equivalent $\mathsf{ARA}(k)$ expression. The converse translation is also possible.

4   Matrices

In this section we show that $(\mathsf{ARA} + \zeta_2)(2)$ is equivalent to a natural version of MATLANG [2]. As a consequence of Corollary 1, we then obtain that also $\mathsf{ARA}(3)$, with database schemas and output relations restricted to arity at most 2, is equivalent to MATLANG. We begin by recalling the definition of this language.

4.1   MATLANG

Let us fix the countable infinite sets **matvar** and **size**, where the latter has a distinguished element $1 \in$ **size**. The elements of **matvar** are called *matrix variables* and the elements of **size** are called *size symbols*.

A *matrix schema* is a function $S : V \to$ **size** $\times$ **size** with $V \subseteq$ **matvar** both finite and nonempty. We write $(\alpha, \beta) \in$ **size** $\times$ **size** also as $\alpha \times \beta$.

MATLANG expressions are recursively defined as follows. At the same time we assign a matrix schema to each MATLANG expression by extending $\mathcal{S}$ from matrix variables to arbitrary MATLANG expressions.

A MATLANG *expression* e over a matrix schema $\mathcal{S}$ is one of the following:

Variable a matrix variable M of $\mathcal{S}$;

Transposition a transposition $(e')^T$, where e' is a MATLANG expression, and $\mathcal{S}(e) := \beta \times \alpha$ if $\mathcal{S}(e') = \alpha \times \beta$;

One-vector a one-vector $\mathbf{1}(e')$, where e' is a MATLANG expression, and $\mathcal{S}(e) := \alpha \times 1$ if $\mathcal{S}(e') = \alpha \times \beta$;

Diagonalization a diagonalization $\mathrm{diag}(e')$, where e' is a MATLANG expression with $\mathcal{S}(e') = \alpha \times 1$, and $\mathcal{S}(e) := \alpha \times \alpha$;

Multiplication a multiplication $e_1 \cdot e_2$, where e_1 and e_2 are MATLANG expressions with $\mathcal{S}(e_1) = \alpha \times \beta$ and $\mathcal{S}(e_2) = \beta \times \gamma$, and $\mathcal{S}(e) := \alpha \times \gamma$;

Addition an addition $e_1 + e_2$, where e_1 and e_2 are MATLANG expressions with $\mathcal{S}(e_1) = \mathcal{S}(e_2)$, and $\mathcal{S}(e) := \mathcal{S}(e_1)$; or

Hadamard product a Hadamard product $e_1 \circ e_2$, where e_1 and e_2 are MATLANGexpressions with $\mathcal{S}(e_1) = \mathcal{S}(e_2)$, and $\mathcal{S}(e) := \mathcal{S}(e_1)$.

A *size assignment* is a function σ that assigns to each size term a strictly positive integer with $\sigma(1) = 1$. Let $\mathcal{M}$ be the set of all matrices over K. We say that $M \in \mathcal{M}$ *conforms* to $\alpha \times \beta \in$ **size** $\times$ **size** by σ if M is a $\sigma(\alpha) \times \sigma(\beta)$-matrix.

If $\mathcal{S} : V \to$ **size** $\times$ **size** is a matrix schema, then an *instance of* $\mathcal{S}$ *with respect to* σ is a function $\mathcal{I} : V \to \mathcal{M}$ such that, for each $M \in V$, the matrix $\mathcal{I}(M)$ conforms to $\mathcal{S}(M)$ by σ.

Remark 6. In practice, a size assignment need only be defined on the size terms that are used in the schema. Thus, it can be finitely specified. While here we have chosen to keep the notion of size assignment and instance separate, it may perhaps be more natural to think of the size assignment as being part of the instance.

$$\mathcal{I}(\text{no\_courses}) = \begin{pmatrix} 5 & 2 & 0 \\ 2 & 1 & 3 \end{pmatrix} \qquad \mathcal{I}(\text{course\_fee}) = \begin{pmatrix} 300 \\ 250 \\ 330 \end{pmatrix}$$

Fig. 2. An example of an instance of a matrix schema.

Example 3. This example is similar to Example 1. Let K be the set of integers and let $\mathcal{S}$ be a matrix schema on $\{\text{no\_courses}, \text{course\_fee}\}$ with $\mathcal{S}(\text{no\_courses}) = \text{student} \times \text{dptm}$ and $\mathcal{S}(\text{course\_fee}) = \text{dptm} \times 1$. Let σ be a size assignment with $\sigma(\text{student}) = 2$ and $\sigma(\text{dptm}) = 3$. An instance $\mathcal{I}$ of $\mathcal{S}$ with respect to σ is shown in Fig. 2.

The semantics for MATLANG is given by the following matrix operations. Let A be an $m \times n$-matrix over K. We define $\mathbf{1}(A)$ to be the $m \times 1$-matrix (i.e., column vector) with $\mathbf{1}(A)_{i,1} = 1$. If $n = 1$ (i.e., A is a column vector), then $\mathrm{diag}(A)$ is the $m \times m$-matrix with $\mathrm{diag}(A)_{i,j}$ equal to $A_{i,1}$ if $i = j$ and to 0 otherwise. If B is an $m \times n$-matrix, then $A \circ B$ denotes the Hadamard product of A and B. In other words, $(A \circ B)_{i,j} = A_{i,j} * B_{i,j}$. Matrix addition and matrix multiplication are as usual denoted by $+$ and $\cdot$, respectively.

Formally, let $\mathcal{S}$ be a matrix schema, let e be a MATLANG expression over $\mathcal{S}$, and let $\mathcal{I}$ be a matrix instance of $\mathcal{S}$. Then the *output* matrix $e(\mathcal{I})$ of e under $\mathcal{I}$ is defined in the obvious way, given the operations just defined. If $e = M$ with M a matrix variable of $\mathcal{S}$, then $e(\mathcal{I})$ is naturally defined to be equal to $\mathcal{I}(M)$.

Remark 7. Matrix addition and the Hadamard product are the pointwise applications of addition and product, respectively. The original definition of MATLANG [2] is more generally defined in terms of an arbitrary set Ω of allowed pointwise functions. So, MATLANG as defined above fixes Ω to $\{+, \cdot\}$. This restriction was also considered by Geerts [4] (who also allows multiplication by constant scalars, but this is not essential).

Also, the original definition of MATLANG fixes K to the field of complex numbers and complex transpose is considered instead of (ordinary) transpose. Of course, transpose can be expressed using complex transpose and pointwise application of conjugation.

Table 1. Symbol table for the simulations between MATLANG and $(\mathsf{ARA} + \zeta_2)(2)$.

Mapping	MATLANG → ARA	ARA → MATLANG
Attributes A/size terms α	row_α, col_α	$\Psi(A)$
Schemas $\mathcal{S}$	$\Gamma(\mathcal{S})$	$\Theta(\mathcal{S})$
Expressions e	$\Upsilon(e)$	$\Phi(e)$
Instances I, relations r/matrices M	$\mathrm{Rel}_{\mathcal{S},\sigma}(I)$, $\mathrm{Rel}_{s,\sigma}(M)$	$\mathrm{Mat}_D(I)$, $\mathrm{Mat}_D(r)$

In the following subsections we provide simulations between MATLANG and $(\mathsf{ARA} + \zeta_2)(2)$. The notations for the different translations that will be given are summarized in Table 1.

4.2   Simulating MATLANG in $(\mathsf{ARA} + \zeta_2)(2)$

For notational convenience, instead of fixing a one-to-one correspondence between **rel** and **matvar**, we assume that **rel** = **matvar**.

Let us now fix injective functions $\mathrm{row} : \mathbf{size} \setminus \{1\} \to \mathbf{att}$ and $\mathrm{col} : \mathbf{size} \setminus \{1\} \to \mathbf{att}$ such that (1) $\mathrm{row}(\alpha)$ and $\mathrm{col}(\alpha)$ are compatible for all $\alpha \in \mathbf{size} \setminus \{1\}$, and (2) the range of row is disjoint from the range of col. To reduce clutter, we also write, for $\alpha \in \mathbf{size} \setminus \{1\}$, $\mathrm{row}(\alpha)$ as row_α and $\mathrm{col}(\alpha)$ as col_α.

Let $s \in \textbf{size} \times \textbf{size}$. We associate to s a relation schema $\Gamma(s)$ with $|\Gamma(s)| \leq 2$ as follows.

$$\Gamma(s) := \begin{cases} \{\text{row}_\alpha, \text{col}_\beta\} & \text{if } s = \alpha \times \beta; \\ \{\text{row}_\alpha\} & \text{if } s = \alpha \times 1; \\ \{\text{col}_\beta\} & \text{if } s = 1 \times \beta; \\ \emptyset & \text{if } s = 1 \times 1, \end{cases}$$

where $\alpha \neq 1 \neq \beta$.

Let $\mathcal{S}$ be a matrix schema on a set of matrix variables V. We associate to $\mathcal{S}$ a database schema $\Gamma(\mathcal{S})$ on V as follows. For $M \in V$, we set $(\Gamma(\mathcal{S}))(M) := \Gamma(\mathcal{S}(M))$.

Let σ be a size assignment. We associate to σ a domain assignment $D(\sigma)$ where, for $\alpha \in \textbf{size}$, $(D(\sigma))(\text{row}_\alpha) := (D(\sigma))(\text{col}_\alpha) := \{1, \ldots, \sigma(\alpha)\}$.

Let $M \in \mathcal{M}$ conform to $s = \alpha \times \beta$ by σ. We associate to M a relation $\text{Rel}_{s,\sigma}(M) : \mathcal{T}_{D(\sigma)}(\Gamma(s)) \to K$ as follows. We have $(\text{Rel}_{s,\sigma}(M))(t) := M_{i,j}$, where (1) i is equal to $t(\text{row}_\alpha)$ if $\alpha \neq 1$ and equal to 1 if $\alpha = 1$; and (2) j is equal to $t(\text{col}_\beta)$ if $\beta \neq 1$ and equal to 1 if $\beta = 1$.

Let $\mathcal{S} : V \to \textbf{size} \times \textbf{size}$ be a matrix schema and let $\mathcal{I}$ be a matrix instance of $\mathcal{S}$ with respect to σ. We associate to $\mathcal{I}$ an instance $\text{Rel}_{\mathcal{S},\sigma}(\mathcal{I})$ of database schema $\Gamma(\mathcal{S})$ with respect to $D(\sigma)$ as follows. For $M \in V$, we set $(\text{Rel}_{\mathcal{S},\sigma}(\mathcal{I}))(M) := \text{Rel}_{\mathcal{S}(M),\sigma}(\mathcal{I}(M))$.

$\mathcal{I}(\text{no\_courses}) =$

row$_{\text{student}}$	col$_{\text{dptm}}$	K
1	1	5
1	2	2
1	3	0
2	1	2
2	2	1
2	3	3

$\mathcal{I}(\text{course\_fee}) =$

row$_{\text{dptm}}$	K
1	300
2	250
3	330

Fig. 3. Matrix instance from Fig. 2 represented as a database instance.

Example 4. Recall $\mathcal{I}$, $\mathcal{S}$, and σ from Example 3. We have that $(\Gamma(\mathcal{S}))$ (no_courses) $= \{\text{row}_{\text{student}}, \text{col}_{\text{dptm}}\}$ and $(\Gamma(\mathcal{S}))(\text{course\_fee}) = \{\text{row}_{\text{dptm}}\}$. The database instance $\text{Rel}_{\mathcal{S},\sigma}(\mathcal{I})$ is shown in Fig. 3.

The next lemma shows that every MATLANG expression can be simulated by an $(\text{ARA} + \zeta_2)(2)$ expression.

Lemma 3. *For each* MATLANG *expression e over a matrix schema $\mathcal{S}$, there exists an $(\text{ARA}+\zeta_2)(2)$ expression $\Upsilon(e)$ over database schema $\Gamma(\mathcal{S})$ such that (1) $\Gamma(\mathcal{S}(e)) = (\Gamma(\mathcal{S}))(\Upsilon(e))$ and (2) for all size assignments σ and matrix instances $\mathcal{I}$ of $\mathcal{S}$ with respect to σ, we have $\text{Rel}_{\mathcal{S}(e),\sigma}(e(\mathcal{I})) = (\Upsilon(e))(\text{Rel}_{\mathcal{S},\sigma}(\mathcal{I}))$.*

Example 5. We continue the running example. In particular, recall $\mathcal{I}$, $\mathcal{S}$, and σ from Example 3. Consider the MATLANG expression $e = \text{no\_courses} \cdot \text{course\_fee}$ over $\mathcal{S}$. We have $\mathcal{S}(e) = \text{student} \times 1$ and

$$e(\mathcal{I}) = \begin{pmatrix} 2000 \\ 1840 \end{pmatrix}; \qquad \text{Rel}_{\mathcal{S}(e),\sigma}(e(\mathcal{I})) = $$

row$_{\text{student}}$	K
1	2000
2	1840

By Lemma 3 and its proof, we have that $\text{Rel}_{\mathcal{S}(e),\sigma}(e(\mathcal{I}))$ is equal to $e'(\text{Rel}_{\mathcal{S},\sigma}(\mathcal{I}))$ with

$$e' = \zeta_{C,2}(\rho_{\varphi_1}(\text{no\_courses}), \rho_{\varphi_2}(\text{course\_fee})),$$

where $\varphi_1(\text{col}_\gamma) = \varphi_2(\text{row}_\gamma) = C \notin \{\text{row}_\alpha, \text{col}_\beta\}$ and φ_1 and φ_2 are the identity otherwise.

4.3   Simulating $(\mathsf{ARA} + \zeta_2)(2)$ in MATLANG

In order to simulate $(\mathsf{ARA} + \zeta_2)(2)$ in MATLANG, we equip **att** with some linear ordering $<$. We remark that $<$ is an ordering on attributes, not on domain elements. Only an ordering on domain elements can have an impact on the expressive power of query languages of query languages [1].

Again we assume that **rel** = **matvar**. Let us fix an injective function $\Psi :$ **att** $\to$ **size** $\setminus \{1\}$.

Let $X \subseteq \{A_1, A_2\}$ be a relation schema for some A_1 and A_2 with $A_1 < A_2$. We associate to X an element $\Theta(X) \in$ **size** $\times$ **size** as follows. We have

$$\Theta(X) := \begin{cases} \Psi(A_1) \times \Psi(A_2) & \text{if } X = \{A_1, A_2\}; \\ \Psi(A) \times 1 & \text{if } X = \{A\} \text{ for some } A; \\ 1 \times 1 & \text{if } X = \emptyset. \end{cases}$$

Let $\mathcal{S}$ a database schema on a set N of relation names of arities at most 2. We associate to $\mathcal{S}$ a matrix schema $\Theta(\mathcal{S})$ on N as follows. For $R \in N$, we set $(\Theta(\mathcal{S}))(R) := \Theta(\mathcal{S}(R))$.

Let D be a domain assignment. We associate to D a size assignment $\sigma(D)$ where, for $A \in$ **att**, $(\sigma(D))(D(A)) = |D(A)|$. If every element in the range of a domain assignment D is of the form $\{1, \ldots, n\}$ for some n, then we say that D is *consecutive*.

Let D be a consecutive domain assignment. Given a relation $r : \mathcal{T}_D(X) \to K$ with $X \subseteq \{A_1, A_2\}$ and $A_1 < A_2$, we associate a matrix $\text{Mat}_D(r)$ conforming to $\Theta(X)$ by $\sigma(D)$ as follows. We define $(\text{Mat}_D(r))_{i,j} := r(t)$, where t is (1) the tuple with $t(A_1) = i$ and $t(A_2) = j$ if $|X| = 2$; (2) the tuple with $t(A) = i$ and $j = 1$ if $X = \{A\}$ for some A; and (3) the unique tuple of $\mathcal{T}_D(X)$ if $X = \emptyset$.

Let $\mathcal{S}$ a database schema on a set N of relation names of arities at most 2, and let $\mathcal{I}$ be a database of $\mathcal{S}$ instance with respect to D. We associate to $\mathcal{I}$ a matrix instance $\text{Mat}_D(\mathcal{I})$ of $\text{Mat}(\mathcal{S})$ with respect to $\sigma(D)$ as follows. For $R \in N$, we set $(\text{Mat}_D(\mathcal{I}))(R) := \text{Mat}_D(\mathcal{I}(R))$.

Example 6. Recall $\mathcal{I}$, $\mathcal{S}$, and D from Example 1. To reduce clutter, assume that **att** $=$ **size** $\setminus \{1\}$ and that Ψ is the identity function. Take student $<$ dptm. We have that $(\Theta(\mathcal{S}))(\text{no\_courses}) = \text{student} \times \text{dptm}$ and $(\Theta(\mathcal{S}))(\text{course\_fee}) = \text{dptm} \times 1$. Consider domain assignment D' and database instance $\mathcal{I}'$ obtained from D and $\mathcal{I}$, respectively, by replacing Alice by 1, Bob by 2, CS by 1, Math by 2, and Bio by 3. Note that D' is consecutive. The instance $\text{Mat}_{D'}(\mathcal{I}')$ is shown in Fig. 2.

The next lemma shows that every $(\mathsf{ARA} + \zeta_2)(2)$ expression can be simulated by a MATLANG expression.

Lemma 4. *For each* $(\mathsf{ARA} + \zeta_2)(2)$ *expression* e *over a database schema* $\mathcal{S}$ *of arity at most 2, there exists a* MATLANG *expression* $\Phi(e)$ *over matrix schema* $\Theta(\mathcal{S})$ *such that (1)* $\Theta(\mathcal{S}(e)) = (\Theta(\mathcal{S}))(\Phi(e))$ *and (2) for all consecutive domain assignments* D *and database instances* $\mathcal{I}$ *with respect to* D*, we have* $\text{Mat}_D(e(\mathcal{I})) = (\Phi(e))(\text{Mat}_D(\mathcal{I}))$.

Example 7. We continue the running example. In particular, recall $\mathcal{I}'$, $\mathcal{S}$, and D' from Examples 1 and 6. Consider the $(\mathsf{ARA} + \zeta_2)(2)$ expression $e = \text{no\_courses} \bowtie \text{course\_fee}$ over $\mathcal{S}$. We have $\mathcal{S}(e) = \{\text{student}, \text{dptm}\}$ and

$$
e(\mathcal{I}') =
\begin{array}{|c|c||c|}
\hline
\text{student} & \text{dptm} & K \\
\hline
1 & 1 & 1500 \\
1 & 2 & 500 \\
1 & 3 & 0 \\
2 & 1 & 600 \\
2 & 2 & 250 \\
2 & 3 & 990 \\
\hline
\end{array}
\quad ; \qquad
\text{Mat}_{D'}(e(\mathcal{I}')) =
\begin{pmatrix} 1500 & 500 & 0 \\ 600 & 250 & 990 \end{pmatrix}.
$$

By Lemma 3 and its proof, we have that $\text{Mat}_{D'}(e(\mathcal{I}'))$ is equal to $e'(\text{Mat}_{D'}(\mathcal{I}'))$ with $e' = \text{no\_courses} \cdot \text{diag}(\text{course\_fee})$.

4.4 Relationship with ARA(3) and Complexity

Corollary 1, Lemmas 3 and 4 together establish the equivalence of MATLANG with the language ARA(3) restricted to database schemas and output relations of arity at most 2.

Theorem 2. *For each* ARA(3) *expression* e *over a database schema* $\mathcal{S}$ *of arity at most 2 and with* $|\mathcal{S}(e)| \leq 2$*, there exists a* MATLANG *expression* e' *such that* $\text{Mat}_D(e(\mathcal{I})) = e'(\text{Mat}_D(\mathcal{I}))$ *for all consecutive domain assignments* D *and instances* $\mathcal{I}$ *with respect to* $\mathcal{S}$ *over* D*.*

Conversely, for each MATLANG *expression* e *over a matrix schema* $\mathcal{S}$*, there exists an* ARA(3) *expression* e' *such that* $\text{Rel}_{\mathcal{S}(e),\sigma}(e(\mathcal{I})) = e'(\text{Rel}_{\mathcal{S},\sigma}(\mathcal{I}))$ *for all size assignments* σ *and matrix instances* $\mathcal{I}$ *of* $\mathcal{S}$ *with respect to* σ*.*

As to complexity, we note that Υ and Φ in Lemmas 3 and 4 can be effectively constructed and of linear length (for fixed schemas; quadratic when the schema is part of the input). We conclude that the direction MATLANG $\to$ $(\mathsf{ARA} + \zeta_2)(2)$ $\to$ ARA(3) in the above result is linear. The direction ARA(3) $\to$ $(\mathsf{ARA} + \zeta_2)(2)$, however, is exponential, see Remark 5.

5   Conclusion

In related work, Yan, Tannen, and Ives consider provenance for linear algebra operators [18]. In that approach, provenance tokens represent not the matrix entries (as in our work), but the matrices themselves. Polynomial expressions (with matrix addition and matrix multiplication) are derived to show the provenance of linear algebra operations applied to these matrices.

Our result that every matrix query expressible in ARA(3) is also expressible in MATLANG provides a partial converse to the observation already made in the original paper [2], to the effect that MATLANG can be expressed in $\mathcal{L}_{\mathrm{Aggr}}(3)$: the relational calculus with summation and numerical functions [9], restricted to three base variables.[3] This observation was made in the extended setting of MATLANG that allows arbitrary pointwise functions (Remark 7). For the language considered here, ARA(3) provides a more appropriate upper bound for comparison, and ARA(3) is still a natural fragment of $\mathcal{L}_{\mathrm{Aggr}}(3)$.

When allowing arbitrary pointwise functions in MATLANG, we actually move beyond the positive relational algebra, as queries involving negation can be expressed. For example, applying the function $x \wedge \neg y$ pointwise to the entries of two $n \times n$ boolean matrices representing two binary relations R and S on $\{1, \ldots, n\}$, we obtain the set difference $R - S$. It is an interesting research question to explore expressibility of queries in MATLANG in this setting. For example, consider the following $\mathcal{L}_{\mathrm{Aggr}}(3)$ query on two matrices M and N:

$$\forall i \exists j \forall k \forall x (M(i, k, x) \rightarrow \exists i\, N(j, i, x))$$

Here, $M(i, k, x)$ means that $M_{i,k} = x$, and similarly for $N(j, i, x)$.

The above query, which does not even use summation, reuses the base variable i and checks whether each row of M, viewed as a set of entries, is included in some row of N, again viewed as a set of entries. We conjecture that the query is not expressible in MATLANG with arbitrary pointwise functions. Developing techniques for showing this is an interesting direction for further research.

Finally, recall that our main result Corollary 1 assumes that K is commutative. It should be investigated whether or not this result still holds in the noncommutative case.

Acknowledgments. We thank Floris Geerts for inspiring discussions. Robert Brijder has been a postdoctoral fellow of the Research Foundation - Flanders (FWO). Jan Van den Bussche was partially supported by the National Natural Science Foundation of China under grant# 61972455.

References

1. Abiteboul, S., Hull, R., Vianu, V.: Foundations of Databases. Addison-Wesley, Reading (1995)

---

[3] $\mathcal{L}_{\mathrm{Aggr}}$ is a two-sorted logic with base variables and numerical variables.

2. Brijder, R., Geerts, F., Van den Bussche, J., Weerwag, T.: On the expressive power of query languages for matrices. In: Kimelfeld, B., Amsterdamer, Y. (eds.) Proceedings of 21st International Conference on Database Theory. LIPIcs, vol. 98, pp. 10:1–10:17. Schloss Dagstuhl-Leibniz Center for Informatics (2018)
3. Van den Bussche, J.: FO³ and the algebra of binary relations. https://databasetheory.org/node/94. Accessed 22 July 2019
4. Geerts, F.: On the expressive power of linear algebra on graphs. In: Barcelo, P., Calautti, M. (eds.) Proceedings of 22nd International Conference on Database Theory. LIPIcs, vol. 127, pp. 7:1–7:19. Schloss Dagstuhl-Leibniz Center for Informatics (2019)
5. Green, T., Karvounarakis, G., Tannen, V.: Provenance semirings. In: Proceedings of 26th ACM Symposium on Principles of Database Systems, pp. 31–40 (2007)
6. Hutchison, D., Howe, B., Suciu, D.: LaraDB: a minimalist kernel for linear and relational algebra computation. In: Afrati, F., Sroka, J. (eds.) Proceedings of 4th ACM SIGMOD Workshop on Algorithms and Systems for MapReduce and Beyond, pp. 2:1–2:10 (2017)
7. Jananthan, H., Zhou, Z., et al.: Polystore mathematics of relational algebra. In: Nie, J.Y., Obradovic, Z., Suzumura, T., et al. (eds.) Proceedings of IEEE International Conference on Big Data, pp. 3180–3189. IEEE (2017)
8. Joglekar, M., Puttagunta, R., Ré, C.: AJAR: aggregations and joins over annotated relations. In: Proceedings of 35th ACM Symposium on Principles of Databases, pp. 91–106. ACM (2016)
9. Libkin, L.: Expressive power of SQL. Theoret. Comput. Sci. **296**, 379–404 (2003)
10. Luo, S., Gao, Z., Gubanov, M., Perez, L., Jermaine, C.: Scalable linear algebra on a relational database system. SIGMOD Rec. **47**(1), 24–31 (2018)
11. Abo Khamis, M., Ngo, H.Q., Rudra, A.: FAQ: questions asked frequently. In: Proceedings of 35th ACM Symposium on Principles of Databases, pp. 13–28. ACM (2016)
12. Abo Khamis, M., Ngo, H.Q., Rudra, A.: Juggling functions inside a database. SIGMOD Rec. **46**(1), 6–13 (2017)
13. Maddux, R.: The origin of relation algebras in the development and axiomatization of the calculus of relations. Stud. Logica. **50**(3/4), 421–455 (1991)
14. Marx, M., Venema, Y.: Multi-Dimensional Modal Logic. APLS, vol. 4. Springer, Dordrecht (1997). https://doi.org/10.1007/978-94-011-5694-3
15. Pratt, V.: Origins of the calculus of binary relations. In: Proceedings of 7th Annual IEEE Symposium on Logic in Computer Science, pp. 248–254 (1992)
16. Tarski, A.: On the calculus of relations. J. Symb. Log. **6**, 73–89 (1941)
17. Tarski, A., Givant, S.: A Formalization of Set Theory Without Variables. AMS Colloquium Publications, vol. 41. American Mathematical Society, Providence (1987)
18. Yan, Z., Tannen, V., Ives, Z.: Fine-grained provenance for linear algebra operators. In: Proceedings of 8th USENIX Workshop on the Theory and Practice of Provenance (2016)

Social Consolidations: Rational Belief in a Many-Valued Logic of Evidence and Peerhood

Yuri David Santos[✉]

University of Groningen, Groningen, The Netherlands
y.david.santos@rug.nl

Abstract. We explore an interpretation of FVEL, a four-valued logic of evidence, where states represent agents, the propositional layer corresponds to the evidence available to these agents, and the relation corresponds to peerhood connections between them. Belief is determined based on the agent's evidence, but also on her peers' evidence. *Consolidation* functions are proposed, which map evidence situations to belief attitudes. We adapt some postulates of Social Choice Theory to our belief formation setting and, with them, we separate rational from irrational consolidations. We define a dynamic operator for addition and removal of evidence, which serves as a basis for some essential dynamic postulates and also for future developments on consolidations that take amounts of evidence into account. Our main technical result is a characterisation of a class of consolidations satisfying most of our rationality postulates.

Keywords: Evidence logics · Epistemic logic · Many-valued logic

1 Introduction

Four-valued epistemic logic (FVEL) [29] was first designed to model scenarios where agents are uncertain about the evidence publicly available. Here we give another interpretation to this logic, where the binary relation represents peerhood connections. Therefore, each state will represent the evidential state of one agent. This puts this work in line with other network logics such as [4,7].

In our setting, agents have four-valued evidence for propositions, embodied by a four-valued valuation function over atoms, which represents only evidence *for* that atom, only evidence *against* it, evidence both *for and against* it, or no evidence at all. Our main goal in this paper is to find rational ways of forming beliefs for these agents, given their own evidence and their peers'. With that in mind, we establish some rationality postulates and check some definitions of belief that respect those postulates, and some that do not.

© Springer Nature Switzerland AG 2020
A. Herzig and J. Kontinen (Eds.): FoIKS 2020, LNCS 12012, pp. 58–78, 2020.
https://doi.org/10.1007/978-3-030-39951-1_4

After that, we introduce a dynamic operator for addition/removal of evidence. This operator is used to axiomatise some of the postulates, but also to define two new ones, which serve to rule out some undesirable consolidations. We then prove that these axioms characterise a class of consolidations satisfying most of the main postulates. Finally, we show how this operator can be used to "count" peers, which in the future can be employed to define consolidations that form beliefs based on the amount of evidence for or against something.[1]

2 Syntax and Semantics

In this section we explore a variant of *four-valued epistemic logic* (FVEL) [29].

2.1 Syntax

Let At be a countable set of atoms. Below, $p \in At$; the classical part of the language, $\mathscr{L}_0$, is represented below by ψ; the propositional part $\mathscr{L}_1$, where $\psi \in \mathscr{L}_0$, is represented by χ; and the complete language $\mathscr{L}$, where $\psi \in \mathscr{L}_0$ and $\chi \in \mathscr{L}_1$, is given by φ:

$$\psi ::= p \mid {\sim}\psi \mid (\psi \wedge \psi) \qquad \chi ::= \psi \mid \neg\chi \mid (\chi \wedge \chi) \mid {\sim}\chi$$

$$\varphi ::= \chi \mid {\sim}\varphi \mid (\varphi \wedge \varphi) \mid \Box\varphi \mid B\psi$$

We abbreviate $\varphi \vee \psi \stackrel{\text{def}}{=} {\sim}({\sim}\varphi \wedge {\sim}\psi)$ and $\Diamond\varphi \stackrel{\text{def}}{=} {\sim}\Box{\sim}\varphi$. We restrict belief to classical propositional formulas ($\mathscr{L}_0$) because formulas with $\neg$ refer to evidence, and we do not want agents forming beliefs about evidence, only about facts.

Formulas such as p are read as *the agent has evidence for* p, whereas $\neg p$ is read as *the agent has evidence against* p, and ${\sim}\varphi$ as *it is not the case that* φ. We read $\Box\varphi$ as φ *holds for all peers* and $B\varphi$ as *the agent believes* φ.[2, 3]

2.2 Semantics

Models are tuples $M = (S, R, V)$, where S is a finite set of agents, R is a binary relation on S representing "peerhood" and $V : At \times S \to \mathscr{P}(\{0, 1\})$ is a four-valued valuation representing agents' evidence: $\{1\}$ is *true* (t), $\{0\}$ is *false* (f),

---

[1] Some proofs at: https://github.com/ydsantos/appendix_scons/blob/master/proofs.
    pdf.

[2] Notice that our language is non-standard in that even though a formula in $\mathscr{L}_1$ has an evidential meaning (such as p meaning *the agent has evidence for* p), under the belief operator B these formulas are read as factual statements (e.g. Bp means that *the agent believes* p and not that *the agent believes that she has evidence for* p).

[3] We chose B (belief) instead of K (knowledge) because we are working with imperfect evidence, which can be misleading. Therefore, our agents can form false beliefs, which violate factivity, a standard requirement for knowledge.

$\{0,1\}$ is *both* (*b*) and $\emptyset$ is *none* (*n*). A satisfaction relation is defined as follows:

$M, s \models p$ iff $1 \in V(p, s)$ $M, s \models \neg p$ iff $0 \in V(p, s)$

$M, s \models {\sim}\varphi$ iff $M, s \not\models \varphi$

$M, s \models (\varphi \wedge \psi)$ iff $M, s \models \varphi$ and $M, s \models \psi$

$M, s \models \neg(\varphi \wedge \psi)$ iff $M, s \models \neg\varphi$ or $M, s \models \neg\psi$

$M, s \models \Box\varphi$ iff for all $t \in S$ s.t. sRt, it holds that $M, t \models \varphi$

$M, s \models \neg{\sim}\varphi$ iff $M, s \models \varphi$ $M, s \models \neg\neg\varphi$ iff $M, s \models \varphi$

Note that the semantics of $\neg$ is defined in a case-by-case fashion (this operator comes from FDE [5]). An extended valuation function $\overline{V}$ can be defined differently for each type of formula. If $\varphi \in \mathscr{L}_1$, then: $1 \in \overline{V}(\varphi, s)$ iff $M, s \models \varphi$; $0 \in \overline{V}(\varphi, s)$ iff $M, s \models \neg\varphi$. Otherwise: $1 \in \overline{V}(\varphi, s)$ iff $M, s \models \varphi$ iff $0 \notin \overline{V}(\varphi, s)$.

As pointed out in [29], this logic can be seen as a modal extension of FDE [5], with the addition of a classical negation. The logic FDE deals with evidence differently than other logics such as intuitionistic logic [18,33]. While both are weaker than classical logic, the concept of justification as existence of constructive proofs is much stronger than what we consider evidence in this paper. In our case, evidence can be misleading, as mentioned before. FDE is more suitable for modelling situations with incomplete and inconsistent evidence, while FVEL extends this logic to a modal setting, enabling us to talk about multiple agents. FVEL also includes a classical negation, which gives it much more expressive power, and many of the definitions and results in this paper make use of this operator ($\sim$). Among other things, it allows us to define formulas discriminating which of the four truth values a formula $\varphi \in \mathscr{L}_1$ has: $\varphi^n \overset{\text{def}}{=} ({\sim}\varphi \wedge {\sim}\neg\varphi)$; $\varphi^f \overset{\text{def}}{=} {\sim}{\sim}({\sim}\varphi \wedge \neg\varphi)$; $\varphi^t \overset{\text{def}}{=} {\sim}{\sim}(\varphi \wedge {\sim}\neg\varphi)$; $\varphi^b \overset{\text{def}}{=} {\sim}{\sim}(\varphi \wedge \neg\varphi)$. In words, a formula φ^x is satisfied ($M, s \models \varphi^x$) iff φ has value $x \in \{t, f, b, n\}$, i.e. $\overline{V}(\varphi, s) = x$.

We say that $\Sigma \models \varphi$ (Σ *entails* φ) when for all models M and states s, if $M, s \models \sigma$ for all $\sigma \in \Sigma$, then $M, s \models \varphi$. We say that $M \models \varphi$ if $M, s \models \varphi$ for all states s of M. And $\models \varphi$ (φ is *valid*) if $M \models \varphi$ for all M; otherwise φ is *invalid*. If $\models {\sim}\varphi$, we say φ is *contradictory*, and if φ is not contradictory nor valid, it is *contingent*. If a formula is valid or contingent, it is *satisfiable*. Call the *truth range* of φ the set $\{x \mid$ there is a model $M = (S, R, V)$ and an $s \in S$ s.t. $\overline{V}(\varphi, s) = x\}$. The following result will be useful for some of the proofs:

Proposition 1. *All formulas in $\mathscr{L}_0$ have one of the following four truth ranges: $\{\{1\}\}$, $\{\{0\}\}$, $\{\{0\}, \{1\}\}$, $\{\emptyset, \{0\}, \{1\}, \{0,1\}\}$. A formula in $\mathscr{L}_1$ can have any truth range in $\mathscr{P}(\mathscr{P}(\{0,1\})) \setminus \emptyset$ except for $\{\emptyset\}$, $\{\{0,1\}\}$, and $\{\emptyset, \{0,1\}\}$.*

The central question of this paper is how to define the semantics for belief based on the evidence, a process we call *consolidation* (see [30]). A key philosophical assumption of this project is that *rational belief is determined by evidence*.

3   Rationality Conditions for Consolidations

3.1   Epistemic Autonomy Versus Epistemic Authority

In social epistemology, there is currently a lot of debate around the topics of peer disagreement and higher-order evidence [6,12,13,17,22–24]. One important question in this debate is: *What should a rational agent do when her peers – who she deems as rational as her – have different opinions on some proposition?* There are many different proposals in the literature as to what to do in this case. Nevertheless, we can roughly categorise them into two main groups: the *equal weight* views [9], and the *steadfast* views [21]. The former tend to consider the agent and her peers to be on equal footing, so if you and your peer disagree on something, your opinion should be something in the middle of both opinions. The latter claim that you are entitled to trust yourself more than you trust your peers – maybe because you have direct access to your evidence, as opposed to mere testimonial access to your peers' evidence, or because of some other reason. In both views, the concept of *peerhood* is preeminent. It is assumed that, in what matters, you and your peers are of equal competence. Evidently, if one's peer is far more competent than oneself in the topic at hand and one knows that, the rational thing to do is to defer to her judgement (but in that case she is *not* your peer). What enables peerhood is the lack of such higher-order knowledge: we usually do not know exactly how competent a peer is, so the reasonable (and modest) thing to do is to assume that the relevant people in the given case are (possibly) as competent as you, except if you have a "defeater" for that belief[4].

3.2   Rationality Postulates

Now we propose and discuss a series of rationality postulates, mostly adapted from postulates from Social Choice Theory (SCT) [2,14,31]. SCT is concerned with determining outcomes of voting from certain voting profiles. The adaptation we make here is in the sense that a rational belief in propositions (atomic or otherwise) will be determined from the evidence possessed by the agent and her peers, so here "voting profiles" become evidence, and "election outcome" becomes belief attitude. Consolidations are not voting procedures, but involve the weighing of inputs to find a suitable outcome.[5]

Regardless of the semantics of B, which is not yet defined, the following function Att serves as a shorthand for the doxastic attitude of an agent s w.r.t. a formula φ (belief, disbelief or abstention):

Definition 1 (Attitude). *Let* Att $: \mathscr{L}_0 \times S \to \{1, 0, -1\}$ *be a function such that:* Att$(\varphi, s) = 1$ *iff* $M, s \models B\varphi$; Att$(\varphi, s) = -1$ *iff* $M, s \models B{\sim}\varphi$; *otherwise* Att$(\varphi, s) = 0$.[6]

---

[4] As a scientist investigating hypothesis H, you consider another scientist also investigating H to be your peer, but not if she committed fraud in the past.

[5] Note, however, that we only make a loose connection to SCT here, not a formal one.

[6] The function Att also depends on a model M, but this will be left implicit. We will usually write Att$'$ if we are referring to another model M', Att$''$ for M'', and so on.

Postulate 1 (Consistency (Con)). *For all models M and $s \in S$: let $\Sigma =$*
$\{\varphi \in \mathscr{L}_0 \mid M, s \models B\varphi\}$. Then $\Sigma \not\models p \wedge \sim p$.

The postulate above is the most important demand on our consolidations: rational belief has to be consistent.

Postulate 2 (Modesty (Mod)). *For all models $M = (S, R, V)$, all $s \in S$,*
and all contingent $\varphi \in \mathscr{L}_0$, there is a model $M' = (S', R', V')$ with $S \subseteq S'$ s.t.
$\mathrm{Att}(\varphi, s) \neq \mathrm{Att}'(\varphi, s)$, where $V|_s = V'|_s$.[7]

Postulate 2 says that it is possible to change an agent's attitude toward a contingent formula just by changing her peerhood connections and the evidence of
her peers. Modesty is adapted from the SCT postulate of *non-dictatorship*: the
outcome of the election is not determined by one single agent. Postulate 3 also
comes from non-dictatorship, but for *Modesty* we think of the agent as her own
dictator.

The plausibility of this postulate hinges on the plausibility of the claim that
regardless of what evidence you have, it is never rational to ignore others' evidence. This, in turn, depends on the outcome of the debate in epistemology
discussed above. In any case, is the format of this postulate adequate? The
restriction to contingent formulas seems justified: if we reject *Logical Omniscience*, it might be acceptable to abstain from judgement on tautologies and
contradictions, but it seems irrational to expect one to be persuaded to aban-
don a belief in a tautology or adhere to a contradiction. Keeping $V|_s$ untouched
captures exactly the idea of not changing one's evidence, but possibly changing
others'. The $S \subseteq S'$ part demands that the original agents be preserved. This
is innocuous, for even if a change in belief demands the removal of a peer, that
can be obtained by removing the connection (changing R); non-peers do not
matter in our setting. A stronger variant of Modesty could be considered, *Strong
Modesty*, where not only is it possible to change the attitude for any formula,
but also any other attitude is possible. This could be plausible, but expecting
a radical change in attitude (for example, from disbelief to belief) for any contingent proposition might require a huge *amount* of evidence, but we are not
representing this *aspect* of evidence here; we do make a step in this direction in
Sect. 5.

Postulate 3 (No Gurus (NG)). *For all agents $s, t \in S$ (with $s \neq t$) and all*
contingent $\varphi \in \mathscr{L}_0$, there is a model $M = (S, R, V)$ s.t. $\mathrm{Att}(\varphi, s) \neq \mathrm{Att}(\varphi, t)$.

This postulate says that for any formula there is a model such that the attitudes
of two agents towards that formula differ, i.e. an agent's opinion is not determined
by anyone else's. This postulate also stems from the postulate of non-dictatorship
in SCT (in a more obvious way). We have that a consolidation satisfying Mod
also satisfies NG (see Proposition 4 later). So if Modesty is plausible, then this
postulate has to be as well. In principle, it might be odd to think that, for
example, two biologists could rationally disagree on whether natural selection

---

[7] We denote by $V|_s$ the restriction of a valuation V to $At \times \{s\}$, with $s \in S$.

happens. This apparent controversy is only superficial, though. If we stick to our key assumption that evidence determines rational belief, then that should be possible *given* they have access to different circles – with one of them possibly possessing misleading evidence.

Postulate 4 (Equal Weight (EW)). *Consider any model $M = (S, R, V)$, any two agents $s, t \in S$, and a valuation V' such that $V'(p, s) = V(p, t)$, $V'(p, t) = V(p, s)$, and $V'(p, u) = V(p, u)$ for all $u \in S \setminus \{s, t\}$, for all $p \in At$. Then if sRt, it holds that $\text{Att}'(\varphi, s) = \text{Att}(\varphi, s)$, for all $\varphi \in \mathscr{L}_0$.*

What this postulate says is that if you swap all your evidence with the evidence of one of your peers, your beliefs do not change: you treat your evidence and your peers' equally. It comes from the SCT postulate of *anonimity*: if we have the same voting profile but swap the voters, the outcome does not change. Again, the plausibility of this postulate depends on your position in the debate of Sect. 3.1.

Postulate 5 (Atom Independence (AI)). *Consider any model $M = (S, R, V)$. For any atom $p \in At$, if V' is a valuation s.t. $V'(p, s) = V(p, s)$ for all $s \in S$, then $\text{Att}(p, s) = \text{Att}'(p, s)$ for all $s \in S$.*

The valuation of one atom should not interfere in the attitudes towards another. This postulate is adapted from *independence of irrelevant alternatives*: the outcome between x and y should only depend on voters opinions w.r.t. x and y; changing the preferences between other candidates does not affect the outcome. A more "local" version of this postulate could be formulated: for any $p \in At$ and $s \in S$, if $V(p, s) = V'(p, s)$ and $V(p, t) = V'(p, t)$ for all t such that sRt, then $\text{Att}(p, s) = \text{Att}'(p, s)$.[8] We can prove that this definition is equivalent to AI.

Let $\preceq$ be the smallest reflexive and transitive relation $\preceq : \mathscr{P}(\{0, 1\}) \times \mathscr{P}(\{0, 1\})$ such that $\{0\} \preceq \emptyset$, $\{0\} \preceq \{0, 1\}$, $\emptyset \preceq \{1\}$ and $\{0, 1\} \preceq \{1\}$. Let $\npreceq$ be the complement of $\preceq$, and define $x \prec y$ iff $x \preceq y$ and $y \npreceq x$.

Postulate 6 (Monotonicity (Mon)). *Consider a model $M = (S, R, V)$ and a V' which coincides with V, except that $V'(p, s) \neq V(p, s)$ for one $s \in S$ and $p \in At$. If $V(p, s) \prec V'(p, s)$, then for all $t \in S$, $\text{Att}(p, t) \leq \text{Att}'(p, t)$. If $V'(p, s) \prec V(p, s)$, then for all $t \in S$, $\text{Att}'(p, t) \leq \text{Att}(p, t)$.*

Postulate 6 states that if the valuation only changes positively/negatively for one atom and one agent, then the attitude towards this atom for any agent should either stay the same, or change according to the same trend (more positive/negative). Monotonicity was adapted from a homonymous SCT postulate: if a profile is altered only by promoting (demoting) one candidate, the outcome should either change only by promoting (demoting) this candidate, or not change.

---

[8] We thank an anonymous reviewer for this suggestion.

Now there is a question of adequacy of the format of this postulate. There is not always a unique way of changing a valuation to produce a certain change in the (extended) valuation of a complex formula, so we limited this postulate to atomic changes. The other question regarding format is why the postulate limits the valuation change to only one atom and one agent. Clearly changing one atom in one direction (according to $\prec$) for more agents, or changing several atoms in this fashion, should preserve monotonicity. These "cumulative" effects are already covered by the postulate as it is.

Postulate 7 (Doxastic Freedom (DF)). *Consider any set of agents S and any function $f : At \times S \rightarrow \{1, -1, 0\}$. Then there is a model $M = (S, R, V)$ such that $\mathrm{Att}(p, s) = f(p, s)$ for all $p \in At$ and $s \in S$.*

DF says that any combination of attitudes towards atoms is possible for any agent. It is adapted from *non-imposition*: every outcome is achievable by some voting profile. This postulate seems somehow connected to AI. However:

Observation 1. *A consolidation satisfying Doxastic Freedom does not necessarily satisfy Atom Independence. The converse also holds.*

Postulate 8 (Consensus (Css)). *If for some agent $s \in S$ and some $\varphi \in \mathscr{L}_0$ we have that $\overline{V}(\varphi, s) = \{1\}$ (or $\{0\}$), and for all $t \in S$ such that sRt: $\overline{V}(\varphi, t) = \{1\}$ (or $\{0\}$), then $\mathrm{Att}(\varphi, s) \neq -1$ (or 1).*

Consensus is derived from the SCT postulate of *unanimity*: if all voters prefer one candidate over another, then so must the outcome. It says that if an agent and all her peers have unambiguous evidence about some atom, then she should not believe contrary to that. We can define *Strong Consensus* in a similar way, but instead of demanding no contrary belief, it demands belief in case of unanimous positive evidence and disbelief in case of unanimous negative evidence.

Observation 2. *A consolidation satisfying Strong Consensus and Consistency also satisfies Consensus.*

Proof. One just has to see that $M, s \models B\varphi$ implies $M, s \not\models B{\sim}\varphi$ for a consolidations satisfying Consistency (and similarly for the $B{\sim}\varphi$ case).   □

Notice that this stronger variant, in combination with Proposition 1, entails a form of *logical omniscience*. We could also have defined the postulate differently by considering unanimity among all agents instead of one agent and her peers, but, again, we are assuming that non-peers are inaccessible/irrelevant.

Postulate 9 (Logical Omniscience (LO)). *For all models M and $s \in S$: if $\Sigma \models \varphi$ and $M, s \models B\sigma$ for all $\sigma \in \Sigma$, then $M, s \models B\varphi$.*

This postulate is not derived from any postulate of SCT. It is debatable whether it should be satisfied or not, but as a normative demand on real agents we consider it too strong. Notice that it implies the knowledge of all validities, as they are consequences of the empty set, and also that the doxastic state has to be consistent or it will be trivialised.

In summary, all the postulates listed in this section are expected to be satisfied by any rational consolidation (call these *core postulates*), except for Mod and EW, whose normative status depend on the reader's philosophical commitments w.r.t. the debate of Sect. 3.1, and LO, which is also part of another long debate [11,19,20,26]. No impossibility theorem à la Arrow [2] ensues, and consolidations satisfying all core postulates are presented. One main difference of our approach that might explain this is that we do not have preference orders over attitudes. Note also that our connection to SCT is not fully formal, our postulates are only inspired by it.

4   Social Consolidations

In this section we will define *consolidation policies*, that is, methods of defining belief from evidence. We expect the most reasonable consolidations to satisfy all the core postulates, and unreasonable ones to violate at least one of them.

4.1   Preliminaries

Before talking about consolidations, we will formally specify what are the possible ones. Now let $\mathbb{M} = \{(M, s) \mid M = (S, R, V)$ is an FVEL model and $s \in S\}$ be the class of all *pointed models*. First, we draw the following definition from the literature on n-bisimulations:

Definition 2 (1-Bisimulation). *Consider two FVEL models $M = (S, R, V)$ and $M' = (S', R', V')$, an $s \in S$ and an $s' \in S'$. We say that $(M, s) \leftrightarrow (M', s')$, read (M, s) is 1-bisimilar to (M', s'), iff:*

atoms *For all $p \in At$, $V(p, s) = V'(p, s')$;*

back *For all $t' \in S'$ s.t. $s'R't'$, there is a $t \in S$ s.t. sRt and $V(p, t) = V'(p, t')$ for all $p \in At$.*

forth *For all $t \in S$ s.t. sRt, there is a $t' \in S'$ s.t. $s'R't'$ and $V(p, t) = V'(p, t')$ for all $p \in At$.*

The purpose of Definition 2 is to determine whether two pointed models have equivalent evidence. Since our relation R of peerhood is not transitive, we assume that our agents only have access to their own evidence and their peers'. So formulas such as $\Box p$ are relevant for consolidation, whereas $\Box\Box p$ is not.

Proposition 2. $(M, s) \leftrightarrow (M', s')$ *implies: $M, s \models \varphi$ iff $M', s' \models \varphi$ for all $\varphi \in \mathscr{L}$ not containing B nor nested $\Box$. The converse also holds for image-finite models (each agent has finitely many peers).*

Proposition 3. *The relation $\leftrightarrow$ is an equivalence relation.*

Then $\leftrightharpoons \subseteq \mathbb{M} \times \mathbb{M}$. Denote by $[M, s]$ the equivalence class of (M, s) under $\leftrightharpoons$, that is, $[M, s] = \{(M', s') \in \mathbb{M} \mid (M, s) \leftrightharpoons (M', s')\}$. Let $\mathbb{M}/\leftrightharpoons$ be the quotient class of $\mathbb{M}$ by $\leftrightharpoons$, that is, the class of equivalence classes of $\mathbb{M}$ under $\leftrightharpoons$. Then, we are interested in the following:

Definition 3. *A consolidation is a function* $\mathbb{C} : \mathbb{M}/\leftrightharpoons \times \mathscr{L}_0 \to \{0, 1\}$. *For any model* $M = (S, R, V)$ *with* $s \in S$, *we set* $M, s \models B\varphi$ *iff* $\mathbb{C}([M, s], \varphi) = 1$.

Proposition 4. *A consolidation satisfying Mod also satisfies NG.*

With these definitions in hand, we will introduce the following:

Definition 4. *We say that a condition is* axiomatisable *when: it holds iff all* $\sigma \in \Sigma$ *are valid, for some* $\Sigma \subseteq \mathscr{L}$. *We say that a condition is* negatively axiomatisable *when: it holds iff all* $\sigma \in \Sigma$ *are invalid, for some* $\Sigma \subseteq \mathscr{L}$.

Proposition 5. *Consistency holds iff for all finite* $\Sigma = \{\sigma_1, ...\sigma_n\} \subseteq \mathscr{L}_0$ *such that* $\Sigma \models p \wedge {\sim}p$, ${\sim}(B\sigma_1 \wedge ... \wedge B\sigma_n)$ *is valid.*

Proof. The logic of $\mathscr{L}_0$ is basically classical propositional logic (as mentioned in [29]), and is, therefore, compact. So for any $\Sigma \models \varphi$ with $\varphi \in \mathscr{L}_0$, there is a finite $\Sigma' \subseteq \Sigma$ such that $\Sigma' \models \varphi$. The case where $\varphi = p \wedge {\sim}p$ is a particular case of this. So all inconsistent subsets of $\mathscr{L}_0$ have a finite inconsistent subset. □

Proposition 6. *Logical Omniscience holds iff for all finite* $\Sigma = \{\sigma_1, ...\sigma_n\} \subseteq \mathscr{L}_0$ *and* $\varphi \in \mathscr{L}_0$ *such that* $\Sigma \models \varphi$, ${\sim}(B\sigma_1 \wedge ... \wedge B\sigma_n \wedge {\sim}B\varphi)$ *is valid.*

Proof. The reasoning is similar to the case for Proposition 5. □

Note that Propositions 5 and 6 follow from compactness of $\mathscr{L}_0$. Now consider the following axioms:

$$\textbf{C1} \quad {\sim}((\varphi^t \wedge \Box\varphi^t) \wedge B{\sim}\varphi) \qquad \textbf{C2} \quad {\sim}((\varphi^f \wedge \Box\varphi^f) \wedge B\varphi)$$

Proposition 7. *A consolidation satisfying Consistency satisfies Consensus iff* **C1** *and* **C2** *are valid.*

4.2   Consolidation Policies

First, we will look at the most straightforward (and naive) possibility: $M, s \models B\varphi$ iff $M, s \models \Box\varphi$. This possibility is appealing because it is familiar and simple. First, let us note that, in order to include the evidence of the agent itself in the consolidation, we have to require the model to be reflexive. This raises the question: is the agent a peer of herself (see [9])? If yes, then we should only work with reflexive models, if not, then only with *anti-reflexive* models (sRs holds for no s). This is not so crucial as we can (and will) use an equivalent definition for anti-reflexive models: $M, s \models B\varphi$ iff $M, s \models \varphi \wedge \Box\varphi$. So we assume that agents are not peers of themselves. We call this latter definition *naive consolidation*.

Proposition 8. *Naive consolidation satisfies Con, Mod, EW, AI, Mon and Strong Css. It does not satisfy DF and LO.*

Surprisingly, naive consolidation only fails one core postulate: Doxastic Freedom. It is surprising because this consolidation actually *ignores all negative evidence*.

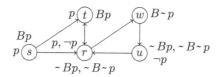

Fig. 1. An example of naive consolidation. Agent s believes p, but not $\sim p$, since all her peers and herself satisfy p (have evidence for p), and not $\sim p$. One of the peers (r) has $\neg p$, but s ignores that. Agent w believes $\sim p$, even though she does not have evidence against p. She believes $\sim p$ only on the grounds that she and r do not have evidence for p. Agent u does not believe p nor $\sim p$, because she does not have evidence for p, but her only peer does.

Another simple consolidation we can analyse is the *sceptical consolidation*, which sets $M, s \not\models B\varphi$ for all $\varphi \in \mathcal{L}_0$. Fortunately this extreme position is blocked by two of our core postulates.

Proposition 9. *Sceptical consolidation satisfies Con, EW, AI, Mon and Css. It does not satisfy in general NG (and therefore Mod), DF and LO.*

Now we will try a more sophisticated definition:

Definition 5. *Call $\mathscr{C}$-consolidations the policies defined by:*

$$M, s \models Bp \qquad \textit{iff } \mathscr{C}(V_p^s, V_{\neg p}^s, V_{\Diamond p}^s, V_{\Diamond \neg p}^s, V_{\Box p}^s, V_{\Box \neg p}^s) = 1$$

$$M, s \models B\sim p \qquad \textit{iff } \mathscr{C}(V_p^s, V_{\neg p}^s, V_{\Diamond p}^s, V_{\Diamond \neg p}^s, V_{\Box p}^s, V_{\Box \neg p}^s) = -1$$

$$M, s \models B\sim\sim\varphi \qquad \textit{iff } M, s \models B\varphi$$

$$M, s \models B(\varphi \wedge \psi) \qquad \textit{iff } M, s \models B\varphi \textit{ and } M, s \models B\psi$$

$$M, s \models B\sim(\varphi \wedge \psi) \qquad \textit{iff } M, s \models B\sim\varphi \textit{ or } M, s \models B\sim\psi$$

where V_χ^t is 1 if $1 \in \overline{V}(\chi, t)$ and 0 otherwise; and $\mathscr{C} : \{0, 1\}^6 \rightarrow \{1, -1, 0\}$ is a function that maps evidence (in this case represented by the six binary parameters) to a belief attitude (1 for belief, -1 for disbelief and 0 for abstention).

What is a good definition for $\mathscr{C}$? As we can see above, the real consolidation effort is only with respect to atomic propositions, while more complex beliefs are formed from those atomic beliefs. Some advantages of this approach are that it uses all evidence available for each atom, the agent still retains some inference power (with which it can derive other beliefs), and avoids malformed definitions, such as: $M, s \models B\varphi$ iff $M, s \models \varphi^t \wedge \Box\varphi^t$; $M, s \models B\text{~}\varphi$ iff $M, s \models \varphi^f \wedge \Box\varphi^f$. In words: the agent believes a formula if she and her peers have only positive evidence for it, and believes its negation if she and her peers have only negative evidence for it. This seems like a good (if too cautious) definition at first sight, but it is actually not well-formed. We can verify whether $B\text{~}\psi$ via the second clause, but also via the first if $\varphi = \text{~}\psi$. And these can sometimes give conflicting results. We avoid that by using $\mathscr{C}$ only to decide belief for literals. Moreover:

Proposition 10. *All $\mathscr{C}$-consolidations satisfy Con and AI.*

Our agents under $\mathscr{C}$-consolidations are not necessarily omniscient, but they present some properties related to unbounded logical power:

Proposition 11. *Consider any $\mathscr{C}$-consolidation, and a maximally consistent set of literals Σ. If $M, s \models B\sigma$ for all $\sigma \in \Sigma$ and $\Sigma \models \varphi$, then $M, s \models B\varphi$.*

Corollary 1. *Any $\mathscr{C}$-consolidation satisfying DF also satisfies NG.*

Proposition 12. *Belief in $\mathscr{C}$-consolidations is closed under* modus ponens: *if $M, s \models B\varphi$ and $M, s \models B\text{~}(\varphi \wedge \text{~}\psi)$, then $M, s \models B\psi$.*

Corollary 2. *Any $\mathscr{C}$-consolidation satisfies Logical Omniscience if we add the following clause to the semantics: if $\models \varphi$, then $M, s \models B\varphi$ (where $\varphi \in \mathscr{L}_0$).*

There are $3^{(2^6)} = 3^{64} \approx 3.43 \times 10^{30}$ consolidation function candidates for $\mathscr{C}$. The combinations $(0, 1, 1)$ for $V^s_{\Diamond p}, V^s_{\Diamond \neg p}, V^s_{\Box p}$ and $(1, 0, 1)$ for $V^s_{\Diamond p}, V^s_{\Diamond \neg p}, V^s_{\Box \neg p}$ are impossible, though, which leaves us with "only" $3^{48} \approx 7.98 \times 10^{22}$ relevantly different candidates. Now we consider some promising possibilities.

Policy I. Our first social consolidation policy is in Fig. 2. In cases of unambiguous evidence, the agent decides for belief or disbelief, accordingly. In the case of conflicting evidence, the agent already has some evidence, and since we want a consistent doxastic state, this entails that the agent will inevitably have to discard some evidence. So, in this case, the mere existence of evidence of one kind from one peer is enough to produce belief. However, when the agent has no evidence at all, even if she decides to abstain there is no waste of evidence, so she will be more demanding to change her view. In this case, unanimity of her peers is needed (see an example in Fig. 3).

Policy II. One might consider that our previous policy still does not justify the different treatment for the problematic evidence cases, and is therefore arbitrary. Hence, we can consider a second policy where the behaviour when the evidence is *none* imitates the case for *both*: consider a decision tree identical to that of Fig. 2 but with the subtree for none (the leftmost subtree) just replaced by that used for both (the rightmost one).

Proposition 13. *Policy I and II satisfy Monotonicity, Doxastic Freedom and Consensus. Modesty and Equal Weight are not satisfied.*

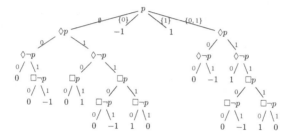

Fig. 2. Decision trees will be used to represent $\mathscr{C}$-consolidations. This one represents $\mathscr{C}$ for Policy I. Nodes are labelled by expressions that are representable with the six parameters for $\mathscr{C}$. The leaves are the outcomes of the consolidation: 1 for belief, -1 for disbelief and 0 for abstention of judgement.

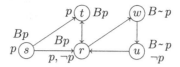

Fig. 3. Policy I applied to the model of Fig. 1. Here all agents except for r and w have unambiguous evidence about p, so they can easily form beliefs without looking at their peers. Agent w has no evidence whatsoever, so by the tree of Fig. 2 she decides to believe $\sim p$ due to her only peer satisfying $\neg p$. Agent r has evidence both for and against p. Since she has a peer with evidence for p, but no peer with evidence against p, she believes p. Note that by Fig. 2 this decision would have been different if r had no evidence at all.

Policy III. The previous policies are in the "steadfast" category. Our agent gives more weight to her own evidence than to others' opinions. We can devise a policy that is more in line with the "equal weight" view. In this case, we consider the relation R to be reflexive, and then "dissolve" the agents' exceptionality in the modal expression. Starting from the consolidation of Fig. 2, we can take its subtree for *both* as the decision tree for this policy (Fig. 4 (left)), ignoring the

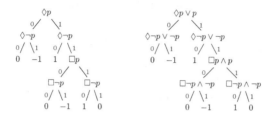

Fig. 4. Decision trees for $\mathscr{C}$ of Policy III for reflexive (left) and anti-reflexive (right) models. Both yield the same beliefs in their respective class of models.

inputs V_φ^s, $V_{\neg\varphi}^s$. This definition makes no distinction between the agent's own evidence and her peers'. We will, however, use the definition of Fig. 4 (right) instead, as we are working with anti-reflexive models. For an example of Policy III, see Fig. 5.

Proposition 14. *Policy III satisfies Mod, EW, Mon, DF and Css.*

5 Dynamics

The dynamic operations we will study use the following models for semantics:

Definition 6. *Consider a model $M = (S, R, V)$. We denote by $M_p^+ = (S, R, V')$ any model s.t. for some $t \in S$, $V'(p, t) = V(p, t) \cup \{1\}$, and $V'(q, r) = V(q, r)$ when $q \neq p$ or $r \neq t$. We define M_p^-, $M_{\neg p}^+$, $M_{\neg p}^-$ analogously, but with $V'(p, t) = V(p, t) \setminus \{1\}$, $V'(p, t) = V(p, t) \cup \{0\}$, $V'(p, t) = V(p, t) \setminus \{0\}$, respectively.*

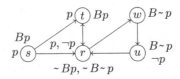

Fig. 5. Policy III applied to the model of Fig. 1. Agent w believes $\sim p$ because she or some peer have $\neg p$, but neither she nor her peer have p. All the other agents have evidence for and against p, either by themselves or via some peer. In this case, if the agent and all her peers have one type of evidence but not the other, a belief is formed. E.g. agent s and her peers have evidence for p but not all of them have $\neg p$, so she settles with belief in p. Agent r, on the other hand, has evidence for and against p (by herself or via a peer), but they are not unanimous about neither, therefore r abstains.

Now, with $l \in \{p, \neg p\}$ for some $p \in At$ and $\circ \in \{+, -\}$, we can define the following operator (with obvious additions to the language).

$$M, s \models [\circ l]\varphi \text{ iff for every model } M_l^\circ \text{ it holds that } M_l^\circ, s \models \varphi$$

So, for example, $M, s \models [+p]\varphi$ can be read as *if evidence for p is added for any agent, φ is the case for s*. A corresponding existential version of this operator can be defined by $\langle\circ l\rangle\varphi \overset{\text{def}}{=} \sim [\circ l]\sim\varphi$, with the expected semantics:

$$M, s \models \langle\circ l\rangle\varphi \text{ iff for some model } M_l^\circ \text{ it holds that } M_l^\circ, s \models \varphi$$

We note the following interactions between modalities:

$$M, s \models \Box[\circ l]\varphi \text{ iff } M, s \models [\circ l]\Box\varphi \qquad M, s \models \Diamond\langle\circ l\rangle\varphi \text{ iff } M, s \models \langle\circ l\rangle\Diamond\varphi$$

Interestingly, we can use the axioms below to define Monotonicity, revealing the hidden dynamic nature of that postulate.

M1 $\sim (Bp \wedge \langle +p \rangle \sim Bp)$

M2 $\sim (Bp \wedge \langle -\neg p \rangle \sim Bp)$

M3 $\sim (\sim B \sim p \wedge \langle +p \rangle B \sim p)$

M4 $\sim (\sim B \sim p \wedge \langle -\neg p \rangle B \sim p)$

M5 $\sim (B \sim p \wedge \langle -p \rangle \sim B \sim p)$

M6 $\sim (B \sim p \wedge \langle +\neg p \rangle \sim B \sim p)$

M7 $\sim (\sim Bp \wedge \langle -p \rangle Bp)$

M8 $\sim (\sim Bp \wedge \langle +\neg p \rangle Bp)$

Proposition 15. *A consolidation satisfying Consistency satisfies Monotonicity iff **M1-M8** are valid.*

Proof. ($\Leftarrow$) If $\sim (Bp \wedge \langle +p \rangle \sim Bp)$ is valid, then for any M, s, it holds that $M, s \not\models Bp$ or $M, s \not\models \langle +p \rangle \sim Bp$, which implies that $M, s \models Bp$ implies $M, s \not\models \langle +p \rangle \sim Bp$. This implies that if $M, s \models Bp$, then there is no M_p^+ such that $M_p^+, s \not\models Bp$. This covers one of the cases of Monotonicity. By analogous reasoning with the other axioms, we get all the other cases.

($\Rightarrow$) The axiom $\sim (Bp \wedge \langle +p \rangle \sim Bp)$ is valid if, for arbitrary M and s, $M, s \models Bp$ implies there is no M_p^+ such that $M_p^+, s \not\models Bp$. Indeed a model M_p^+ satisfies the condition $V(p, t) \preceq V'(p, t)$ for some t (by Definition 6). In this case Monotonicity implies that $\mathrm{Att}'(p, s) \geq \mathrm{Att}(p, s)$. So indeed, if $M, s \models Bp$, which by Consistency means that $\mathrm{Att}(p, s) = 1$, we can only have $\mathrm{Att}'(p, s) = 1$, so $M_p^+, s \models Bp$. So the semantic conditions for **M1** are satisfied. Notice that the case for **M2** is similar, because a model $M_{\neg p}^-$ also satisfies $V(p, t) \preceq V'(p, t)$ for some t. The cases for the other axioms are similar.   □

We can do something similar for AI, where $l \in \{q, \neg q\}$ and $q \neq p$:

AI1 $\sim (Bp \wedge \langle \circ l \rangle \sim Bp)$

AI2 $\sim (B \sim p \wedge \langle \circ l \rangle \sim B \sim p)$

AI3 $\sim (\sim Bp \wedge \langle \circ l \rangle Bp)$

AI4 $\sim (\sim B \sim p \wedge \langle \circ l \rangle B \sim p)$

Proposition 16. *For image-finite models and a finite At, a consolidation satisfies Atom Independence iff **AI1-AI4** are valid. For infinite At, validity of **AI1-AI4** do not imply Atom Independence.*

Proof. ($\Leftarrow$) Suppose **AI1-AI4** are valid. If our models are image-finite and At is finite, then for any two models M and M', if there is a p such that for all $s \in S$ we have $V(p, s) = V'(p, s)$, then there is a finite sequence: $M, M_{l_1}^{\circ_1}, (M_{l_1}^{\circ_1})_{l_2}^{\circ_2}, ..., M'$, where $l_1, l_2, ...$ do not involve p. If $\mathrm{Att}(p, s) \neq \mathrm{Att}'(p, s)$ (for M and M', respectively), then there is one M_i in this sequence such that $\mathrm{Att}_i(p, s) \neq \mathrm{Att}_{i+1}(p, s)$. But if **AI1-AI4** are valid, this is not possible.

($\Rightarrow$) Assume that Atom Independence is satisfied, and $Bp \wedge \langle \circ l \rangle \sim Bp$ is satisfiable. Then there is a M_l° and s such that $M_l^\circ, s \not\models Bp$, while $M, s \models Bp$. But then $V_l^\circ(p, t) = V(p, t)$ for all t, but $\mathrm{Att}(p, s) \neq \mathrm{Att}_l^\circ(p, s)$, and therefore Atom Independence does not hold. Contradiction. Therefore **AI1** is valid. The other cases are similar.

Now we show a consolidation which satisfies **AI1-AI4** but violates Atom Independence (in a setting with infinite $At)$. First, we will need to define some preliminary notions. Let M, s have a *p-canonical* valuation iff $V(p, s) = \{1\}$ and $V(p, t) = \{1\}$ for all t with sRt, and $V(q, s) = \{0\}$ and $V(q, t) = \{0\}$ for all t with sRt, for all $q \neq p$. The *p-canonical* model of M, s is a pointed model $M^\star, s$, where the valuation of $M^\star$ is such that $M^\star, s$ has a *p*-canonical valuation. For two pointed models M, s and M', s which differ only in V, define the *distance* between them to be the size of the sequence (similar to the one built in the first part of this proof) needed to go from M to M'. If no such sequence exists, the distance is infinite. We can easily show that (*) if $M, s \leftrightarrows M', s'$, then M, s is at a finite distance from its *p*-canonical model iff M', s' is at a finite distance from its *p*-canonical model. Now define a consolidation $\mathbb{C}$ as follows: $M, s \models Bp$ iff M, s is at a finite distance from its *p*-canonical model, and $M, s \not\models B\varphi$ for all non-atomic φ. This consolidation respects Definition 4, due to (*). Moreover, this definition violates Atom Independence, for if we take a *p*-canonical M, s (with $\text{Att}(p, s) = 1$) and change the valuation of infinitely many atoms (without changing p) to obtain $M^\star, s$, this new pointed model is not at a finite distance from its *p*-canonical model M, s, and therefore $\text{Att}^\star(p, s) \neq 1$. This violates Atom Independence. Axioms **AI1** to **AI4**, however, are valid. Suppose $M, s \models Bp$. Then M, s is at a finite distance from its *p*-canonical model. For $M, s \models \langle ol \rangle \sim Bp$ to be satisfied, there needs to be a M_l°, s such that $M_l^\circ, s \models \sim Bp$. But that would mean that M_l° is at an infinite distance from its *p*-canonical model. This is impossible, for M, s is *p*-canonical and M_l° only differs from it in one atom for one agent. $\qquad\square$

The following formula means that *there is an agent other than myself such that if we add/remove evidence l for her, φ holds* (where $l \in \{p, \neg p\}$, for some $p \in At$):

$$\langle\langle ol \rangle\rangle\varphi \stackrel{\text{def}}{=} (p^t \wedge \langle ol \rangle(p^t \wedge \varphi)) \vee (p^f \wedge \langle ol \rangle(p^f \wedge \varphi)) \vee (p^b \wedge \langle ol \rangle(p^b \wedge \varphi)) \vee (p^n \wedge \langle ol \rangle(p^n \wedge \varphi))$$

The two following postulates could have been defined before, but now we can define them less cumbersomely:

ES1	$Bp \wedge \langle +\neg p \rangle \sim Bp$	**ES3**	$Bp \wedge \langle -p \rangle \sim Bp$
ES2	$B\sim p \wedge \langle -\neg p \rangle \sim B\sim p$	**ES4**	$B\sim p \wedge \langle +p \rangle \sim B\sim p$
SS1	$Bp \wedge \langle\langle +\neg p \rangle\rangle \sim Bp$	**SS3**	$Bp \wedge \langle\langle -p \rangle\rangle \sim Bp$
SS2	$B\sim p \wedge \langle\langle -\neg p \rangle\rangle \sim B\sim p$	**SS4**	$B\sim p \wedge \langle\langle +p \rangle\rangle \sim B\sim p$

Postulate 10 (Evidence Sensitivity (ES)). *ES1-ES4 are satisfiable.*

Postulate 11 (Social Sensitivity (SS)). *SS1-SS4 are satisfiable.*

Observation 3. *A consolidation satisfying SS also satisfies ES.*

Now from Propositions 5, 7, 15–16 and Postulate 10–11, we get our main technical result:

Corollary 3. *A consolidation satisfies Consistency, Monotonicity, Consensus, Evidence Sensitivity and Social Sensitivity iff:* $\sim(B\sigma_1 \wedge \dots \wedge B\sigma_n)$ *is valid, for all finite* $\Sigma = \{\sigma_1, \dots \sigma_n\} \subseteq \mathcal{L}_0$ *such that* $\Sigma \models p \wedge \sim p$; **M1-M8**, **C1-C2** *are valid; and* **ES1-ES4**, **SS1-SS4** *are satisfiable.*

Atom Independence can be included (with its respective axioms **AI1-AI4**) if we apply the restrictions of Proposition 16. The significance of Corollary 3 is that it characterises a class of consolidations satisfying almost all core postulates. We conjecture that Doxastic Freedom and No Gurus are not axiomatisable (nor negatively so). A hint of why that might be the case for NG is that it is equivalent to saying that there is a model s.t.: $(M, s \models B\varphi$ and $M, t \not\models B\varphi)$ or $(M, s \models B\sim\varphi$ and $M, t \not\models B\sim\varphi)$ or $(M, s \not\models B\varphi$ and $M, t \models B\varphi)$ or $(M, s \not\models B\sim\varphi$ and $M, t \models B\sim\varphi)$. Our language, however, can only talk of belief from an agent's perspective, or modally (e.g. $\Diamond B\varphi$ – *there is a peer who believes* φ).

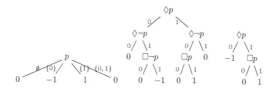

Fig. 6. Anti-social consolidation (left), Policy IV (center), and Policy V (right).

Figure 6 defines three more $\mathscr{C}$-consolidations which will show the importance of the new postulates. First, Social Sensitivity is the only core postulate to rule out anti-social consolidation, an unacceptable function that only takes the agent's own evidence into account.

Proposition 17. *Anti-social consolidation satisfies Mon, DF, Css and ES. Mod, EW and SS are not satisfied.*

Now it can be speculated that Evidence Sensitivity can be forced by a combination of other postulates, such as Strong Modesty, Atom Independence and Monotonicity. Policy IV satisfies all those postulates:

Proposition 18. *Policy IV satisfies Strong Modesty, Monotonicity, Doxastic Freedom, Consensus and Social Sensitivity. It does not satisfy Equal Weight.*

But that logical connection between those postulates does not hold. Interestingly, Policy IV violates Equal Weight, but this time not by the agents not giving enough importance to their peers, but by failing to appreciate their own evidence.

Policy V, which is just a modified version of naive consolidation that satisfies Doxastic Freedom, violates Evidence Sensitivity, because, as its cousin, it completely ignores negative evidence. What Evidence Sensitivity enforces is exactly this: that all evidence is taken into account at least in some occasions.

Proposition 19. *Policy V satisfies Strong Mod, Mon, DF and Css. It does not satisfy EW and ES.*

Proposition 20. *Policies I, II and III satisfy Social Sensitivity. Naive and sceptical consolidations do not satisfy Evidence Sensitivity.*

A summary of the consolidations appears in Table 1. But the main conclusion is that indeed the straightforward definitions such as naive and sceptical consolidations are very unsatisfactory, and the best ones (the only ones satisfying all core postulates) are Policies I-IV, depending on whether one adheres to equal weight or steadfast views.

Table 1. Postulates satisfied by consolidations. A.-S. is anti-social consolidation.

	Naive	Scept.	A.-S.	Pol. I	Pol. II	Pol. III	Pol. IV	Pol. V
AI	✓	✓	✓	✓	✓	✓	✓	✓
Mon	✓	✓	✓	✓	✓	✓	✓	✓
Css	✓	✓	✓	✓	✓	✓	✓	✓
NG	✓		✓	✓	✓	✓	✓	✓
DF		✓	✓	✓	✓	✓	✓	✓
ES		✓	✓	✓	✓	✓	✓	
SS			✓	✓	✓	✓	✓	
Mod	✓					✓	✓	✓
EW	✓	✓				✓		

The $[ol]$ operators make the language more expressive, so we cannot use reduction axioms to obtain equivalent non-dynamic formulas. With these operators we gain the power to *count peers*[9]. Let us abbreviate $\langle ol \rangle...\langle ol \rangle$, repeated n times, by $\langle ol \rangle^n$, with $\langle ol \rangle^0 \varphi \stackrel{\text{def}}{=} \varphi$. Then, with $l \in \{p, \neg p\}$ for some $p \in At$, we have:

$$M, s \models \sim \langle -l \rangle^n \square \sim l \qquad \text{iff } s \text{ has more than } n \text{ peers satisfying } l$$

$$M, s \models \sim \langle +l \rangle^n \square l \qquad \text{iff } s \text{ has more than } n \text{ peers not satisfying } l$$

$$M, s \models \langle -l \rangle^n \square \sim l \qquad \text{iff } s \text{ has at most } n \text{ peers satisfying } l$$

$$M, s \models \langle +l \rangle^n \square l \qquad \text{iff } s \text{ has at most } n \text{ peers not satisfying } l$$

We can abbreviate those formulas by formulas such as $[>n]x$ and $[\leq n]x$, meaning *agent has more than n peers satisfying x* and *agent has at most n peers satisfying x*, respectively, where $x \in \{p, \neg p, \sim p, \sim \neg p\}$, for $p \in At$. We can also define $[=n]x \stackrel{\text{def}}{=} [\leq n]x \wedge [>n-1]x$, with $n \geq 1$, meaning that *the agent has exactly n peers satisfying x*. For $n = 0$, define $[=0]x \stackrel{\text{def}}{=} \square \sim x$. Now $[=n]l \wedge [=m]\sim l$,

---

[9] See [1,3,4,25] for modal logics with notions of counting.

where $l \in \{p, \neg p\}$, indicates that the agent has exactly $n + m$ peers in total. Since for any $n \in \mathbb{N}$ there are exactly $n + 1$ binary sums that equal n, we can define $[[= n]]$, meaning that an *agent has exactly n peers in total*, via a finite disjunction $([=n]p \wedge [=0] \sim p) \vee ([=n-1]p \wedge [=1] \sim p) \vee ... \vee ([=0]p \wedge [=n] \sim p)$.

Notice that our counting abilities are limited to $\neg$-literals (like p and $\neg p$) and their $\sim$-negations, since our base modalities $[\circ l]$ deal only with $\neg$-literals. This indicates that a consolidation taking amounts of evidence into account would have to work on the atomic level, just as our $\mathscr{C}$-consolidations, but the development of such consolidations will be left for future work.

6   Related Work

Now we briefly put our work in context with other belief formation/update theories. There are similar works, but in general our multi-agent perspective plus the qualitative and "modal" processing of evidence set our approach apart.

The term "consolidation" employed here is inspired by the homonymous belief revision operator [15,16], where an inconsistent belief base is transformed into a consistent one; likewise, our consolidations must respect the Consistency postulate. One of the most obvious differences between our approach and belief revision is that we are dealing with a multi-agent setting.

As for Bayesianism, the Bayesian update rule tells us how to update our beliefs, but not how to form them – those are the priors, which are usually allowed to be arbitrary. Our models, in principle, seem to be more in line with objective Bayesianism, which is a controversial position, but more research is needed in order to make a more rigorous comparison.

Dempster-Shafer theory of evidence [8,32] is a generalisation of probability theory where probabilities can be assigned not only to events but also to sets of events. This theory offers rules for combinations of probability assignments, which in a way can be seen as a kind of consolidation operation.

One of the main differences between our modelling and theories as Dempster-Shafer's and Bayesianism is that the latter have a clear quantitative take on evidence. Our framework employs a more limited modal language, where such quantitative statements are not even expressible (although we lay the groundwork for such possibility in Sect. 5). In our models, features such as unanimity and existence of at least one peer with some evidence play important roles, whereas in the other two theories mentioned above these notions are not straightforwardly expressible. Our paper illustrates that there are *some* sensible rationality constraints for formation of evidence-based beliefs even in a limited modal setting, but on the other hand shows the limitations of such a framework and gives the next step towards a quantitative, many-valued modal logical approach to the consolidation problem.

This modal/qualitative perspective is also one of the main differences between our models and opinion diffusion models such as [4]. Although our system is very much in the spirit of other works in opinion dynamics and aggregation and social choice theory (see e.g. [10]), our setup and treatment of evidence is unique.

This contribution does not attempt to offer a *better* formalism for multi-agent evidence-based beliefs, but to highlight how a many-valued modal logic can be used for such a task, bringing an entirely new perspective to this field.

7   Conclusions and Future Work

In this paper we took a many-valued modal logic (FVEL) and showed how it can be used to model networks of peers, where each one may have different evidence for each atomic proposition, including conflicting and incomplete evidence. We showed that in this setup, there is a question of *consolidation*: how to form beliefs given some evidence? We delineated formally a reasonable class of possible consolidations (Definition 3), using a concept similar to bisimulation. Then, we proposed postulates that have to be satisfied in order for a consolidation to be rational, and we showed that (i) they are enough to block many inadequate consolidations and (ii) they are not too strong, as they are jointly satisfiable.

Moreover, we have defined one dynamic operator with the aim of adding and removing evidence. We showed that this operator is useful to formalise some postulates inside the language, and also proposed two important new postulates formulated as axioms containing this operator, without which some unreasonable consolidations would be allowed. With these axioms, we characterised a class of consolidations satisfying most core postulates – with the exception of two which are not axiomatisable. Finally, we showed that this dynamic operator makes the language strictly more expressive, giving it the ability to "count peers", and how this lays the groundwork for quantitative consolidations that take amounts of evidence into account – but the development of those are left for future work.

A complete tableau system for FVEL is found in [29], and [27,28] give an axiomatisation for a language similar to it. Since we use a different version of FVEL, a calculus for it is still missing. Given that we already presented axioms for most postulates, an axiomatic system is preferable. It remains to be seen, however, if such axiomatisation is possible, given that some postulates are not axiomatisable and others are only "negatively" so. Considering that we have not defined a unique belief operator but only constrained the possibilities for such an operator, a complete axiomatisation for our variant of FVEL will probably require one particular consolidation to be chosen. Although we have not talked about public announcements, which in this setting are operations that *remove peers not satisfying some conditions regarding evidence, higher-order evidence or even beliefs*, we know that not all of the reduction axioms of [29] apply here.

Finally, in the consolidations presented here, the agents form beliefs based on their evidence and *their peers' evidence*. Another possibility is to make the evidence private to each agent, so that they have to resort only to their own evidence and their peers' *opinions*.

Acknowledgements. Special thanks to Barteld Kooi and Rineke Verbrugge for important suggestions. I would also like to thank the anonymous reviewers for very relevant comments. Research supported by Ammodo KNAW project "Rational Dynamics and Reasoning".

References

1. Areces, C., Hoffmann, G., Denis, A.: Modal logics with counting. In: Dawar, A., de Queiroz, R. (eds.) WoLLIC 2010. LNCS (LNAI), vol. 6188, pp. 98–109. Springer, Heidelberg (2010). https://doi.org/10.1007/978-3-642-13824-9_9
2. Arrow, K.J.: Social Choice and Individual Values. Wiley, New York (1951)
3. Baltag, A., Christoff, Z., Hansen, J.U., Smets, S.: Logical models of informational cascades. Stud. Logic **47**, 405–432 (2013)
4. Baltag, A., Christoff, Z., Rendsvig, R., Smets, S.: Dynamic epistemic logics of diffusion and prediction in social networks. In: Proceedings of the 12th Conference on Logic and the Foundations of Game and Decision Theory (2016)
5. Belnap, N.: A useful four-valued logic. In: Dunn J.M., Epstein G., (eds.) Modern Uses of Multiple-valued Logic, vol. 2, pp. 5–37. Springer, Dordrecht (1977). https://doi.org/10.1007/978-94-010-1161-7_2
6. Christensen, D.: Higher-order evidence. Philos. Phenomenol. Res. **81**(1), 185–215 (2010)
7. Christoff, Z., Hansen, J.U.: A logic for diffusion in social networks. J. Appl. Logic **13**(1), 48–77 (2015)
8. Dempster, A.P.: A generalization of Bayesian inference. J. Roy. Stat. Soc.: Ser. B (Methodol.) **30**(2), 205–232 (1968)
9. Elga, A.: Reflection and disagreement. Noûs **41**(3), 478–502 (2007)
10. Endriss, U., Grandi, U.: Graph aggregation. Artif. Intell. **245**, 86–114 (2017)
11. Fagin, R., Halpern, J.: Belief, awareness, and limited reasoning. Artif. Intell. **34**(1), 39–76 (1987)
12. Foley, R.: Self-trust and the authority of others. In: Foley, R. (ed.) Intellectual Trust in Oneself and Others, pp. 83–130. Cambridge University Press, Cambridge (2001)
13. Fricker, E.: Testimony and epistemic autonomy. In: Lackey, J., Sosa, E. (eds.) The Epistemology of Testimony, pp. 225–250. Oxford University Press, Oxford (2006)
14. Gibbard, A.: Manipulation of voting schemes: a general result. Econometrica **41**(4), 587–601 (1973)
15. Hansson, S.O.: Belief base dynamics. Ph.D. thesis, Uppsala University (1991)
16. Hansson, S.O.: Semi-revision. J. Appl. Non-Class. Logics **7**(1–2), 151–175 (1997)
17. Hardwig, J.: Epistemic dependence. J. Philos. **82**(7), 335–349 (1985)
18. Heyting, A.: Intuitionism: an Introduction, vol. 41. Elsevier (1966)
19. Hintikka, J.: Impossible possible worlds vindicated. In: Saarinen, E. (ed.) Game-Theoretical Semantics, vol. 5, pp. 367–379. Springer, Dordrecht (1979). https://doi.org/10.1007/978-1-4020-4108-2_13
20. Hintikka, J.: Knowledge and Belief: An Introduction to the Logic of the Two Notions, Contemporary Philosophy, vol. 4. Cornell University Press, Ithaca (1962)
21. Kelly, T.: Peer disagreement and higher order evidence. In: Goldman, A.I., Whitcomb, D. (eds.) Social Epistemology: Essential Readings, pp. 183–217. Oxford University Press (2010)
22. Lasonen-Aarnio, M.: Higher-order evidence and the limits of defeat. Philos. Phenomenol. Res. **88**(2), 314–345 (2014)
23. Lehrer, K.: Social information. The Monist **60**(4), 473–487 (1977)
24. Martini, C., Sprenger, J., Colyvan, M.: Resolving disagreement through mutual respect. Erkenntnis **78**(4), 881–898 (2013)
25. Pacuit, E., Salame, S.: Majority logic. In: Dubois, D., Welty, C., Williams, M.A. (eds.) Principles of Knowledge Representation and Reasoning, KR 2004, vol. 4, pp. 598–605. AAAI Press (2004)

26. Rantala, V.: URN models: a new kind of non-standard model for first-order logic. J. Symbolic Logic **4**, 455–474 (1975)
27. Rivieccio, U.: Algebraic semantics for bilattice public announcement logic. In: Studia Logica Proceedings Trends in Logic XIII. Springer, Heidelberg (2014)
28. Rivieccio, U.: Bilattice public announcement logic. In: Goré, R., Kooi, B., Kurucz, A. (eds.) AiML, vol. 10, pp. 459–477. College Publications (2014)
29. Santos, Y.D.: A dynamic informational-epistemic logic. In: Madeira, A., Benevides, M. (eds.) DALI 2017. LNCS, vol. 10669, pp. 64–81. Springer, Cham (2018). https://doi.org/10.1007/978-3-319-73579-5_5
30. Santos, Y.D.: Consolidation of belief in two logics of evidence. In: Blackburn, P., Lorini, E., Guo, M. (eds.) LORI 2019. LNCS, vol. 11813, pp. 57–70. Springer, Heidelberg (2019). https://doi.org/10.1007/978-3-662-60292-8_5
31. Satterthwaite, M.A.: Strategy-proofness and arrow's conditions: existence and correspondence theorems for voting procedures and social welfare functions. J. Econ. Theory **10**(2), 187–217 (1975)
32. Shafer, G.: A Mathematical Theory of Evidence, vol. 42. Princeton University Press, Princeton (1976)
33. Troelstra, A.S., Van Dalen, D.: Constructivism in Mathematics, vol. 1. North Holland, Amsterdam (1988)

ASPARTIX-V19 - An Answer-Set Programming Based System for Abstract Argumentation

Wolfgang Dvořák$^{(\boxtimes)}$ [ID], Anna Rapberger [ID], Johannes P. Wallner [ID], and Stefan Woltran [ID]

Institute of Logic and Computation, TU Wien, Vienna, Austria
{dvorak,arapberg,wallner,woltran}@dbai.tuwien.ac.at

Abstract. We present ASPARTIX-V, a tool for reasoning in abstract argumentation frameworks that is based on answer-set programming (ASP), in its 2019 release. ASPARTIX-V participated in this year's edition of the International Competition on Computational Models of Argumentation (ICCMA'19) in all classical (static) reasoning tasks. In this paper we discuss extensions the ASPARTIX suite of systems has undergone for ICCMA'19. This includes incorporation of recent ASP language constructs (e.g. conditional literals), domain heuristics within ASP, and multi-shot methods. In particular, with this version of ASPARTIX-V we partially deviate from an earlier focus on monolithic approaches (i.e., one-shot solving via a single ASP encoding) to further enhance performance. We also briefly report on the results achieved by ASPARTIX-V in ICCMA'19.

Keywords: Abstract argumentation · Argumentation system · Answer-set programming

1 Introduction

Abstract argumentation frameworks (AFs) as introduced by Dung [4] are a core formalism for many problems and applications in the field of formal argumentation. In a nutshell, AFs formalize statements as arguments together with a relation denoting conflicts between arguments. Semantics of these AFs give a handle to resolve the conflicts between statements by selecting coherent subsets of the arguments. This selection is solely based on the relation between the arguments and considers arguments as abstract entities. Several different semantics to select coherent subsets of arguments have already been proposed by Dung [4] but numerous other semantics have been introduced later on which lead to a multitude of argumentation semantics (see [1]).

A prominent line of research in the field of computational argumentation has focused on implementations of reasoning procedures for abstract argumentation

© Springer Nature Switzerland AG 2020
A. Herzig and J. Kontinen (Eds.): FoIKS 2020, LNCS 12012, pp. 79–89, 2020.
https://doi.org/10.1007/978-3-030-39951-1_5

(see, e.g., [2]) and cumulated in the biennial International Competition on Computational Models of Argumentation (ICCMA)[1] which has been established in 2015. There are two kinds of approaches to such systems. First, the direct approach of implementing dedicated algorithms for argumentation problems which are often based one some kind of labelling propagation (see, e.g., [20]). Second, the reduction-based approach where the argumentation problem is encoded in some other formalism for which sophisticated solvers already exist. Prominent target formalisms for the later are answer-set programming (ASP) [17,18] and propositional logic with SAT-solving technology; see [3] for an overview.

In this paper we consider the ASPARTIX[2] system that exploits ASP technology to solve argumentation reasoning problems and describe the ASPARTIX-V (Answer Set Programming Argumentation Reasoning Tool - Vienna) version in its 2019 edition which is dedicated to the reasoning tasks of ICCMA'19. We discuss the specifics of ASPARTIX-V19 and differences to earlier versions of ASPARTIX. This includes incorporation of recent ASP language constructs (e.g. conditional literals), domain heuristics within ASP, and multi-shot methods. In particular, with this version of ASPARTIX-V we partially deviate from an earlier focus on monolithic approaches (i.e., one-shot solving via a single ASP encoding) to further enhance performance. Moreover, we give a first analysis of the results achieved by ASPARTIX-V in ICCMA'19.

In the remainder of the paper we first recall the necessary argumentation background and the tracks of this years ICCMA competition. We then give an overview on the ASPARTIX system and explain the aim of our ASPARTIX-V19 edition. In the main part we discuss technical specifics of the ASPARTIX-V19 edition. Finally, we briefly discuss the performance of our system at ICCMA'19.

2   Preliminaries

In this section we briefly introduce the necessary background on abstract argumentation and discuss the tracks of the ICCMA'19 competition.

2.1   Abstract Argumentation

Let us introduce argumentation frameworks [4] and recall the semantics relevant for this work (for a comprehensive introduction, see [1]).

Definition 1. *An* argumentation framework (AF) *is a pair* $F = (A, R)$ *where* A *is a finite set of arguments and* $R \subseteq A \times A$ *is the attack relation. The pair* $(a, b) \in R$ *means that* a *attacks* b, *and we say that a set* $S \subseteq A$ *attacks (in* F) *an argument* b *if* $(a, b) \in R$ *for some* $a \in S$. *An argument* $a \in A$ *is* defended *(in* F) *by a set* $S \subseteq A$ *if each* b *with* $(b, a) \in R$ *is attacked by* S *in* F.

---

[1] www.argumentationcompetition.org.

[2] www.dbai.tuwien.ac.at/research/argumentation/aspartix/.

Semantics for argumentation frameworks are defined as functions σ which assign to each AF $F = (A, R)$ a set $\sigma(F) \subseteq 2^A$, with each set $S \in \sigma(F)$ called an extension. We consider for σ the functions cf, grd, stb, adm, com, $ideal$, prf, sem and stg which stand for conflict-free, grounded, stable, admissible, complete, ideal, preferred, semi-stable and stage extensions, respectively. Towards the definition of these semantics we introduce the following notation. For a set $S \subseteq A$, we denote the set of arguments attacked by (resp. attacking) S in F as $S_F^+ = \{x \mid S \text{ attacks } x \text{ in } F\}$ (resp. $S_F^- = \{x \mid x \text{ attacks some } s \in S \text{ in } F\}$), and define the *range of* S in F as $S_F^\oplus = S \cup S_F^+$.

We are now prepared to give the formal definitions of the abstract argumentation semantics we will consider.

Definition 2. *Let* $F = (A, R)$ *be an AF. A set* $S \subseteq A$ *is* conflict-free (*in* F), *if there are no* $a, b \in S$, *such that* $(a, b) \in R$. $cf(F)$ *denotes the collection of conflict-free sets of* F. *For a conflict-free set* $S \in cf(F)$, *it holds that*

- $S \in stb(F)$, *if each* $a \in A \setminus S$ *is attacked by* S *in* F;
- $S \in adm(F)$, *if each* $a \in S$ *is defended by* S *in* F;
- $S \in com(F)$, *if* $S \in adm(F)$ *and each* $a \in A$ *defended by* S *in* F *is contained in* S;
- $S \in grd(F)$, *if* $S \in com(F)$ *and there is no* $T \subset S$ *such that* $T \in com(F)$;
- $S \in prf(F)$, *if* $S \in adm(F)$ *and there is no* $T \supset S$ *such that* $T \in adm(F)$;
- $S \in ideal(F)$, *if* S *is a* $\subseteq$-*maximal admissible set that is contained in each preferred extension of* F;
- $S \in sem(F)$, *if* $S \in adm(F)$ *and there is no* $T \in adm(F)$ *with* $S_R^\oplus \subset T_R^\oplus$;
- $S \in stg(F)$, *if there is no* $T \in cf(F)$, *with* $S_R^\oplus \subset T_R^\oplus$.

Notice that $grd(F)$, $ideal(F)$ respectively, always yields a unique extension, the grounded, ideal respectively, extension of F.

Example 1. Consider the AF $F = (A, R)$, with arguments $A = \{a, b, c, d, e\}$ and attacks $R = \{(a, b), (c, b), (c, d), (d, c), (d, e), (e, e)\}$. The graph representation of F is as follows.

Considering the extensions of F, we have $stb(F) = stg(F) = sem(F) = \{\{a, d\}\}$. The admissible sets of F are $\emptyset$, $\{a\}$, $\{c\}$, $\{d\}$, $\{a, c\}$ and $\{a, d\}$ and thus the set of preferred extensions is $prf(F) = \{\{a, c\}, \{a, d\}\}$ and the complete extensions are $\{a\}$, $\{a, c\}$ and $\{a, d\}$. Finally, the grounded extension is $\{a\}$ and coincides with the ideal extension. $\diamond$

2.2   Tracks of ICCMA'19

ICCMA'19[3] is the third edition of the International Competition on Computational Models of Argumentation (ICCMA) and had two types of tracks, the classical tracks and the novel dynamic tracks. In the classical tracks the solver

---

[3] https://www.iccma2019.dmi.unipg.it/.

is given an argumentation framework and has to solve a specific reasoning task while in the dynamic tracks the solver is given an initial argumentation frameworks and a list of updates to that framework and the reasoning task has to be evaluated after each update to the framework. As the ASPARTIX-V system supports only the classical tracks we will focus on these tracks here.

For the classical tracks ICCMA'19 considers the following four reasoning tasks, that correspond to the standard reasoning problems studied in the literature (see, e.g., [7]).

- DC-σ: Decide Credulous acceptance of an argument w.r.t. a semantics σ: Given $F = (A, R)$, $a \in A$ decide whether $a \in E$ for some extension $E \in \sigma(F)$.
- DS-σ: Decide Skeptical acceptance of an argument w.r.t. a semantics σ: Given $F = (A, R)$, $a \in A$ decide whether $a \in E$ for all extensions $E \in \sigma(F)$.
- SE-σ: compute Some σ-Extension: Given $F = (A, R)$ return some $E \in \sigma(F)$.
- EE-σ: Enumerate all σ-Extensions: Given $F = (A, R)$ enumerate all $E \in \sigma(F)$.

For σ, seven semantics were considered, namely complete, preferred, stable, semi-stable, stage, grounded and ideal. This resulted in a total number of 24 classical tracks, as for $\sigma \in \{ideal, grd\}$ (the semantics with a unique extension) we have DC-σ = DS-σ and SE-σ = EE-σ.

3   The ASPARTIX System and Its V19 Edition

The ASPARTIX system was one of the first systems that supported efficient reasoning for a broad collection of abstract argumentation semantics starting with the work of Gaggl et al. (see, e.g., [11]) and has been continuously expanded and improved since then (see, e.g., [8–10,13,21]). However, the system is not limited to abstract argumentation frameworks but also supports enhancements of AFs by, e.g., preferences or recursive attacks. It is thus frequently used as reference system in the literature.

ASPARTIX is based on answer-set programming (ASP) and the idea of characterizing argumentation semantics via ASP encodings. With such an encoding of a semantics one can easily apply state-of-art systems for ASP to solve diverse reasoning tasks or to enumerate all extensions of a given AF. Given an AF as input, in the apx format of ICCMA, ASPARTIX delegates the main reasoning to an answer set programming solver (e.g., [15]), with answer set programs encoding the argumentation semantics and reasoning tasks. The basic workflow is shown in Fig. 1, i.e., the AF is given in apx format (facts in the ASP language), and the AF semantics and reasoning tasks are encoded via ASP rules, possibly utilizing further ASP language constructs. For more information on the ASPARTIX system and its derivatives in general the interested reader is referred to the systems web-page:

www.dbai.tuwien.ac.at/research/argumentation/aspartix/

In this work we shall focus on ASPARTIX-V19 which is a derivative of ASPARTIX tuned towards the tracks of ICCMA'19. That is, ASPARTIX-V19

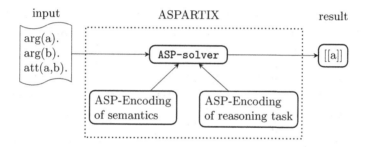

Fig. 1. Basic workflow of ASPARTIX

is restricted to AFs and supports all the standard tasks of ICCMA'19, i.e. credulous/skeptical acceptance and computing all/some extension(s) for complete, preferred, stable, semi-stable, stage, grounded, and ideal semantics. In the following we highlight specifics of the current version and in particular differences to prior versions. In this instance of the argumentation competition, the software systems were collected as docker containers. The competition version ASPARTIX-V19 is available at

https://hub.docker.com/r/aspartix19/aspartix19-repo.

In this competition version of the ASPARTIX system we deviate from classical ASPARTIX design virtues. First, while traditional ASPARTIX encodings are modular in the sense that fixed encodings for semantics can be combined with the generic encodings of reasoning tasks, we use semantics encodings specific to a reasoning task. Second, when appropriate, we apply multi-shot methods for reasoning, which is in contrast to the earlier focus on so-called monolithic encodings, where one uses a single ASP-encoding and runs the solver only once (as illustrated in Fig. 1). Third we make use of advanced features of the ASP-language, and utilize clingo v5.3.0 and v4.4.0[4] [15].

Next, we list and overview some of the ASP-techniques novel to the ASPARTIX system. First, we exploit the concept of conditional literals [14, Section 3.1.11], which has first been applied for ASP-encodings of argumentation semantics in [13]. For example we simplified the encoding of grounded semantics (cf. Listing 1.1). Moreover, conditional literals enable us to give ASPARTIX style encodings of the translations from AF semantics to ASP semantics provided in [22]. Second, we exploit clingo domain heuristics [16] (see also [14, Chapter 10]), in order to compute subset-maximal extensions while only specifying constraints for the base semantics [12].

4   Implementation Details

When not stated otherwise, for a supported semantics we provide an ASP-encoding such that when combined with an AF in the `apx` format the answer-sets

---

[4] https://potassco.org/.

of the program are in a one-to-one correspondence with the extensions of the AF. Given an answer-set of such an encoding the corresponding extension is given by the in(·) predicate, i.e., an argument a is in the extensions iff in(a) is in the answer-set. With such an encoding we can exploit a standard ASP-solver to compute some extension (SE) by computing an answer-set; enumerate all extensions (EE) by enumerating all answer-sets; decide credulous acceptance (DC) of an argument a by adding the constraint ← in(a) to the program and testing whether the program is satisfiable, i.e., a is credulously accepted if there is at least one answer set; and decide skeptical acceptance (DS) of an argument a by adding the constraint ← *not* in(a) to the program and testing whether the program is unsatisfiable, i.e., a is skeptically accepted if there is no answer set.

4.1   Conditional Literals

We make use of the *conditional literal* [14]. In the head of a disjunctive rule literals may have conditions, e.g. consider the head of rule "$p(X) : q(X) \leftarrow$". Intuitively, this represents a head of disjunctions of atoms $p(a)$ where also $q(a)$ is true. Rules might as well have conditions in their body, e.g. consider the body of rule "$\leftarrow p(X) : q(X)$", which intuitively represents a conjunction of atoms $p(a)$ where also $q(a)$ is true.

A bottleneck of previous encodings for grounded semantics was the grounding step of the solver, i.e., the instantiation of variables with constants typically produces large programs. By utilizing conditional literals we were able to provide a compact encoding (cf. Listing 1.1) with significant smaller grounded programs.

Listing 1.1. Encoding for grounded semantics (using conditional literals)

```
in(X) ← arg(X), defeated(Y) : att(Y,X).
defeated(X) ← arg(X), in(Y), att(Y,X).
```

Moreover, conditional literals allow for an ASPARTIX style implementation of the translations from argumentation framework to grounded logic programs provided in [22]. For example consider our one line encoding of stable semantics in Listing 1.2 and the encoding of preferred semantics in Listing 1.3.

Listing 1.2. Encoding for stable semantics (using conditional literals)

```
in(Y) ← arg(Y), not in(X) : att(X,Y).
```

Listing 1.3. Encoding for preferred semantics (using conditional literals)

```
defended(X) | defeated(X) ← arg(X).
defended(X) ← arg(X), defeated(Y) : att(Y,X).
defeated(X) ← defended(Y), att(Y,X).
← defended(X), not defeated(Y), att(Y,X).
← defeated(X), not defended(Y) : att(Y,X).
in(X) ← defended(X), not defeated(X).
```

4.2   Domain Heuristics

Clingo provides an option to specify user-specific domain heuristics in the ASP-program which guides the ASP-solver. In particular one can define heuristics in order to select the answer-sets that are subset-maximal/minimal w.r.t. a specified predicate. Inspired by [12] we use such heuristics to compute preferred extensions by utilizing an encoding for complete semantics and identifying the subset-maximal answer-sets w.r.t. the in($\cdot$) predicate (cf. Listing 1.4). Moreover, we use domain heuristics and three-valued labelling-based characterizations of complete semantics via the predicates in($\cdot$), out($\cdot$), and undec($\cdot$) in order to compute the subset-maximal ranges of complete and conflict-free sets, i.e. we compute the subset-minimal answer-sets w.r.t. the undec($\cdot$) predicate. This can be exploited for computing some semi-stable or stage extensions. However, the domain heuristics only return one witnessing answer-set for each minima and thus this technique is not directly applicable to the corresponding enumerations tasks (we would miss some extensions if several extensions have the same range). In the next section we present a multi-shot method addressing this problem.

Listing 1.4. Encoding for preferred semantics (using domain heuristics)

```
%% Complete labellings
in(X) | out(X) | undec(X) ← arg(X).
in(X) ← arg(X), out(Y) : att(Y,X).
out(X) ← in(Y), att(Y,X).
← in(X), not out(Y), att(Y,X).
← out(X), not in (Y): att(Y,X).
← in(X), out(X).
← undec(X), out(X).
← undec(X), in(X).
%% We now apply heuristics to get the complete labeling with subset-maximal in(.) set
#heuristic in(X) : arg(X). [1,true]
```

4.3   Multi-shot Methods

We utilize multi-shot strategies and pre-processing of the AF for several semantics and reasoning tasks. In the current section, we briefly describe these methods.

For credulous and skeptical reasoning with complete, preferred, grounded, and ideal semantics we do not need to consider the whole framework but only those arguments that have a directed path to the query argument (notice that this does not hold true for stable, semi-stable and stage semantics). We perform pre-processing on the given AF that removes arguments without a directed path to the queried argument before starting the reasoning with an ASP-solver.

For computing the ideal extension we follow a two-shot strategy that is inspired by algorithms proposed earlier for ideal semantics [5,6]. That is, we first use an encoding for complete semantics and the brave reasoning mode of clingo to compute all arguments that are credulously accepted/attacked w.r.t.

preferred semantics. Second, we use the outcome of the first call together with an encoding that computes a fixed-point corresponding to the ideal extension. For reasoning with ideal semantics we use an encoding for ideal sets and perform credulous reasoning on ideal sets in the standard way.

Semi-stable extensions correspond to those complete labellings for which the set of undecided arguments is subset-minimal. In our approach, we utilize an encoding for complete semantics extended by an undec(·) predicate and process the answer-sets. We check whether models without an undec(·) predicate have been computed; in that case, semi-stable extensions coincide with stable extensions. In the other case, we compute all subset-minimal sets among all undecided sets using the set class in python and return the corresponding models.

For enumerating stage extensions we use a multi-shot strategy. First we use the domain heuristics to compute the maximal ranges w.r.t. naive semantics (as each range maximal conflict-free set is also subset-maximal it is sufficient to only consider naive sets, i.e. subset-maximal conflict free sets). Second, for each of the maximal ranges we start another ASP-encoding which computes conflict-free sets with exactly that range (this is equivalent to computing stable extension of a restricted framework). Each of these extensions corresponds to a different stage extension of the AF.

For reasoning with semi-stable and stage semantics we use a multi-shot strategy similar to that for enumerating the stage extensions. First we use domain heuristics to compute the maximal ranges w.r.t. complete or naive semantics. In the second step we iterate over these ranges and perform skeptical (credulous) reasoning over complete extensions (conflict-free sets) with the given range. For skeptical acceptance, we answer negatively as soon as a counterexample to a positive answer is found when iterating the extensions; otherwise, after processing all maximal ranges we answer with YES. Analogously, for credulous acceptance, we check in each iteration whether we can report a positive answer; otherwise, after processing all maximal ranges, we return NO.

5   Discussion

We next briefly discuss the performance of our system at ICCMA'19 (detailed results of the competition are published at https://www.iccma2019.dmi.unipg. it/results/results-main.html). The competition was dominated by the μ-toksia system by Niskanen and Järvisalo [19], an optimized system based on modern SAT-solving technology which won all the tracks of the competition and only failed to solve two of the benchmark instances in the given time-limit of 600 seconds.

The ASPARTIX-V19 system scored third in the overall evaluation of the competition, scored second in 8 of the 24 tracks and scored second in the aggregated evaluation of complete and stable semantics. Moreover, for 16 tracks ASPARTIX-V19 solved all instances of the competition within the given time-limit. Noteworthy, ASPARTIX-V19 was to only system to solve the enumeration task under stage semantics for the n256p3q08n.apx instance.

The ICCMA'19 results also reported different kinds of errors in the results of the ASPARTIX-system, which we investigated and shall discuss in the following. This errors include wrong results, malformed output, crashed computations and for enumeration tasks incomplete list of extensions which are not due to a timeout. The affected tasks are skeptical acceptance under preferred and semi-stable semantics, credulous acceptance under semi-stable semantics, stage and ideal semantics and enumeration of semi-stable and stage semantics.

The main reason for these errors seems to be side-effects of concurrent calls to the solver. Towards understanding the erroneous results, we performed additional experiments. For these experiments we considered all skeptical and credulous acceptance instances of the competition where ASPARTIX-V19 returned an erroneous result or crashed and reran the ASPARTIX-V19 docker on these instances in an isolated setting. For all but one instance we got the correct results. In this isolated setting ASPARTIX-V19 only reported one wrong result for skeptical reasoning with semi-stable semantics on the `Small-result-b86.apx` instance. This seems to be due to a bug in the used ASP solver, which can be resolved by using an earlier version of the solver (we got correct results with clingo 4.4.0). We maintain an updated and extended version of ASPARTIX-V19, available at the systems web-page[5]. For the enumeration tasks we investigated selected instances with erroneous/incomplete results and again got correct results when running them in an isolated setting and on the other hand could generate erroneous results by concurrent calls to the solver.

From our development work and the results achieved in the international competition, we conclude that (i) a performance increase was achieved by utilizing advanced language features of ASP, across multiple reasoning tasks covering several levels of complexity of the polynomial hierarchy (e.g., argumentative reasoning tasks considered in the ICCMA range from polynomial-time decidable to being complete for a class on the second level of the polynomial hierarchy), (ii) said language features, furthermore, provide means for compact and accessible modeling of problem shortcuts in the ASP language, however care needs to be taken when designing systems that interface ASP solvers, and (iii) while our prototype was outperformed by the SAT based approach of μ-toksia, performance of ASPARTIX-V19 does not lag behind for several cases. Indeed, as witnessed by the uniquely solved instance only by ASPARTIX-V19, certain shortcuts included in ASPARTIX-V19 can lead to complementary performance for families of instances.

Acknowledgments. The authors are grateful to a reviewer for suggesting directions for further improvements in the encodings.

This work has been funded by the Austrian Science Fund (FWF): P30168-N31, W1255-N23, and I2854.

---

[5] https://www.dbai.tuwien.ac.at/research/argumentation/aspartix/dung.html#iccma_interface.

References

1. Baroni, P., Caminada, M., Giacomin, M.: Abstract argumentation frameworks and their semantics. In: Baroni, P., Gabbay, D., Giacomin, M., van der Torre, L. (eds.) Handbook of Formal Argumentation, Chapt. 4. College Publications, London (2018)
2. Cerutti, F., Gaggl, S.A., Thimm, M., Wallner, J.P.: Foundations of implementations for formal argumentation. In: Baroni, P., Gabbay, D., Giacomin, M., van der Torre, L. (eds.) Handbook of Formal Argumentation, Chapt. 15. College Publications, London (2018). Also available as an article in the IfCoLog Journal of Logics and their Applications 4(8), 2623–2706
3. Charwat, G., Dvořák, W., Gaggl, S.A., Wallner, J.P., Woltran, S.: Methods for solving reasoning problems in abstract argumentation - a survey. Artif. Intell. **220**, 28–63 (2015)
4. Dung, P.M.: On the acceptability of arguments and its fundamental role in non-monotonic reasoning, logic programming and n-person games. Artif. Intell. **77**(2), 321–358 (1995)
5. Dunne, P.E.: The computational complexity of ideal semantics. Artif. Intell. **173**(18), 1559–1591 (2009)
6. Dunne, P.E., Dvořák, W., Woltran, S.: Parametric properties of ideal semantics. Artif. Intell. **202**, 1–28 (2013)
7. Dvořák, W., Dunne, P.E.: Computational problems in formal argumentation and their complexity. In: Baroni, P., Gabbay, D., Giacomin, M., van der Torre, L. (eds.) Handbook of Formal Argumentation, Chapt. 14. College Publications, London (2018). Also available as an article in the IfCoLog Journal of Logics and their Applications 4(8), 2557–2622
8. Dvořák, W., Gaggl, S.A., Wallner, J.P., Woltran, S.: Making use of advances in answer-set programming for abstract argumentation systems. In: Tompits, H., Abreu, S., Oetsch, J., Pührer, J., Seipel, D., Umeda, M., Wolf, A. (eds.) INAP/WLP -2011. LNCS (LNAI), vol. 7773, pp. 114–133. Springer, Heidelberg (2013). https://doi.org/10.1007/978-3-642-41524-1_7
9. Dvořák, W., Gaggl, S.A., Linsbichler, T., Wallner, J.P.: Reduction-based approaches to implement Modgil's extended argumentation frameworks. In: Eiter, T., Strass, H., Truszczyński, M., Woltran, S. (eds.) Advances in Knowledge Representation, Logic Programming, and Abstract Argumentation. LNCS (LNAI), vol. 9060, pp. 249–264. Springer, Cham (2015). https://doi.org/10.1007/978-3-319-14726-0_17
10. Dvořák, W., Greßler, A., Woltran, S.: Evaluating SETAFs via answer-set programming. In: Thimm, M., Cerutti, F., Vallati, M. (eds.) CEUR Workshop Proceedings of the SAFA Co-Located with COMMA 2018, vol. 2171, pp. 10–21. CEUR-WS.org (2018)
11. Egly, U., Gaggl, S.A., Woltran, S.: Answer-set programming encodings for argumentation frameworks. Argument Comput. **1**(2), 147–177 (2010)
12. Faber, W., Vallati, M., Cerutti, F., Giacomin, M.: Enumerating preferred extensions using ASP domain heuristics: the ASPrMin solver. In: Modgil, S., Budzynska, K., Lawrence, J. (eds.) Proceedings of the COMMA, Frontiers in Artificial Intelligence and Applications, vol. 305, pp. 459–460. IOS Press (2018)
13. Gaggl, S.A., Manthey, N., Ronca, A., Wallner, J.P., Woltran, S.: Improved answer-set programming encodings for abstract argumentation. Theory Pract. Logic Program. **15**(4–5), 434–448 (2015)

14. Gebser, M., et al.: Potassco guide version 2.2.0 (2019). https://github.com/potassco/guide/releases/tag/v2.2.0

15. Gebser, M., Kaminski, R., Kaufmann, B., Schaub, T.: Clingo = ASP + control: Preliminary report. CoRR, abs/1405.3694 (2014)

16. Gebser, M., et al.: Domain-specific heuristics in answer set programming. In: des Jardins, M., Littman, M.L. (eds.) Proceedings of the AAAI, pp. 350–356. AAAI Press (2013)

17. Marek, V.W., Truszczyński, M.: Stable models and an alternative logic programming paradigm. In: Apt, K.R., Marek, V.W., Truszczynski, M., Warren, D.S. (eds.) The Logic Programming Paradigm - A 25-Year Perspective, pp. 375–398. Springer, Heidelberg (1999)

18. Niemelä, I.: Logic programming with stable model semantics as a constraint programming paradigm. Ann. Math. Artif. Intell. **25**(3–4), 241–273 (1999)

19. Niskanen, A., Järvisalo, M.: μ-toksia participating in ICCMA 2019 (2019). https://www.iccma2019.dmi.unipg.it/papers/ICCMA19_paper_11.pdf

20. Nofal, S., Atkinson, K., Dunne, P.E.: Algorithms for decision problems in argument systems under preferred semantics. Artif. Intell. **207**, 23–51 (2014)

21. Ronca, A., Wallner, J.P., Woltran, S.: ASPARTIX-V: utilizing improved ASP encodings (2015). http://argumentationcompetition.org/2015/pdf/paper_11.pdf

22. Sakama, C., Rienstra, T.: Representing argumentation frameworks in answer set programming. Fundam. Inform. **155**(3), 261–292 (2017)

Proper Hierarchies in Polylogarithmic Time and Absence of Complete Problems

Flavio Ferrarotti[1]([✉]), Senén González[1], Klaus-Dieter Schewe[2], and José María Turull-Torres[3]

[1] Software Competence Center Hagenberg, Hagenberg, Austria
{flavio.ferrarotti,senen.gonzalez}@scch.at
[2] Zhejiang University, UIUC Institute, Haining, China
kd.schewe@intl.zju.edu.cn
[3] Universidad Nacional de La Matanza, Buenos Aires, Argentina
jturull@unlam.edu.ar

Abstract. The polylogarithmic time hierarchy structures sub-linear time complexity. In recent work it was shown that all classes $\tilde{\Sigma}_m^{plog}$ or $\tilde{\Pi}_m^{plog}$ ($m \in \mathbb{N}$) in this hierarchy can be captured by semantically restricted fragments of second-order logic. In this paper the descriptive complexity theory of polylogarithmic time is taken further showing that there are strict hierarchies inside each of the classes of the hierarchy. A straightforward consequence of this result is that there are no complete problems for these complexity classes, not even under polynomial time reductions.

Keywords: Strict hierarchies · Polylogarithmic time · Complexity theory · Complete problems · Finite models · Logic

1 Introduction

Computations with sub-linear time complexity have not been studied intensively. However, such computations appear rather naturally, e.g. in the area of circuits. Mix Barrington studied the complexity of circuits [9] characterizing a class of families of constant-depth quasi-polynomial size AND/OR-circuits. In particular, he proved that the class of Boolean queries computable by the class of DTIME[$(\log n)^{O(1)}$] DCL-uniform families of Boolean circuits of unbounded fan-in, size $2^{(\log n)^{O(1)}}$ and depth $O(1)$ coincides with the class of Boolean queries expressible in a fragment SO^b of second-order logic. As used in his study, the

The research reported in this paper results from the project *Higher-Order Logics and Structures* supported by the Austrian Science Fund (FWF: **[I2420-N31]**). It has also been partly supported by the Austrian Ministry for Transport, Innovation and Technology, the Federal Ministry for Digital and Economic Affairs, and the Province of Upper Austria in the frame of the COMET center SCCH.

© Springer Nature Switzerland AG 2020
A. Herzig and J. Kontinen (Eds.): FoIKS 2020, LNCS 12012, pp. 90–105, 2020.
https://doi.org/10.1007/978-3-030-39951-1_6

complexity class DTIME$[2^{(\log n)^{O(1)}}]$ is known as *quasipolynomial time*. Furthermore, the fastest known algorithm for checking graph isomorphisms is in quasipolynomial time [1].

In [3] we started a deeper investigation of sub-linear time computations emphasising complexity classes DPolyLogTime and NPolyLogTime of decision problems that can be solved deterministically or non-deterministically with a time complexity in $O(\log^k n)$ for some k, where n is as usual the size of the input. We extended these complexity classes to a complete hierarchy, the *polylogarithmic time hierarchy*, analogous to the polynomial time hierarchy, and for each class Σ_m^{plog} or Π_m^{plog} ($m \in \mathbb{N}$) in the hierarchy we defined a fragment of semantically restricted second-order logic capturing it [5,6]. While the hierarchy as a whole captures the same class of problems studied by Mix Barrington, the various classes of the hierarchy provide fine-grained insights into the nature of decision problems decidable in sub-linear time.

With these promising results the natural question occurs, whether there are complete problems in the hierarchy, and what would be an appropriate notion of reduction to define complete problems. Note that for the somehow related complexity class PolyLogSpace it is known since long time that it does not have complete problems.

In this paper we address this problem. We show that for none of the classes $\tilde{\Sigma}_m^{plog}$ and $\tilde{\Pi}_m^{plog}$ ($m \in \mathbb{N}$) in the polylogarithmic time hierarchy there exists a complete problem. It turns out that this result is a rather simple consequence of the existence of proper hierarchies inside each of the classes $\tilde{\Sigma}_m^{plog}$ and $\tilde{\Pi}_m^{plog}$. Note that a similar approach shows the non-existence of complete problems for PolyLogSpace, but the corresponding proof exploits theorems by Hartmanis et al. that cannot be applied to our case, as these theorems (which are well known in complexity theory as the space and time hierarchy theorems) require at least linear time.

The remainder of this paper is organized as follows. Section 2 summarizes the necessary preliminaries for our investigation introducing the complexity classes of the polylogarithmic time hierarchy. This is complemented in Sect. 3 by reviewing SO^{plog}, the polylogarithmically-restricted fragment of second-order logic that is used to define subsets capturing the complexity classes $\tilde{\Sigma}_m^{plog}$ and $\tilde{\Pi}_m^{plog}$. Section 4 introduces concrete decision problems that we use to show the existence of proper hierarchies inside $\tilde{\Sigma}_m^{plog}$ and $\tilde{\Pi}_m^{plog}$. We use the capturing logics to define these problems that are parametrised by $k \in \mathbb{N}$, and the various different values for k give rise to the hierarchies. Theorems showing that we obtain proper hierarchies inside $\tilde{\Sigma}_m^{plog}$ and $\tilde{\Pi}_m^{plog}$ are proven in Sect. 5. Then the non-existence of complete problems arises as a rather straightforward consequence, as we will show in Sect. 6. We conclude with a brief summary in Sect. 7.

2 Polylogarithmic Time Complexity Classes

The sequential access that Turing machines have to their tapes makes it impossible to compute anything in sub-linear time. Therefore, logarithmic time complexity classes are usually studied using models of computation that have random access to their input. As this also applies to the poly-logarithmic complexity classes studied in this paper, we adopt a Turing machine model that has a *random access* read-only input, similar to the log-time Turing machine in [10].

In the following, $\log n$ always refers to the binary logarithm of n, i.e., $\log_2 n$. With $\log^k n$ we mean $(\log n)^k$.

A *random-access Turing machine* is a multi-tape Turing machine with (1) a read-only (random access) *input* of length $n+1$, (2) a fixed number of read-write *working tapes*, and (3) a read-write input *address-tape* of length $\lceil \log n \rceil$.

Every cell of the input as well as every cell of the address-tape contains either 0 or 1 with the only exception of the $(n+1)$st cell of the input, which is assumed to contain the endmark ◁. In each step the binary number in the address-tape either defines the cell of the input that is read or if this number exceeds n, then the $(n+1)$st cell containing ◁ is read.

Example 2.1. Let polylogCNFSAT be the class of satisfiable propositional formulae in conjunctive normal form with $c \leq \lceil \log n \rceil^k$ clauses, where n is the length of the formula. Note that the formulae in polylogCNFSAT tend to have few clauses and many literals. We define a random-access Turing machine M which decides polylogCNFSAT. The alphabet of M is $\{0, 1, \#, +, -\}$. The input formula is encoded in the input tape as a list of $c \leq \lceil \log n \rceil^k$ indices, each index being a binary number of length $\lceil \log n \rceil$, followed by c clauses. For every $1 \leq i \leq c$, the i-th index points to the first position in the i-th clause. Clauses start with $\#$ and are followed by a list of literals. Positive literals start with a $+$, negative with a $-$. The $+$ or $-$ symbol of a literal is followed by the ID of the variable in binary. M proceeds as follows: (1) Using binary search with the aid of the "out of range" response ◁, compute n and $\lceil \log n \rceil$. (2) Copy the indices to a working tape, counting the number of indices (clauses) c. (3) Non-deterministically guess c input addresses $a_1, \ldots, a_c$, i.e., guess c binary numbers of length $\lceil \log n \rceil$. (4) Using c 1-bit flags, check that each $a_1, \ldots, a_c$ address falls in the range of a different clause. (5) Check that each $a_1, \ldots, a_c$ address points to an input symbol $+$ or $-$. (6) Copy the literals pointed by $a_1, \ldots, a_c$ to a working tape, checking that there are *no* complementary literals. (7) Accept if all checks hold.

Let L be a language accepted by a random-access Turing machine M. Assume that for some function f on the natural numbers, M makes at most $O(f(n))$ steps before accepting an input of length n. If M is deterministic, then we write $L \in \mathrm{DTIME}(f(n))$. If M is non-deterministic, then we write $L \in \mathrm{NTIME}(f(n))$. We define the classes of deterministic and non-deterministic poly-logarithmic time computable problems as follows:

$$\mathrm{DPolyLogTime} = \bigcup_{k,c \in \mathbb{N}} \mathrm{DTIME}((\log n)^k \cdot c)$$

$$\mathrm{NPolylogTime} = \bigcup_{k,c \in \mathbb{N}} \mathrm{NTIME}((\log n)^k \cdot c)$$

The non-deterministic random-access Turing machine in Example 2.1 clearly works in polylog-time. Therefore, polylogCNFSAT $\in$ NPolylogTime.

Recall that an alternating Turing machine comes with a set of states Q that is partitioned into subset $Q_\exists$ and $Q_\forall$ of so-called existential and universal states. Then a configuration c is accepting iff

- c is in a final accepting state,
- c is in an existential state and there exists a next accepting configuration, or
- c is in a universal state, there exists a next configuration and all next configurations are accepting.

In analogy to our definition above we can define a *random-access alternating Turing machine*. The languages accepted by such a machine M, which starts in an existential state and makes at most $O(f(n))$ steps before accepting an input of length n with at most m alternations between existential and universal states, define the complexity class ATIME$(f(n), m)$. Analogously, we define the complexity class ATIME$^{op}(f(n), m)$ comprising languages that are accepted by a random-access alternating Turing machine that starts in a universal state and makes at most $O(f(n))$ steps before accepting an input of length n with at most $m - 1$ alternations between universal and existential states. With this we define

$$\tilde{\Sigma}_m^{plog} = \bigcup_{k,c\in\mathbb{N}} \text{ATIME}[(\log n)^k \cdot c, m] \qquad \tilde{\Pi}_m^{plog} = \bigcup_{k,c\in\mathbb{N}} \text{ATIME}^{op}[(\log n)^k \cdot c, m].$$

The poly-logarithmic time hierarchy is then defined as PLH $= \bigcup_{m\geq 1} \tilde{\Sigma}_m^{plog}$. Note that $\tilde{\Sigma}_1^{plog} = $ NPolylogTime holds.

Remark 2.1. Note that a simulation of a NPolylogTime Turing machine M by a deterministic machine N requires checking all computations in the tree of computations of M. As M works in time $(\log n)^{O(1)}$, N requires time $2^{\log n^{O(1)}}$. This implies NPolylogTime $\subseteq$ DTIME$(2^{\log n^{O(1)}})$, which is the complexity class called quasipolynomial time of the fastest known algorithm for graph isomorphism [1], which further equals the class DTIME$(n^{\log n^{O(1)}})$[1].

3   Logics for Polylogarithmic Time

The descriptive complexity of the polylogarithmic time complexity classes described in the previous section, has been recently studied in deepth in [3–7], where precise logical characterization of those classes were presented. The logics used in those characterizations are quite useful to think and describe the problems used in this paper to prove proper hierarchies inside polylogarithmic time. In this section we describe these logics and the results regarding their correspondence with the different polylogarithmic time complexity classes.

---

[1] This relationship appears quite natural in view of the well known relationship NP $=$ NTIME$(n^{O(1)}) \subseteq$ DTIME$(2^{n^{O(1)}}) = $ EXPTIME.

The capturing results for polylogarithmic time hold over ordered structures. A finite ordered σ-structure $\mathbf{A}$ is a finite structure of vocabulary $\sigma \cup \{<\}$, where $\leq \notin \sigma$ is a binary relation symbol and $<^{\mathbf{A}}$ is a linear order on A. Every finite ordered structure has a corresponding isomorphic structure, whose domain is an initial segment of the natural numbers. Thus, we assume, as usual, that $A = \{0, 1, \ldots, n-1\}$, where n is the cardinality $|A|$ of A. In the case of non-deterministic polylogarithmic time complexity, the capturing results also assume that σ includes SUCC, BIT and constants for $\log n$, the minimum, second and maximum elements. In every structure $\mathbf{A}$, the symbol SUCC is interpreted by the successor relation corresponding to the $<^{\mathbf{A}}$ ordering. The constant symbols 0, 1 and max are in turn interpreted as the minimum, second and maximum elements under the $<^{\mathbf{A}}$ ordering and the constant $logn$ as $\lceil \log |A| \rceil$. Finally, BIT is interpreted by the following binary relation:

$$\mathrm{BIT}^{\mathbf{A}} = \{(i, j) \in A^2 \mid \text{Bit } j \text{ in the binary representation of } i \text{ is } 1\}.$$

W.l.o.g., we assume that all structures have at least *three* elements. This results in a cleaner presentation, avoiding trivial cases which would unnecessarily complicate some formulae.

Let us start with DPolylogTime. This class is captured by the *index logic* introduced in [6]. Index logic is two-sorted; variables of the first sort range over the domain of the input structure. Variables of the second sort range over an initial segment of the natural numbers; this segment is bounded by the logarithm of the size of the input structure. Thus, the elements of the second sort represent the bit positions needed to address elements of the first sort. Index logic includes full fixpoint logic on the second sort. Quantification over the first sort, however, is heavily restricted. Specifically, a variable of the first sort can only be bound using an address specified by a subformula that defines the positions of the bits of the address that are set. This "indexing mechanism" lends index logic its name.

The following result confirms that the problems that can be described in the index logic are in *DPolylogTime* and vice versa.

Theorem 3.1 ([6]). *Index logic captures DPolylogTime over ordered structures.*

Regarding nondeterministic polylogarithmic time, the restricted second-order logic SO^{plog} defined in [3–5] captures the polylogarithmic-time hierarchy, with its quantifier prenex fragments Σ_m^{plog} and Π_m^{plog} capturing the corresponding levels $\tilde{\Sigma}_m^{plog}$ and $\tilde{\Pi}_m^{plog}$ of this hierarchy, respectively.

SO^{plog} is a fragment of second-order logic where second-order quantification range over relations of polylogarithmic size and first-order quantification is restricted to the existential fragment of first-order logic plus universal quantification over variables under the scope of a second-order variable.

Formally, we can inductively define the syntax of SO^{plog} as follows:

– Every formula in the existential fragment of first-order logic with equality is a SO^{plog} formula.

- If X is a second-order variable of arity r, and $t_1, \ldots, t_r$ are first-order terms, then both $X(t_1, \ldots, t_r)$ and $\overline{X(t_1, \ldots, t_r)}$ are SO^{plog} formulae.
- If φ and ψ are SO^{plog} formulae, then $(\varphi \wedge \psi)$ and $(\varphi \vee \psi)$ are SO^{plog} formulae.
- If φ is a SO^{plog} formula, X is a second-order variable of arity r and $\bar{x}$ is an r-tuple of first-order variables, then $\forall \bar{x}(X(\bar{x}) \rightarrow \varphi)$ is SO^{plog} formula.
- If φ is a SO^{plog} formula and x is a first-order variable, then $\exists x \varphi$ is a SO^{plog} formula.
- If φ is a SO^{plog} formula and X is a second-order variable, then both $\exists X \varphi$ and $\forall X \varphi$ are SO^{plog} formulae.

The most significant restriction of SO^{plog} is in its semantics. In addition to its arity, each second-order variable X is associated with another non-negative integer, its *exponent*, and it is required that any X of arity r and exponent k is interpreted on a structure of domain A as an r-ary relation *of cardinality smaller or equal than* $\log^k |A|$. Otherwise, the semantics of SO^{plog} follows the standard semantics of second-order logic.

As usual, the fragments Σ_m^{plog} (resp. Π_m^{plog}) are defined by considering SO^{plog} formulae with m alternating blocks of second-order quantifiers in quantifier prenex (Skolem) normal form, starting with an existential (resp. universal) block. Note that by Lemma 3 in [4], for every SO^{plog} formula φ there is an equivalent formula φ' that is in quantifier prenex normal form. In the following we will assume that the reader is familiar with the techniques that can be applied to transform arbitrary SO^{plog} formulae into equivalent formulae in Skolem normal form. Those techniques are detailed in the proof of Lemma 3 in Appendix B in [4].

The following result characterizes precisely the expressive power of SO^{plog} in terms of the nondeterministic polylogarithmic time hierarchy. Note that in particular, existential SO^{plog} captures $NPolylogTime$.

Theorem 3.2 ([3,5]). *Over ordered structures with successor relation,* BIT *and constants for* $\log n$*, the minimum, second and maximum elements,* Σ_m^{plog} *captures* $\tilde{\Sigma}_m^{plog}$ *and* Π_m^{plog} *captures* $\tilde{\Pi}_m^{plog}$ *for all* $m \geq 1$.

4 Problems that Lead to Proper Hierarchies

Here we introduce the decision problems that we use in the next section to show the existence of proper hierarchies of polylogarithmic-time. In addition, for the nondeterministic classes we give a precise definition of these problems in terms of the logic SO^{plog} studied in [3–5] and discussed in the previous section.

From now on we work with the class of structures known as *word models* (see for instance [2]). Let π be the vocabulary $\{<, R_0, R_1\}$, where $<$ is a binary relation symbol and R_0, R_1 are unary relation symbols. We can identify any binary string (word) $w = a_1 \ldots a_n$ in $\{0,1\}^+$ with a π-structure (word model) $\mathbf{A}_w$, where the cardinality of the domain A of $\mathbf{A}_w$ equals the length of w, $<^{\mathbf{A}_w}$ is a linear order in A, $R_0^{\mathbf{A}_w}$ contains the positions in w carrying a 0, and $R_1^{\mathbf{A}_w}$ contains the positions in w carrying a 1.

Problem 4.1 (InitialZeros^k). The problem InitialZeros^k consists on deciding (over word models of signature π) the language of binary strings which have a prefix of at least $\lceil \log n \rceil^k$ consecutive zeros, where n is the length of the string.

Problem 4.2 (ConseqZeros^k). Let ConseqZeros^k denote the problem of deciding the language of binary strings which have at least $\lceil \log n \rceil^k$ consecutive bits set to 0, where n is the length of the string. This can be expressed formally in SO^{plog} as follows:

$$\exists X(|X| = \log^k n \wedge SEQ(X) \wedge \forall x(X(x) \rightarrow R_0(x))),$$

where X is of arity 1 and exponent k, the expression $|X| = \log^k n$ denotes the sub-formula which defines that the cardinality of X is $\lceil \log n \rceil^k$, and $SEQ(X)$ denotes the sub-formula expressing that the elements of X are a contiguous subsequence of the order $<$.

The sub-formula expressing $|X| = \log^k n$ can be written as follows:

$$\exists Y \bar{x}(Y(\bar{x}) \wedge \bar{x} = \bar{0} \wedge \forall \bar{y}(Y(\bar{y}) \rightarrow (SUCCk(\bar{y}, \overline{\log n}) \vee \exists \bar{z}(Y(\bar{z}) \wedge SUCCk(\bar{y}, \bar{z}))) \\ \wedge |X| = |Y|)$$

where Y is of arity k and exponent k, $\bar{x}, \bar{y}, \bar{z}$ denote k-tuples of first-order variables, $SUCCk(\bar{y}, \bar{z})$ denotes a sub-formula expressing that $\bar{z}$ is the immediate successor of $\bar{y}$ in the lexicographical order of k-tuples, and $|X| = |Y|$ expresses that X and Y have equal cardinality. $SUCCk(\bar{y}, \bar{z})$ can be expressed by a quantifier-free SO^{plog} formula (for details refer to $SUCC_k$ in Sect. 4 in [4]). In turn, $|X| = |Y|$ can be expressed by an existential SO^{plog} formula using second order variables of arity $k + 1$ and exponent k (for details refer to Sect. 3.1 in [4]).

Finally, $SEQ(X)$ can be expressed in SO^{plog} as follows:

$$\forall x(X(x) \rightarrow \exists y(SUCC(x, y) \vee \forall z(X(z) \rightarrow z < x)))$$

The whole formula for ConseqZeros^k can then be rewritten in Skolem normal form as a formula in Σ_1^{plog} with second order variables of exponent k.

Problem 4.3 (NoConseqZeros^k). Let NoConseqZeros^k denote the problem of deciding the language of binary strings which do *not* have greater than or equal $\lceil \log n \rceil^k$ consecutive bits set to 0, where n is the length of the string. Since syntactically the negation of a formula in SO^{plog} is not always a formula in SO^{plog}, we cannot just negate the formula for ConseqZeros^k in Problem 4.2 to get the SO^{plog} formula for NoConseqZeros^k. We can nevertheless define NoConseqZeros^k as follows:

$$\forall X(|X| = \log^k n \wedge SEQ(X) \rightarrow \exists x(X(x) \wedge R_1(x)))$$

This is equivalent to:

$$\forall X(\neg(|X| = \log^k n) \vee \neg SEQ(X) \vee \exists x(X(x) \wedge R_1(x))).$$

It follows that the negations of the sub-formulae $|X| = \log^k n$ that we defined in Problem 4.2 is in Π_1^{plog}. Regarding $\neg SEQ(X)$, it can be written in SO^{plog} as

$$\exists xyz(X(x) \wedge \neg X(y) \wedge X(z) \wedge x < y < z).$$

We then get that the formula for NoConseqZeros^k can be rewritten in Skolem normal form as a formula in Π_1^{plog} with second order variables of exponent k.

Problem 4.4 (ExactlyOnce^k). Let ExactlyOnce^k denote the problem of deciding the language of binary strings which contain the substring $0^{\lceil \log n \rceil^k}$ exactly once, i.e., s is in ExactlyOnce^k iff $0^{\lceil \log n \rceil^k}$ is a substring of s and every other substring of s is not $0^{\lceil \log n \rceil^k}$. This can be expressed formally in SO^{plog} by combining the formulae for ConseqZeros^k and NoConseqZeros^k (see Problems 4.2 and 4.3, respectively) as follows:

$$\exists X(|X| = \log^k n \wedge SEQ(X) \wedge \forall x(X(x) \rightarrow R_0(x))$$
$$\wedge \forall Y(Y = X \vee \neg(|Y| = \log^k n) \vee \neg SEQ(Y) \vee \exists x(X(x) \wedge R_1(x)))),$$

Clearly, all second order variables in the formula need maximum exponent k and the formula itself can be rewritten in Skolem normal form as a formula in Σ_2^{plog}.

The formulae expressing the following two problems can be well understood as formulae defining B-trees where the leaves are pointers to positions in the input string.

Problem 4.5 (AtLeastBlocks$_l^k$). Let AtLeastBlocks$_l^k$ for $k, l \geq 0$ denote the problem of deciding the language of binary strings with at least $(\lceil \log n \rceil^k)^l$ non-overlapping adjacent substrings of the form $0^{\lceil \log n \rceil^k}$ where n is the length of the string, or equivalently, the language of binary strings which have at least $(\lceil \log n \rceil^k)^{l+1}$ consecutive bits set to 0.

If $l = 0$ then this is equivalent to ConseqZeros^k and, as discussed in Problem 4.2, it can be expressed in Σ_1^{plog}.

If $l = 1$, we can express AtLeastBlocks$_l^k$ in SO^{plog} as follows:

$$\exists X \forall xy \exists Z(|X| = \log^k n \wedge SEQP(X)$$
$$\wedge (X(x, y) \rightarrow (|Z| = \log^k n \wedge SEQ(Z) \wedge min(Z) = x \wedge max(Z)$$
$$= y \wedge \forall z(Z(z) \rightarrow R_0(z))))).$$

Here $SEQP(X)$ denotes the sub-formula expressing that X is a set of ordered pairs that form a sequence where every consecutive (a_1, a_2) and (b_1, b_2) in the sequence satisfy that a_2 is the immediate predecessor of b_1 in the order $<$. This is clearly expressible by a SO^{plog} formula free of second-order quantification. The sub-formulae $min(Z) = x$ and $max(Z) = y$ have the obvious meaning and again can easily be expressed in SO^{plog} without using second-order quantification. The whole sentence can be transformed into an equivalent sentence in Σ_3^{plog}.

Finally, for every $l \geq 2$, we can express AtLeastBlocks$_l^k$ in SO^{plog} with formulae of the form:

$$\exists X_1 \forall x_1 y_1 \exists X_2 \forall x_2 y_2 \cdots \exists X_l \forall x_l y_l \exists Z(|X_1| = \log^k n \wedge SEQP(X_1) \wedge$$
$$(X_1(x_1, y_1) \rightarrow$$
$$\quad (|X_2| = \log^k n \wedge SEQP(X_2) \wedge minp(X_2) = x_1 \wedge maxp(X_2) = y_1 \wedge$$
$$\quad\quad \cdots \wedge (X_{l-1}(x_{l-1}, y_{l-1}) \rightarrow$$
$$\quad\quad\quad (|X_l| = \log^k n \wedge SEQP(X_l) \wedge minp(X_l) = x_{l-1} \wedge maxp(X_l) = y_{l-1} \wedge$$
$$\quad\quad\quad (X_l(x_l, y_l) \rightarrow$$
$$\quad\quad\quad\quad (|Z| = \log^k n \wedge SEQ(Z) \wedge min(Z) = x_l \wedge max(Z) = y_l \wedge$$
$$\quad\quad\quad\quad \forall z(Z(z) \rightarrow R_0(z)))))) \cdots))).$$

The sub-formulae of the form $minp(X) = x$ (resp. $maxp(X) = x$) express that x is the smallest first element (resp. biggest second element) of any tuple in X and is easily expressible in SO^{plog} by a formula free of second-order quantifiers. We can rewrite the whole formula as a $\Sigma_{2 \cdot l + 1}^{plog}$ formula.

Problem 4.6 (ExactlyBlocks$_l^k$). Let ExactlyBlocks$_l^k$ for $k, l \geq 0$ denote the problem of deciding the language of binary strings with exactly $(\lceil \log n \rceil^k)^l$ non-overlapping adjacent substrings of the form $0^{\lceil \log n \rceil^k}$ where n is the length of the string, or equivalently, the language of binary strings which contain the substring $0^{(\lceil \log n \rceil^k)^{l+1}}$ exactly once.

If $l = 0$ then this is equivalent to ExactlyOnce^k and, as discussed in Problem 4.4, it can be expressed in Σ_2^{plog}.

If $l = 1$, we can express ExactlyBlocks$_l^k$ in SO^{plog} as follows:

$$\exists X \forall xy \exists Z(|X| = \log^k n \wedge SEQP(X) \wedge$$
$$\quad (X(x, y) \rightarrow (|Z| = \log^k n \wedge SEQ(Z) \wedge min(Z) = x \wedge max(Z) = y \wedge$$
$$\quad\quad \forall z(Z(z) \rightarrow R_0(z)) \wedge$$
$$\quad\quad \forall X' \exists x' y' \forall Z'(X' = X \vee \neg(|X'| = \log^k n) \vee \neg SEQP(X') \vee$$
$$\quad\quad\quad (X'(x', y') \wedge (Z' = Z \vee \neg(|Z'| = \log^k n) \vee \neg SEQ(Z') \vee$$
$$\quad\quad\quad \neg(min(Z') = x') \vee \neg(max(Z') = y') \vee$$
$$\quad\quad\quad \exists z'(Z'(z') \wedge R_1(z')))))))).$$

It is not difficult to see that this formula can be rewritten as a Σ_4^{plog} formula.

Finally, for every $l \geq 2$, we can express ExactlyBlocks$_l^k$ in SO^{plog} with formulae of the form:

$$\exists X_1 \forall x_1 y_1 \exists X_2 \forall x_2 y_2 \cdots \exists X_l \forall x_l y_l \exists Z(|X_1| = \log^k n \wedge SEQP(X_1) \wedge$$
$$(X_1(x_1, y_1) \rightarrow$$
$$\quad (|X_2| = \log^k n \wedge SEQP(X_2) \wedge minp(X_2) = x_1 \wedge maxp(X_2) = y_1 \wedge$$
$$\quad\quad \cdots \wedge (X_{l-1}(x_{l-1}, y_{l-1}) \rightarrow$$
$$\quad\quad\quad (|X_l| = \log^k n \wedge SEQP(X_l) \wedge minp(X_l) = x_{l-1} \wedge maxp(X_l) = y_{l-1} \wedge$$
$$\quad\quad\quad (X_l(x_l, y_l) \rightarrow$$
$$\quad\quad\quad\quad (|Z| = \log^k n \wedge SEQ(Z) \wedge min(Z) = x_l \wedge max(Z) = y_l \wedge$$
$$\quad\quad\quad\quad \forall z(Z(z) \rightarrow R_0(z)) \wedge$$

$$\forall X_1' \exists x_1' y_1' \forall X_2' \exists x_2' y_2' \cdots \forall X_l' \exists x_l' y_l' \forall Z' (X_1' = X_1 \vee$$
$$\neg(|X_1'| = \log^k n) \vee \neg SEQP(X_1') \vee (X_1'(x_1', y_1') \wedge (X_2' = X_2 \vee$$
$$\neg(|X_2'| = \log^k n) \vee \neg SEQP(X_2') \vee \neg minp(X_2') = x_1' \vee$$
$$\neg maxp(X_2') = y_1' \vee (\cdots \vee (X_{l-1}'(x_{l-1}', y_{l-1}') \wedge (X_l' = X_l \vee$$
$$\neg(|X_l'| = \log^k n) \vee \neg SEQP(X_l') \vee \neg(minp(X_l') = x_{l-1}') \vee$$
$$\neg(maxp(X_l') = y_{l-1}') \vee (X_l'(x_l', y_l') \wedge (Z' = Z \vee$$
$$\neg(|Z'| = \log^k n) \vee \neg SEQ(Z') \vee \neg(min(Z') = x_l') \vee$$
$$\neg(max(Z') = y_l') \vee \exists z'(Z'(z') \wedge R_1(z')))))) \cdots))))))) \cdots))).$$

We can rewrite formulae of this form as $\Sigma_{2 \cdot l+2}^{plog}$ formulae.

5 Proper Hierarchies in Polylogarithmic Time

We now present the key results of the paper showing that all the polylogarithmic complexity classes defined in Sect. 2, including every level of the polylogarithmic time hierarchy, contain proper hierarchies defined in terms of the smallest degree of the polynomial required for the decision problems introduced in the previous section.

In order to relate the problems described in the previous section using logics to the polylogarithmic complexity classes defined in terms of random-access Turing machines, we adhere to the usual conventions concerning binary encoding of finite structures [8]. That is, if $\sigma = \{R_1^{r_1}, \ldots, R_p^{r_p}, c_1, \ldots, c_q\}$ is a vocabulary, and $\mathbf{A}$ with $A = \{0, 1, \ldots, n-1\}$ is an ordered structure of vocabulary σ. Each relation $R_i^{\mathbf{A}} \subseteq A^{r_i}$ of $\mathbf{A}$ is encoded as a binary string $\text{bin}(R_i^{\mathbf{A}})$ of length n^{r_i} where 1 in a given position indicates that the corresponding tuple is in $R_i^{\mathbf{A}}$. Likewise, each constant number $c_j^{\mathbf{A}}$ is encoded as a binary string $\text{bin}(c_j^{\mathbf{A}})$ of length $\lceil \log n \rceil$. The encoding of the whole structure $\text{bin}(\mathbf{A})$ is simply the concatenation of the binary strings encodings its relations and constants. The length $\hat{n} = |\text{bin}(\mathbf{A})|$ of this string is $n^{r_1} + \cdots + n^{r_p} + q \lceil \log n \rceil$, where $n = |A|$ denotes the size of the input structure $\mathbf{A}$. Note that $\log \hat{n} \in O(\lceil \log n \rceil)$, so $\text{NTIME}[\log^k \hat{n}] = \text{NTIME}[\log^k n]$ (analogously for DTIME). Therefore, we can consider random-access Turing machines, where the input is the encoding $\text{bin}(\mathbf{A})$ of the structure $\mathbf{A}$ followed by the endmark $\triangleleft$.

The following simple lemmas are useful to prove our hierarchy theorems. They show that the problems in the previous section can be expressed by random-access machines working in the required levels of the hierarchy theorems.

Lemma 5.1. InitialZeros^k (see Problem 4.1) is in $\text{DTIME}(\log^k n)$.

Proof. Assume the input tape encodes a word model $\mathbf{A}$ of signature π, i.e., a binary string. A deterministic random-access Turing machine can in deterministic time $O(\log n)$ calculate and write in its index-tape the address of the first bit in the encoding of $R_0^{\mathbf{A}}$. Then it only needs to check whether this bit and the subsequent $\lceil \log n \rceil^k - 1$ bits in the input-tape are 1. If that is the case, then the machine accepts the input. Clearly, this process takes time $O(\log^k n)$. □

Lemma 5.2. ConseqZeros^k *(see Problem 4.2) is in* NTIME($\log^k n$).

Proof. Assume the input tape encodes a word model **A** of signature π. A random-access Turing machine M can non-deterministically guess a position i in the input tape which falls within the cells encoding $R_0^{\mathbf{A}}$. This takes time $O(\log n)$. Then M can check (working deterministically) in time $O(log^{k+1}n)$ whether each cell of the input tape between positions i and $i + log^{k+1}n$ has a 0. $\qquad\square$

Lemma 5.3. NoConseqZeros^k *(see Problem 4.3) is in* ATIME^{op}($\log^k n, 1$).

Proof. Assume the input tape encodes a word model **A** of signature π. In a universal state, a random-access alternating Turing machine M can check whether for all cell in some position i in the input tape which falls in a position encoding $R_0^{\mathbf{A}}$ and is at distance at least $\lceil \log n \rceil^k$ from the end of the encoding, there is a position between positions i and $i + \lceil \log n \rceil^k - 1$ with 0. Each of these checking can be done deterministically in time $O(log^k n)$. Therefore this machine decides NoConseqZeros^k in ATIME^{op}($\log^k n, 1$). $\qquad\square$

Lemma 5.4. ExactlyOnce^k *(see Problem 4.4) is in* ATIME($\log^k n, 2$).

Proof. We only need to combine the machines that decide ConseqZeros^k and NoConseqZeros^k in Lemmas 5.2 and 5.3, respectively. An alternating random-access Turing machine M can decide ExactlyOnce^k as follows: Assume the input tape encodes a word model **A** of signature π. Let s and t be the cells that mark the beginning and end of the encoding of $R_0^{\mathbf{A}}$. These cells can be calculated by M in $DTIME(\log n)$. First M checks in an existential state whether there is a position i in the input tape which fall between s and $t - \lceil \log n \rceil^k + 1$ such that each cell between positions i and $i + \lceil \log n \rceil^k - 1$ has a 1. Then M switches to a universal state and checks whether for all cell in some position j between s and $t - \lceil \log n \rceil^k + 1$ of the input tape other than position i, there is a cell between positions j and $j + \lceil \log n \rceil^k - 1$ with 0. If these two checks are successful, then the input string belongs to ExactlyOnce^k. We already saw in Lemmas 5.2 and 5.3 that both checks can be done in time $O(\log^k n)$. $\qquad\square$

In order to get tighter upper bounds, in the previous lemmas we explicitly defined the random-access Turing machines that decide the problems. For the following two lemmas we use the upper bounds resulting from the proof of Theorem 3.2 instead, since there seems to be no better upper bounds for these cases. Thus, Lemmas 5.5 and 5.6 follow from the facts that: (a) to evaluate the SO^{plog} formulae in Problems 4.5 and 4.6 for AtLeastBlocks$_l^k$ and ExactlyBlocks$_l^k$, respectively, the machine needs (as shown in the proof of Theorem 3.2 in [4]) to "guess" $\lceil \log n \rceil^k$ addresses, each of length $\lceil \log n \rceil$; and (b) the formula for AtLeastBlocks$_l^k$ and ExactlyBlocks$_l^k$ are in $\Sigma_{2 \times l+1}^{plog}$ and $\Sigma_{2 \times l+2}^{plog}$, respectively.

Lemma 5.5. AtLeastBlocks$_l^k$ *(see Problem 4.5) is in* ATIME($\log^{k+1} n, 2 \cdot l + 1$).

Lemma 5.6. ExactlyBlocks$_l^k$ *(see Problem 4.6) is in* ATIME($\log^{k+1} n, 2 \cdot l + 2$).

We can now prove our first hierarchy theorem which shows that there is a strict hierarchy of problems inside DPolylogTime.

Theorem 5.1. *For every $k > 1$, $DTIME(\log^k n) \subsetneq DTIME(\log^{k+1} n)$.*

Proof. Lemma 5.1 proves that $InitialZeros^{k+1} \in DTIME(\log^{k+1} n)$. Regarding the lower bound, we will show that $InitialZeros^{k+1}$ (see Problem 4.1) is *not* in $DTIME(\log^k n)$.

Let us assume for the sake of contradiction that there is a deterministic random-access Turing machine M that decides $InitialZeros^{k+1}$ in time $\lceil \log n \rceil^k \cdot c$, for some constant $c \geq 1$. Take a string s of the form 0^n such that $\lceil \log n \rceil^{k+1} > \lceil \log n \rceil^k \cdot c$. Let $\mathbf{A}$ be its corresponding word model. Since the running time of M on input $\mathbf{A}$ is strictly less than $\lceil \log n \rceil^{k+1}$, then there must be at least one position i among the first $\lceil \log n \rceil^{k+1}$ cells in the encoding of $R_0^{\mathbf{A}}$ in the input tape that was not read in the computation of $M(\mathbf{A})$. Define a string $s' = 0^i 10^{n-i-1}$ and a corresponding word model $\mathbf{B}$. Clearly, the output of the computations of $M(\mathbf{A})$ and $M(\mathbf{B})$ are identical. This contradicts the assumption that M decides $InitialZeros^{k+1}$, since it is not true that the first $\lceil \log n \rceil^{k+1}$ bits of s' are 0.   □

Our second hierarchy theorem shows that there is also a strict hierarchy of problem inside NPolylogTime.

Theorem 5.2. *For every $k > 1$, $NTIME(\log^k n) \subsetneq NTIME(\log^{k+1} n)$.*

Proof. Lemma 5.2 proves that $ConseqZeros^{k+1} \in NTIME(\log^{k+1} n)$. Regarding the lower bound, we will show that $ConseqZeros^{k+1}$ (see Problem 4.2) is *not* in $NTIME(\log^k n)$.

Let us assume for the sake of contradiction that there is a nondeterministic random-access Turing machine M that decides $ConseqZeros^{k+1}$ in time $\lceil \log n \rceil^k \cdot c$, for some constant $c \geq 1$. Take a binary string s of the form $0^{\lceil \log n \rceil^{k+1}} 1^{n - \lceil \log n \rceil^{k+1}}$ such that $\lceil \log n \rceil^{k+1} > \lceil \log n \rceil^k \cdot c$. Let $\mathbf{A}$ be its corresponding word model. Since M accepts $\mathbf{A}$, then there is at least one computation ρ of M which accepts $\mathbf{A}$ in at most $\lceil \log n \rceil^k \cdot c$ steps. Then there must be at least one position i among the first $\lceil \log n \rceil^{k+1}$ cells in the encoding of $R_0^{\mathbf{A}}$ in the input tape that was not read during computation ρ. Define a string $s' = 0^i 10^{\lceil \log n \rceil^{k+1} - i - 1} 1^{n - \lceil \log n \rceil^{k+1}}$ and a corresponding word model $\mathbf{B}$. Clearly, the accepting computation ρ of $M(\mathbf{A})$ is also an accepting computation of $M(\mathbf{B})$. This contradicts the assumption that M decides $ConseqZeros^{k+1}$, since it is not true that there are $\lceil \log n \rceil^{k+1}$ consecutive zeros in s'.   □

The following theorem shows that there is a strict hierarchy of problems inside the first level of the $\tilde{\Pi}_m^{plog}$ hierarchy.

Theorem 5.3. *For every $k > 1$, $ATIME^{op}(\log^k n, 1) \subsetneq ATIME^{op}(\log^{k+1} n, 1)$.*

Proof. Lemma 5.3 proves that $NoConseqZeros^{k+1} \in ATIME^{op}(\log^{k+1} n, 1)$. Regarding the lower bound, we will show that $NoConseqZeros^{k+1}$ (see Problem 4.3) is *not* in $ATIME^{op}(\log^k n, 1)$.

Let us assume for the sake of contradiction that there is an alternating random-access Turing machine M that decides NoConseqZeros^{k+1} using only universal states and in time $\lceil \log n \rceil^k \cdot c$, for some constant $c \geq 1$. Take a binary string s of the form $0^{\lceil \log n \rceil^{k+1}} 1^{n - \lceil \log n \rceil^{k+1}}$ such that $\lceil \log n \rceil^{k+1} > \lceil \log n \rceil^k \cdot c$. Let $\mathbf{A}$ be its corresponding word model. From our assumption that M decides NoConseqZeros^{k+1}, we get that there is a rejecting computation ρ of $M(\mathbf{A})$. Since every computation of M which rejects $\mathbf{A}$ must do so reading at most $\lceil \log n \rceil^k \cdot c$ cells, then there must be at least one position i among the first $\lceil \log n \rceil^{k+1}$ cells in the encoding of $R_0^{\mathbf{A}}$ in the input tape that was not read during computation ρ. Define a string $s' = 0^i 1 0^{\lceil \log n \rceil^{k+1} - i - 1} 1^{n - \lceil \log n \rceil^{k+1}}$ and a corresponding word model $\mathbf{B}$. Clearly, the rejecting computation ρ of $M(\mathbf{A})$ is also a rejecting computation of $M(\mathbf{B})$. This contradicts the assumption that M decides NoConseqZeros^{k+1}, since s' do not have $\lceil \log n \rceil^{k+1}$ consecutive bits set to 0 and should then be accepted by all computations of M. $\qquad \square$

The following theorem shows that there is a strict hierarchy of problems inside the second level of the $\tilde{\Sigma}_m^{plog}$ hierarchy.

Theorem 5.4. *For every $k > 1$, ATIME$(\log^k n, 2) \subsetneq$ ATIME$(\log^{k+1} n, 2)$.*

Proof. Lemma 5.4 proves that ExactlyOnce$^{k+1} \in$ ATIME$(\log^{k+1} n, 2)$. Regarding the lower bound, we will show that ExactlyOnce^{k+1} (see Problem 4.4) is *not* in ATIME$(\log^k n, 2)$.

We assume for the sake of contradiction that there is an alternating random-access Turing machine M that decides ExactlyOnce^{k+1} in ATIME$(\log^k n, 2)$. We further assume, w.l.o.g., that every final state of M is universal. Let M work in time $\lceil \log n \rceil^k \cdot c$ for some constant c. Take a binary string s of the form $0^{\lceil \log n \rceil^{k+1}} 1 0^{\lceil \log n \rceil^{k+1}} 1^{n - 2 \cdot \lceil \log n \rceil^{k+1} - 1}$ such that $\lceil \log n \rceil^{k+1} > \lceil \log n \rceil^k \cdot c$. Let $\mathbf{A}$ be its corresponding word model. From our assumption that M decides ExactlyOnce^{k+1}, we get that there is a rejecting computation ρ of $M(\mathbf{A})$. Since every computation of M which rejects $\mathbf{A}$ must do so reading at most $\lceil \log n \rceil^k \cdot c$ cells, then there must be a position i among the first $\lceil \log n \rceil^{k+1}$ cells in the encoding of $R_0^{\mathbf{A}}$ in the input tape that was not read during computation ρ. Define a string $s' = 0^i 1 0^{\lceil \log n \rceil^{k+1} - i - 1} 1 0^{\lceil \log n \rceil^{k+1}} 1^{n - 2 \cdot \lceil \log n \rceil^{k+1} - 1}$ and a corresponding word model $\mathbf{B}$. Clearly, the rejecting computation ρ of $M(\mathbf{A})$ is also a rejecting computation of $M(\mathbf{B})$. This contradicts the assumption that M decides ExactlyOnce^{k+1}, since s' has exactly one substring $0^{\lceil \log n \rceil^{k+1}}$ and should then be accepted by all computations of M. $\qquad \square$

The following result, together with Theorems 5.2 and 5.4, shows that there is a proper hierarchy of problems for every level of the polylogarithmic time hierarchy $\tilde{\Sigma}_m^{plog}$.

Theorem 5.5. *For $m > 2$ and $k > 1$, it holds that ATIME$(\log^k n, m) \subsetneq$ ATIME$(\log^{k+2} n, m)$.*

Proof. Since $m > 2$, we have that Lemma 5.5 proves that if m is odd, then AtLeastBlocks$_{(m-1)/2}^{k+1}$ is in ATIME($\log^{k+2} n, m$). Likewise, Lemma 5.6 proves that if m is even, then ExactlyBlocks$_{(m-2)/2}^{k+1}$ is in ATIME($\log^{k+2} n, m$). Regarding the lower bounds, it is easy to see (given our previous results in this section) that: (a) for odd m, AtLeastBlocks$_{(m-1)/2}^{k+1}$ is *not* in ATIME($\log^k n, m$), and (b) for even m, ExactlyBlocks$_{(m-2)/2}^{k+1}$ is also *not* in ATIME($\log^k n, m$). Note that if m is odd, then we can prove (a) by contradiction following a similar argument than in the proof of the lower bound for Theorem 5.2. Likewise, if m is even, then we can prove (b) by contradiction following a similar argument than in the proof of Theorem 5.4. □

It is clear that by taking the complements of the problems AtLeastBlocks$_l^k$ and ExactlyBlocks$_l^k$, a similar result holds for each level of the Π_m^{plog} hierarchy.

Theorem 5.6. *For $m = 2$ and every $k > 1$, it holds that* ATIME$^{op}(\log^k n, m) \subsetneq$ ATIME$^{op}(\log^{k+1} n, m)$. *Moreover, For every $m > 2$ and every $k > 1$, it holds that* ATIME$^{op}(\log^k n, m) \subsetneq$ ATIME$^{op}(\log^{k+2} n, m)$.

6   On Polylogarithmic-Time and Complete Problems

In this section we investigate whether the concept of complete problem can somehow be applied to the complexity classes DPolylogTime and NPolylogTime. That is, we want to know whether we can isolate the most difficult problems inside these sublinear time complexity classes. The first step towards this objective is to find a suitable concept of many-one reducibility (m-reducibility for short).

It is quite clear that m-reductions with sublinear time bounds do not work. Consider for instance DPolylogTime reductions. Assume there is a complete problem P for the class NPolylogTime under DPolylogTime reductions. Let P' belong to NPolylogTime and let M be a deterministic random-access Turing machine that reduces P' to P in time $c' \cdot log^{k'} n$ for some constant c'. Then the output of M given an instance of P' of length n has maximum length $c' \cdot \log^{k'} n$. This means that, given an input of length n for P' and its reduction, the random-access Turing machine that computes the complete problem P can actually compute $P(s)$ in time $O((\log \log n)^k)$ for some fixed k. This is already highly unlikely. If as one would expect there are more than a single complete problem for the class, then we could keep applying reductions from one problem to the other, infinitely reducing the time required to compute the original problem.

Let us then consider the standard concept of Karp reducibility, i.e., deterministic polynomially bounded many-one reducibility, so that we can avoid the obvious problem described in the previous paragraph. Rather surprisingly, there is no complete problems for DPolylogTime and NPolylogTime, even under these rather expensive reductions for the complexity classes at hand.

Theorem 6.1. *DPolylogTime does* not *have complete problems under deterministic polynomially bounded many-one reductions.*

Proof. We prove it by contradiction. Assume that there is such a complete problem P. Since P is in DPolylogTime, then there is a random-access Turing machine M which computes P in time $O(\log^k n)$ for some fixed k. Thus P belongs to DTIME($\log^k n$). Let us take the problem InitialZeros^{k+1} of deciding the language of binary strings which have a prefix of at least $\lceil \log n \rceil^{k+1}$ consecutive zeros. Since P is complete for the whole class DPolylogTime, there must be a function $f : \{0,1\}^* \to \{0,1\}^*$, computable in polynomial-time, such that $x \in$ InitialZeros^{k+1} iff $f(x) \in P$ holds for all $x \in \{0,1\}^*$. It then follows that the size of $f(x)$ is polynomial in the size of x. Let $|f(x)| = |x|^{k'}$, we get that the machine M which computes the complete problem P can also decide InitialZeros^{k+1} in time $O(\log^k n^{k'}) = O((k' \cdot \log n)^k) = O(\log^k n)$. This contradicts the fact that InitialZeros$^{k+1} \notin$ DTIME($\log^k n$) as shown in the proof of Theorem 5.1.   □

Using a similar proof than in the previous theorem for DPolylogTime, we can prove that the same holds for NPolylogTime. In fact, we only need to replace the problem InitialZeros^{k+1} by ConseqZeros^{k+1} and the reference to Theorem 5.1 by a reference to Theorem 5.2 in the previous proof, adapting the argument accordingly.

Theorem 6.2. *NPolylogTime does* not *have complete problems under deterministic polynomially bounded many-one reductions.*

Moreover, using the problems AtLeastBlocks$_l^k$ and ExactlyBlocks$_l^k$ together with its complements and Theorems 5.5 and 5.6, it is easy to prove that the same holds for every individual level of the polylogarithmic time hierarchy.

Theorem 6.3. *For every $m \geq 1$, Σ_m^{plog} and Π_m^{plog} do* not *have complete problems under deterministic polynomially bounded many-one reductions.*

7   Concluding Remarks

In this paper we showed that none of the classes $\tilde{\Sigma}_m^{plog}$ and $\tilde{\Pi}_m^{plog}$ ($m \in \mathbb{N}$) in the polylogarithmic time hierarchy has a complete problem. This result follows from the existence of proper hierarchies inside each of the classes. The proof that such hierarchies exist is constructive by defining concrete problems parameterized by $k \in \mathbb{N}$ for each class. For the definition of these concrete problems we exploit the logics capturing $\tilde{\Sigma}_m^{plog}$ and $\tilde{\Pi}_m^{plog}$, respectively. We expect that these results can be taken further towards an investigation of the strictness of the polylogarithmic time hierarchy as such. We also expect that similar strict hierarchies can be defined in terms of subsets of formulae in Σ_m^{plog} and Π_m^{plog}. Notice that the latter does not follow directly from the strict hierarchies proven in this paper, since in the proofs of the characterization results for the polylogarithmic-time hierarchy [3,5], there is *not* an exact correspondence between the exponents in the polylogarithmic functions that bound the time complexity of the machines and the exponents in the restricted second-order variables of the SO^{plog} formulae that define the machines.

References

1. Babai, L.: Graph isomorphism in quasipolynomial time. In: Proceedings of the Forty-Eighth Annual ACM Symposium on Theory of Computing (STOC 2016), pp. 684–697 (2016)
2. Ebbinghaus, H.-D., Flum, J.: Finite Model Theory. Perspectives in Mathematical Logic. Springer, Heidelberg (1995)
3. Ferrarotti, F., González, S., Schewe, K.-D., Turull Torres, J.M.: The polylog-time hierarchy captured by restricted second-order logic. In: 20th International Symposium on Symbolic and Numeric Algorithms for Scientific Computing, SYNASC 2018, Timisoara Romania, 20–23 September 2018, pp. 133–140. IEEE (2018)
4. Ferrarotti, F., González, S., Schewe, K.-D., Turull Torres, J.M.: The polylog-time hierarchy captured by restricted second-order logic. CoRR, abs/1806.07127 (2018)
5. Ferrarotti, F., González, S., Schewe, K.-D., Turull Torres, J.M.: A restricted second-order logic for non-deterministic poly-logarithmic time. Logic J. IGPL (2019, to appear)
6. Ferrarotti, F., González, S., Turull Torres, J.M., Van den Bussche, J., Virtema, J.: Descriptive complexity of deterministic polylogarithmic time. In: Iemhoff, R., Moortgat, M., de Queiroz, R. (eds.) WoLLIC 2019. LNCS, vol. 11541, pp. 208–222. Springer, Heidelberg (2019). https://doi.org/10.1007/978-3-662-59533-6_13
7. Ferrarotti, F., González, S., Turull Torres, J.M., Van den Bussche, J., Virtema, J.: Descriptive complexity of deterministic polylogarithmic time and space. Submitted for publication (2019)
8. Immerman, N.: Descriptive complexity. Graduate Texts in Computer Science. Springer, New York (1999)
9. Mix Barrington, D.A.: Quasipolynomial size circuit classes. In: Proceedings of the Seventh Annual Structure in Complexity Theory Conference, Boston, Massachusetts, USA, 22–25 June 1992, pp. 86–93. IEEE Computer Society (1992)
10. Mix Barrington, D.A., Immerman, N., Straubing, H.: On uniformity within NC^1. J. Comput. Syst. Sci. 41(3), 274–306 (1990)

Diversity, Dependence and Independence

Pietro Galliani[1]($\boxtimes$) and Jouko Väänänen[2]

[1] Free University of Bozen-Bolzano, Bolzano, Italy
Pietro.Galliani@unibz.it
[2] University of Helsinki, Helsinki, Finland
jouko.vaananen@helsinki.fi

Abstract. We introduce the concepts of dependence and independence in a very general framework. We use a concept of *rank* to study dependence and independence. By means of the rank we identify (total) *dependence* with inability to create more diversity, and (total) *independence* with the presence of maximum diversity. We show that our theory of dependence and independence covers a variety of dependence concepts, for example the seemingly unrelated concepts of linear dependence in algebra and dependence of variables in logic.

Keywords: Dependence Logic · Matroids · Independence · Team Semantics

1 Introduction

Our starting point is very general. Suppose we have a set M of objects. We want to make sense of the concept that a subset $x \subseteq M$ **depends** on another subset $y \subseteq M$, or that a subset $x \subseteq M$ is **independent** of another subset $y \subseteq M$. To accomplish this in the most general sense, we define the concept of **diversity** of a set $x \subseteq M$. The idea is that a small set has less diversity than a bigger set, hence our diversity function is monotone. Also, the diversity of X arises from properties of the individual elements, hence our diversity function satisfies certain further conditions. The connection between diversity and dependence arises from the idea that dependence reduces diversity, and respectively independence preserves diversity. If x totally determines y, then adding y to x does not increase the diversity of x at all. On the other hand, if x and y are independent, then putting them together means simply adding the diversities together: nothing is lost, because there is no interaction between x and y.

Because of the generality of our approach, according to which M is just a set of objects about which we a priori know nothing, we do not *define* the diversity function explicitly, but rather give a few conditions it ought to satisfy. The point is that on the basis of these conditions we can introduce natural notions of **dependence** and **independence** with a variety of applications.

The concepts of dependence and independence occur widely in science. Exact study of these concepts has taken place at least in four different contexts:

- **Mathematics:** Dependence and independence are fundamental concepts in algebra: linear dependence in linear algebra and algebraic dependence in field theory. In both cases independence is defined as the lack of dependence: elements $\{x_1, \ldots, x_n\}$ are *independent* if no x_i is dependent on the rest. Whitney [15] and van der Waerden [14]

© Springer Nature Switzerland AG 2020
A. Herzig and J. Kontinen (Eds.): FoIKS 2020, LNCS 12012, pp. 106–121, 2020.
https://doi.org/10.1007/978-3-030-39951-1_7

pointed out the similarity between these two notions of dependence and proposed axioms that cover both cases. Whitney suggested the name *matroid* for the general dependence structure inherent in algebra, giving rise to *matroid theory*, nowadays a branch of discrete mathematics.

- **Computer science:** *Functional* dependence [1] is a fundamental concept of database theory. The design and analysis of so called relational databases is often based on a careful study of the functional dependencies between attributes of various parts of the database. The more general *multivalued* dependencies are analogous to what we call independence relations between attributes.
- **Statistics and probability theory:** Dependence and independence of events (or random variables) is the basis of probability theory and statistical analysis of data.
- **Logic:** Dependence of a variable on another is the basic concept in quantification theory. In *Dependence Logic* [12] this concept is separated from quantification, making it possible, as in *Independence-Friendly Logic* [10], to write formulas with more complicated dependence relations between variables than what first order logic allows. *Independence Logic* [5], likewise, extends First Order Logic by an atom $x \perp y$ that states that the tuples of quantified variables x and y are chosen independently, in the sense that every possible choice of x and of y may occur together. These logics – and the generalization of Tarski's Semantics used for their analysis, commonly called *Team Semantics* – have lead in the last decade to a considerable amount of research regarding logics augmented by various *notions of dependence and independence*, in the first order case but also in the propositional case [16], in the modal case [7,13], in the temporal case [9], and recently even in probabilistic cases [3,6,8].

As it turns out, these concepts of dependence and independence arise from one particular more general concept—diversity—that simultaneously generalizes all the above cases and satisfies the same axioms in each case. It is the purpose of this paper to introduce this concept and suggest thereby a wide-ranging general theory of dependence.

2   Diversity Rank in a General Setting

We now define the concept of *rank* in an entirely general setting. We use the notation xy to denote the union $x \cup y$ of subsets x and y of a fixed set M. The following is the key definition of this work:

Definition 1. *Suppose M is an arbitrary set and $\mathcal{M} \subseteq \mathcal{P}(M)$ such that $\mathcal{M}$ is closed under (finite) unions and contains the finite subsets of M. A function $x \mapsto \|x\|$ from $\mathcal{M}$ to $\mathbb{R}^+ \cup \{0\}$ is called a* diversity rank function *on M if it satisfies the following conditions for all $x, y, z \in \mathcal{M}$:*

R1: $\|\emptyset\| = 0$;
R2: $\|x\| \leq \|xy\| \leq \|x\| + \|y\|$;
R3: *If* $\|xy\| = \|x\|$ *then* $\|xyz\| = \|xz\|$;
R4: *If* $\|xyz\| = \|x\| + \|yz\|$ *then* $\|xy\| = \|x\| + \|y\|$.[1]

---

[1] Since set union is commutative, it also follows that if $\|xyz\| = \|x\| + \|yz\|$ then $\|xz\| = \|x\| + \|z\|$.

Intuitively, the rank of x is the amount of "diversity" or "variation" that x contains. For example, if x is a sequence of vectors in a vector space, the amount of diversity in x is revealed by the dimension of the subspace spanned by x. If x is the set of attributes in a relation schema in a given database schema, the amount of diversity in x is revealed by the maximum number of different tuples (records) that may exist in the corresponding relation in a given database instance that can be considered as valid for that schema. Finally, if x is simply a word in a finite alphabet, a possible measure of the amount of diversity in x is the number of different letters in x, so that for example the diversity of "abbab" is 2 and the diversity of "abcdda" is 4.

It is obvious that we have to require **R1** and **R2**. The empty set cannot manifest any diversity, more elements means more diversity, and the amount of diversity manifesting in two sets taken together is at most the sum of the amounts of diversity occurring in each of them separately.

The axioms **R3** and **R4** are less intuitive. In brief, Axiom **R3** states that if adding y to x does not increase the amount of diversity of x (that is, y is "trivial" given x), then adding it to xz does not increase the amount of diversity of xz (that is, y is also "trivial" given xz) either; and Axiom **R4** states that if adding yz to x increases the amount of diversity of the maximum amount possible (that is, yz is "maximally non-trivial" given x) then adding y to x also increases the amount of diversity of the maximum amount possible (that is, y is also "maximally non-trivial" given x). **R3** and **R4** (as well as the right part of **R2**) would follow at once if we assumed that our rank-function is *submodular*, in the sense that it satisfies the condition

SUBM: $\|xyz\| + \|z\| \le \|xz\| + \|yz\|$:

Proposition 1. *Every function from subsets of some set M to non-negative real numbers satisfying R1, the left part of R2 and SUBM is a diversity rank function in the sense of Definition 1.*

Proof. Let $\|x\|$ satisfy **R1**, the left part of **R2** and **SUBM**. We need to show that $\|x\|$ also satisfies the right part of **R2** as well as **R3** and **R4**.

Choosing $z = \emptyset$, we obtain immediately from submodularity that $\|xy\| + \|\emptyset\| \le \|x\| + \|y\|$. But by **R1** we know that $\|\emptyset\| = 0$, and subadditivity (that is, the right part of **R2**) follows.

As for **R3**, suppose that $\|xy\| = \|x\|$. Then by **SUBM**, $\|xyz\| \le \|xy\| + \|xz\| - \|x\| = \|x\| + \|xz\| - \|x\| = \|xz\|$; but on the other hand $\|xz\| \le \|xyz\|$ by the left part of **R2** and so $\|xyz\| = \|xz\|$ as required.

Finally, **R4** holds. Indeed, by **SUBM** we know that $\|xyz\| + \|y\| \le \|xy\| + \|yz\|$. Thus, if $\|xyz\| = \|x\| + \|yz\|$ we have at once that $\|x\| + \|yz\| + \|y\| \le \|xy\| + \|yz\|$, that is, that $\|x\| + \|y\| \le \|xy\|$. But $\|xy\| \le \|x\| + \|y\|$ by the right part of **R2**, which we already proved, and hence $\|xy\| = \|x\| + \|y\|$ as required.

As we will soon see, however, submodularity would be too strong a requirement for our intended application.

A direct consequence of the above result is that our diversity rank functions generalize matroids:

Definition 2 (Matroid). *A matroid over some finite set E may be defined in terms of a rank function*[2] *from subsets of E to non-negative integers satisfying the following conditions for all subsets x and y of E:*

M1 $r(x) \leq |x|$;
M2 $r(xy) + r(x \cap y) \leq r(x) + r(y)$;
M3 *If* $|y| = 1$ *then* $r(x) \leq r(xy) \leq r(x) + 1$.

Corollary 1. *Let r be the rank function of a matroid over some finite set E. Then r is a diversity rank function over E.*

Proof. **R1** holds for r. Indeed, by **M1**, $r(\emptyset) \leq |\emptyset| = 0$, and therefore the only possibility is that $r(\emptyset) = 0$.

The left part of **R2** holds at once because of **M3** (this can be shown by easy induction on the number of elements in $y \backslash x$).

Moreover, **SUBM** also holds of r. Indeed, since $xyz = xzyz$, we have by **M2** that $r(xyz) + r(xz \cap yz) \leq r(xz) + r(yz)$. But $z \subseteq xz \cap yz$, and so by the left part of **R2** (which we already proved) $r(z) \leq r(xz \cap yz)$ and hence $r(xyz) + r(z) \leq r(xz) + r(yz)$, as required.

The conclusion then follows, since because of Proposition 1 the right part of **R2** as well as **R3** and **R4** are also true of r.

Given a notion of diversity, it is easy to define dependence and independence in terms of *minimal* and *maximal* diversity contributions:

Definition 3. *Suppose M is a set and $\| \cdot \|$ a rank-function on M. We can now define dependence relations on $\mathcal{P}(M)$ (with respect to $\| \cdot \|$) as follows:*

– **Dependence:** *y (totally) depends on x, in symbols $=(x, y)$, if $\|xy\| = \|x\|$.*
– **Constancy:** *x is constant, in symbols $=(x)$, if $\|x\| = 0$.*
– **Independence:** *x and y are independent, in symbols $x \perp y$, if $\|x\| + \|y\| = \|xy\|$.*

The idea is that $=(x, y)$ holds under a rank-function if the amount of diversity inherent in x in terms of the rank-function does not increase when y is added. Simply put, x determines y, so no new diversity occurs. $=(x)$, on the other hand, holds if x has no diversity at all; and $x \perp y$ holds if the diversity inherent in x is so unrelated to the diversity inherent in y that when the two are put together into xy, the diversity is the sum of the diversity of x and the diversity of y: no loss of diversity occurs because there is—intuitively—no connection between x and y.

---

[2] Equivalently, it is possible to define matroids in terms of its *independent sets* (that is, the x such that $r(x) = |x|$), in terms of *circuits* (maximal independent sets), in terms of *bases* (minimal non-independent sets), or *closure operations*. All these definitions are equivalent. We refer the reader to [11] for more details.

3 Examples

Let us now consider some examples of our definitions, in order to get a better feel of their applicability and consequences:

Example 1 (Constant diversity). As an extreme case we have the constant rank $\|x\| = c$, c constant, for all $x \subseteq M$, with the exception that $\|\emptyset\| = 0$. If $c = 0$, there is no diversity, every set depends on every other set and is also independent of every other set. If $c \neq 0$ then every set y is still dependent on any non-empty set x, because $\|xy\| = c = \|x\|$, and every set x is still independent from the empty set $\emptyset$, because $\|x\emptyset\| = \|x\| = \|x\| + \|\emptyset\|$; but two non-empty sets x and y are not independent, because $\|xy\| = c \neq c + c = \|x\| + \|y\|$.

Example 2 (Singular diversity). Let $a_0 \in M$ be fixed. Let

$$\|x\| = \begin{cases} 1, & \text{if } a_0 \in x, \\ 0, & \text{otherwise.} \end{cases}$$

In this case a chosen individual a_0 is the only "diversity" there is. A set has diversity 1 iff it contains a_0. In this case y depends on x if

$$a_0 \in y \rightarrow a_0 \in x$$

and two sets x and y are independent if at most one of them contains a_0. So dependence reduces in this case to implication and independence to the Sheffer stroke (also known as NAND).

Example 3 (Two-valued diversity). Suppose $\|\{a\}\|$ is either 0 or 1 for all $a \in M$, and

$$\|x\| = \max\{\|\{a\}\| : a \in x\}$$

for all $x \subseteq M$.

In this case some chosen individuals are declared to have diversity 1 and the rest are of diversity 0. A set has diversity 1 if it contains one of those chosen individuals. In this case diversity is an on/off thing, either it exists (1) or it does not (0), and a set has diversity if it includes some singleton that has it. In an extreme case $\|\{a\}\| = 0$ for *all* singletons $\{a\}$, $a \in M$, and we have constant diversity: every set has diversity 0. In another extreme case $\|\{a\}\| = 1$ for *all* singletons $\{a\}$, $a \in M$, and we are again in constant diversity: every non-empty set has diversity 1.

According to this diversity notion, $=(x, y)$ if and only if $\exists a \in x$ s.t. $\|\{a\}\| = 1 \Rightarrow \exists b \in y$ s.t. $\|\{b\}\| = 1$; and x is independent from y if and only if at most one of x and y contain an element c with $\|\{c\}\| = 1$.

Example 4 (Uniform diversity). Suppose

$$\|x\| = |x|.$$

This is the bold choice of taking the cardinality of the (finite) set as the measure of diversity. Dependence means inclusion: y is (totally) dependent on x if and only if

$|xy| = |x|$, that is, if and only if $y \subseteq x$. Independence is disjointness: x and y are independent if and only if $|xy| = |x| + |y|$, that is, if and only if $x \cap y = \emptyset$. Note that if M has at least three elements a, b, c, then independence is not equivalent to the failure of dependence both ways, as consideration of $\{a, b\}$ and $\{b, c\}$ reveals.

Example 5 (Coverage diversity). Suppose U is a finite set and we have a set $A_m \subseteq U$ for each $m \in M$. For $a_1 \ldots a_n \in M$, let

$$\|\{a_1, \ldots, a_n\}\| = |A_{a_1} \cup \ldots \cup A_{a_n}|.$$

We can think of each A_m as "data", about the element m of M. The more data we have the more diversity we give to the element, and the diversity of a set is obtained by simply putting together all the data we have. In this simple example the data is not thought to be specific to the m in M, so the data about different m is just lumped together. For example, if a and b are two genera, such as *Astragalus* and *Angylocalyx*, the diversity of $\{a, b\}$ in a set U of data about species (e.g. in some location) is obtained by counting how many different species of *Astragalus* or *Angylocalyx* there are in U.

According to this diversity notion, y is dependent on x if and only if $\bigcup\{A_{a_i} : a_i \in y\} \subseteq \bigcup\{A_{b_i} : b_i \in x\}$, that is, every data point corresponding to some element of y also corresponds to some element of x; and y is independent from x if and only if $\bigcup\{A_{a_i} : a_i \in y\} \cap \bigcup\{A_{b_i} : b_i \in x\} = \emptyset$, that is, if no data point corresponds to some element of x *and* to some element of y.

Example 6 (Relational diversity). Suppose X is a nonempty, finite set of *variable assignments* s from a finite set V of variables to a set A of elements (in the language of Dependence and Independence Logic, such a X is said to be a *team* over A with domain V).[3] Given some $x = \{v_1 \ldots v_n\} \subseteq V$, let[4]

$$\|x\| = \log(\#\mathrm{rows}_X(v_1 \ldots v_n)).$$

where

$$\#\mathrm{rows}_X(v_1 \ldots v_n) = |\{(s(v_1), \ldots, s(v_n)) : s \in X\}|$$

is the number of different values that $x = v_1 \ldots v_n$ takes in X.

We can think of each $s \in X$ as an "observation", or "data", about the possible values that the variables in V can take. The more different observations we have the more diversity we give to the element. Note the difference with the coverage diversity, where the data was not specific to the element of A. Here what matters is the relationships of the different observations to each other. Thus

$$\|\{v\}\| = \log|\{s(v) : s \in X\}|,$$

that is, the diversity rank of a single element v of V is the (logarithm of the) number of different observations about v. The diversity of a pair $\{v, w\}$ is the (logarithm of the) number of different combinations of observations of v and w. For example, if v and

---

[3] In general, in Dependence and Independence Logic teams do not necessarily have to be finite, but we will focus on the finite case in this example.

[4] In this work, log will always represent the base-2 logarithm.

w are two genera, the diversity of $\{v, w\}$ in a set X of observations is calculated by counting how many different pairs of observations of a specimen of v and a specimen of w there are in X.

The dependence relation arising from the relational diversity rank is the usual functional dependence relation of database theory and dependence logic. Why? By definition, $=(x, y)$ if and only if $\log(\#\text{rows}_X(xy)) = \log(\#\text{rows}_X(x))$, that is, if and only if $\#\text{rows}_X(xy) = \#\text{rows}_X(x)$. This can be the case if and only if any two $s, s' \in X$ which differ with respect to xy differ already on x alone, or, by contrapositive, if and only if any two $s, s' \in X$ which are the same with respect to x are also the same with respect to y. This is precisely the usual notion of functional dependence.

The independence relation arising from the relational diversity rank is also the independence relation of Independence Logic [5]. Indeed, $x \perp y$ if and only if

$$\log(\#\text{rows}_X(x)) + \log(\#\text{rows}_X(y)) = \log(\#\text{rows}_X(xy)),$$

or, in other words, if and only if

$$\#\text{rows}_X(xy) = \#\text{rows}_X(x) \cdot \#\text{rows}_X(y).$$

This is the case if and only if the projection of X along xy is the Cartesian product of its projections along x and along y, that is, if and only if every possible value for x and every possible value for y can occur together in X – or, more formally, if and only if for every $s, s' \in X$ there exists some $s'' \in X$ such that $s''(x) = s(x)$ and $s''(y) = s'(y)$.

It may be instructive to verify that the relational diversity notion of rank satisfies our axioms:

R1 Since $\#\text{rows}(\emptyset) = |\{()\}| = 1$ for any choice of X, where $()$ represents the empty tuple, we have that $\|\emptyset\| = 0$ as required.

R2 Since $\#\text{rows}(x) \leq \#\text{rows}(xy)$ and the logarithm is a monotone function, we have at once that $\|x\| \leq \|xy\|$; and since $\#\text{rows}(xy) \leq \#\text{rows}(x) \cdot \#\text{rows}(y)$, we have at once that $\|xy\| \leq \|x\| + \|y\|$.

R3 If $\|xy\| = \|x\|$, $\#\text{rows}(xy) = \#\text{rows}(x)$ and hence every possible value of x occurs together with only one possible value of y. But then every possible value of xz occurs together with only one possible value of y, and hence $\#\text{rows}(xyz) = \#\text{rows}(xz)$ and $\|xyz\| = \|xz\|$;

R4 If $\|xyz\| = \|x\| + \|yz\|$, it must be the case that $\#\text{rows}(xyz) = \#\text{rows}(x) \cdot \#\text{rows}(yz)$, and hence that every possible value of x occurs together with every possible value for yz. But then in particular every possible value for x occurs together with every possible value for y, and so $\#\text{rows}(xy) = \#\text{rows}(x) \cdot \#\text{rows}(y)$ and $\|xy\| = \|x\| + \|y\|$.

Note, however, that differently from the other examples given this notion of relational diversity is *not* submodular, as the following counterexample, which we owe to (Tong Wang, personal communication) shows:

Proposition 2. *Relational diversity fails to satisfy **SUBM**.*

Proof. Consider the relation

$$
\begin{array}{ccc}
v_1 & v_2 & v_3 \\
\hline
1 & 1 & 1 \\
1 & 1 & 2 \\
2 & 1 & 1 \\
1 & 2 & 1 \\
2 & 1 & 2 \\
\end{array}
$$

Then $\#\text{rows}(v_1 v_2 v_3) = 5$, $\#\text{rows}(v_2) = 2$, and $\#\text{rows}(v_1 v_2) = \#\text{rows}(v_2 v_3) = 3$. Thus, $\#\text{rows}(v_1 v_2 v_3) \cdot \#\text{rows}(v_2) = 10 > 9 = \#\text{rows}(v_1 v_2) \cdot \#\text{rows}(v_2 v_3)$, and hence $\|v_1 v_2 v_3\| + \|v_2\| > \|v_1 v_2\| + \|v_2 v_3\|$.

Example 7 (Algebraic diversity). Suppose that V is a vector space and that h maps M into V. We get a rank function by letting for $x \subseteq M$:

$\quad \|x\| =$ the dimension of the subspace generated by $\{h(a) : a \in x\}$.

Submodularity **SUBM** follows at once from the known fact that if U and V are vector subspaces,
$$
\dim(U \cup V) = \dim(U) + \dim(V) - \dim(U \cap V).
$$

In this context, it is not hard to verify that V is dependent on U if and only if every $\dim(U \cup V) = \dim(U)$, that is, if and only if every vector of V is a linear combination of vectors in U; and that, on the other hand, U and V are independent if and only if $\dim(U \cup V) = \dim(U) + \dim(V)$, that is, if and only if no nonzero vector belongs in both the subspaces generated by U and V.

Likewise, if F is a field, we get a rank function by letting for $x \subseteq M$ and letting h map M into F instead:

$\quad \|x\| =$ the transcendence degree of the subfield generated by $\{h(a) : a \in x\}$.

This gives rise to the concepts of algebraic dependence and independence.

As mentioned in the Introduction, this notion of rank defines a matroid (in fact, it was one of the original motivations for the development of Matroid Theory); and thus, by Corollary 1, it is also a diversity rank function according to our definition.

Example 8 (Entropy). Let us think of the individuals of M as *discrete random variables* $v_1, v_2, \ldots$ over some probability space and with outcomes in some finite set A.[5] Then for any $x = \{v_1 \ldots v_k\} \subseteq M$ we can define $\|x\|$ as the joint entropy $H(x)$ of $v_1 \ldots v_k$, that is, as

$$
- \sum_{(m_1 \ldots m_k) \in A^k} P(v_1 \ldots v_k = m_1 \ldots m_k) \log P(v_1 \ldots v_k = m_1 \ldots m_k).
$$

---

[5] Nothing in this example hinges on A being the same for all $v \in M$, but we will assume so for simplicity.

This definition clearly satisfies rule **R0**, since the entropy of the only possible distribution over the empty space is zero; moreover, it is not hard to convince oneself that it is monotone and submodular. In brief, this can be shown by considering the *conditional entropy* $H(y|x) = H(xy) - H(x)$.

Indeed, it can be proved (see any Information Theory textbook, for instance Theorem 2.2.1 of [2]) that the conditional entropy $H(y|x)$ is always non-negative[6], from which we have at once the left part of **R2**; and furthermore (see e.g. Theorem 2.6.5 of [2])[7] that $H(x|yz) \leq H(x|z)$, from which we obtain at once that $H(xyz) - H(yz) \leq H(xz) - H(z)$, that is, Axiom **SUBM**.

From Proposition 1, we can conclude at once that entropy is an example of a diversity rank function. y depends on x according to the entropy diversity rank if and only if $H(xy) = H(x)$, that is, if and only if the relative entropy of y given x is 0, or in other words if the value of y is completely determined by the value of x; and x and y are independent according to this rank if and only if they are independent tuples of random variables, that is, $P(x = m, y = m') = P(x = m)P(y = m')$ for all possible choices of values a and b for x and y.

4   From Diversity to Dependence

We have already identified *dependence* $=(x, y)$ of y on x with xy not contributing any diversity to x, in the sense that $\|xy\| = \|x\|$.

It is easy to verify that any notion of dependence thus defined satisfies the following axioms:

Proposition 3. *The following always hold:*

1. **Reflexivity:** $=(xy, x)$.
2. **Augmentation:** $=(x, y)$ *implies* $=(xz, yz)$.
3. **Transitivity:** *If* $=(x, y)$ *and* $=(y, z)$, *then* $=(x, z)$.

Proof

Reflexivity: Clearly $\|xyx\| = \|xy\|$. Therefore, $=(xy, x)$.

Augmentation: Suppose that $\|xy\| = \|x\|$. Then, by **R3**, $\|xyz\| = \|xz\|$; and therefore, $\|xzyz\| = \|xz\|$, or, in other words, $=(xz, yz)$.

Transitivity: Suppose that $\|x\| = \|xy\|$ and $\|y\| = \|yz\|$. Again, by **R3**, from $\|x\| = \|xy\|$ we get that $\|xz\| = \|xyz\|$; and similarly, from $\|y\| = \|yz\|$ we get that $\|xy\| = \|xyz\|$. By the transitivity of equality, we can conclude that $\|xz\| = \|xy\|$. But we have as an hypothesis that $\|xy\| = \|x\|$, and therefore we can conclude that $\|x\| = \|xz\|$, or, in other words, that $=(x, z)$.

---

[6] More precisely, this theorem shows that $H(xy) - H(x) = -\sum_m P(x = m) \sum_{m'} P(y = m'|x = m) \log P(y = m'|x = m)$, and the right hand side is straightforwardly seen to be non-negative.

[7] Strictly speaking, this theorem states that $H(x) - H(x|y) \geq 0$, but if we consider the above inequality with respect to distributions already conditioned on z the result follows at once.

We can use the above rules as axioms of a proof system for inferring the consequences of a set of dependence statements:

Definition 4. *Let Σ be a finite set of relations of the form $=(z, w)$ for $z, w \subseteq M$, and let also $x, y \subseteq M$. Then $\Sigma \vdash =(x, y)$ if it is possible to derive $=(x, y)$ from Σ through applications of the rules of Reflexivity, Augmentation and Transitivity.*

According to these axioms, a dependency notion is entirely defined even if we only consider singletons on the right-hand side of it:

Corollary 2. *Let Σ be a finite set of relations of the form $=(z, w)$ for $z, w \subseteq M$, and let also $x, y \subseteq M$. Then $\Sigma \vdash =(x, y)$ if and only if $\Sigma \vdash =(x, \{m\})$ for all $m \in y$.*

Proof. By Reflexivity, if $m \in y$ then it is always the case that $\Sigma \vdash =(y, \{m\})$. If $\Sigma \vdash =(x, y)$, by Transitivity it is thus the case that $\Sigma \vdash =(x, \{m\})$ for all such m.

Conversely, suppose that $\Sigma \vdash =(x, \{m\})$ for all $m \in y$. Then, in order to reach our conclusion that $\Sigma \vdash =(x, y)$, it suffices to verify that whenever $\Sigma \vdash =(x, y_1)$ and $\Sigma \vdash =(x, y_2)$ it is also the case that $\Sigma \vdash =(x, y_1 y_2)$.

This is easily shown: if $\Sigma \vdash =(x, y_1)$, by Augmentation we have that $\Sigma \vdash =(x, xy_1)$ (remember that in our notation $xx = x \cup x = x$), and if $\Sigma \vdash =(x, y_2)$ again by Augmentation we have that $\Sigma \vdash =(xy_1, y_1 y_2)$, and an application of Transitivity gives us $\Sigma \vdash =(x, y_1 y_2)$. The conclusion follows at once.

The following is essentially proved in [1], albeit in the special case of relations and functional dependences:

Theorem 1. *Let M be a set and Σ is a finite set of relations of the form $=(z, w)$ and let also $x, y \subseteq M$. The following are equivalent:*

1. *$=(x, y)$ holds under any rank-function on M, under which Σ holds.*
2. *$=(x, y)$ holds under any rank-function $\mathcal{P}(M) \to \{0, 1\}$, under which Σ holds.*
3. *$=(x, y)$ holds under any relational diversity rank-function, under which Σ holds.*
4. *$=(x, y)$ follows from Σ by the rules of Proposition 3.*

Proof. Trivially, (1) implies (2) and (3), and (4) implies (1). We demonstrate that (2) and (3) both separately imply (4). Let us first assume (2). Suppose $=(x, y)$ does not follow from Σ by the rules of Proposition 3. Let V be the set of $m \in M$ such that $=(x, \{m\})$ follows from Σ by the above rules. By Corollary 2, for all $w \subseteq M$ we have that $\Sigma \vdash =(x, w)$ if and only if $w \subseteq V$.

Let W be all the remaining elements of M. Note that y is not a subset of V, since $\Sigma \not\vdash =(x, y)$, and therefore $W \neq \emptyset$. Let us define a rank-function on M by letting for $m \in M$:

$$\|\{m\}\| = \begin{cases} 0, & \text{if } m \in V \\ 1, & \text{if } m \in W. \end{cases}$$

and otherwise

$$\|\{m_1, \ldots, m_n\}\| = \max\{\|\{m_1\}\|, \ldots, \|\{m_n\}\|\}.$$

Note that $\|xy\| = 1$, while $\|x\| = 0$. Thus the relation $=(x, y)$ does not hold under this rank-function. Suppose then $=(z, w) \in \Sigma$. If $z \subseteq V$, this means that $\Sigma \vdash =(x, z)$; and then, by Transitivity, $\Sigma \vdash =(x, w)$ and so $w \subseteq V$ as well. So $\|zw\| = \|z\| = 0$ and $=(z, w)$ holds. On the other hand, if $z \nsubseteq V$, then $\|z\| = 1$. So $\|zw\| = \|z\| = 1$, whence $=(z, w)$ holds again.

Let us then assume (3). We proceed as above. Let X consist of the two functions $\{s_1, s_2\}$, where $s_1(m) = 0$ for all $m \in M$, $s_2(m) = 0$ for $m \in V$ and $s_2(m) = 1$ for $m \in W$. We get the same rank as above, so we are done.

Since – as we saw – functional dependence is exactly the dependency notion generated by the relational dependency rank function, we get that

Corollary 3 (Armstrong). *A functional dependence follows semantically, in all databases, from a given set of functional dependencies if and only if it follows by the rules of Proposition 3.*

The proof of Theorem 1 shows that Armstrong's completeness theorem for functional dependence is actually a more general completeness theorem of dependence relations arising from diversity ranks.

5   From Diversity to Independence

We shall now study the properties of the notions of independence arising from our diversity ranks. Let us recall that, according to our definition, x and y are independent $(x \perp y)$ if and only if $\|xy\| = \|x\| + \|y\|$.

Proposition 4. *The following always hold:*

1. **Empty Set:** $x \perp \emptyset$.
2. **Symmetry:** *If* $x \perp y$, *then* $y \perp x$.
3. **Decomposition:** *If* $x \perp yz$, *then* $x \perp y$.
4. **Mixing:** *If* $x \perp y$ *and* $xy \perp z$, *then* $x \perp yz$.
5. **Constancy:** *If* $z \perp z$ *then* $z \perp x$.[8]

Proof. Let us prove that these axioms follow from our notion of independence:
Empty Set: Since $\|\emptyset\| = 0$, $\|x\| + \|\emptyset\| = \|x\| + 0 = \|x\| = \|x\emptyset\|$.
Symmetry: Follows at once from the commutativity of sum and union. If $\|x\| + \|y\| = \|xy\|$ then $\|y\| + \|x\| = \|xy\| = \|yx\|$.
Decomposition: Suppose that $x \perp yz$, that is, $\|x\| + \|yz\| = \|xyz\|$.
    By **R4**, we then have that $\|xy\| = \|x\| + \|y\|$ and $x \perp y$.
Mixing: Suppose that $\|xy\| = \|x\| + \|y\|$ and $\|xyz\| = \|xy\| + \|z\|$. We need to prove that $\|xyz\| = \|x\| + \|yz\|$.
    Begin by observing that $\|x\| + \|y\| + \|z\| = \|xy\| + \|z\| = \|xyz\|$. But by **R2** $\|yz\| \leq \|y\| + \|z\|$, and therefore $\|x\| + \|yz\| \leq \|x\| + \|y\| + \|z\| = \|xyz\|$.
    On the other hand, again by **R2**, $\|xyz\| \leq \|x\| + \|yz\|$, and so in conclusion $\|xyz\| = \|x\| + \|yz\|$, as required.

---

[8] If one is uninterested in independence statements $x \perp y$ in which x and y overlap, this axiom can be removed. Our proof of Theorem 2 then reduces essentially to the proof in [4].

Constancy: If $z \perp z$ then $\|z\| = \|z\| + \|z\|$, and hence $\|z\| = 0$. But then by **R2** $\|x\| \leq \|xz\| \leq \|x\| + \|z\| = \|x\|$, and thus $\|xz\| = \|x\| + \|z\|$ and $z \perp x$.

Definition 5. *Let Σ be a finite set of relations of the form $z \perp w$ for $z, w \subseteq M$, and let also $x, y \subseteq M$. Then $\Sigma \vdash x \perp y$ if it is possible to derive $x \perp y$ from Σ through applications of the rules Empty Set, Symmetry, Decomposition, Mixing and Constancy.*

The following derived rule will be useful:

Corollary 4 (Constancy Augmentation). *Let Σ be a finite set of relations of the form $z \perp w$ for $z, w \subseteq M$, and suppose that $\Sigma \vdash u \perp u$ and $\Sigma \vdash x \perp y$. Then $\Sigma \vdash xu \perp y$*

Proof. By Constancy, if $\Sigma \vdash u \perp u$ then $\Sigma \vdash u \perp xy$, and so by Symmetry $\Sigma \vdash xy \perp u$. If furthermore $\Sigma \vdash x \perp y$, by Symmetry $\Sigma \vdash y \perp x$; and thus, by Mixing, $\Sigma \vdash y \perp xu$, and by Symmetry once more $\Sigma \vdash xu \perp y$ as required.

The following is a variation of the proof in [4], with the added complication that we do not require the left- and right-hand sides of an independence statement to be disjoint and generalized from probabilistic independence to our more general setting. First of all, we will show that that the above axioms are complete for statements of the form $\{m\} \perp \{m\}$:

Lemma 1 (Completeness of Independence Axioms wrt Constancy Statements). *Let M be a finite set and let $m \in M$. Then the following conditions are equivalent:*

1. *$\{m\} \perp \{m\}$ holds under any rank-function on M under which Σ holds.*
2. *$\{m\} \perp \{m\}$ holds under any relational diversity rank-function under which Σ holds.*
3. *$\{m\} \perp \{m\}$ follows from Σ by the rules of Proposition 4.*

Proof. Trivially (1) implies (2) and (3) implies (1). Let us verify that (2) implies (3). Suppose that $\{m\} \perp \{m\}$ does not follow from Σ by the above rules. Then let V contain all $m' \in M$ such that $\Sigma \vdash \{m'\} \perp \{m'\}$ and let S be a team with domain M over $\{0, 1\}$ (that is, a set of functions from M to $\{0, 1\}$) that contains all $s : M \to \{0, 1\}$ such that $s(m') = 0$ for all $m' \in V$.

Now let $\| \cdot \| = \log(\#\text{rows}_S(\cdot))$ be the relational diversity rank-function induced by S: as already discussed, such a rank-function satisfies an independence statement $z \perp w$ if and only if any possible values of z and w in S may occur together, or, in other words, if and only if for all $s, s' \in S$ there exists some $s'' \in S$ that agrees with s on z and with s' on w. In particular, for the S given, this means that $z \perp w$ is satisfied if and only if $z \cap w \subseteq V$. Thus, S does not satisfy $\{m\} \perp \{m\}$, since by assumption $m \notin V$.

On the other hand, S satisfies all statements of Σ. Indeed, consider any $z \perp w \in \Sigma$. By Decomposition and Symmetry, every element of $m_0 \in M$ which is in both z and w is such that $\Sigma \vdash \{m_0\} \perp \{m_0\}$, that is, such that $m_0 \in V$. Thus, $z \cap w \subseteq V$ and therefore $z \perp w$ is satisfied by (the relational diversity rank-function corresponding to) S, as required.

In conclusion, from the assumption that $\{m\} \perp \{m\}$ does not follow from Σ according to the rules we were able to find a relational diversity rank-function that satisfies Σ but not $\{m\} \perp \{m\}$. Thus (2) implies (3), and this concludes the proof.

Now we can generalize the completeness result to arbirary independence statements:

Theorem 2 (Completeness of the Independence Axioms). *Let M be a finite set. Then the following conditions are equivalent:*

1. *$x \perp y$ holds under any rank-function on M under which Σ holds.*
2. *$x \perp y$ holds under any relational diversity rank-function under which Σ holds.*
3. *$x \perp y$ follows from Σ by the rules of Proposition 4.*

Proof. We adapt the proof of [4] into our framework. Trivially (1) implies (2) and (3) implies (1). So we prove only that (2) implies (3). Suppose $x \perp y$ follows semantically from Σ but does not follow by the above rules. Without loss of generality, we can assume that Σ is closed under the rules. We may assume that x and y are minimal, that is, if $x' \subseteq x$ and $y' \subseteq y$ and at least one containment is proper, then if $x' \perp y'$ follows from Σ semantically (which it does, because Decomposition and Symmetry are semantically valid) then $x' \perp y' \in \Sigma$.

Suppose $x = \{a_1, \ldots, a_l\}$ and $y = \{b_1, \ldots, b_{l'}\}$. Without loss of generality, $l \geq 1$ and $l' \geq 1$, because otherwise $x \perp y$ would be in Σ via the Empty Set Rule.

Now let $z = \{c_1, \ldots, c_q\}$ enumerate all the $c_i \in M$ for which $\{c_i\} \perp \{c_i\} \in \Sigma$. By the minimality of $x \perp y$, it follows at once that z does not intersect x or y: indeed, if for instance x were of the form $x'\{c\}$ for some $c \in z$ then we would have by the minimality of $x \perp y$ that $x' \perp y \in \Sigma$, and thus – by the Constancy Augmentation Rule and the fact that $\{c\} \perp \{c\} \in \Sigma$ – we could conclude that $x \perp y$ is also in Σ, contrarily to our premises. Since we already saw that our axioms are complete with respect to constancy expressions $\{c\} \perp \{c\}$, this also implies that x and y are disjoint. Indeed, suppose that $m \in x \cap y$. Then it would follow semantically from Σ that $\{m\} \perp \{m\}$ (because it follows semantically from Σ that $x \perp y$ and because Symmetry and Decomposition are semantically valid), which would imply that $m \in z$, which would contradict the already verified fact that x and y cannot intersect with z.

Then let $w = \{d_1, \ldots, d_k\}$ be the set of the remaining elements of M.

We construct a team S with domain $xyzw$ over $\{0, 1\}$ (that is, a set of functions from $xyzw$ to $\{0, 1\}$) as follows: we take to S every $s : xyzw \to \{0, 1\}$ which satisfies

$$s(a_1) = \text{the number of ones in } s[\{a_2, \ldots, a_l, b_1, \ldots, b_{l'}\}] \bmod 2$$

and such that $s(c_i) = 0$ for all $c_i \in z$. We use the relational diversity rank

$$\|\{v_1, \ldots, v_n\}\| = \log |\{(s(v_1), \ldots, s(v_n)) : s \in S\}|,$$

according to which, as shown previously, S satisfies $u \perp u'$ if and only if for every $s, s' \in S$ there exists some $s'' \in S$ that agrees with s on u and with s' on u' It is easy to see that $x = (a_1 \ldots a_l)$ and $y = (b_1 \ldots b_{l'})$ take all 2^l (respectively $2^{l'}$) possible values in S, and hence that $\|x\| = l$ and $\|y\| = l'$.

Claim 1: $x \perp y$ is not true according to [the relational diversity rank function induced by] S. Consider the following two assignments in S:

	a_1	other a_i	b_1	other b_i	other
s	1	0	1	0	0
s'	0	0	0	0	0

If s'' agrees with s on $x = \{a_1 \ldots a_l\}$ and with s' on $\{b_1 \ldots b_{l'}\}$, then $s'' \notin S$. Thus, it is not true that $x \perp y$ according to S.

Claim 2: [The relational diversity rank function induced by] S satisfies all the independence atoms in Σ. Suppose $u' \perp u'' \in \Sigma$.

If $u' = \emptyset$, there is nothing to prove, because every diversity rank satisfies trivially $\emptyset \perp u''$ for any choice of u''. Likewise, if $u'' = \emptyset$ then all ranks (and in particular the one induced by S) satisfy $u' \perp u''$.

Furthermore, without loss of generality, we can assume that $u' \cap z = u'' \cap z = \emptyset$. Indeed, suppose for instance that u' were of the form $u'_0\{z_i\}$ for some $z_i \in z$: then z_i takes only constant value 0 in S, and hence S satisfies $u' \perp u''$ if and only if S satisfies $u'_0 \perp v''$.

Now suppose that $u' \cap u'' = t \neq \emptyset$. Then, since $u' \perp u'' \in \Sigma$, by Decomposition and Symmetry we have that $\{e\} \perp \{e\} \in \Sigma$ for all $e \in t$, and thus that $t \subseteq z$. But we already said that we can assume u' and u'' do not intersect with z, and so we can also assume that $u' \cap u'' = \emptyset$.

Thus, without loss of generality, we can assume that $u' \neq \emptyset$, $u'' \neq \emptyset$, $u' \cap z = u'' \cap z = \emptyset$ and $u' \cap u'' = \emptyset$. We need to prove that S satisfies $u' \perp u''$.

If $u'u''$ does not cover all of xy then S satisfies $u' \perp u''$, because we can fix parity on the variable in xy which does not occur in u' nor in u''. So let us assume $u'u''$ covers all of xy. Thus $u' \supseteq x'y'$ and $u'' \supseteq x''y''$, where $x'x'' = x$ and $y'y'' = y$, and by Decomposition and Symmetry $x'y' \perp x''y'' \in \Sigma$.

But by the minimality of $x \perp y$ among the statements that follow semantically from Σ but are not in it, we then have that $x' \perp y'$ is in Σ. Thus, using the Mixing axiom, from $x'y' \perp x''y'' \in \Sigma$ we can derive that $x' \perp x''y'y''$ is in Σ as well. Once more, by the minimality of $x \perp y$ we have that $x'' \perp y'y'' \in \Sigma$, and so - by Mixing and Symmetry - that $x'x'' \perp y'y''$ is in Σ. But this contradicts our assumption that $x \perp y \notin \Sigma$.

5.1 Dependence/Independence Axioms

At this point, it would be natural to ask whether there are any rules that govern the interaction between dependence and independence in our framework. Here we will consider only two simple such axioms:

Constancy Equivalence: $x \perp x$ if and only if $=(\emptyset, x)$;
Propagation: If $x \perp y$ and $=(y, z)$ then $x \perp yz$.

Both of these can be shown to follow easily from our notion of rank.

Constancy Equivalence: Suppose that $x \perp x$: then, by definition, $\|x\| + \|x\| = \|xx\|$. But on the other hand, $xx = x$: and therefore, $\|x\| = 0$ and $\|x\| = \|\emptyset x\| = \|\emptyset\| = 0$. Conversely, suppose that $=(\emptyset, x)$: then $\|x\| = \|\emptyset x\| = \|\emptyset\| = 0$, and therefore $\|x\| + \|x\| = 0 = \|xx\|$.

Propagation: Suppose that $\|x\| + \|y\| = \|xy\|$ and that $\|yz\| = \|y\|$. From the second hypothesis, by **R3**, we can show that $\|xy\| = \|xyz\|$; and therefore, in the first hypothesis we can replace $\|y\|$ with $\|yz\|$ and $\|xy\|$ with $\|xyz\|$, thus obtaining

$$\|x\| + \|yz\| = \|xyz\|.$$

Therefore, $x \perp yz$, as required.

6 Conclusions

In this work, we showed how many distinct notions of dependence and independence, having their origin in different branches of mathematics, may be treated as instances of the same framework, which can be seen as a generalization of matroid theory that allows for non-integer ranks and that weakens the submodularity condition. In this framework, y is said to be dependent on x if adding it to x does not increase at all the amount of diversity, while y is independent from x if adding it to x increases *maximally* the amount of diversity.

Despite this considerable amount of generality, this framework is nonetheless powerful enough to prove non-trivial results - including, in particular, completeness theorems for the corresponding dependence and independence notions which adapt to the entire setting the completeness theorems by Armstrong and by Geiger-Paz-Pearl for functional dependence and for probabilistic independence respectively.

The natural next step would consist in investigating further the properties of this formalism, in particular with respect to the interaction between independence and dependence statements. Combinatorial properties of this system would also be worth investigating, as would be the study of possible operations that *combine* different diversity rank functions. This could also contribute to the logical study of notions of dependence and independence in the context of Team Semantics, in particular providing a unifying approach for the different variants (e.g. probabilistic, modal, propositional, ...) of it.

Acknowledgments. We thank the reviewers for a number of helpful comments and suggestions. The research of the second author was partially supported by grant 322795 of the Academy of Finland, and a grant of the Faculty of Science of the University of Helsinki.

References

1. Armstrong, W.W.: Dependency structures of data base relationships. Inf. Process. **74**, 580–583 (1974)
2. Cover, T.M., Thomas, J.A.: Entropy, relative entropy and mutual information. Elem. Inf. Theory **2**, 1–55 (1991)
3. Durand, A., Hannula, M., Kontinen, J., Meier, A., Virtema, J.: Probabilistic team semantics. In: Ferrarotti, F., Woltran, S. (eds.) FoIKS 2018. LNCS, vol. 10833, pp. 186–206. Springer, Cham (2018). https://doi.org/10.1007/978-3-319-90050-6_11
4. Geiger, D., Paz, A., Pearl, J.: Axioms and algorithms for inferences involving probabilistic independence. Inf. Comput. **91**(1), 128–141 (1991)
5. Grädel, E., Väänänen, J.: Dependence and independence. Stud. Logica. **101**(2), 399–410 (2013)

6. Hannula, M., Hirvonen, Å., Kontinen, J., Kulikov, V., Virtema, J.: Facets of distribution identities in probabilistic team semantics. In: Calimeri, F., Leone, N., Manna, M. (eds.) JELIA
   2019. LNCS (LNAI), vol. 11468, pp. 304–320. Springer, Cham (2019). https://doi.org/10.
   1007/978-3-030-19570-0_20
7. Hella, L., Luosto, K., Sano, K., Virtema, J.: The expressive power of modal dependence
   logic. In: Advances in Modal Logic, vol. 10, pp. 294–312. Coll. Publ., London (2014)
8. Hyttinen, T., Paolini, G., Väänänen, J.: A logic for arguing about probabilities in measure
   teams. Arch. Math. Logic **56**(5–6), 475–489 (2017)
9. Krebs, A., Meier, A., Virtema, J., Zimmermann, M.: Team semantics for the specification and
   verification of hyperproperties. In: 43rd International Symposium on Mathematical Foundations of Computer Science, LIPIcs. Leibniz International Proceedings of the Information,
   vol. 117, pages Art. No. 10, 16. Schloss Dagstuhl. Leibniz-Zent. Inform., Wadern (2018)
10. Mann, A.L., Sandu, G., Sevenster, M.: Independence-Friendly Logic. London Mathematical
    Society Lecture Note Series, vol. 386. Cambridge University Press, Cambridge (2011). A
    game-theoretic approach
11. Oxley, J.G.: Matroid Theory, vol. 3. Oxford University Press, New York (2006)
12. Väänänen, J.: Dependence Logic. London Mathematical Society Student Texts, vol. 70.
    Cambridge University Press, Cambridge (2007)
13. Väänänen, J.: Modal Dependence Logic. New Perspectives on Games and Interaction. Texts
    Log. Games, vol. 4, pp. 237–254. Amsterdam Univ. Press, Amsterdam (2008)
14. van der Waerden, B.L.: Moderne Algebra. Springer, Berlin (1940)
15. Whitney, H.: On the abstract properties of linear dependence. Amer. J. Math. **57**(3), 509–533
    (1935)
16. Yang, F., Väänänen, J.: Propositional logics of dependence. Ann. Pure Appl. Logic **167**(7),
    557–589 (2016)

Towards Probabilistic Reasoning in Type Theory - The Intersection Type Case

Silvia Ghilezan[1,2], Jelena Ivetić[1], Simona Kašterović[1(✉)], Zoran Ognjanović[2], and Nenad Savić[3]

[1] Faculty of Technical Sciences, University of Novi Sad, Trg Dositeja Obradovića 6, 21000 Novi Sad, Serbia
{gsilvia,jelenaivetic,simona.k}@uns.ac.rs
[2] Mathematical Institute SANU, Kneza Mihaila 36, 11000 Belgrade, Serbia
zorano@mi.sanu.ac.rs
[3] Institute of Computer Science, University of Bern, Neubrueckstrasse 10, 3012 Bern, Switzerland
savic@inf.unibe.ch

Abstract. The development of different probabilistic models of uncertainty has been inspired by the rapid progress in various fields, e.g. in AI, probabilistic programming, etc. Lambda calculus is a universal model of computation suitable to express programming languages concepts. Hence, different methods for probabilistic reasoning in lambda calculus have been investigated. In this paper, we develop a formal model for probabilistic reasoning about lambda terms with intersection types, which is a combination of lambda calculus and probabilistic logic. The language of lambda calculus with intersection types is endowed with a probabilistic operator. We propose a semantics based on the possible world approach. An infinitary axiomatization is given for this system and it is proved to be sound with respect to the proposed semantics.

Keywords: Probabilistic reasoning · Lambda calculus · Intersection types · Kripke-style semantics · Soundness

1 Introduction

Over the last decades the interest in probabilistic programming has been rapidly growing due to the role reasoning about uncertain knowledge has in computer science and artificial intelligence. In order to formalize uncertain reasoning, different approaches have been investigated.

One of the approaches, completely different from ours, consists in taking a probabilistic choice as primitive and obtaining probabilistic computation, which has proved to be extremely applicable and useful in various areas, such as

This work was supported by the Serbian Ministry of Education and Science through projects ON174026, III 044006 and by the Swiss National Science Foundation grant 200021_165549.

© Springer Nature Switzerland AG 2020
A. Herzig and J. Kontinen (Eds.): FoIKS 2020, LNCS 12012, pp. 122–139, 2020.
https://doi.org/10.1007/978-3-030-39951-1_8

robotics [27], machine learning [26], and natural language processing [22]. This approach was used in [8–10,19].

On the other hand, our approach consists in extending the language of typed lambda calculus with probabilistic operators. We are interested in probabilistic reasoning in type theory. More precisely, we want to define a logical framework where we can express the probability that a term (program) has a certain type (behavior).

As many others mathematical stories, the story about probabilistic logic begins in ancient times. In order to argue about legal, medical or political questions, sophists Corax and Tisias in Plato's Phaedrus and Aristotle's Rhetoric used the notion of probability [21]. Many mathematicians were interested in the strong connection between mathematical logic and probability theory, as early as Leibnitz, Lambert and Boole. Until the 1960s the ideas about probability logic did not have a lot in common with proof-theoretical and model-theoretical results of Gödel, Alfred Tarski, Leon Henkin, Abraham Robinson, Saul Kripke and others. The papers by Gaifman [11], Keisler [20], Hamblin [15] and Hailperin [14] changed this situation. In 1980s the development of application of reasoning about uncertain knowledge in economics, artificial intelligence, computer science and philosophy resulted in growing interest in probability logics and numerous publications. Nils Nilsson in his seminal work [23] started the modern development of probability logic. Following the approaches from [18,24,25] we extend the lambda calculus with intersection types and give a sound infinitary axiomatization.

Alonzo Church proposed the lambda calculus in the early 1930s, as a simple formal system capable of expressing all effectively computable functions, and equivalent to Turing machines. The first formalism Church proposed was untyped. Later Church introduced the system with simple types in [6]. Intersection types [5,7] were introduced in the lambda calculus as an extension of the simple types in order to overcome the limitations of the simple (functional) types and to completely characterize the termination of reduction (strong normalization) in the lambda calculus. For example, term $\lambda x.xx$ is not typable in simply typed lambda calculus, but it is typable in lambda calculus with intersection types. In lambda calculus with intersection types, we can infer the following typing statement $\vdash \lambda x.xx : (\sigma \cap (\sigma \rightarrow \tau)) \rightarrow \tau$. Moreover, intersection types gave rise to filter models which ensure completeness of the type assignment in [3].

Our motivation for the formalization of the system which captures probabilistic reasoning over programs (typed lambda terms) was increased application of reasoning about uncertain knowledge in computer science and artificial intelligence. In order to overcome some limitations of probabilistic logic over simply typed lambda calculus, we consider lambda calculus with intersection types.

In the paper, we develop a formal model for probabilistic reasoning about lambda terms with intersection types, which is a combination of lambda calculus and probabilistic logic. We extend the language of lambda calculus with intersection types with a probabilistic operator $P_{\geqslant s}$ and obtain formulas of the form $P_{\geqslant s}M : \sigma$, which have the following meaning:

"A probability that the term M has the type σ is equal to or greater than s."

A semantics we propose is based on the possible world approach. We gave an infinitary axiomatization for this system and as a main result we prove it to be sound with respect to the proposed semantics. Our future goal is to prove that this axiomatization is strongly complete, that is that every consistent set is satisfiable.

The paper is organized as follows: in Sect. 2 we recall some basic notions of the probabilistic logic LPP_2, the lambda calculus with intersection types and filter models. Then, in Sect. 3 we introduce the syntax of the logic $P\Lambda^\cap$ and propose the semantics. We give an axiomatization of logic $P\Lambda^\cap$ in Sect. 4 and prove it to be sound. In Sect. 5 we introduce the probabilistic logic with a finitary axiomatization. Finally, Sect. 6 contains some concluding remarks.

2   Preliminaries

In this section, we recall some basic notions of the probabilistic logic LPP_2 [25], lambda calculus [1], intersection types [5,7], lambda models [1,17] and revisit the soundness and completeness result for the intersection type assignment proved in [3].

2.1   Probabilistic Logic

Let $S = [0,1] \cap \mathbb{Q}$ be the set of all rational numbers from the real unit interval $[0,1]$. The alphabet of the logic LPP_2 consists of

- primitive propositions from the countable set $P = \{p, q, r, \ldots, p_1, \ldots\}$,
- classical propositional connectives $\neg$, and $\wedge$,
- a list of probability operators $P_{\geqslant}s$ for every $s \in S$.

The set For_C of all classical propositional formulas over the set P is the smallest set containing all primitive propositions which is closed under Boolean connectives. Classical propositional formulas are generated by the following grammar:

$$\mathsf{For}_C \quad \alpha ::= p \mid \alpha \wedge \alpha \mid \neg\alpha \tag{1}$$

Other propositional Boolean connectives $\Rightarrow$, $\vee$, $\Leftrightarrow$ are defined as usual. The formulas from the set For_C will be denoted by $\alpha, \beta, \ldots$, indexed if necessary. Probabilistic formulas are defined as follows:

Definition 1. *If $\alpha \in \mathsf{For}_C$ and $s \in S$, then $P_{\geqslant s}\alpha$ is a basic probabilistic formula. The intuitive meaning of this formula is "the probability that α is true is greater than or equal to s". The set For_P of all probabilistic formulas is the smallest set containing all basic probabilistic formulas which is closed under Boolean connectives. Probabilistic formulas are generated by the following grammar:*

$$\mathsf{For}_P \quad A ::= P_{\geqslant s}\alpha \mid A \wedge A \mid \neg A \tag{2}$$

The set of all LPP_2-formulas is $\mathsf{For}_{LPP_2} = \mathsf{For}_C \cup \mathsf{For}_P$.

The formulas from the sets For_P and For_{LPP_2} will be denoted by $A, B, \ldots$, and $\varphi, \psi, \ldots$, respectively, and indexed if necessary.

We use the following abbreviations to introduce other probabilistic operators:

$$P_{<s}\alpha \text{ stands for } \neg P_{\geqslant s}\alpha,$$
$$P_{\leqslant s}\alpha \text{ stands for } P_{\geqslant 1-s}\neg\alpha,$$
$$P_{>s}\alpha \text{ stands for } \neg P_{\leqslant s}\alpha,$$
$$P_{=s}\alpha \text{ stands for } P_{\geqslant s}\alpha \wedge \neg P_{>s}\alpha.$$

Note that mixing of pure propositional formulas and probabilistic formulas and nested probability operators are not allowed in LPP_2. Nesting of probability operators allowed in some other probabilistic logics [25] is beyond the scope of this paper.

The semantics for LPP_2 is based on the possible world approach.

Definition 2. *An LPP_2-model is a structure $\mathcal{M} = (W, H, \mu, v)$ where:*

- *W is a nonempty set of objects called worlds,*
- *H is an algebra of subsets of W,*
- *μ is a finitely additive probability measure, $\mu : H \to [0, 1]$, and*
- *$v : W \times P \to \{true, false\}$ provides a two-valued valuation of primitive propositions, for each world $w \in W$.*

For each world $w \in W$, we extend the truth valuation $v(w, \cdot)$ to all classical propositional formulas from For_C in the usual way.

Let $\mathcal{M}$ be an LPP_2-structure. By $[\alpha]_{\mathcal{M}}$, we will denote the set of all worlds in which α is true, i.e. $\{w \in W \mid w \models \alpha\}$. If $\mathcal{M}$ is clear from the context, we omit the subscript $\mathcal{M}$ from $[\alpha]_{\mathcal{M}}$ and write $[\alpha]$.

Definition 3. *An LPP_2-model $\mathcal{M} = (W, H, \mu, v)$ is measurable if $[\alpha]_{\mathcal{M}} \in H$ for every formula $\alpha \in \mathsf{For}_C$. The class of all measurable LPP_2-models is denoted by $LPP_{2,\text{Meas}}$.*

Definition 4 (Satisfiability relation). *The satisfiability relation $\models \subseteq LPP_{2,\text{Meas}} \times \mathsf{For}_{LPP_2}$ is defined in the following way:*

- *$\mathcal{M} \models \alpha$ if and only if for every $w \in W$, $v(w, \alpha) = true$;*
- *$\mathcal{M} \models P_{\geqslant s}\alpha$ if and only if $\mu([\alpha]) \geqslant s$;*
- *$\mathcal{M} \models \neg A$ if and only if it is not the case that $\mathcal{M} \models A$;*
- *$\mathcal{M} \models A \wedge B$ if and only if $\mathcal{M} \models A$ and $\mathcal{M} \models B$.*

Definition 5. *Let $A \in \mathsf{For}_{LPP_2}$ be a formula and $F \subseteq \mathsf{For}_{LPP_2}$*

- *A is satisfiable if there is a $LPP_{2,\text{Meas}}$-model $\mathcal{M}$ such that $\mathcal{M} \models A$;*
- *A set of formulas F is satisfiable if there is a $LPP_{2,\text{Meas}}$-model $\mathcal{M}$ such that $\mathcal{M} \models A$ for every $A \in F$;*
- *A is valid if for every $LPP_{2,\text{Meas}}$-model $\mathcal{M}$, $\mathcal{M} \models A$;*
- *A is a semantical consequence of a set of formulas F, denoted by $F \models A$, if it holds that A is satisfied in a model $\mathcal{M}$ (denoted by $\mathcal{M} \models A$) whenever F is satisfied in that model (denoted by $\mathcal{M} \models F$)*

The axiom system Ax_{LPP_2} for the logic LPP_2 includes the following set of axiom schemes:

(1) all instances of the classical propositional tautologies (atoms are any LPP_2-formulas),
(2) $P_{\geqslant 0}\alpha$,
(3) $P_{\leqslant r}\alpha \Rightarrow P_{<s}\alpha$, $s > r$,
(4) $P_{<s}\alpha \Rightarrow P_{\leqslant s}\alpha$,
(5) $(P_{\geqslant r}\alpha \wedge P_{\geqslant s}\beta \wedge P_{\geqslant 1}(\neg\alpha \vee \neg\beta)) \Rightarrow P_{\geqslant \min\{1, r+s\}}(\alpha \vee \beta)$,
(6) $(P_{\leqslant r}\alpha \wedge P_{<s}\beta) \Rightarrow P_{<r+s}(\alpha \vee \beta)$, $r + s \leqslant 1$.

and inference rules

$$(1)\ \frac{\varphi \quad \varphi \Rightarrow \psi}{\psi} \quad (2)\ \frac{\alpha}{P_{\geqslant 1}\alpha} \quad (3)\ \frac{\{A \Rightarrow P_{\geqslant s - \frac{1}{k}}\alpha \mid k \geqslant \frac{1}{s}\}}{A \Rightarrow P_{\geqslant s}\alpha}$$

Classical propositional logic is a sublogic of LPP_2 because of Axiom scheme (1), and Rule (1). Thus, all instances of classical propositional tautologies are $LPP_{2;\text{Meas}}$-valid. By Axiom (2) sets of worlds that satisfy classical formulas have non-negative measures. As a consequence (by using $\neg\alpha$ instead of α we obtain $P_{\leqslant 1}\alpha$), the upper bound of probabilities is 1. Axioms (3) and (4) imply monotonicity of measures, while Axiom (5) and Axiom (6) guarantee finite additivity of measures.

Rule (1) is the classical Modus Ponens. Rule (2) corresponds to necessitation in modal logics. It can be applied to the classical propositional formulas only, since iterations of probability operators are not allowed in LPP_2. Note that Rule (3) is the only infinitary rule in the system. It has a countable set of assumptions and one conclusion. Rule (3) corresponds to Archimedean axiom for real numbers. Intuitively it says that if the probability is arbitrary close to s, then it is at least s.

Definition 6. *A formula φ is deducible from a set T of formulas (denoted by $T \vdash_{LPP_2} \varphi$) if there is a sequence $\varphi_0, \ldots, \varphi_n$ (n is a finite or countable ordinal) of For_{LPP_2}-formulas, such that*

– $\varphi_n = \varphi$, and
– every φ_i, $i \leqslant n$, is an axiom-instance, or $\varphi_i \in T$, or φ_i is derived by an inference rule applied to some previous members of the sequence.

A proof for φ from T is the corresponding sequence of formulas.

Using soundness of classical propositional logic and the properties of probability measures, soundness of LPP_2 logic was proved in [25].

Theorem 1 (Soundness). *The axiomatic system Ax_{LPP_2} is sound with respect to the class of $LPP_{2,\text{Meas}}$-models.*

Definition 7. *A set T of formulas is* consistent *if there is at least a formula from For_C, and at least a formula from For_P that are not deducible from T, otherwise T is* inconsistent.

In [25], the authors also prove strong completeness (every consistent set of formulas is satisfiable) of LPP_2 logic.

Theorem 2 (Strong completeness of LPP_2). *Every consistent set of LPP_2-formulas T is $LPP_{2,\mathsf{Meas}}$-satisfiable.*

The proof can be found in [25], Chapter 3. It consists of the following main steps: proving Deduction theorem, showing how to extend a consistent set of formulas to a maximal consistent set and constructing a canonical model, using a maximal consistent set.

2.2 Intersection Type Assignment

Intersection types [5, 7] were introduced in the lambda calculus as an extension of the simple types in order to overcome the limitations of the simple (functional) types. Indeed, lambda calculus with intersection types has two unique properties which do not hold in other type systems. First, it completely characterizes the termination of reduction, a.k.a strong normalization, in lambda calculus (e.g. [12]). Second, its type assignment is sound and complete with respect to the filter model, which was proven in the seminal paper by Barendregt et al. [3].

The set of untyped *terms (λ-terms) Λ* is defined by the following grammar:

$$M ::= x \mid \lambda x.M \mid MM \tag{3}$$

where x belongs to a countable set of λ-term variables, V_Λ. The letters $M, N, \ldots$ will denote arbitrary λ-terms. The operator λx is a binder and the set of *free variables* of a λ-term M is defined as usual. The β-reduction is the rewriting rule $(\lambda x.M)N \to_\beta M[N/x]$. The definition and main properties of β-reduction (and $\beta\eta$-reduction) can be found in [1,16].

Intersection types are generated by the following grammar:

$$\sigma ::= a \mid \omega \mid \sigma \to \sigma \mid \sigma \cap \sigma \tag{4}$$

where a belongs to a countable set of propositional variables $\mathsf{V}_{\mathsf{Type}}$, and ω is a constant.

The set of all intersection types is denoted by $\mathsf{Type}^\cap$ and will be ranged over by $\sigma, \tau, \ldots, \sigma_1, \ldots$.

The *pre-order* $\leqslant$ is a binary relation on the set of all intersection types defined in the following way:

(1) $\sigma \leqslant \sigma$; (2) $\sigma \leqslant \tau, \tau \leqslant \rho$ then $\sigma \leqslant \rho$;
(3) $(\sigma \to \rho) \cap (\sigma \to \tau) \leqslant (\sigma \to \rho \cap \tau)$; (4) $\omega \leqslant \omega \to \omega$;
(5) $\sigma \leqslant \rho, \sigma \leqslant \tau$ then $\sigma \leqslant \rho \cap \tau$; (6) $\sigma \cap \tau \leqslant \sigma, \sigma \cap \tau \leqslant \tau$;
(7) $\sigma \leqslant \sigma_1, \tau \leqslant \tau_1$, then $\sigma_1 \to \tau \leqslant \sigma \to \tau_1$, (8) $\sigma \leqslant \omega$.

The induced equivalence relation $\sim$ is defined by

$$\sigma \sim \tau \quad \text{if and only if } \sigma \leqslant \tau \text{ and } \tau \leqslant \sigma.$$

A *lambda statement* is an expression of the form $M : \sigma$, with $M \in \Lambda$ and $\sigma \in \mathsf{Type}^\cap$. The type σ is the predicate and the term M is the subject of the statement. *A Declaration (basic statement)* is a lambda statement with a term-variable as subject, i.e. $x : \sigma$. A *basis (context)*, denoted by Γ, is a set of basic statements with distinct term variables. It can be infinite.

Definition 8. *Intersection (extended) type assignment, $\Lambda^\cap$, is defined as follows:*

$$[x : \sigma]$$

$$\frac{M : \sigma \to \tau \quad N : \sigma}{MN : \tau} \ (\to_E) \qquad \begin{array}{c} \vdots \\ \dfrac{M : \tau}{\lambda x.M : \sigma \to \tau} \ (\to_I) \end{array}$$

$$\frac{M : \sigma \cap \tau}{M : \sigma} \ (\cap_E) \qquad \frac{M : \sigma \cap \tau}{M : \tau} \ (\cap_E)$$

$$\frac{M : \sigma \quad M : \tau}{M : \sigma \cap \tau} \ (\cap_I) \qquad \frac{}{M : \omega} \ (\omega)$$

$$\frac{M : \sigma \quad \sigma \leqslant \tau}{M : \tau} \ (\leqslant)$$

The rule $(\to_E)$, represents rule for the typing of application and has the following meaning: if a term M has type $\sigma \to \tau$, then it can be applied to some term N of type σ and the the resulting term MN has type τ. According to the rule $\to_I$, if we can derive that term M has type τ whenever variable x has type σ, then lambda abstraction $\lambda x.M$, which represents a function, has type $\sigma \to \tau$. The explanation of other rules as well as the motivation for introducing intersection types for lambda calculus and the pre-order relation on the set of all types, can be found in [2].

If $M : \sigma$ is derivable from a set of premises $\Gamma = \{x_1 : \sigma_1, \ldots, x_n : \sigma_n\}$ by the rules given in Definition 8, it is denoted by $\Gamma \vdash M : \sigma$.

There are several attempts to give intersection type assignment *à la* Church, nevertheless none of them became standard. This is the reason why we work with type assignments *à la* Curry. For more details on the two versions of type systems we refer the reader to [4].

Lambda Models. In this section we recall the notion of the lambda model and the *interpretation of terms* in it, which is given in [1], §5.2.

A lambda model for untyped lambda calculus is of the form $\langle D, \cdot, [\![\,]\!] \rangle$, where $\langle D, \cdot \rangle$ is an applicative structure and $[\![\,]\!]$ is an interpretation map which assigns to each term M and each map ρ ($\rho : \mathsf{V}_\Lambda \to D$), an element $[\![M]\!]_\rho$, of D such that

(i) $[\![x]\!]_\rho = \rho(x)$;
(ii) $[\![MN]\!]_\rho = [\![M]\!]_\rho \cdot [\![N]\!]_\rho$;
(iii) if $\sigma(x) = \rho(x)$ for all x free in M, then $[\![M]\!]_\sigma = [\![M]\!]_\rho$;
(iv) $[\![\lambda x.M]\!]_\rho \cdot a = [\![M]\!]_{\rho[x:=a]}$;

(v) if $M =_\beta N$, then $[\![M]\!]_\rho = [\![N]\!]_\rho$.

The interpretation of intersection types in a lambda model is defined as follows.

Definition 9. *Let $\mathcal{M} = \langle D, \cdot, [\![\ ]\!] \rangle$ be a λ-model.*

(i) If $\rho : V_\Lambda \to D$ is the valuation of term variables in D, then $[\![M]\!]_\rho \in D$ is the interpretation of $M \in \Lambda$ in $\mathcal{M}$ via ρ.

(ii) Let $\xi : V_{\text{Type}} \to \mathcal{P}(D)$ be a valuation of type variables. The interpretation of $\sigma \in \text{Type}^\cap$ in $\mathcal{M}$ via ξ, denoted by $[\![\sigma]\!]_\xi \in \mathcal{P}(D)$, is defined as follows:
- $[\![a]\!]_\xi = \xi(a)$;
- $[\![\sigma \to \tau]\!]_\xi = \{d \in D \mid \forall e \in [\![\sigma]\!]_\xi,\ d \cdot e \in [\![\tau]\!]_\xi\}$;
- $[\![\omega]\!]_\xi = D$;
- $[\![\sigma \cap \tau]\!]_\xi = [\![\sigma]\!]_\xi \cap [\![\tau]\!]_\xi$.

(iii) The satisfiability relation and semantical consequence are defined as follows:
- $\mathcal{M}, \rho, \xi \models M : \sigma$ *if and only if* $[\![M]\!]_\rho \in [\![\sigma]\!]_\xi$;
- $\mathcal{M}, \rho, \xi \models \Gamma$ *if and only if* $\mathcal{M}, \rho, \xi \models x : \sigma$ *for all* $x : \sigma \in \Gamma$;
- $\Gamma \models M : \sigma$ *if and only if* $(\forall \mathcal{M}, \rho, \xi \models \Gamma)\ \mathcal{M}, \rho, \xi \models M : \sigma$.

Soundness of intersection type assignment is proved with respect to the notion of lambda models (e.g. [5]). However completeness can not be proved.

Theorem 3 (Soundness of $\Lambda^\cap$). $\Gamma \vdash M : \sigma \Rightarrow \Gamma \models M : \sigma$.

For proving completeness of the type assignment, the notion of filters is introduced in [3].

Definition 10. *(i) A subset $d \subseteq \text{Type}^\cap$ is a filter if and only if*
- $\omega \in d$;
- *if $\sigma, \tau \in d$ then $\sigma \cap \tau \in d$;*
- *if $\tau \in d$ and $\sigma \geqslant \tau$ then $\sigma \in d$.*

(ii) Let $\mathcal{F} = \{d \mid d$ is a filter$\}$, then for $d_1, d_2 \in \mathcal{F}$, we define $d_1 \cdot d_2 = \{\tau \in \text{Type} \mid \exists \sigma \in d_2, \sigma \to \tau \in d_1\}$.

It is shown in [3] that $\langle \mathcal{F}, \cdot, [\![\ ]\!] \rangle$ is a lambda model, called the *filter lambda model*, where $[\![\ ]\!]$ is explicitly defined as follows:

(i) If $\rho : V_\Lambda \to \mathcal{F}$ is a valuation in $\mathcal{F}$, then $\Gamma_\rho = \{x : \sigma \mid \sigma \in \rho(x)\}$.
(ii) If $M \in \Lambda$, $[\![M]\!] = \{\sigma \mid \Gamma_\rho \vdash M : \sigma\}$.

Completeness of the intersection type assignment with respect to the class of filter lambda models was proved in [3].

Theorem 4 (Strong completeness of $\Lambda^\cap$). $\Gamma \models M : \sigma \Rightarrow \Gamma \vdash M : \sigma$.

3   Probabilistic Logical System for Lambda Terms with Intersection Types P$\Lambda^\cap$

We define the probabilistic logical system for lambda terms with intersection types P$\Lambda^\cap$ as a probabilistic logic over the lambda calculus with intersection types $\Lambda^\cap$. In this section, we introduce the syntax and propose the semantics of P$\Lambda^\cap$.

3.1   Syntax of P$\Lambda^\cap$

Let S be the set of rational numbers from $[0, 1]$. The *alphabet* of the logic P$\Lambda^\cap$ consists of

- all symbols needed to define lambda terms with intersection types, given in Sect. 2.2,
- the classical propositional connectives $\neg$ and $\wedge$,
- the list of probability operators $P_{\geqslant s}$, for every $s \in$ S.

Other classical propositional connectives $\Rightarrow$, $\vee$, $\Leftrightarrow$ being defined as usual.

Basic Formulas. The set For$_B$ of all *basic formulas* over the set of term-variables V_Λ is the smallest set containing all lambda statements $M : \sigma$ with $M \in \Lambda$ and $\sigma \in$ Type$^\cap$, which is closed under the Boolean connectives. Basic formulas are generated by the following grammar:

$$\text{For}_B \quad \alpha ::= M : \sigma \mid \alpha \wedge \alpha \mid \neg \alpha. \tag{5}$$

The set of all basic formulas is denoted by For$_B$ and basic formulas will be denoted by $\alpha, \beta, \ldots$, possibly indexed.

Probabilistic Formulas. If $\alpha \in$ For$_B$ and $s \in$ S, then a *basic probabilistic formula* is any formula of the form $P_{\geqslant s}\alpha$.

The set of all probabilistic formulas, denoted by For$_P$, is the smallest set containing all basic probabilistic formulas which is closed under Boolean connectives.

Probabilistic formulas are generated by the following grammar:

$$\text{For}_P \quad \phi ::= P_{\geqslant s}\alpha \mid \phi \wedge \phi \mid \neg \phi. \tag{6}$$

The formulas from the set For$_P$ will be denoted by $\phi, \psi, \ldots$, possibly with subscripts.

Formulas of P$\Lambda^\cap$. The language of P$\Lambda^\cap$ consists of both basic formulas and probabilistic formulas

$$\text{For}_{P\Lambda^\cap} = \text{For}_B \cup \text{For}_P. \tag{7}$$

The set of formulas For$_{P\Lambda^\cap}$ will be ranged over by $\mathfrak{A}, \mathfrak{A}_1, \mathfrak{A}_2, \ldots$. We use the same abbreviations as in Sect. 2.1 to introduce other probabilistic operators ($P_{<s}$, $P_{>s}$, $P_{\leqslant s}$ and $P_{=s}$).

We also denote both $\alpha \wedge \neg \alpha$ and $\phi \wedge \neg \phi$ by $\perp$ (and dually for $\top$).

Similarly as in Sect. 2.1 and [13], where we introduced a formal model for reasoning about probabilities of simply typed lambda terms, neither mixing of basic formulas and probabilistic formulas, nor nested probability operators is allowed.

For example, the following two expressions are not (well defined) formulas of the logic $\mathsf{P}\wedge^\cap$:

$$(x : \sigma) \wedge P_{\geqslant \frac{1}{2}}(x : \sigma \cap \tau), \qquad P_{\geqslant \frac{1}{3}} P_{\geqslant \frac{1}{2}}(x : \sigma \cap \tau).$$

3.2   Semantics of $\mathsf{P}\wedge^\cap$

Similarly as in [13], the semantics for $\mathsf{P}\wedge^\cap$ will be based on the possible-world approach, with an important difference that worlds are now filter models, defined in Sect. 2.2.

Definition 11 ($\mathsf{P}\wedge^\cap$-structure). *A $\mathsf{P}\wedge^\cap$-structure is a tuple $\mathcal{M} = \langle W, \rho, \xi, H, \mu \rangle$, where:*

  (i) W is a nonempty set of worlds, where each world is the filter model, i.e. for every $w \in W$, $w = \langle \mathcal{F}_w, \cdot_w, [\![\ ]\!]_w \rangle$;
  (ii) $\rho : V_\Lambda \times \{w\} \longrightarrow \mathcal{F}_w$, $w \in W$;
  (iii) $\xi : V_{\mathsf{Type}} \times \{w\} \longrightarrow \mathcal{P}(\mathcal{F}_w)$, $w \in W$;
  (iv) H is an algebra of subsets of W, i.e. $H \subseteq \mathcal{P}(W)$ such that
      – $W \in H$,
      – if $U, V \in H$, then $W \setminus U \in H$ and $U \cup V \in H$;
  (v) μ is a finitely additive probability measure defined on H, i.e.
      – $\mu(W) = 1$,
      – if $U \cap V = \varnothing$, then $\mu(U \cup V) = \mu(U) + \mu(V)$,
      for all $U, V \in H$.

The elements of H are called *measurable sets of worlds*. We will write $\rho_w(x)$, instead of $\rho(x, w)$ and similarly for ξ.

Definition 12. *We say that a lambda statement $M : \sigma$ holds in a world w, denoted by $w \models M : \sigma$, if and only if*

$$[\![M]\!]_\rho^w \in [\![\sigma]\!]_\xi^w, \tag{8}$$

where $[\![M]\!]_\rho^w$ is the interpretation of a term M in a world w via ρ, and $[\![\sigma]\!]_\xi^w$ is the interpretation of a type σ in a world w via ξ. We define interpretation of Boolean combinations of lambda statements in the usual way:

  – $\omega \models \alpha \wedge \beta$ if and only if $\omega \models \alpha$ and $\omega \models \beta$;
  – $\omega \models \neg \alpha$ if and only if $\omega \nvDash \alpha$.

If $\mathcal{M}$ is a $\mathsf{P}\wedge^\cap$-structure, the set of all worlds in which α is true, $\{w \in W \mid w \models \alpha\}$, is denoted by $[\alpha]_\mathcal{M}$. We will omit the subscript $\mathcal{M}$ from $[\alpha]_\mathcal{M}$ and write $[\alpha]$ if $\mathcal{M}$ is clear from the context.

Definition 13 (Measurable structure). *A structure $\mathcal{M}$ is measurable if $[\alpha]_{\mathcal{M}} \in H$ for every $\alpha \in$ For$_B$. The class of all measurable structures of the logic $P\Lambda^{\cap}$ will be denoted by $P\Lambda^{\cap}_{\text{Meas}}$.*

Definition 14 (Satisfiability relation). *The satisfiability relation $\models\subseteq$ $P\Lambda^{\cap}_{\text{Meas}} \times$ For$_{P\Lambda^{\cap}}$ is defined in the following way:*

- $\mathcal{M} \models M : \sigma$ *iff* $w \models M : \sigma$, *for all* $w \in W$;
- $\mathcal{M} \models P_{\geqslant s}\alpha$ *iff* $\mu([\alpha]) \geqslant s$;
- $\mathcal{M} \models \neg\mathfrak{A}$ *iff it is not the case that* $\mathcal{M} \models \mathfrak{A}$;
- $\mathcal{M} \models \mathfrak{A}_1 \wedge \mathfrak{A}_2$ *iff* $\mathcal{M} \models \mathfrak{A}_1$ *and* $\mathcal{M} \models \mathfrak{A}_2$.

Definition 15 (Formula satisfiability). *Let $\mathfrak{A} \in$ For$_{P\Lambda^{\cap}}$ be a formula and $F \subseteq$ For$_{P\Lambda^{\cap}}$*

- $\mathfrak{A}$ *is satisfiable if there is a $P\Lambda^{\cap}_{\text{Meas}}$-model $\mathcal{M}$ such that $\mathcal{M} \models \mathfrak{A}$;*
- $\mathfrak{A}$ *is valid if for every $P\Lambda^{\cap}_{\text{Meas}}$-model $\mathcal{M}$, $\mathcal{M} \models \mathfrak{A}$;*
- *A set of formulas F is satisfiable if there is a $P\Lambda^{\cap}_{\text{Meas}}$-model $\mathcal{M}$ such that $\mathcal{M} \models \mathfrak{A}$ for every $\mathfrak{A} \in F$;*
- $\mathfrak{A}$ *is a semantical consequence of a set of formulas F, denoted by $F \models \mathfrak{A}$, if it holds that $\mathfrak{A}$ is satisfied in a model $\mathcal{M}$ (denoted by $\mathcal{M} \models \mathfrak{A}$) whenever F is satisfied in that model (denoted by $\mathcal{M} \models F$).*

In order to clarify the above notions we give an example.

Example 1. Let $\Gamma_1 = \{x : \sigma \to \tau, y : \sigma\}$, $\Gamma_2 = \{y : \sigma\}$ and $\Gamma_3 = \{z : \rho\}$. Consider the following $P\Lambda^{\cap}$-model with three worlds, i.e. let $\mathcal{M} = \langle W, \rho, \xi, H, \mu \rangle$, such that:

- $W = \{w_1, w_2, w_3\}$ and for all $i \in \{1, 2, 3\}$, $w_i = \langle \mathcal{F}, \cdot, [\![\ ]\!] \rangle$, defined as in Definition 10.
- $H = \mathcal{P}(W)$;
- $\mu(\{w_i\}) = \frac{1}{3}$, $i = 1, 2, 3$,

Let the valuations ρ and ξ be defined in the following way:

- $\rho_{w_i}(x) = \{\sigma \mid \Gamma_i \vdash x : \sigma\}$, for all $x \in V_\Lambda$ and $i \in \{1, 2, 3\}$;
- $\xi_{w_i}(a) = \{d \in \mathcal{F} \mid a \in d\}$, for all $a \in V_{\text{Type}}$ and $i \in \{1, 2, 3\}$.

It can be proved that for all $\sigma \in$ Type, $[\![\sigma]\!]^{w_i}_{\xi_{w_i}} = \{d \in \mathcal{F} \mid \sigma \in d\}$, see [3]. In [3] it is proved that $\{\sigma \mid \Gamma \vdash M : \sigma\}$ is a filter, for any term M. Now, we have that $\rho_{w_1}(x) = \{\sigma \mid \Gamma_1 \vdash x : \sigma\}$ is a filter which contains $\sigma \to \tau$, since $\Gamma_1 \vdash x : \sigma \to \tau$. Thus, $\rho_{w_1}(x) \in [\![\sigma \to \tau]\!]^{w_1}_{\xi_{w_1}}$. From the latter and Definition 12 we conclude $w_1 \models x : \sigma \to \tau$. Similarly, we obtain $w_1 \models y : \sigma$, $w_2 \models \neg(x : \sigma \to \tau)$, $w_2 \models y : \sigma$, $w_3 \models \neg(x : \sigma \to \tau)$ and $w_3 \models \neg(y : \sigma)$ (Figure 1). Finally, we can conclude that $M \models P_{=\frac{1}{3}}(x : \sigma \to \tau)$, $M \models P_{=\frac{2}{3}}(y : \sigma)$ and $M \models P_{\geqslant\frac{1}{3}}(xy : \tau)$. ∎

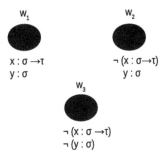

W_1

$x : \sigma \rightarrow \tau$
$y : \sigma$

W_2

$\neg (x : \sigma \rightarrow \tau)$
$y : \sigma$

W_3

$\neg (x : \sigma \rightarrow \tau)$
$\neg (y : \sigma)$

Fig. 1. An illustration of the Example 1

The compactness theorem which states that "if every finite subset of F is satisfiable, then F is satisfiable" does not hold for $\mathsf{P}\Lambda^{\cap}$. We show this in the following example.

Example 2. Consider the set

$$F = \{\neg P_{=0}\alpha\} \cup \{P_{<\frac{1}{n}}\alpha \mid n \quad is \quad a \quad positive \quad integer\} \tag{9}$$

Every finite subset of F is clearly $\mathsf{P}\Lambda^{\cap}_{\mathsf{Meas}}$-satisfiable, but the set F itself is not, since there is no real number greater than 0 and smaller than all positive rationals due to the Archimedean property of real numbers[1]. Therefore, the compactness theorem does not hold for $\mathsf{P}\Lambda^{\cap}$.

4 The Axiomatization $Ax_{\mathsf{P}\Lambda^{\cap}}$

In this section, an axiom system for the logic $\mathsf{P}\Lambda^{\cap}$, denoted by $Ax_{\mathsf{P}\Lambda^{\cap}}$, is introduced. This axiom system was obtained from axiom system of probability logic LPP_2 and intersection type assignment system.

Axiom Schemes

(1) all instances of the classical propositional tautologies, (atoms are λ-statements or any $\mathsf{P}\Lambda^{\cap}$-formulas),
(2) $P_{\geqslant 0}\alpha$,
(3) $P_{\leqslant r}\alpha \Rightarrow P_{<s}\alpha, s > r$,
(4) $P_{<s}\alpha \Rightarrow P_{\leqslant s}\alpha$,
(5) $(P_{\geqslant r}\alpha \wedge P_{\geqslant s}\beta \wedge P_{\geqslant 1}(\neg \alpha \vee \neg \beta)) \Rightarrow P_{\geqslant min\{1, r+s\}}(\alpha \vee \beta)$,
(6) $(P_{\leqslant r}\alpha \wedge P_{<s}\beta) \Rightarrow P_{<r+s}(\alpha \vee \beta), r + s \leqslant 1$,
(7) $P_{\geqslant 1}(\alpha \Rightarrow \beta) \Rightarrow (P_{\geqslant s}\alpha \Rightarrow P_{\geqslant s}\beta)$.

The meaning of the first six axioms is explained in Sect. 2.1. In order to better understand Axiom 1, we give an instance of axiom, that can be obtained from it. The formula $(\alpha \Rightarrow \beta) \Leftrightarrow (\neg \beta \Rightarrow \neg \alpha)$ is tautology of classical propositional

---

[1] For any real number $\epsilon > 0$ there exists an $n \in \mathbb{N}$ such that $\frac{1}{n} < \epsilon$.

Table 1. Inference Rules I

$$[x : \sigma]$$

$$(1) \; \frac{M : \sigma \to \tau \qquad N : \sigma}{MN : \tau} \; (\to_E)$$

$$(2) \; \frac{\begin{array}{c} \vdots \\ M : \tau \end{array}}{\lambda x.M : \sigma \to \tau} \; (\to_I)$$

$$(3) \; \frac{M : \sigma \cap \tau}{M : \sigma} \; (\cap_E) \qquad (4) \; \frac{M : \sigma \cap \tau}{M : \tau} \; (\cap_E)$$

$$(5) \; \frac{M : \sigma \qquad \Gamma \vdash M : \tau}{M : \sigma \cap \tau} \; (\cap_I) \qquad (6) \; \frac{}{M : \omega} \; (\omega)$$

$$(7) \; \frac{M : \sigma \qquad \sigma \leqslant \tau}{M : \tau} \; (\leqslant)$$

Table 2. Inference rules II

$$(1) \; \frac{\mathfrak{A}_1 \qquad \mathfrak{A}_1 \Rightarrow \mathfrak{A}_2}{\mathfrak{A}_2} \; (MP) \qquad (2) \; \frac{\alpha}{P_{\geqslant 1}\alpha} \qquad (3) \; \frac{\left\{ \phi \Rightarrow P_{\geqslant s - \frac{1}{k}}\alpha \mid k \geqslant \frac{1}{s} \right\}}{\phi \Rightarrow P_{\geqslant s}\alpha}$$

logic. If α and β are replaced with $P_{\geqslant s}(x : \sigma)$ and $P_{\geqslant r}(x : \sigma)$, we get $(P_{\geqslant s}(x : \sigma) \Rightarrow P_{\geqslant r}(x : \sigma)) \Leftrightarrow (\neg P_{\geqslant r}(x : \sigma) \Rightarrow \neg P_{\geqslant s}(x : \sigma))$. Using abbreviations for probabilistic operators, we obtain the formula $(P_{\geqslant s}(x : \sigma) \Rightarrow P_{\geqslant r}(x : \sigma)) \Leftrightarrow (P_{<r}(x : \sigma) \Rightarrow P_{<s}(x : \sigma))$, which is an axiom of the logical system $\mathsf{P\Lambda}^\cap$. Axiom 7 ensures that equivalent formulas have equal measures.

Inference Rules I are given in Table 1. These rules correspond to inference of lambda terms with intersection types and they can be applied only on lambda statements.

Inference Rules II are given in Table 2. They are also discussed in Sect. 2.1.

Similarly as in [13], we introduce a notion of proof, consistent set, maximal consistent set and deductively closed set.

Definition 16. *Let T be a set of $\mathsf{P\Lambda}^\cap$-formulas and $\mathfrak{A}$ a formula.*

1. *$T \vdash_{\mathsf{P\Lambda}^\cap} \mathfrak{A}$ means that there exists a sequence $\mathfrak{A}_0, \dots, \mathfrak{A}_n$ (n is finite or countable ordinal) of formulas, such that $\mathfrak{A}_n = \mathfrak{A}$ and for all $i \leqslant n$, $\mathfrak{A}_i$ is an axiom-instance, or $\mathfrak{A}_i \in T$, or $\mathfrak{A}_i$ is a $\mathsf{P\Lambda}^\cap$-formula which can be derived by some inference rule from Tables 1 and 2 applied on some previous members of the sequence.*
2. *Instead of $\varnothing \vdash_{\mathsf{P\Lambda}^\cap} \mathfrak{A}$ we write $\vdash_{\mathsf{P\Lambda}^\cap} \mathfrak{A}$. Any formula $\mathfrak{A}$ such that $\vdash_{\mathsf{P\Lambda}^\cap} \mathfrak{A}$ will be called a theorem.*
3. *T is consistent if*
   (1) there is at least a formula $\alpha \in \mathsf{For_B}$ and a formula $\phi \in \mathsf{For_P}$ that are not deducible from T and
   (2) for every lambda statement $M : \sigma$ that is in T, if x_i, $i \in I$ are all free variables of M, then all basic statements of the form $x_i : \tau_i$,

needed to adequately derive $M : \sigma$ according to Inference Rules I, i.e. $x_i : \sigma_i, \ldots, x_n : \sigma_n \vdash M : \sigma$, are also in T.[2]

Otherwise, T is inconsistent;

4. T is a maximally consistent set if it is consistent and:
   (1) for every $\alpha \in \mathsf{For_B}$, if $T \vdash_{\mathsf{P\wedge^\cap}} \alpha$, then $\alpha \in T$ and $P_{\geqslant 1}\alpha \in T$
   (2) for every $\phi \in \mathsf{For_P}$, either $\phi \in T$ or $\neg\phi \in T$.
5. T is deductively closed if for every $\mathfrak{A} \in \mathsf{For_{P\wedge^\cap}}$, if $T \vdash_{\mathsf{P\wedge^\cap}} \mathfrak{A}$, then $\mathfrak{A} \in T$.

If $T \vdash_{\mathsf{P\wedge^\cap}} \bot$, then the condition (1) from the definition of a consistent set does not hold, and T is inconsistent. The opposite does not hold. Inconsistency of set T does not imply $T \vdash_{\mathsf{P\wedge^\cap}} \bot$. However, the first condition is equivalent to $T \vdash_{\mathsf{P\wedge^\cap}} \bot$. Hence if T is inconsistent due to the fact that condition (1) does not, although condition (2) does hold, then $T \vdash_{\mathsf{P\wedge^\cap}} \bot$. Note that it is not required that for every $\alpha \in \mathsf{For_{B^\cap}}$, either α or $\neg\alpha$ belongs to a maximal consistent set (as it is done for formulas from $\mathsf{For_{P^\cap}}$). It can be proved that, otherwise, in our canonical model, for each α we would have $P_{=1}\alpha$ or $P_{=0}\alpha$, which trivializes the probabilities.

Recall Example 2. It shows that we need an infinitary axiomatization in order to obtain strong completeness for our formal model. Inconsistency of F cannot be proved by a finitary proof, since only finite number of members of F can be used in such a proof, while every finite subset of F is satisfiable and consistent. Hence, for every finitary axiomatization there are consistent sets of formulas, which are unsatisfiable. Thus, strong completeness can not be obtained for any finite axiomatic system.

Theorem 5 (Deduction theorem). *Let T be a set of* $\mathsf{P\wedge^\cap}$*-formulas and* $\phi, \psi \in \mathsf{For_P}$*. If $T \cup \{\phi\} \vdash_{\mathsf{P\wedge^\cap}} \psi$ then $T \vdash_{\mathsf{P\wedge^\cap}} \phi \Rightarrow \psi$.*

Proof. Analogous to the proof of Deduction theorem in [13].

Theorem 6 (Soundness of $\mathsf{P\wedge^\cap}$). *The axiomatic system $Ax_{\mathsf{P\wedge^\cap}}$ is sound with respect to the class of $\mathsf{P\wedge^\cap_{Meas}}$-models.*

Proof. Soundness is a consequence of soundness of intersection type assignment with respect to the (filter) lambda models and soundness of probabilistic logic LPP_2 with respect to the semantics presented in Sect. 2.1. The idea is to prove that that every instance of an axiom scheme holds in every model and that the inference rules preserve the validity. The proof that every instance of an axiom scheme holds in every model and that inference rules in Table 2 preserve validity proceed as in the probabilistic logical system for simply typed lambda terms, introduced in [13]. This is due to the fact that there is a certain similarity between the structures of the model proposed in Sect. 3.2 and the one proposed in [13]. An important difference between those models is that worlds of models we propose in the paper are filter models, while in [13] worlds are term models. Inference rules from Table 1 are validity-preserving because of the soundness of intersection type assignment. For example, let us consider rule

---

[2] Note that the notion of a consistent set is different than usual. We have one additional condition, namely condition (2).

$$(1) \quad \frac{M : \sigma \to \tau \qquad N : \sigma}{MN : \tau} \ (\to_E)$$

Suppose that model $\mathcal{M} = \langle W, \rho, \xi, H, \mu \rangle$ satisfies premises of the rule, i.e. $\mathcal{M} \models M : \sigma \to \tau$ and $\mathcal{M} \models: \sigma$. From Definition 14, it follows that for every world $w \in W$, $w \models M : \sigma \to \tau$ and $w \models N : \sigma$. This is equivalent to the following:

$$[\![M]\!]_\rho^w \in [\![\sigma \to \tau]\!]_\xi^w \text{ and } [\![N]\!]_\rho^w \in [\![\sigma]\!]_\xi^w, \tag{10}$$

by Definition 12. Using the definition of the interpretation of a functional type, i.e. $[\![\sigma \to \tau]\!]_\xi$, and the property of the interpretation map in a lambda model, $[\![MN]\!]_\rho = [\![M]\!]_\rho \cdot [\![N]\!]_\rho$, we can conclude $[\![MN]\!]_\rho^w \in [\![\tau]\!]_\xi^w$, that is $w \models MN : \tau$. Since the latter holds for every world $w \in W$, we obtain $\mathcal{M} \models MN : \tau$. Thus, this rule preserves validity. Next, let us consider the rule

$$(3) \quad \frac{M : \sigma \cap \tau}{M : \sigma} \ (\cap_E)$$

If model $\mathcal{M} = \langle W, \rho, \xi, H, \mu \rangle$ satisfy premise, i.e. $\mathcal{M} \models M : \sigma \cap \tau$, then for every world $w \in W$, $w \models M : \sigma \cap \tau$ holds. By Definition 12 and the definition of the interpretation of an intersection type ($[\![\sigma \cap \tau]\!]_\xi$), we obtain $[\![M]\!]_\rho^w \in [\![\sigma \cap \tau]\!]_\xi^w = [\![\sigma]\!]_\xi^w \cap [\![\tau]\!]_\xi^w$. Hence, we can conclude $[\![M]\!]_\rho^w \in [\![\sigma]\!]_\xi^w$. Since the latter holds for every world $w \in W$, it follows that $\mathcal{M} \models M : \sigma$. Thus, this rule also preserves validity. The proofs that other rules are validity-preserving can be obtained similarly.

Next example illustrates the difference between the probabilistic logic over simply typed lambda calculus and probabilistic logic over lambda calculus with intersection types.

Example 3. Let us consider term $\lambda x.xx$. As we already mentioned in Introduction, in lambda calculus with intersection types we can infer $\vdash \lambda x.xx : (\sigma \cap (\sigma \to \tau)) \to \tau$ for any types σ and τ. Thus, in $\mathsf{PA}^\cap$ logic, we can infer $\vdash P_{\geqslant 1} \lambda x.xx : (\sigma \cap (\sigma \to \tau)) \to \tau$. Because of the soundness of $\mathsf{PA}^\cap$ logic we conclude that every model $\mathcal{M}$ satisfies formula $P_{\geqslant 1} \lambda x.xx : (\sigma \cap (\sigma \to \tau)) \to \tau$. On the other hand, in probabilistic logic over simply typed lambda calculus, introduced in [13] this does not hold.

One interesting direction for further work would be the study of probabilistic logic over some lambda calculus where for some s, $P_{=s}M : \sigma$ is always true and $P_{>s}M : \sigma$ is not.

5   The Logic $\mathsf{PA}^{\mathsf{Fr}(n)}$

In this section, we introduce countably many more logics (for each positive integer n, one logic) which are similar to logic $\mathsf{PA}^\cap$. The main difference is that in semantics we allow only that $\mu : H \to Range$, where $Range = \{0, \frac{1}{n}, \ldots, \frac{n-1}{n}, 1\}$. We will call these semantics $\mathsf{PA}_{\mathsf{Meas}}^{\mathsf{Fr}(n)}$-models. That change in semantics provides the change in Definition 16, i.e. the proofs are *finite* sequences of formulas.

For $s \in [0,1)$, let $s^+ = \min\{r \in Range \mid s < r\}$, and if $s \in (0,1]$, let $s^- = \max\{r \in Range \mid s > r\}$.

The axiomatization of the logic $\mathsf{PA}^{\mathsf{Fr}(n)}$ includes all the axioms from Sect. 4, plus one more axiom:

(8) $P_{>s}\alpha \rightarrow P_{\geqslant s^+}\alpha$,

and, the only infinitary rule, i.e. Rule (3) from group Inference Rules II, is not included. Therefore, this gives us a finite axiomatization.

It is straightforward to prove that Deduction theorem holds and that the axiomatization is sound with respect to the proposed semantics.

Theorem 7 (Soundness of $P\Lambda^{Fr(n)}$). *The axiomatic system of logic $P\Lambda^{Fr(n)}$ is sound with respect to the class of $P\Lambda_{Meas}^{Fr(n)}$-models.*

Proof. The proof is similar to the proof of soundness for $P\Lambda^\cap$ logic. Again, we prove that every instance of an axiom scheme holds in every model and that the inference rules preserve the validity. The difference with respect to the proof of Theorem 6 is that we need to prove that Axiom (8) holds in every model and we do not consider the infinitary rule. Using the definition of the satisfiability of a probabilistic formula $P_{\geqslant s}\alpha$ in a model and the definition of s^+, we can easily obtain that Axiom (8) holds in every model.

Completeness is still an open problem.

In Example 2, we showed that the logic $P\Lambda^\cap$ is not compact. The next theorem states that logics $P\Lambda^{Fr(n)}$ are compact.

Theorem 8 *(Compactness theorem for $P\Lambda^{Fr(n)}$).* *Let C be any class of models considered in this section, and let T be any set of formulas. If every finite subset of T is C-satisfiable, then T is also C-satisfiable.*

Proof. Suppose that T is not C-satisfiable. Then, it is not $Ax_{P\Lambda^{Fr(n)}}$-consistent, so $T \vdash \perp$. Because the axiomatic system $Ax_{P\Lambda^{Fr(n)}}$ is finite, there must be a finite set $T_1 \subseteq T$, such that $T_1 \vdash \perp$. Contradiction with the fact that every finite subset of T is both C-satisfiable and $Ax_{P\Lambda^{Fr(n)}}$-consistent.

6 Conclusion

In this paper, we introduced a formal model for probabilistic reasoning in lambda calculus with intersection types. We extended with probabilistic operators the language of lambda calculus with intersection types. The main result is the proof that the proposed infinitary axiomatization is sound with respect to the measurable possible world semantics.

In order to overcome some limitations of probabilistic logic over simply typed lambda calculus, we consider lambda calculus with intersection types. Probabilistic reasoning is introduced in simply typed lambda calculus with the equality rule in [13]. Here we have introduced probabilistic reasoning in lambda calculus with intersection types, which is a pure type system without equality. Furthermore, the lambda models employed in the construction of the Kripke-style models in [13] are term models, whereas in this paper filter models of lambda calculus are the basis for this construction. Our further goal is to push forward probabilistic reasoning in type theory and to consider probabilistic extensions of other typed lambda calculi.

References

1. Barendregt, H.P.: The Lambda Calculus: Its Syntax and Semantics. North Holland Publishing Company, Amsterdam (1984)
2. Barendregt, H.P., Dekkers, W., Statman, R.: Lambda Calculus with Types. Perspectives in Logic. Cambridge University Press, Cambridge (2013). http://www.cambridge.org/de/academic/subjects/mathematics/logic-categories-and-sets/lambda-calculus-types
3. Barendregt, H., Coppo, M., Dezani-Ciancaglini, M.: A filter lambda model and the completeness of type assignment. J. Symb. Logic 48(4), 931–940 (1983). https://doi.org/10.2307/2273659
4. Barendregt, H.P.: Lambda calculi with types. In: Handbook of Logic in Computer Science. vol. 2, pp. 117–309. Oxford University Press Inc, New York (1992). http://dl.acm.org/citation.cfm?id=162552.162561
5. Ben-Yelles, C.B.: Type assignment in the lambda-calculus: syntax and semantics. Ph.D. thesis, Department of Pure Mathematics, University College of Swansea, September 1979
6. Church, A.: A formulation of the simple theory of types. J. Symb. Logic 5(2), 56–68 (1940). https://doi.org/10.2307/2266170
7. Coppo, M., Dezani-Ciancaglini, M.: A new type assignment for λ-terms. Arch. Math. Log. 19(1), 139–156 (1978). https://doi.org/10.1007/BF02011875
8. Crubillé, R., Dal Lago, U.: On probabilistic applicative bisimulation and call-by-value λ-calculi. In: Shao, Z. (ed.) ESOP 2014. LNCS, vol. 8410, pp. 209–228. Springer, Heidelberg (2014). https://doi.org/10.1007/978-3-642-54833-8_12
9. Crubillé, R., Dal Lago, U., Sangiorgi, D., Vignudelli, V.: On applicative similarity, sequentiality, and full abstraction. In: Meyer, R., Platzer, A., Wehrheim, H. (eds.) Correct System Design. LNCS, vol. 9360, pp. 65–82. Springer, Cham (2015). https://doi.org/10.1007/978-3-319-23506-6_7
10. Dal Lago, U., Sangiorgi, D., Alberti, M.: On coinductive equivalences for higher-order probabilistic functional programs. In: POPL, pp. 297–308. ACM (2014)
11. Gaifman, H.: Concerning measures in first order calculi. Israel J. Math. 2(1), 1–18 (1964). https://doi.org/10.1007/BF02759729
12. Ghilezan, S.: Strong normalization and typability with intersection types. Notre Dame J. Form. Logic 37(1), 44–52 (1996). https://doi.org/10.1305/ndjfl/1040067315
13. Ghilezan, S., Ivetić, J., Kašterović, S., Ognjanović, Z., Savić, N.: Probabilistic reasoning about simply typed lambda terms. In: Artemov, S., Nerode, A. (eds.) LFCS 2018. LNCS, vol. 10703, pp. 170–189. Springer, Cham (2018). https://doi.org/10.1007/978-3-319-72056-2_11
14. Hailperin, T.: Best possible inequalities for the probability of a logical function of events. Am. Math. Monthly 72(4), 343–359 (1965). http://www.jstor.org/stable/2313491
15. Hamblin, C.L.: The modal "probably". Mind 68(270), 234–240 (1959). https://doi.org/10.1093/mind/LXVIII.270.234
16. Hindley, J.R.: Basic Simple Type Theory. Cambridge Tracts in Theoretical Computer Science 42. Cambridge University Press, Cambridge (1997). http://gen.lib.rus.ec/book/index.php?md5=3BB4134A46F16E81D2D16744850F44EA
17. Hindley, J.R., Longo, G.: Lambda-calculus models and extesionality. Math. Logic Q. 26, 289–310 (1980)

18. Ikodinović, N., Ognjanović, Z., Rašković, M., Marković, Z.: First-order probabilistic logics and their applications. In: Zbornik radova, Subseries Logic in Computer Science, vol. 18, no. 26, pp. 37–78. Matematički institut(2015)
19. Kašterović, S., Pagani, M.: The discriminating power of the let-in operator in the lazy call-by-name probabilistic lambda-calculus. In: 4th International Conference on Formal Structures for Computation and Deduction, FSCD 2019, 24–30 June 2019, Dortmund, Germany, pp. 26:1–26:20 (2019). https://doi.org/10.4230/LIPIcs.FSCD.2019.26
20. Keisler, H.: Hyperfinite models of adapted probability logic. Ann. Pure Appl. Logic **31**, 71–86 (1986). https://doi.org/10.1016/0168-0072(86)90063-1. http://www.sciencedirect.com/science/article/pii/0168007286900631
21. Kraus, M.: Early Greek probability arguments and common ground in dissensus. In: Ontario Society for the Study of Argumentation (OSSA) Proceedings, pp. 1–11. OSSA Conference Archive (2007)
22. Manning, C.D., Schütze, H.: Foundations of Statistical Natural Language Processing. MIT Press, Cambridge (1999)
23. Nilsson, N.J.: Probabilistic logic. Artif. Intell. **28**(1), 71–87 (1986). https://doi.org/10.1016/0004-3702(86)90031-7
24. Ognjanović, Z., Rašković, M., Marković, Z.: Probability logics. In: Zborik radova, Subseries logic in computer science, vol. 12, no. 20, pp. 35–111. Matematički institut (2009)
25. Ognjanović, Z., Rašković, M., Marković, Z.: Probability Logics: Probability-Based Formalization of Uncertain Reasoning. Springer, Cham (2016). https://doi.org/10.1007/978-3-319-47012-2
26. Pearl, J.: Probabilistic Reasoning in Intelligent Systems - Networks of Plausible Inference. Morgan Kaufmann series in representation and reasoning. Morgan Kaufmann, San Mateo (1989)
27. Thurn, S.: Exploring Artificial Intelligence in the New Millennium, chap. Robotic Mapping: A Survey, pp. 1–35. Morgan Kaufmann Publishers Inc., San Francisco (2003), http://dl.acm.org/citation.cfm?id=779343.779345

Measuring Inconsistency in a General Information Space

John Grant[1] and Francesco Parisi[2(✉)]

[1] University of Maryland at College Park, College Park, USA
grant@cs.umd.edu
[2] DIMES Department, University of Calabria, Rende, Italy
fparisi@dimes.unical.it

Abstract. AI systems often need to deal with inconsistent information. For this reason since the early 2000s some AI researchers have developed ways to measure the amount of inconsistency in a knowledge base. By now there is a substantial amount of research about various aspects of inconsistency measuring. The problem is that most of this work applies only to knowledge bases formulated as sets of formulas in propositional logic. Hence this work is not really applicable to the way that information is actually stored. The purpose of this paper is to extend inconsistency measuring to real world information. We first define the concept of *general information space* which encompasses various types of databases and scenarios in AI systems. Then, we show how to transform any general information space to an *inconsistency equivalent* propositional knowledge base, and finally apply propositional inconsistency measures to find the inconsistency of the general information space. Our method allows for the direct comparison of the inconsistency of different information spaces, even though the data is presented in different ways. We demonstrate the transformation on three general information spaces: a relational database, a graph database, and a Blocks world scenario, where we apply several inconsistency measures after performing the transformation.

Keywords: Inconsistency measurement · General information space · Inconsistency measures

1 Introduction

As AI systems may need to deal with inconsistency, some AI researchers started in the early 2000s to develop ways of measuring the inconsistency of a propositional knowledge base, that is, a set of formulas in propositional logic. By now a substantial amount of work has been done along these lines, which includes for instance the approaches developed in [18, 20–24, 26, 28, 34, 35, 38]. A survey on the topic can be found in [13].

But information in many cases is not restricted to propositional logic formulas. In this paper we show that much of the work done for measuring inconsistency in propositional logic knowledge bases can be applied to measure inconsistency in more complex frameworks, where real world information is actually stored. Our approach is as follows.

ⓒ Springer Nature Switzerland AG 2020
A. Herzig and J. Kontinen (Eds.): FoIKS 2020, LNCS 12012, pp. 140–156, 2020.
https://doi.org/10.1007/978-3-030-39951-1_9

- We first introduce the concept of a *general information space* that covers various types of databases and AI frameworks. A general information space consists of a *framework*, such as a database schema, a set of *information units*, such as tuples in a relation, and a set of *restrictions*, such as integrity constraints.
- Then, we formulate a *transformation* from a general information space to a propositional knowledge base in such a way that the inconsistencies are preserved. As expected, the propositional knowledge base resulting from the transformation does not capture the rich content of the general information space, but it does capture all of its inconsistencies in a precise way.
- Finally, after the transformation we apply a propositional inconsistency measure to the just obtained propositional knowledge base to find the inconsistency of the general information space.

Hence, our approach allows for measuring the inconsistency of any information space and the method can be applied in a uniform manner over a variety of information scenarios. To the best of our knowledge, this is the first approach that lifts the idea of inconsistency measure from propositional knowledge bases to a range of different frameworks used for storing real world data. This makes it possible to apply all the results about inconsistency measures for propositional knowledge bases to a wide range of applications. We believe that this is a significant advance for the whole idea of inconsistency measurement that has been developed for propositional knowledge bases, and explored in several other individual settings such as software specifications [27], databases [2,25,31], and ontologies [39,40], among others.

The plan of this paper is as follows. We start by giving some basic information and examples of inconsistency measures for propositional knowledge bases. Then, we define the concept of a general information space. This is followed by the steps of the transformation and the equivalence of the two for inconsistencies. Then, specific examples illustrate what happens for a relational database [4], a graph database [33], and a Blocks world scenario [17]. We also show that this method allows for the important task of evaluating and comparing the inconsistency [10] even for such different ways of storing information.

2 Brief Background on Inconsistency Measures for Propositional Knowledge Bases

The idea of an inconsistency measure is to assign a nonnegative number to a knowledge base that measures its inconsistency. We start with a propositional language of formulas composed from a countable set of atoms, the fundamental propositions, and the connectives $\wedge$ (*conjunction*), $\vee$ (*disjunction*), and $\neg$ (*negation*). We write $\mathscr{K}$ for the set of all propositional knowledge bases (KBs), i.e. the set of all finite sets of formulas in the language. We write K for an individual KB. 2^X is the set of all subsets (the power set) of any set X. In general, an inconsistency measure assigns each KB a nonnegative real number or infinity.

Definition 1. *A function* $I : \mathscr{K} \to \mathbb{R}_\infty^{\geq 0}$ *is an* inconsistency measure *if the following conditions hold for all* $K, K' \in \mathscr{K}$:

1. *Consistency.* $I(K) = 0$ *iff* K *is consistent.*
2. *Monotony. If* $K \subseteq K'$, *then* $I(K) \le I(K')$.

Consistency and Monotony are called (rationality) postulates. Postulates are desirable properties for inconsistency measures. Consistency means that all and only consistent KBs get measure 0. Monotony means that the enlargement of a KB cannot decrease its measure. Monotony is not appropriate for relative measures [19] where the ratio of inconsistency may decrease with the addition of consistent information; however, it is appropriate for the measures we consider in this paper.

For a knowledge base K, $\mathsf{MI}(K)$ is the set of minimal inconsistent subsets of K. Also, if $\mathsf{MI}(K) = \{M_1,...,M_n\}$ then $\mathsf{Problematic}(K) = M_1 \cup ... \cup M_n$, and $\mathsf{Free}(K) = K \setminus \mathsf{Problematic}(K)$. A *free* formula is not involved in an essential way in any inconsistency, while a *problematic* formula is so involved in at least one inconsistency. As an example, for $K_{ex} = \{a_1, a_2, a_3, a_4, \neg a_1 \vee \neg a_2, \neg a_2 \vee \neg a_3, a_4 \wedge a_5\}$, $\mathsf{MI}(K_{ex}) = \{\{a_1, a_2, \neg a_1 \vee \neg a_2\}, \{a_2, a_3, \neg a_2 \vee \neg a_3\}\}$, and $\mathsf{Problematic}(K_{ex}) = \{a_1, a_2, a_3, \neg a_1 \vee \neg a_2, \neg a_2 \vee \neg a_3\}$, while a_4 and $a_4 \wedge a_5$ are free formulas.

A classical interpretation i for K assigns each atom a that appears in a formula of K the truth value T or F, that is, $i : \mathsf{Atoms}(K) \to \{T, F\}$. However, there is an important propositional inconsistency measure, I_C, that uses 3 truth values: T, F, and B, where B indicates inconsistency. This measure uses Priest's 3-valued logic. This interpretation uses an ordering on the truth values where $F < B < T$ and $\wedge$ computes the minimum value while $\vee$ computes the maximum value; also $\neg(B) = B$. So, for example, $B \wedge F = F$ and $B \vee F = B$. Then, an interpretation i satisfies a formula iff the truth-value of the formula for i is T or B.

Now we are ready to define the propositional inconsistency measures we will consider in this paper. Although we can use any of the inconsistency measures that have been formulated for propositional knowledge bases, we will present ones that involve in some way the minimal inconsistent subsets. As was shown in [5] many important inconsistency measures can be defined using only the structure of the minimal inconsistent subsets and these are particularly relevant in view of the transformation. Below the definition we briefly explain the meanings of these measures.

Definition 2 (Propositional Inconsistency Measures). *For a knowledge base K, the inconsistency measures I_B, I_M, $I_\#$, I_P, I_H, and I_{nc} are such that*

- $I_B(K) = 1$ *if K is inconsistent and $I_B(K) = 0$ if K is consistent.*
- $I_M(K) = |\mathsf{MI}(K)|$.
- $I_\#(K) = \begin{cases} 0 & \text{if } K \text{ is consistent,} \\ \sum_{X \in \mathsf{MI}(K)} \frac{1}{|X|} & \text{otherwise.} \end{cases}$
- $I_P(K) = |\mathsf{Problematic}(K)|$.
- $I_H(K) = \min\{|X| \mid X \subseteq K \text{ and } \forall M \in \mathsf{MI}(K)(X \cap M \ne \varnothing)\}$.
- $I_{nc}(K) = |K| - \max\{n \mid \forall K' \subseteq K : |K'| = n \text{ implies that } K' \text{ is consistent }\}$.
- $I_C(K) = \min\{|i^{-1}(B)| \text{ such that } i \text{ satisfies every formula in } K\}$.

We explain the measures as follows. I_B is also called the drastic measure [20]: it simply distinguishes between consistent and inconsistent KBs. I_M counts the number of minimal inconsistent subsets [20]. $I_\#$ also counts the number of minimal inconsistent subsets, but it gives larger sets a smaller weight; the reason is that when a minimal inconsistent set contains more formulas than another minimal inconsistent set, the former is intuitively less inconsistent than the latter [20]. I_P counts the number of formulas that contribute essentially to one or more inconsistencies [11]. Finally, I_C counts the minimal number of atoms that must be assigned the truth-value B in the three-valued logic by an interpretation that satisfies every formula in the KB [11]. I_H counts the minimal number of formulas whose deletion makes the set consistent [12]. I_{nc} uses the largest number such that all sets with that many formulas are consistent [6]. For the KB K_{ex}, we have that $I_B(K_{ex}) = 1$ as it is inconsistent; $I_M(K_{ex}) = 2$ as there are 2 minimal inconsistent subsets; $I_\#(K_{ex}) = \frac{1}{3} + \frac{1}{3} = \frac{2}{3}$ as both minimal inconsistent subsets consist of 3 formulas; $I_P(K_{ex}) = 5$ as there are 5 problematic formulas; $I_H(K_{ex}) = 1$ as deleting a_2 suffices to make K_{ex} consistent; $I_{nc}(K_{ex}) = 7 - 2 = 5$ as 2 is the largest number such that all subsets of size 2 are consistent; and $I_C(K_{ex}) = 1$ as the following interpretation satisfies all the formulas: $i(a_1) = i(a_3) = i(a_4) = i(a_5) = T$, $i(a_2) = B$.

3   General Information Spaces

Our goal is to lift the idea of inconsistency measure from propositional knowledge bases to more complex cases that are useful in AI and databases. For this reason we now define the concept of a general information space that encompasses many such cases.

Definition 3. *A general information space* $S = \langle F, U, C \rangle$ *is a triple where F is the framework for the information, U is a set of information units, and C is a set of requirements that U must satisfy, where the following hold:*

A1 (Consistency of individual information units). The set of information units, U, simply gives some information and each unit is itself consistent.

A2 (Consistency of requirements). There are no inconsistencies among requirements. All inconsistencies arise from the interaction of U and C.

A3 (Procedure for finding violations of the requirements). For every requirement, there has to be a procedure that finds all violations of that requirement.

A relational database is an example of a general information space. Here the framework is the database schema as well as the language used to describe the database and the domains for the constants. The information units are the tuples in the relations of the schema. The set of requirements is the set of constraints.

A graph database is another example. Here the framework is the type of information stored in the vertices and edges. The information units are the information given by the vertices and edges. The requirements are the graph constraints.

A Blocks world is a general information space as well. Here the framework is the type of information stored about blocks such as block color and which block is on top of another block. The elements of the domain are also part of the framework. The information units describe the Blocks world such as giving the colors of the blocks and which block is on top of another block. The requirements are the rules of the Blocks world such as that a green block cannot be on top of a blue block.

Another example is a board game configuration. Here the framework is information about the board and the pieces. The information units describe the position of each piece and some action or actions. The requirements are the rules of the game, such as how the pieces may move and what constitutes a winning position.

Thus a general information space encompasses many ways in which information is stored and presented; we give three concrete examples later.

It is worth noting that A1, A2, and A3 hold in many real world scenarios, such as those mentioned earlier. For instance, for relational databases, tuples are units of information that are usually assumed to be consistent when considered alone (without interacting with the integrity constraints), integrity constraints are usually satisfiable (there exists a database instance that satisfies them), and procedures for checking inconsistency are well-known for large classes of integrity constraints [1, 37].

A requirement for a general information space is really a constraint but we use this terminology to indicate that there need not be a formal language in which the requirement is presented and we will use English in some examples. In many cases a positive number, called the *arity*, can be associated with each requirement that indicates *the minimal number of information units that together violate the requirement and thereby cause an inconsistency*. We will also explain later that we can also handle an additional case where the arity is set to 0.

Next we give some details concerning requirements for relational databases, and illustrate the notion of arity for them.[1] Consider a relational database with two relations, a binary relation R_1 and a ternary relation R_2. As is usual we call a requirement a constraint. Consider the constraint: $\neg R_1(1, 2)$. This constraint specifically excludes the tuple $(1, 2)$ from R_1. Hence the existence of that tuple with the constraint is inconsistent. Thus, the arity of this constraint is 1. Consider the more general constraint: $\forall x_1 x_2 x_3 [R_1(x_1, x_2) \wedge R_1(x_1, x_3) \rightarrow x_2 = x_3]$. This states the functional dependency of the second attribute of R_1 on the first. This would be violated, for instance, by the two tuples: $(1, 2)$ and $(1, 3)$ in R_1 causing an inconsistency. So the arity of this constraint is 2. Next consider the constraint: $\forall x_1 x_2 [R_1(x_1, x_2) \rightarrow \exists x_3 x_4 (R_2(x_2, x_3, x_4))]$. This states the inclusion dependency that the elements in the second column of R_1 are included in the first column of R_2. A violation of this constraint is caused by a single tuple in R_1 whose second element is not in the first column of R_2. This means that the arity is 1. It is worth pointing out an important difference between this case and the previous two cases. Take the first constraint, $\neg R_1(1, 2)$, the functional dependency case is similar. The existence in R_1 of the tuple $(1, 2)$, that is, $R_1(1, 2)$ together with the constraint $\neg R_1(1, 2)$ form an inconsistent set of formulas. But in the inclusion dependency the existence of some tuple in R_1, say $R_1(1, 2)$ together with the inclusion dependency is not inconsistent: there may be a tuple in R_2, say $R_2(2, 3, 4)$ that would satisfy the constraint. Hence for this type of constraint the context is needed, such as what tuples are in R_2. In any case, all the usual database constraints such as the various types of dependencies, denial and key constraints are included in our concept of a requirement.

There is also a somewhat different type of constraint. Consider that the requirement $R_1(1, 2)$ requires the tuple $(1, 2)$ to be in R_1. So in this case it is a lack of the tuple

---

[1] We assume the reader is familiar with the classical notions of database scheme, relation scheme, relation instance, and integrity constraint. We will recall these notions in Sect. 5.1.

that violates the requirement and causes an inconsistency. A more general example is $\exists x_1 x_2 R_2(1, x_1, x_2)$ which states that there must be a tuple in the R_2 relation whose first element is 1. Note how there is some similarity between this constraint and the inclusion dependency given above: in both cases a required tuple is not in the database. The difference is that for an inclusion dependency we can point to a tuple in R_1 that is the source of the problem, i. e. without that tuple there would not be a constraint violation. But this example is purely existential; no deletion from the database would negate the violation. Hence the arity of such a constraint is 0 and the constraint is inconsistent with respect to U.

4   Transforming a General Information Space to a Propositional Knowledge Base

We now show how any general information space can be transformed to a propositional knowledge base in such a way that all the violations of the requirements are inconsistencies in the knowledge base. Note however that the transformation also loses some information: there is no way to go back from the propositional knowledge base to the original general information space. In fact many different information spaces representing different phenomena may be transformed to the same knowledge base. But the transformation is appropriate if we are interested in measuring inconsistency.

Definition 4 (Transformation). *The transformation from a general information space $S = \langle F, U, C \rangle$ to a propositional KB K_S is as follows.*

- *Let $A_U = \{a_1, \ldots, a_{|U|}\}$ be a set of $|U|$ propositional atoms.*
- *Define a bijective function $f : U \to A_U$ that assigns a distinct propositional atom to each information unit in U.*
- *Let $B_C = \{b_1, \ldots, b_{|C|}\}$ be another set of $|C|$ propositional atoms.*
- *Define a bijective function $h : C \to B_C$ that assigns a distinct propositional atom to each requirement in C.*
- *Let $\mathscr{F}_S$ be the set of propositional formulas using $A_U \cup B_C$.*
- *Define a function $g : C \to \mathscr{F}_S$ as follows: For each requirement $c \in C$ do as follows.*
    (1) If there is no violation of the requirement, then set $g(c) = h(c)$. Otherwise, there is at least one violation of c.
    (2) If the arity of c is greater than 0, then a minimal inconsistency is formed by one or more information units together with c. Find all such sets, say $M_c = \{U_1, \ldots, U_k\}$ and suppose that $|U_i| = n$. Let $U_i = \{u_i^1, \ldots, u_i^n\}$ (where each u_i^j is an information unit). Define $\rho(U_i) = \neg f(u_i^1) \vee \ldots \vee \neg f(u_i^n)$ which is a propositional logic formula. Then, define
    $$g(c) = \left(\bigwedge_{U_i \in M_c} \rho(U_i) \right) \wedge h(c).$$

    (3) When the arity of c is 0, define $g(c) = \neg h(c) \wedge h(c)$.
- *Define $K_S = \{f(u) \mid u \in U\} \cup \{g(c) \mid c \in C\}$.*

Clearly g is one-to-one because each $g(c)$ is identified with its corresponding unique $h(c)$. Next we show the equivalence between the violation of the requirements C for S and the minimal inconsistent subsets of K_S. A requirement violation causes an inconsistency for S but we need a definition for it.

Definition 5 (Inconsistency of a general information space). *An* inconsistency *of S consists of one of two cases:*

(1) The arity of the requirement c is a positive number k. In this case an inconsistency of S is a set of k information units, $\{u_1, \ldots, u_k\}$, that violates c. We write such an inconsistency as $\{u_1, \ldots, u_k, c\}$.

(2) The arity of the requirement c is 0. If c is violated by S, there is an inconsistency written as $\{c, \neg c\}$.

Then we define $\mathsf{Inc}(S)$ as the set of inconsistencies of S.

Theorem 1. *A general information space S and its transformation to a propositional knowledge base K_S are equivalent for inconsistencies in the sense that there is a bijection $m : \mathsf{Inc}(S) \to \mathsf{MI}(K_S)$. Furthermore, for $M \in \mathsf{Inc}(S)$, $|M| = |m(M)|$.*

Proof. ($\Rightarrow$) Let $M \in \mathsf{Inc}(S)$. Every inconsistency of S contains exactly one requirement.

Suppose the arity of the requirement c is a positive integer k. In this case the inconsistency contains k elements from U, say $u_1, \ldots, u_k$, and a requirement $c \in C$. Then, according to the construction, $g(c)$ is a propositional formula in CNF, one of whose conjuncts is $\neg f(u_1) \vee \cdots \vee \neg f(u_k)$. Hence $\{f(u_1), \ldots, f(u_k), g(c)\} \in \mathsf{MI}(K_S)$. Thus, in this case $m(\{u_1, \ldots, u_k, c\}) = \{f(u_1), \ldots, f(u_k), g(c)\}$. It is clear from the construction that $|M| = |m(M)|$.

In the case where the arity of c is 0, c is inconsistent with respect to U. Then, according to the construction $g(c) = \neg h(c) \wedge h(c)$ which is also a minimal inconsistent set (of size 1) in K_S. Thus, in this case $m(\{\neg c, c\}) = \{\neg h(c), h(c)\}$.

($\Leftarrow$) Let $M \in \mathsf{MI}(K_S)$. Based on the structure of the transformation with the information units transformed to atoms and only the constraints transformed to formulas that may involve negation, there are two cases. In the first case $M = \{a_1, \ldots, a_k, g(c)\}$ where $\neg a_1 \vee \cdots \vee \neg a_k$ is a conjunct in $g(c)$. As g is one-to-one, g^{-1} exists and $g^{-1}(g(c)) = c$. Therefore $m^{-1}(M) = \{f^{-1}(a_1), \ldots, f^{-1}(a_k), c\} \in \mathsf{Inc}(S)$. The other case is where $M = \{\neg b_i \wedge b_i\}$ for some i, $1 \leq i \leq |C|$. This means that $m^{-1}(M) = \{g^{-1}(M)\}$ which is exactly a constraint that is inconsistent (with respect to U). In both cases the construction gives $|M| = |m^{-1}(M)|$.

Our approach to measuring the inconsistency of a general information space according to a propositional inconsistency measure I_x is to apply the inconsistency measure to the transformed space, that is, we define $I_x(S) = I_x(K_S)$.

In the rest of this paper, we consider several examples to illustrate this process. In each case we first describe the general information space, then do the transformation, and finally compute the inconsistency measures according to the propositional inconsistency measures of Definition 2.

5 Examples of Instantiation

We now show how the transformation and the calculation of the inconsistency measures work on three general information spaces: a relational database, a graph database, and a Blocks world scenario. For the relational database case, we start by recalling the formal definitions of database scheme, database instance, and integrity constraint, and then illustrate how these concepts correspond to the components of a general information space. Then an example is provided. For the other two cases, we directly give the examples.

5.1 A Relational Database as a General Information Space

A *relation scheme* $R(A_1, \ldots, A_n)$ consists of a relation name R and a sorted list of attribute names $A_1, \ldots, A_n$, where each attribute A_i (with $i \in [1..n]$) has associated a domain $DOM(A_i)$. A *database scheme* $\mathscr{DS}$ is a nonempty finite set of relation schemes. A *tuple* $\mathbf{t} = (v_1, \ldots, v_n)$ over $R(A_1, \ldots, A_n)$ is a mapping assigning to each attribute A_i of R a value $v_i \in DOM(A_i)$. A *relation instance* is a set of tuples over a given relation scheme, and a *database instance* D is a set of relation instances over a given database scheme. We use u_i for terms. An atom over a database scheme $\mathscr{DS}$ is an expression of one of the following forms: (a) $R(u_1, \ldots, u_n)$ where R is a relation scheme in $\mathscr{DS}$, or (b) $u_i \circ u_j$ where $\circ \in \{=, \neq, >, <, \geq, \leq\}$. An *integrity constraint* over $\mathscr{DS}$ is any (function-free) first-order sentence over the database scheme $\mathscr{DS}$. For a database scheme $\mathscr{DS}$ and a set $\mathscr{C}$ of integrity constraints over $\mathscr{DS}$, an instance D of $\mathscr{DS}$ is said to be *consistent* w.r.t. $\mathscr{C}$ (or, equivalently, $\mathscr{C}$ is *satisfied* by D, $\mathscr{C}$ is not *violated* by D) iff $D \models \mathscr{C}$ in the standard model-theoretic sense.

The components of the general information space $S = \langle F, U, C \rangle$ for a relational database instance D over the database scheme $\mathscr{DS}$ with a set $\mathscr{C}$ of integrity constraints are as follows. The framework F is the database scheme $\mathscr{DS}$ and the (function-free) first-order language using a set of uninterpreted constants and predicate symbols for relation names, as well as domains of the attributes for the evaluation of constants. The set U of information units is the instance D (the set of the tuples in the relation instances), and the set C of requirements is the set $\mathscr{C}$ of integrity constraints.

We now provide an example. Let the framework F be the database scheme consisting of the relation schemes *Asset(SN, DateLoaned, Employee, Date Returned)* whose instance contains the serial number, the loan date, the employee' identifier, and the returned date of assets provided by a company to the employees, about whom information is stored in two relations: *Employee(SSN, Name, HiringDate)*, and *Family(SSN, Child, Project)*. Here U, the database instance, is shown in the usual tabular form in Fig. 1. Altogether there are 13 information units (tuples) that for convenience we name t_i, with $1 \leq i \leq 13$.

There are 8 requirements in C as given below both as first-order logic formulas and in English.

- $c_1 = \forall x_1 \ldots x_4 [Asset(x_1, x_2, x_3, x_4) \rightarrow x_2 \leq x_4]$, stating that, for every asset, the loan date must predate the return date.
- $c_2 = \forall x_1 \ldots x_7 [Asset(x_1, x_2, x_3, x_4) \wedge Asset(x_1, x_5, x_6, x_7) \rightarrow (x_2 = x_5 \wedge x_3 = x_6 \wedge x_4 = x_7)]$, i.e. the constraint that the serial number is a key for *Asset*.

Asset

Atom	SN	DateLoaned	Employee	DateReturned	Tuple
a_1	999	2015-02-01	123456789	2016-03-15	t_1
a_2	999	2015-02-01	123456789	2018-12-31	t_2
a_3	999	2013-06-15	222222222	2017-12-31	t_3
a_4	888	2016-12-01	222222222	2013-12-01	t_4
a_5	555	2014-07-01	333333333	2013-06-20	t_5
a_6	666	2014-07-01	333333333	2015-09-10	t_6
a_7	777	2014-07-01	333333333	2014-05-21	t_7

Employee

Atom	ID	Name	HiringDate	Tuple
a_8	333333333	Robert	1980-01-01	t_8
a_9	444444444	William	1975-06-01	t_9
a_{10}	123456789	William	1975-06-01	t_{10}

Family

Atom	ID	Child	Project	Tuple
a_{11}	123456789	Steve	Q1	t_{11}
a_{12}	123456789	Mary	Q2	t_{12}
a_{13}	123456789	Steve	Q2	t_{13}

Fig. 1. Instances of *Asset*, *Employee*, and *Family*

- $c_3 = \forall x_1 \ldots x_8 [\ Asset(x_1,x_2,x_3,x_4) \wedge Asset(x_5,x_2,\ x_3,x_6) \wedge\ Asset(x_7,x_2,x_3,x_8) \rightarrow (x_1 = x_5 \vee x_1 = x_7 \vee x_5 = x_7)]$, stating the numerical dependency [15,16] *DateLoaned, Employee* $\rightarrow^2 SN$ whose meaning is that for every date and employee there can be at most 2 assets loaned.
- $c_4 = \forall x_1 \ldots x_5 [\ Employee(x_1,x_2,x_3) \wedge Employee(x_1,\ x_4,x_5) \rightarrow (x_2 = x_4 \wedge x_3 = x_5)]$, stating that *ID* is a key for *Employee*.
- $c_5 = \forall x_1 \ldots x_4 [\ Employee(x_1,x_2,x_3) \wedge Employee(x_4,\ x_2,x_3) \rightarrow x_1 = x_4]$, that is, the pair of attributes *Name* and *HiringDate* also form a key for *Employee*.
- $c_6 = \forall x_1 \ldots x_6 [Asset(x_1,x_2,x_3,x_4) \rightarrow \exists\ x_5,x_6\ Employee\ (x_3,x_5,x_6)]$ i.e., the inclusion dependency *Asset*[*Employee*] $\subseteq$ *Employee*[*ID*].
- $c_7 = \forall x_1 \ldots x_5 [\ Family(x_1,x_2,x_3) \wedge Family(x_1,x_4,\ x_5) \rightarrow Family(x_1,x_2,x_5)]$, i.e. the multivalued dependency [8] *Family: ID* $\rightarrow\rightarrow$ *Child*.
- $c_8 = \exists x_1 \ldots x_6 [\ Family(x_1,x_2,x_3) \wedge Family(x_4,x_5,\ x_6) \wedge x_1 \neq x_4)]$ stating that there must be at least two employees referenced in the *Family* relation.

Transformation to a Propositional Knowledge Base. We now show how the transformation from a general information space to a propositional knowledge base is applied for this relational database.

1. $A_U = \{a_1, \ldots, a_{13}\}$ corresponding to the 13 tuples,
2. $f(t_i) = a_i$ for all i, $1 \le i \le 13$.
3. $B_C = \{b_1, \ldots, b_8\}$ corresponding to the 8 constraints.
4. $h(c_i) = b_i$ for all i, $1 \le i \le 8$.
5. $\mathscr{F}_S$ is the set of propositional formulas using $A_U \cup B_C$.
6. Now we show the mapping g by going over the constraints one at a time.

   c_1 The arity of c_1 is 1. The 3 tuples t_4, t_5, and t_7 each violate c_1. Hence, $g(c_1) = \neg a_4 \wedge \neg a_5 \wedge \neg a_7 \wedge b_1$.

   c_2 The arity of c_2 is 2. The 3 tuples t_1, t_2, and t_3 all have the same serial number but are not identical. Hence, $g(c_2) = (\neg a_1 \vee \neg a_2) \wedge (\neg a_1 \vee \neg a_3) \wedge (\neg a_2 \vee \neg a_3) \wedge b_2$.

   c_3 The arity of c_3 is 3. t_5, t_6, and t_7 together violate this constraint. Hence, $g(c_3) = (\neg a_5 \vee \neg a_6 \vee \neg a_7) \wedge b_3$.

   c_4 This constraint is satisfied. Hence, $g(c_4) = b_4$.

   c_5 The arity of c_5 is 2. It is violated by the pair t_9 and t_{10}. Hence, $g(c_5) = (\neg a_9 \vee \neg a_{10}) \wedge b_5$.

   c_6 The arity of c_6 is 1. It is violated separately by t_3 and t_4. Hence, $g(c_6) = \neg a_3 \wedge \neg a_4 \wedge b_6$.

   c_7 The arity of c_7 is 2. It is violated by the pair t_{11} and t_{12}. Hence, $g(c_7) = (\neg a_{11} \vee \neg a_{12}) \wedge b_7$.

   c_8 The arity of c_8 is 0 and it is violated by U. Hence, $g(c_8) = \neg b_8 \wedge b_8$.

7. Therefore $K_S = \{a_1, \ldots, a_{13}, \neg a_4 \wedge \neg a_5 \wedge \neg a_7 \wedge b_1, (\neg a_1 \vee \neg a_2) \wedge (\neg a_1 \vee \neg a_3) \wedge (\neg a_2 \vee \neg a_3) \wedge b_2, (\neg a_5 \vee \neg a_6 \vee \neg a_7) \wedge b_3, b_4, (\neg a_9 \vee \neg a_{10}) \wedge b_5, \neg a_3 \wedge \neg a_4 \wedge b_6, (\neg a_{11} \vee \neg a_{12}) \wedge b_7, \neg b_8 \wedge b_8\}$.

The Calculation of the Inconsistency Measures. Below are the results of calculating the inconsistency measures of the relational database example. We use the fact that
$MI(K_S) = \{\{a_4, \neg a_4 \wedge \neg a_5 \wedge \neg a_7 \wedge b_1\}, \{a_5, \neg a_4 \wedge \neg a_5 \wedge \neg a_7 \wedge b_1\}, \{a_7, \neg a_4 \wedge \neg a_5 \wedge \neg a_7 \wedge b_1\}, \{a_1, a_2, (\neg a_1 \vee \neg a_2) \wedge (\neg a_1 \vee \neg a_3) \wedge (\neg a_2 \vee \neg a_3) \wedge b_2\}, \{a_1, a_3, (\neg a_1 \vee \neg a_2) \wedge (\neg a_1 \vee \neg a_3) \wedge (\neg a_2 \vee \neg a_3) \wedge b_2\}, \{a_2, a_3, (\neg a_1 \vee \neg a_2) \wedge (\neg a_1 \vee \neg a_3) \wedge (\neg a_2 \vee \neg a_3) \wedge b_2\}, \{a_5, a_6, a_7, (\neg a_5 \vee \neg a_6 \vee \neg a_7) \wedge b_3\}, \{a_9, a_{10}, (\neg a_9 \vee \neg a_{10}) \wedge b_5\}, \{a_3, \neg a_3 \wedge \neg a_4 \wedge b_6\}, \{a_4, \neg a_3 \wedge \neg a_4 \wedge b_6\}, \{a_{11}, a_{12}, (\neg a_{11} \vee \neg a_{12}) \wedge b_7\}, \{\neg b_8 \wedge b_8\}\}$.

- $I_B(S) = 1$ as K_S is inconsistent.
- $I_M(S) = 12$ as there are 12 minimal inconsistent subsets for K_S.
- $I_\#(S) = 1 + 5 \times \frac{1}{2} + 5 \times \frac{1}{3} + \frac{1}{4} = \frac{65}{12}$ as there is one minimal inconsistent subset of size 1, 5 of size 2, 5 of size 3, and 1 of size 4 in K_S.
- $I_P(S) = 11 + 7 = 18$ as 11 atoms (i.e., tuples) plus 7 propositional formulas (i.e., constraints) are problematic in K_S.
- $I_H(S) = 7$ as the deletion of the 7 formulas of $g(c_i)$ for all i, $1 \le i \le 3$ and $5 \le i \le 8$ makes K_S consistent and there is no set of smaller cardinality that accomplishes the same.
- $I_{nc}(S) = 21$ as the set $\{\neg b_8 \wedge b_8\}$ has size 1 and is inconsistent.
- $I_C(S) = 8$ as there must be at least 8 atoms, for example a_2, a_3, a_4, a_5, a_7, a_9, a_{11}, and b_8, that must be given the value B for a 3-valued interpretation in order to satisfy all the formulas.

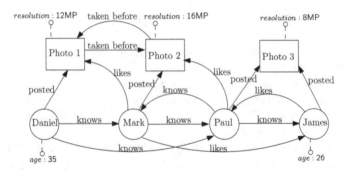

Fig. 2. Graph database instance

5.2 A Graph Database as a General Information Space

We start by giving the components of $S = \langle F, U, C \rangle$. It will be helpful to look at the graph for this example in Fig. 2. We consider a general form of graph databases where vertices may be associated with properties and edges may be labeled [32,33]. The framework F consists of basic information about the vertices and the edges of the graph, that is, the sets of *vertex names*, *edge labels*, and *vertex properties*. Each vertex property has an associated domain, which is a set of values that can be assigned to the property. For instance, a property for the vertices of our graph database example is *type*, whose domain includes person and media. In Fig. 2, the vertices indicated by circles represent people and the vertices indicated by rectangles represent media objects. Moreover, property *age* is associated with people and *resolution* with media objects (their domains are obvious). The edges represent relationships between vertices whose meaning is given by the labels.

The data units are the vertices and the edges. For convenience we number them so that each unit is some u_i starting with the square vertices: $u_1 \ldots u_3$, then the circular vertices: $u_4 \ldots u_7$, and finally the edges: $u_8 \ldots u_{22}$.

u_1 : (Photo 1, 12MP)	u_2 : (Photo 2, 16MP)
u_3 : (Photo 3, 8MP)	u_4 : (Daniel, 35)
u_5 : (Mark)	u_6 : (Paul)
u_7 : (James, 26)	u_8 : (Daniel, posted, Photo 1)
u_9 : (Daniel, knows, Mark)	u_{10} : (Daniel, knows, Paul)
u_{11} : (Mark, likes, Photo 1)	u_{12} : (Mark, posted, Photo 2)
u_{13} : (Mark, knows, Paul)	u_{14} : (Mark, likes, James)
u_{15} : (Paul, knows, Mark)	u_{16} : (Paul, likes, Photo 2)
u_{17} : (Paul, posted, Photo 3)	u_{18} : (Paul, knows, James)
u_{19} : (James, likes, Paul)	u_{20} : (James, posted, Photo 3)
u_{21} : (Photo 1, taken before, Photo 2)	u_{22} : (Photo 2, taken before, Photo 1)

Finally we get to the requirements that are constraints on the graph; we write them in English.

- c_1: Every circular vertex must have an associated age value.
- c_2: Every rectangular vertex must have an associated resolution.
- c_3: There may not be a cycle on rectangular vertices.
- c_4: There cannot be 2 edges with the label "posted" going to the same rectangular vertex.
- c_5: For every edge between circular vertices that has the label "likes" there must be another edge with the label "knows".

Transformation to a Propositional Knowledge Base. We now show how the transformation from a general information space to a propositional knowledge base is applied for this graph database.

1. $A_U = \{a_1, \ldots, a_{22}\}$ corresponding to the 7 vertices and 15 edges.
2. $f(u_i) = a_i$ for all i, $1 \leq i \leq 22$.
3. $B_C = \{b_1, \ldots, b_5\}$ corresponding to the 5 constraints.
4. $h(c_i) = b_i$ for all i, $1 \leq i \leq 5$.
5. $\mathscr{F}_S$ is the set of propositional formulas using $A_U \cup B_C$.
6. Now we show the mapping g by going over the constraints one at a time.

   c_1 The arity of c_1 is 1. The two nodes u_5 and u_6 each violate c_1. Hence, $g(c_1) = \neg a_5 \wedge \neg a_6 \wedge b_1$.

   c_2 This constraint is satisfied. Hence, $g(c_2) = b_2$.

   c_3 This constraint does not have a fixed arity because a cycle does not have a fixed number of elements. However, if it is violated its arity is greater than zero. It is violated by the pair of edges u_{21} and u_{22}. Hence, $g(c_3) = (\neg a_{21} \vee \neg a_{22}) \wedge b_3$.

   c_4 The arity of c_4 is 2. It is violated by the pair of edges u_{17} and u_{20}. Hence, $g(c_4) = (\neg a_{17} \vee \neg a_{20}) \wedge b_4$.

   c_5 The arity of c_5 is 1. The two edges u_{14} and u_{19} each violate c_5. Hence, $g(c_5) = \neg a_{14} \wedge \neg a_{19} \wedge b_5$.

7. Therefore $K_S = \{a_1, \ldots, a_{22}, \neg a_5 \wedge \neg a_6 \wedge b_1, b_2, (\neg a_{21} \vee \neg a_{22}) \wedge b_3, (\neg a_{17} \vee \neg a_{20}) \wedge b_4, \neg a_{14} \wedge \neg a_{19} \wedge b_5\}$.

The Calculation of the Inconsistency Measures. Below are the results of calculating the inconsistency measures of the graph database example. We use the fact that $\mathsf{MI}(K_S) = \{\{a_5, \neg a_5 \wedge \neg a_6 \wedge b_1\}, \{a_6, \neg a_5 \wedge \neg a_6 \wedge b_1\}, \{a_{21}, a_{22}, (\neg a_{21} \vee \neg a_{22}) \wedge b_3\}, \{a_{17}, a_{20}, (\neg a_{17} \vee \neg a_{20}) \wedge b_4\}, \{a_{14}, \neg a_{14} \wedge \neg a_{19} \wedge b_5\}, \{a_{19}, \neg a_{14} \wedge \neg a_{19} \wedge b_5\}\}$.

- $I_B(S) = 1$ as K_S is inconsistent.
- $I_M(S) = 6$ as there are 6 minimal inconsistent subsets for K_S.
- $I_\#(S) = 4 \times \frac{1}{2} + 2 \times \frac{1}{3} = \frac{8}{3}$ as there are 4 minimal inconsistent subsets of size 2 and 2 minimal inconsistent subsets of size 3 for K_S.
- $I_P(S) = 8 + 4 = 12$ as 8 atoms (i.e., vertices and edges) plus 4 propositional formulas (i.e., the transformations of the constraints) are problematic in K_S.
- $I_H(S) = 4$ as the deletion of the 4 formulas: $g(c_1)$, $g(c_3)$, $g(c_4)$, and $g(c_5)$ makes K_S consistent and there is no smaller cardinality set that accomplishes the same.
- $I_{nc}(S) = 27 - 1 = 26$ as there is a minimal inconsistent subset of size 2.
- $I_C(S) = 6$ as a 3-valued interpretation must give at least a_5, a_6, a_{14}, a_{19}, one of a_{21} and a_{22}, and one of a_{17} and a_{20} the value B to satisfy all the formulas.

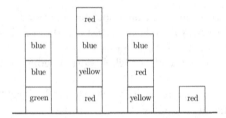

Fig. 3. Blocks world configuration

5.3   A Blocks World Configuration as a General Information Space

Blocks-world planning has been widely investigated in AI, as it captures several aspects of planning systems [17]. In this case, the components of $S = \langle F, U, C \rangle$ are as follows. The framework indicates that there is a finite number of colored blocks of the same size in stacks on a table, which is large enough to hold all (i.e., the number of stacks can be equal to number of blocks).

A blocks world configuration is shown in Fig. 3.

We indicate the data units, which are the stack and the colors of the block in them by writing $st_{i,j} : color$ to indicate that the block in stack i in the jth position has that color. Here the first position means directly on the table, the second position means on top of the first block, and so on.

The data units in this example are as follows:

$st_{11} : green$	$st_{12} : blue$	$st_{13} : blue$	
$st_{21} : red$	$st_{22} : yellow$	$st_{23} : blue$	$st_{24} : red$
$st_{31} : yellow$	$st_{32} : red$	$st_{33} : blue$	$st_{41} : red$

So there are 4 stacks with 3, 4, 3, and 1 blocks respectively.

Finally we get to the requirements that are the constraints in this Blocks world; we write them in English.

- c_1: No blue block can be on top of another blue block.
- c_2: There cannot be a yellow block that has a red block below it and a red block above it.
- c_3: There cannot be a red block on the table (i.e. at the bottom of a stack).
- c_4: No stack has both a green block and a blue block.
- c_5: At least one of the blocks is purple.
- c_6: There must be a blue block in at least 3 stacks.

Transformation to a Propositional Knowledge Base. We now show how the transformation from a general information space to a propositional knowledge base is applied for this blocks world example.

1. $A_U = \{a_1, \ldots, a_{11}\}$ corresponding to the 11 blocks.
2. $f(st_{11}) = a_1$, $f(st_{12}) = a_2$, $f(st_{13}) = a_3$, $f(st_{21}) = a_4$, $f(st_{22}) = a_5$, $f(st_{23}) = a_6$, $f(st_{24}) = a_7$, $f(st_{31}) = a_8$, $f(st_{32}) = a_9$, $f(st_{33}) = a_{10}$, $f(st_{41}) = a_{11}$,

3. $B_C = \{b_1, \ldots, b_6\}$ corresponding to the 6 constraints.
4. $h(c_i) = b_i$ for all i, $1 \leq i \leq 6$.
5. $\mathscr{F}_S$ is the set of propositional formulas using $A_U \cup B_C$.
6. Now we show the mapping g by going over the constraints one at a time.

   c_1 The arity of c_1 is 2. The two blocks st_{12} and st_{13} together violate c_1. Hence,
   $g(c_1) = (\neg a_2 \vee \neg a_3) \wedge b_1$.

   c_2 The arity of c_2 is 3. The 3 blocks that together violate this constraint are st_{21}, st_{22},
   and st_{24}. Hence, $g(c_2) = (\neg a_4 \vee \neg a_5 \vee \neg a_7) \wedge b_2$.

   c_3 The arity of c_3 is 1. The blocks st_{21} and st_{41} both violate this constraint. Hence,
   $g(c_3) = \neg a_4 \wedge \neg a_{11} \wedge b_3$.

   c_4 The arity of c_4 is 2. The blocks st_{11} and st_{12} as well as the blocks st_{11} and st_{13}
   violate this constraint. Hence, $g(c_4) = (\neg a_1 \vee \neg a_2) \wedge (\neg a_1 \vee \neg a_3) \wedge b_4$.

   c_5 The arity of c_5 is 0. There is no purple block in any stack. Hence, $g(c_5) = \neg b_5 \wedge$
   b_5.

   c_6 This constraint is satisfied. Hence, $g(c_6) = b_6$.

7. Therefore $K_S = \{a_1, \ldots, a_{11}, (\neg a_2 \vee \neg a_3) \wedge b_1, (\neg a_4 \vee \neg a_5 \vee \neg a_7) \wedge b_2, \neg a_4 \wedge \neg a_{11} \wedge$
   $b_3, (\neg a_1 \vee \neg a_2) \wedge (\neg a_1 \vee \neg a_3) \wedge b_4, \neg b_5 \wedge b_5, b_6\}$.

The Calculation of the Inconsistency Measures. Below are the results of calculating
the inconsistency measures of the Blocks world example. We use the fact that $\text{MI}(K_S) =$
$\{\{a_2, a_3, (\neg a_2 \vee \neg a_3) \wedge b_1\}, \{a_4, a_5, a_7, (\neg a_4 \vee \neg a_5 \vee \neg a_7) \wedge b_2\}, \{a_4, \neg a_4 \wedge \neg a_{11} \wedge$
$b_3\}, \{a_{11}, \neg a_4 \wedge \neg a_{11} \wedge b_3\}, \{a_1, a_2, (\neg a_1 \vee \neg a_2) \wedge (\neg a_1 \vee \neg a_3) \wedge b_4\}, \{a_1, a_3, (\neg a_1 \vee$
$\neg a_2) \wedge (\neg a_1 \vee \neg a_3) \wedge b_4\}, \{\neg b_5 \wedge b_5\}\}$.

- $I_B(S) = 1$ as K_S is inconsistent.
- $I_M(S) = 7$ as there are 7 minimal inconsistent subsets for K_S.
- $I_\#(S) = 1 + 2 \times \frac{1}{2} + 3 \times \frac{1}{3} + 1 \times \frac{1}{4} = \frac{13}{4}$ as there is 1 minimal inconsistent subset of
  size 1, 2 minimal inconsistent subsets of size 2, 3 minimal inconsistent subsets of
  size 3, and 1 minimal inconsistent subset of size 4 for K_S.
- $I_P(S) = 7 + 5 = 12$ as 7 atoms (i.e., colored block locations) plus 5 propositional
  formulas (i.e., the transformations of the requirements) are problematic in K_S.
- $I_H(S) = 5$ as the deletion of the 5 formulas: $g(c_1)$, $g(c_2)$, $g(c_3)$, $g(c_4)$, and $g(c_5)$
  makes K_S consistent and there is no smaller cardinality set that accomplishes the
  same.
- $I_{nc}(S) = 17$ as there is a minimal inconsistent subset of size 1.
- $I_C(S) = 5$ as a 3-valued interpretation that satisfies all the formulas must give a_4,
  a_{11}, b_5, and at least 2 other atoms, for example, a_1 and a_2 the value B.

6   Conclusions and Future Work

As inconsistency in real-world information systems can not be easily avoided, many
inconsistency-tolerant approaches have been developed to live with inconsistency [9],
and provide appropriate mechanisms to handle inconsistent data [3,7]. A key issue in
such situations is measuring the amount of inconsistency to assess its nature and understand the degree of the dirtiness of data.

In this paper, we developed a general approach for measuring inconsistency in general information spaces which encompasses various ways in which information is stored in real-world systems. An important advantage of defining inconsistency measures for general information spaces is its wide range of use. Consider an inconsistency measure defined specifically for relational databases. Such a measure allows for comparing two relational databases and determining if one is less inconsistent than the other one or if they have the same inconsistency. But using the concept of a general information space and the uniformity of the definition of an inconsistency measure allows for comparing the inconsistency of a relational database and a graph database or a Block worlds scenario.

For instance, we can compare the inconsistency measures obtained for the 3 examples using the 6 inconsistency measures we have considered. These inconsistency measures measure different aspects of the inconsistencies. Hence we find that for some measures one of the examples is more inconsistent than another but for a different measure it is the reverse. So for comparing these general information spaces we need to decide which aspect we really want to measure. In particular, measuring the number of inconsistent subsets, that is, using I_M, is a good way to get a general sense of the amount of inconsistency. Using that measure we find that the answer is largest for the first example, the relational database, where $I_M(S) = 12$. This measure gives almost the same result for the graph database example, 6, and the Blocks world example, 7.

For the sake of the presentation we have focused on showing how our approach works with the inconsistency measures listed in Definition 2. But it is important to observe that the transformation creates a propositional knowledge base; hence all propositional inconsistency measures ever proposed are applicable.

There are several directions that we plan to explore for future work. For one, we will provide more instantiations of our framework. Recently, there has been some work on measuring inconsistency in spatio-temporal databases [14]. Interestingly, a spatio-temporal database can be viewed as a general information space where information units are the possible positions of objects in time and the requirements are entailed by the spatio-temporal atoms along with the spatio-temporal constraints (e.g. an object cannot be in two different places at the same time). However, how to encode probabilistic spatio-temporal knowledge bases [29,30], and more in general probabilistic information, into a general information space needs further investigation that may lead to define a concept of *probabilistic* general information space.

We plan to consider complexity issues as well. One aspect is the complexity of the transformation itself. The second aspect is the complexity of the calculation of the inconsistency measures. For general propositional knowledge bases this calculation is not polynomial [36]. However, the result of the transformation is a set of formulas that must have a specific form and may have lower complexity. Finally, we also plan to study what aspects of inconsistency the various inconsistency measures actually measure to determine which ones are the most appropriate to use for general information spaces.

References

1. Foto, N.A., Kolaitis, P.G.: Repair checking in inconsistent databases: algorithms and complexity. In: Proceedings of International Conference on Database Theory (ICDT), pp. 31–41 (2009)
2. Bertossi, L.E.: Repair-based degrees of database inconsistency. In: Proceedings of International Conference on Logic Programming and Nonmonotonic Reasoning (LPNMR), pp. 195–209 (2019)
3. Bertossi, L.E., Hunter, A., Schaub, T.: Introduction to inconsistency tolerance. In: Inconsistency Tolerance, pp. 1–14 (2005)
4. Codd, E.F.: The Relational Model for Database Management, Version 2. Addison-Wesley, Boston (1990)
5. De Bona, G., Grant, J., Hunter, A., Konieczny, S.: Towards a unified framework for syntactic inconsistency measures. In: Proceedings of the Thirty-Second AAAI Conference on Artificial Intelligence, (AAAI), pp. 1803–1810 (2018)
6. Doder, D., Raskovic, M., Markovic, Z., Ognjanovic, Z.: Measures of inconsistency and defaults. Int. J. Approximate Reasoning 51(7), 832–845 (2010)
7. Eiter, T., Fink, M., Schüller, P., Weinzierl, A.: Finding explanations of inconsistency in multi-context systems. Artif. Intell. 216, 233–274 (2014)
8. Fagin, R.: Multivalued dependencies and a new normal form for relational databases. ACM Trans. Database Syst. 2(3), 262–278 (1977)
9. Gabbay, D.M., Hunter, A.: Making inconsistency respectable: a logical framework for inconsistency in reasoning. In: Proceedings of International Workshop on Fundamentals of Artificial Intelligence Research (FAIR), pp. 19–32 (1991)
10. Grant, J., Hunter, A.: Measuring inconsistency in knowledgebases. J. Intell. Inf. Syst. 27(2), 159–184 (2006)
11. Grant, J., Hunter, A.: Measuring consistency gain and information loss in stepwise inconsistency resolution. In: Proceedings of European Conference Symbolic and Quantitative Approaches to Reasoning with Uncertainty (ECSQARU), pp. 362–373 (2011)
12. Grant, J., Hunter, A.: Distance-based measures of inconsistency. In: Proceedings of ECSQARU, pp. 230–241 (2013)
13. Grant, J., Martinez, M.V.: Measuring Inconsistency in Information. College Publications (2018)
14. Grant, J., Martinez, M.V., Molinaro, C., Parisi, F.: On measuring inconsistency in spatio-temporal databases. In: Grant, J., Martinez, M.V. (eds.) Measuring Inconsistency in Information, volume 73 of Studies in Logic, pp. 313–342. College Publications (2018)
15. Grant, J., Minker, J.: Inferences for numerical dependencies. Theoret. Comput. Sci. 41, 271–287 (1985)
16. Grant, J., Minker, J.: Normalization and axiomatization for numerical dependencies. Inf. Control 65(1), 1–17 (1985)
17. Gupta, N., Nau, D.S.: Complexity results for blocks-world planning. In: Proceedings of the 9th National Conference on Artificial Intelligence (AAAI), pp. 629–633 (1991)
18. Hunter, A.: Measuring inconsistency in knowledge via quasi-classical models. In: Proceedings of National Conference on Artificial Intelligence and Conference on Innovative Applications of Artificial Intelligence (AAAI/IAAI), pp. 68–73 (2002)
19. Hunter, A., Konieczny, S.: Approaches to measuring inconsistent information. In: Inconsistency Tolerance, pp. 191–236 (2005)
20. Hunter, A., Konieczny, S.: Measuring inconsistency through minimal inconsistent sets. In: Proceedings of International Conference on Principles of Knowledge Representation and Reasoning (KR), pp. 358–366 (2008)

21. Hunter, A., Konieczny, S.: On the measure of conflicts: Shapley inconsistency values. Artif. Intell. **174**(14), 1007–1026 (2010)
22. Jabbour, S., Ma, Y., Raddaoui, B., Sais, L.: Quantifying conflicts in propositional logic through prime implicates. Int. J. Approximate Reasoning **89**, 27–40 (2017)
23. Knight, K.: Measuring inconsistency. J. Philos. Logic **31**(1), 77–98 (2002)
24. Ma, Y., Qi, G., Xiao, G., Hitzler, P., Lin, Z.: An anytime algorithm for computing inconsistency measurement. In: Proceedings of International Conference on Knowledge Science, Engineering and Management (KSEM), pp. 29–40 (2009)
25. Martinez, M.V., Pugliese, A., Simari, G.I., Subrahmanian, V.S., Prade, H.: How dirty is your relational database? An axiomatic approach. In: Proceedings of European Conference Symbolic and Quantitative Approaches to Reasoning with Uncertainty (ECSQARU), pp. 103–114 (2007)
26. McAreavey, K., Liu, W., Miller, P.C.: Computational approaches to finding and measuring inconsistency in arbitrary knowledge bases. Int. J. Approximate Reasoning **55**(8), 1659–1693 (2014)
27. Mu, K., Jin, Z., Lu, R., Liu, W.: Measuring inconsistency in requirements specifications. In: Proceedings of European Conference on Symbolic and Quantitative Approaches to Reasoning with Uncertainty (ECSQARU), pp. 440–451 (2005)
28. Kedian, M., Liu, W., Jin, Z., Bell, D.A.: A syntax-based approach to measuring the degree of inconsistency for belief bases. Int. J. Approximate Reasoning **52**(7), 978–999 (2011)
29. Parisi, F., Grant, J.: Knowledge representation in probabilistic spatio-temporal knowledge bases. J. Artif. Intell. Res. **55**, 743–798 (2016)
30. Parisi, F., Grant, J.: On repairing and querying inconsistent probabilistic spatio-temporal databases. Int. J. Approximate Reasoning **84**, 41–74 (2017)
31. Parisi, F., Grant, J.: Inconsistency measures for relational databases. CoRR, abs/1904.03403 (2019)
32. Parisi, F., Park, N., Pugliese, A., Subrahmanian, V.S.: Top-k user-defined vertex scoring queries in edge-labeled graph databases. ACM Trans. Web (TWEB) **12**(4), 211–2135 (2018)
33. Robinson, I., Webber, J., Eifrem, E.: Graph Databases. O'Reilly Media Inc., Sebastopol (2013)
34. Thimm, M.: On the expressivity of inconsistency measures. Artif. Intell. **234**, 120–151 (2016)
35. Thimm, M.: Stream-based inconsistency measurement. Int. J. Approximate Reasoning **68**, 68–87 (2016)
36. Thimm, M., Wallner, J.P.: On the complexity of inconsistency measurement. Artif. Intell. **275**, 411–456 (2019)
37. Vardi, M.Y.: The complexity of relational query languages (extended abstract). In Proceedings of Symposium on Theory of Computing (STOC), pp. 137–146 (1982)
38. Xiao, G., Ma, Y.: Inconsistency measurement based on variables in minimal unsatisfiable subsets. In Proceedings of 20th European Conference on Artificial Intelligence (ECAI), pp. 864–869 (2012)
39. Zhang, X., Wang, K., Wang, Z., Ma, Y., Qi, G., Feng, Z.: A distance-based framework for inconsistency-tolerant reasoning and inconsistency measurement in dl-lite. Int. J. Approximate Reasoning **89**, 58–79 (2017)
40. Zhou, L., Huang, H., Qi, G., Ma, Y., Huang, Z., Qu, Y.: Measuring inconsistency in dl-lite ontologies. In: Proceedings of International Conference on Web Intelligence (WI), pp. 349–356 (2009)

Parameterised Complexity
of Model Checking and Satisfiability
in Propositional Dependence Logic

Yasir Mahmood$^{(\boxtimes)}$ and Arne Meier

Leibniz Universität Hannover, Institut für Theoretische Informatik,
Hannover, Germany
{mahmood,meier}@thi.uni-hannover.de

Abstract. In this paper, we initiate a systematic study of the parameterised complexity in the field of Dependence Logics which finds its origin in the Dependence Logic of Väänänen from 2007. We study a propositional variant of this logic (PDL) and investigate a variety of parameterisations with respect to the central decision problems. The model checking problem (MC) of PDL is NP-complete (Ebbing and Lohmann, SOFSEM 2012). The subject of this research is to identify a list of parameterisations (formula-size, formula-depth, treewidth, team-size, number of variables) under which MC becomes fixed-parameter tractable. Furthermore, we show that the number of disjunctions or the arity of dependence atoms (dep-arity) as a parameter both yield a paraNP-completeness result. Then, we consider the satisfiability problem (SAT) which classically is known to be NP-complete as well (Lohmann and Vollmer, Studia Logica 2013). There we are presenting a different picture: under team-size, or dep-arity SAT is paraNP-complete whereas under all other mentioned parameters the problem is in FPT. Finally, we introduce a variant of the satisfiability problem, asking for teams of a given size, and show for this problem an almost complete picture.

Keywords: Propositional dependence logic · Parameterised complexity · Model checking · Satisfiability

1 Introduction

The logics of dependence and independence are a recent innovation studying such central formalisms occurring in several areas of research: computer science, logic, statistics, game theory, linguistics, philosophy, biology, physics, and social choice theory [18]. Väänänen [35] initiated this subfield of research in 2007, and nowadays, it is a vibrant area of study [1]. Its focus widened from initially first-order dependence logic further to modal logic [36], temporal logics [24,25], probabilistic logics [12], logics for independence [23], inclusion logics [16,22], multi-team semantics [11], and poly-team semantics [20].

Funded by German Research Foundation (DFG), project ME 4279/1-2.

© Springer Nature Switzerland AG 2020
A. Herzig and J. Kontinen (Eds.): FoIKS 2020, LNCS 12012, pp. 157–174, 2020.
https://doi.org/10.1007/978-3-030-39951-1_10

Table 1. (Left) An example database with 4 attributes and universe size 15. (Right) An encoding with $\lceil \log_2(3) \rceil + \lceil \log_2(3) \rceil + \lceil \log_2(5) \rceil + \lceil \log_2(4) \rceil$ many propositional variables.

Instructor	Room	Time	Course	$i_1 i_2$	$r_1 r_2$	$t_1 t_2 t_3$	$c_1 c_2$
Antti	A.10	09.00	Logic	00	11	110	11
Antti	A.10	11.00	Statistics	00	11	111	00
Antti	B.20	15.00	Algebra	00	00	000	01
Jonni	C.30	10.00	LAB	01	01	001	10
Juha	C.30	10.00	LAB	10	01	001	10
Juha	A.10	13.00	Statistics	10	11	010	00

In this paper, we study a sub-logic of the modal variant which is called propositional dependence logic ($\mathcal{PDL}$) [21,37]. The main concept also in this logic, the *dependence atom* $\mathsf{dep}(P;Q)$, intuitively states that the variables $p \in P$ functionally determine the values of the variables $q \in Q$. As functional dependence only makes sense on sets of assignments, which Väänänen called *teams*, team-semantics are the heart of the satisfaction relation $\models$ in this logic. Formally, a team T is a set of classical propositional assignments $t \colon \mathrm{VAR} \to \{0,1\}$, and $T \models \mathsf{dep}(P;Q)$ if and only if for all $t, t' \in T$, we have that t and t' agree on the variables in P implies t and t' agree on variables in Q.

The model checking question (MC), given a team T and a $\mathcal{PDL}$-formula φ, asks if $T \models \varphi$ is true. The satisfiability problem (SAT), given a $\mathcal{PDL}$-formula φ, asks for the existence of a team T such that $T \models \varphi$. It is known that MC as well as SAT are **NP**-complete by Ebbing and Lohmann [14], respectively, by Lohmann and Vollmer [27]. These authors classify the complexity landscape of even operator-fragments of $\mathcal{PDL}$ yielding a deep understanding of these problems from a classical complexity point of view. For an overview of how other atoms (e.g., inclusion, or independence) influence the complexity of these problems consider the tables in the work Hella et al. [22].

Example 1. We illustrate an example from relational databases providing understanding of team logics. Table 1 depicts a database which can be expressed in $\mathcal{PDL}$ via binary encoding of the possible entries for the attributes. The set of rows then corresponds to a team T. The database satisfies two functional dependencies:

$\mathsf{dep}(\{\mathtt{Room}, \mathtt{Time}\}; \{\mathtt{Course}\})$ and $\mathsf{dep}(\{\mathtt{Instructor}, \mathtt{Time}\}; \{\mathtt{Room}, \mathtt{Course}\})$.

Whereas, it does not satisfy $\mathsf{dep}(\{\mathtt{Room}, \mathtt{Time}\}; \{\mathtt{Instructor}\})$ as witnessed by the tuples (Juha, C.30, 10, LAB) and (Jonni, C.30, 10, LAB). Formally, we have that

$$T \models \mathsf{dep}(\{\mathtt{Room}, \mathtt{Time}\}; \{\mathtt{Course}\}) \wedge \mathsf{dep}(\{\mathtt{Instructor}, \mathtt{Time}\}; \{\mathtt{Room}, \mathtt{Course}\}),$$

but

$$T \not\models \mathsf{dep}(\{\mathtt{Room}, \mathtt{Time}\}; \{\mathtt{Instructor}\}).$$

Notice that in propositional logic, we cannot express a table of so many values. As a result, we need to binary encode the values of each column separately. This might cause a logarithmic blow-up (by binary encoding the universe values for each column) in the parameter values, for example, it influences the number of variables. Furthermore, one also has to rewrite variables in the occurring formulas accordingly. For instance, as in Table 1, for $\mathsf{dep}(\{\mathtt{Room}, \mathtt{Time}\}; \{\mathtt{Instructor}\})$ this would yield the formula $\mathsf{dep}(\{r_1, r_2, t_1, t_2, t_3\}; \{i_1, i_2\})$. The parameters discussed in this paper correspond to the already encoded values. This means that there is no need in considering this blow-up as in this example.

Often, when a problem is shown to be intrinsic hard, a possible way to further unravel the true reasons for the intractability is the framework of parameterised complexity theory [10]. Here, one aims for a more fine-grained complexity analysis involving the study of parameterisations and how they pin causes for intractability substantially. One distinguishes two runtimes of a different quality: $f(k) \cdot p(|x|)$ versus $p(|x|)^{f(k)}$, where f is an arbitrary computable function, p is a polynomial, $|x|$ the input length and k the value of the parameter. Clearly, both runtimes are polynomial in x for each fixed k but the first one is much better as the polynomial degree is independent of the parameter's value. Problems that can be solved with algorithms running in a time of the first kind are said to be fixed-parameter tractable (or **FPT**). Whereas, problems of category two are in the complexity class **XP**. It is known that $\mathbf{FPT} \subsetneq \mathbf{XP}$ [15]. Whenever runtimes of the form $f(k) \cdot p(|x|)$ are considered with respect to nondeterministic machines, one studies the complexity class $\mathbf{paraNP} \supseteq \mathbf{FPT}$. In between these two classes a presumably infinite **W**-hierarchy is contained: $\mathbf{FPT} \subseteq \mathbf{W[1]} \subseteq \mathbf{W[2]} \subseteq \cdots \subseteq \mathbf{paraNP}$. It is unknown whether any of these inclusions is strict. Showing $\mathbf{W[1]}$-hardness of a problem intuitively corresponds to being intractable in the parameterised world.

The area of research of parameterised problems is tremendously growing and often provides new insights into the inherent difficulty of the studied problems [9]. However, the area of dependence logic is rather blank with respect to this direction of research, only Meier and Reinbold [30] investigated the (parameterised) enumeration complexity of a fragment of $\mathcal{PDL}$ recently. As a subject of this research, we want to initiate and to further push a study of the parameterised complexity of problems in these logics.

Applications. The teams in the team semantic bear a close resemblance with the relations studied in relational database theory. Moreover, dependence atoms are analogous to functional dependencies in the context of database systems. The MC problem for dependence logic, for example, is equivalent to determining whether a relation in the database satisfies a functional dependency. The teams of $\mathcal{PDL}$ also relate to theinformation states of inquisitive logic [7]; a semantic framework for the study of the notion of meaning and information exchange among agents.

Contributions. We study a wealth of parameters, also relevant from the perspective of database theory. Specifically, the parameter team-size corresponds to the

Table 2. Complexity classification overview showing the results of the paper with pointers to the theorems. All results are completeness results. The question mark symbol means that the precise complexity is unknown.

Parameter	MC	SAT	m-SAT
formula-tw	**paraNP**[14]	**FPT**[22]	?
formula-team-tw	**FPT**[19]	see above	see above
team-size	**FPT**[16]	**paraNP**[21]	**paraNP**[26]
formula-size	**FPT**[18]	**FPT**[24]	**FPT**[25]
formula-depth	**FPT**[18]	**FPT**[24]	**FPT**[25]
#variables	**FPT**[18]	**FPT**[24]	**FPT**[25]
#splits	**paraNP**[17]	**FPT**[23]	?
dep-arity	**paraNP**[15]	**paraNP**[21]	**paraNP**[26]

number of entries in the database and #variables is the number of attributes. The parameter formula-tw denotes how much interleaving is present among the attributes in the query and dep-arity bounds the size of functional dependencies in the query. Furthermore, the parameter formula-team-tw bounds the interleaving between a query and the database, formula-size limits the size of the query, formula-depth restricts the nesting depth of the query, and #splits controls the unions in relational algebra queries. With respect to all parameters, we study MC and SAT. Furthermore, we introduce a satisfiability variant m-SAT, which has an additional unary input $m \in \mathbb{N}$, and asks for a satisfying team of size exactly m.

In Table 2, we give an overview of our results. In this article, we prove dichotomies for MC and SAT: depending on the parameter the problem is either fixed-parameter tractable or **paraNP**-complete. Only the satisfiability variant under the parameters formula-tw and #splits resist a complete classification and are left for further research.

Related Work. The notion of treewidth is due to Robertson and Seymour [32]. The study of the complexity of bounded treewidth query evaluation is a vibrant area of research [3–6,13,19]. As stated earlier, the formulas of dependence logic correspond to the functional dependencies in the database context. Bläsius et al. [2] study the parameterised complexity of dependency detection. The problem is defined as, given a database T and a positive integer k whether there is a non-trivial functional dependency of size (dep-arity in our notion) at most k that is satisfied by T. These authors prove that this problem is **W**[2]-complete.

Organisation of the Article. At first, we introduce some required notions and definitions in (parameterised) complexity theory, dependence logic, and propositional logic. Then we study the parameterised complexity of the model checking problem. We proceed with the satisfiability problem and study a variant of it. Finally, we conclude and discuss open questions. For results marked with a ($\star$) their proof can be found in the related full version [29].

2   Preliminaries

In this paper, we assume familiarity with standard notions in complexity theory [31] such as the classes **NP** and **P**.

2.1   Parameterised Complexity

We will recapitulate some relevant notion of parameterised complexity theory, now. For a broader introduction consider the textbook of Downey and Fellows [10], or Flum and Grohe [15]. A parameterised problem (PP) $\Pi \subseteq \Sigma^* \times \mathbb{N}$ consists of tuples (x, k), where x is called the *instance* and k the *(value of the) parameter*.

Definition 2 (Fixed-parameter tractable and paraNP). *Let Π be a PP over $\Sigma^* \times \mathbb{N}$. We say that Π is* fixed-parameter tractable *(or is in the class* **FPT***) if there exists a deterministic algorithm $\mathcal{A}$ deciding Π in time $f(k) \cdot |x|^{O(1)}$ for every input $(x, k) \in \Sigma^*$, where f is a computable function. If $\mathcal{A}$ is a nondeterministic algorithm instead, then Π belongs to the class* **paraNP***.*

Let P be a PP over $\Sigma^* \times \mathbb{N}$. Then the *$\ell$-slice* of P, for $\ell \geq 0$, is the set $P_\ell := \{ x \mid (x, \ell) \in P \}$. It is customary to use the notation $O^\star(f(k))$ to denote the runtime dependence only on the parameter and to ignore the polynomial factor in the input. We will use the following result from parameterised complexity theory to prove **paraNP**-hardness results.

Proposition 3 ([15, Theorem 2.14]). *Let P be a PP. If there exists an $\ell \geq 0$ such that P_ℓ is* **NP***-complete, then P is* **paraNP***-complete.*

Moreover, we will use the following folklore result to get several upper bounds.

Proposition 4. *Let Q be a problem such that (Q, k) is in* **FPT** *and let ℓ be another parameter such that $k \leq f(\ell)$ for some computable function f, then (Q, ℓ) is also in* **FPT***.*

2.2   Propositional Dependence Logic

Let VAR be a countably infinite set of variables. The syntax of propositional dependence logic ($\mathcal{PDL}$) is defined via the following EBNF:

$$\varphi ::= \top \mid \bot \mid x \mid \neg x \mid \varphi \vee \varphi \mid \varphi \wedge \varphi \mid \mathsf{dep}(X; Y) \mid \neg\mathsf{dep}(X; Y),$$

where $\top$ is *verum*, $\bot$ is *falsum*, $x \in$ VAR is a variable, $X, Y \subset$ VAR are finite sets of variables, $\mathsf{dep}(\cdot; \cdot)$ is called the *dependence atom*, and the disjunction $\vee$ is also called *split-junction*. Observe that we only consider atomic negation. We let $\mathcal{PL}$ be defined as the $\mathcal{PDL}$-formulas without $\mathsf{dep}(\cdot; \cdot)$. Finally, the set X in $\mathsf{dep}(X; Y)$ can be empty, giving rise to formulas of the form $\mathsf{dep}(; Y)$. To simplify the notation, when either set in the arguments of $\mathsf{dep}(X; Y)$ is singleton then we write, for example, $\mathsf{dep}(x; y)$ instead of $\mathsf{dep}(\{x\}; \{y\})$.

Definition 5 (Team semantics). *Let φ, ψ be $\mathcal{PDL}$-formulas and $P, Q \subseteq \mathrm{VAR}$ be two finite sets of variables. A team T is a set of assignments $t \colon \mathrm{VAR} \to \{0, 1\}$. Furthermore, we define the satisfaction relation $\models$ as follows, where $T \models \top$ is always true, $T \models \bot$ is never true, and $T \models \neg\mathsf{dep}(P; Q)$ iff $T = \emptyset$:*

$$
\begin{aligned}
T &\models x &\quad \textit{iff} &\quad \forall t \in T : t(x) = 1 \\
T &\models \neg x &\quad \textit{iff} &\quad \forall t \in T : t(x) = 0 \\
T &\models \varphi \wedge \psi &\quad \textit{iff} &\quad T \models \varphi \text{ and } T \models \psi \\
T &\models \varphi \vee \psi &\quad \textit{iff} &\quad \exists T_1 \exists T_2 (T_1 \cup T_2 = T) : T_1 \models \varphi \text{ and } T_2 \models \psi \\
T &\models \mathsf{dep}(P; Q) &\quad \textit{iff} &\quad \forall t, t' \in T : \bigwedge_{p \in P} t(p) = t'(p) \text{ implies } \bigwedge_{q \in Q} t(q) = t'(q)
\end{aligned}
$$

Observe that for formulas of the form $\mathsf{dep}(; Q)$ the team has to be constant with respect to Q. That is why such atoms are called *constancy atoms*. Note that in literature there exist two semantics for the split-junction operator: *lax* and *strict* semantics (e.g., Hella et al. [22]). Strict semantics requires the "splitting of the team" to be a partition whereas lax semantics allow an "overlapping" of the team. We use lax semantics here. Notice that the computational complexity for SAT and MC in $\mathcal{PDL}$ are the same irrespective of the considered semantics. Furthermore, our proofs work for both semantics. Also further note that allowing an unrestricted negation operator dramatically increases the complexity of SAT in this logic to **ATIME-ALT(exp, poly)** (alternating exponential time with polynomially many alternations) as shown by Hannula et al. [21]. That is one reason why we stick to atomic negation.

In the following, we define three well-known formula properties which are relevant to results in the paper. A formula ϕ is *flat* if, given any team T, we have that $T \models \phi \iff \{s\} \models \phi$ for every $s \in T$. A logic $\mathcal{L}$ is *downwards closed* if for every $\mathcal{L}$-formula ϕ and team T, if $T \models \phi$ then for every $P \subseteq T$ we have that $P \models \phi$. A formula ϕ is *2-coherent* if for every team T, we have that $T \models \phi \iff \{s_i, s_j\} \models \phi$ for every $s_i, s_j \in T$. The classical $\mathcal{PL}$-formulas are flat. This also implies that for $\mathcal{PL}$-formulas, the truth value is evaluated under each assignment individually, consequently, the semantics is the usual Tarski semantic. Moreover, $\mathcal{PDL}$ is downwards closed and every dependence atom is 2-coherent.

2.3   Representation of Inputs as Graphs

As we will consider specific structural parameters, we need to agree on a representation of formulas, respectively, teams. Classically, propositional formulas were represented via different kinds of graphs (e.g., Gaifman graph, primal graph) [33]. However, in this setting usually CNF-formulas are considered. Coping with this restriction, Lück et al. [28] defined syntax circuits for temporal logic formulas that also allow arbitrary formulas. In our setting, we continue this direction and define the syntax (or formula) circuit with respect to a $\mathcal{PDL}$-formula.

An important observation regarding the graph representation for the $\mathcal{PDL}$-formulas is due to Grädel [17]. In the usual setting for logics with team semantics, we take the syntax tree and not the associated circuit, that is, we distinguish between different occurrences of the same subformula. The reason for this choice is that a formula $\phi \vee \phi$ is not equivalent to ϕ, and in its evaluation, different teams are entitled to the two occurrences of ϕ in the formula. Consequently, the well-formed formulas of $\mathcal{PDL}$ can be seen as binary trees with leaves as atomic subformulas (variables and dependence atoms).

Example 6. The team $\{00, 01, 10, 11\}$ satisfies $\mathsf{dep}(x; y) \vee \mathsf{dep}(x; y)$, even though it does not satisfy $\mathsf{dep}(x; y)$.

Notice that according to the graph representation of formulas as trees, the treewidth (Definition 8) of a $\mathcal{PDL}$-formula is already 1. As a consequence, the 1-slice of each problem is **NP**-hard and both problems (MC and SAT) are **paraNP**-complete when parameterised by the treewidth of the syntax tree of a $\mathcal{PDL}$-formula. For this reason we consider the syntax circuit rather than the syntax tree as a graph structure.

Given an instance $\langle T, \Phi \rangle$ of the model checking problem, where Φ is a $\mathcal{PDL}$-formula with propositional variables $\{x_1, \ldots, x_n\} \subseteq \mathrm{VAR}$ and $T = \{s_1, \ldots s_m\}$ is a team of assignments $s_i \colon \mathrm{VAR} \to \{0, 1\}$. Then we consider the graph-structure $\mathcal{A}_{T,\Phi}$ with vocabulary $\tau_{T,\Phi}$ and represent the formula by its syntax circuit. Henceforth, we write $\mathcal{A}$ instead of $\mathcal{A}_{T,\Phi}$ when it is clear that our input instance is T, Φ and define the vocabulary as

$$\tau_{T,\Phi} := \{\mathsf{SF}^1, \succcurlyeq^2, r, \mathsf{VAR}^1, \mathsf{NEG}^2, \mathsf{CONJ}^3, \mathsf{DISJ}^3, \mathsf{DEP}^3,$$
$$\mathsf{inTeam}^1, \mathsf{isTrue}^2, \mathsf{isFalse}^2, c_1, \ldots, c_m\},$$

where the superscripts denote the arity of each relation. The set of vertices A of the graph is $\mathrm{SF}(\Phi) \cup \{c_1^{\mathcal{A}}, \ldots, c_m^{\mathcal{A}}\}$, where $\mathrm{SF}(\Phi)$ denotes the set of subformulas of Φ.

- SF and VAR are unary relations representing 'is a subformula of Φ' and 'is a variable in Φ' respectively.
- $\succcurlyeq$ is a binary relation such that $\phi \succcurlyeq^{\mathcal{A}} \psi$ iff ψ is the immediate subformula of ϕ and r is a constant symbol representing Φ.
- The set $\{c_1, \ldots, c_m\}$ encodes the team T, where each c_i is interpreted as $c_i^{\mathcal{A}} \in \mathcal{A}$ and each c_i corresponds to an assignment $s_i \in T$ for $i \leq m$.
- $\mathsf{inTeam}(c)$ is true if and only if $c \in \{c_1, \ldots, c_m\}$.
- isTrue and isFalse relate variables with the team elements. $\mathsf{isTrue}(c, x)$ (resp., $\mathsf{isFalse}(c, x)$) is true if and only if x is mapped 1 (resp., 0) by the assignment interpreted by c.

The remaining relations interpret how subformulas are related to each other.

Definition 7 (Gaifman graph). *Given a team T and a $\mathcal{PDL}$-formula Φ, the Gaifman graph $G_{T,\Phi} = (A, E)$ of the $\tau_{T,\Phi}$-structure $\mathcal{A}$ is defined as*

$$E := \{\{u, v\} \mid u, v \in A, u \text{ and } v \text{ share a tuple in a relation in } \mathcal{A}\}.$$

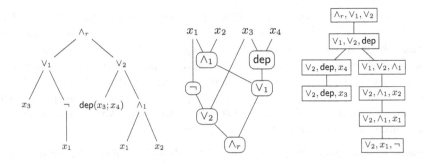

Fig. 1. An example syntax tree (left) with the corresponding circuit graph (middle) and a tree decomposition (right) for $(x_3 \vee \neg x_1) \wedge [(\mathsf{dep}(x_3; x_4)) \vee (x_1 \wedge x_2)]$.

Definition 8 (Treewidth). *The* tree-decomposition *of a given graph* $G = (V, E)$ *is a tree* $T = (B, E_T)$, *where the vertex set* $B \subseteq \mathcal{P}(V)$ *is called* bags *and* E_T *is the edge relation such that the following is true: (i)* $\bigcup_{b \in B} = V$ *(ii) for every* $\{u, v\} \in E$ *there is a bag* $b \in B$ *with* $u, v \in b$, *and (iii) for all* $v \in V$ *the restriction of* T *to* v *(the subtree with all bags containing* v) *is connected. The* width *of a given tree-decomposition* $T = (B, E_T)$ *is the size of the largest bag minus one:* $\max_{b \in B} |b| - 1$. *The* treewidth *of a given graph* G *is the minimum over all widths of tree-decompositions of* G.

Observe that if G is a tree then the treewidth of G is one. Intuitively, one can say that treewidth accordingly is a measure of tree-likeliness of a given graph.

Example 9. Figure 1 represents a graph (in middle) with a tree-decomposition (on the right). Since the largest bag is of size 3, the graph has a treewidth of 2.

2.4 Considered Parameterisations

We consider eight different parameters for all three problems (MC, SAT and m-SAT). These include formula-tw, formula-team-tw, team-size, formula-size, #variables, formula-depth, #splits and dep-arity. All these parameters arise naturally in the problems we study. #splits denotes the number of times a split junction ($\vee$) appears in a formula and #variables denotes the total number of propositional variables. The parameter formula-depth is the depth of the syntax tree of Φ, that is, the length of the longest path from root to any leaf in the tree. Arity of a dependence atom $\mathsf{dep}(P; Q)$ is the length of P and dep-arity is the maximum arity of a dependence atom in the input formula.

Regarding treewidth, notice first that for the MC problem, we can also include the assignment-variable relation in the graph representation. This yields two treewidth notions formula-tw and formula-team-tw, the name emphasises whether the team is also part of the graph. formula-tw is the treewidth of the syntax circuit of input formula alone whereas formula-team-tw comes from the syntax circuit when team elements are also part of the representation. Clearly, formula-team-tw is only relevant for the MC problem.

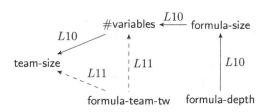

Fig. 2. The relationship among different parameters. The direction of arrow in $p \leftarrow q$ implies that bounding q results in bounding p. The dotted line indicates that the parameter bounds either (minimum) of the given two. Li means the Lemma i.

The following lemma proves relationships between several of the aforementioned parameters (also see Fig. 2). The notation $\kappa(T, \Phi)$ stands for the parameter κ of the input instance (T, Φ).

Lemma 10. *Given a team T and a formula Φ then*

1. $\text{team-size}(T, \Phi) \leq 2^{\#\text{variables}(T,\Phi)}$
2. $\text{team-size}(T, \Phi) \leq 2^{\text{formula-size}(T,\Phi)}$
3. $\text{formula-size}(T, \Phi) \leq 2^{2 \cdot \text{formula-depth}(T,\Phi)}$

Proof. If a $\mathcal{PDL}$-formula Φ has m variables then there are 2^m many assignments and the maximum size for a team is 2^m. As a result, we have team-size $\leq$ $2^{\#\text{variables}}$. Furthermore, the number of variables in a $\mathcal{PDL}$-formula Φ is bounded by the formula-size and as a consequence, we have $2^{\#\text{variables}} \leq 2^{\text{formula-size}}$. This proves the second claim.

If a formula Φ has formula-depth $= d$ then there are $\leq 2^d$ leaves in the (binary) syntax tree of Φ and $\leq 2^d$ internal nodes. Then formula-size $\leq 2^{2d}$ is true. $\square$

Now we prove the following non-trivial lemma stating that treewidth of the structure $\mathcal{A}_{T,\Phi}$ bounds either the team size or the number of variables. This implies that bounding the treewidth of the structure also bounds either of the two parameters. Recall that we talk about the formula-team-tw of the Gaifman graph underlying the structure $\mathcal{A}_{T,\Phi}$ that encodes the MC question.

Lemma 11. *Let $\langle T, \Phi \rangle$ be a given MC instance. Then the following relationship between parameters is true,*

$$\text{formula-team-tw}(T, \Phi) \geq \min\{\text{team-size}(T, \Phi), \#\text{variables}(T, \Phi)\}$$

Proof. We prove that if formula-team-tw is smaller than the two then such a decomposition must have cycles and hence cannot be a tree decomposition. The proof idea uses the fact that in the Gaifman graph $\mathcal{A}_{T,\Phi}$, every team element is related to each variable. As a consequence, in any tree decomposition, the assignment-variable relations 'isTrue' and 'isFalse' cause some bag to have their size larger than either the team size or the number of variables (based on which of the two values is smaller). We consider individual bags corresponding to an edge in the Gaifman graph due to the relations from $\tau_{T,\Phi}$. Let $\{x_1, \ldots, x_n\}$ be the variables that also appear as leaves in the formula tree $\langle \mathrm{SF}(\Phi), \succcurlyeq, \Phi \rangle$.

Consider a minimal tree decomposition $\langle B_T, \prec \rangle$ for the Gaifman graph of $\mathcal{A}$. Denote by B_{x_i, c_j} the bag that covers the edge between a variable x_i and an assignment-element c_j, that is, either $\mathsf{isTrue}(x_i, c_j)$ or $\mathsf{isFalse}(x_i, c_j)$ is true. Moreover, denote by B_{x_i, α_r} the bag covering the edges between a variable x_i and its immediate $\succcurlyeq$-predecessor α_r. Recall that in the formula part of the Gaifman graph, there is a path from each variable x_i to the formula Φ due to $\succcurlyeq$. This implies that there exists a minimal path between any pair of variables in the Gaifman graph, and this path passes through some subformula Ψ of Φ. Let $B_{x, \alpha_1}, B_{\alpha_1, \alpha_2}, \ldots B_{\alpha_n, \Psi}, \ldots B_{\Psi, \beta_n}, \ldots B_{\beta_2, \beta_1}, B_{\beta_1, y}$ be the sequence of bags that covers $\succcurlyeq^{\mathcal{A}}$-edges between x and y. Without loss of generality, we assume that all these bags are distinct. Now, for any pair x, y of variables, the bags B_{x, c_i} and B_{y, c_i} contain c_i for each $i \leq m$ and as a consequence, we have either of the following two cases.

- The two bags are equal, that is $B_{x, c_i} = B_{y, c_i}$ and as a consequence, we have $|B_{x, c_i}| \geq 3$ because B_{x, c_i} contains at least x, y, c_i. Moreover, if this is true (otherwise case two applies) for each pair of variables, then there is a single bag B_{c_i} that contains all the variables and the element c_i. This means the maximum bag size must be larger than the total number of variables, a contradiction.

- Every path between B_{x, c_i} and B_{y, c_i} contains c_i. We know that if a B_{x, α_1}-$B_{\beta_1, y}$-path between x and y due to $\succcurlyeq$ exist, then the bags B_{x, c_i} and B_{y, c_i} cannot be incident because this will produce a cycle, a contradiction again. Now, for two different assignment-elements c_i, c_j, consider the bags B_{y, c_i}, B_{y, c_j}. If these two bags are incident then B_{x, c_j}, B_{y, c_j} cannot be incident and the path between B_{x, c_j}, B_{y, c_j} must contain c_j. Notice that both B_{y, c_i}, B_{y, c_j} and B_{x, c_j}, B_{y, c_j} cannot be incident since this would, again, create a cycle. Consequently, the only possible case is that either B_{y, c_i} and B_{y, c_j} are not incident and every path between these bags contains y, or B_{x, c_j} and B_{y, c_j} are not incident and every path between these bags contains c_j. Also see Fig. 3 explaining this situation.

Finally, since this is true for all the variables and all the elements c_i with $i \leq m$ this proves that either there is a bag that contains all the variables, or there is one that contains all c_i's. The remaining case that there are cycles in the tree decomposition is not applicable. This proves the claim and completes the proof to the lemma. $\square$

The following corollary is immediate due to previous lemma.

Corollary 12. *Let $\Phi \in \mathcal{PDL}$ and T be a team. Then* formula-team-tw(T, Φ) *bounds* team-size(T, Φ).

Proof. If formula-team-tw $\geq$ #variables then bounding formula-team-tw bounds #variables which in turn bounds team-size because team-size $\leq 2^{\text{#variables}}$. Otherwise we already have formula-team-tw $\geq$ team-size according to Lemma 11. $\square$

Fig. 3. The rectangles represent bags corresponding to a variable-assignment relation. If the c_i-bags do not contain c_j-nodes, then there can be only either dotted or dashed edges between the bags to avoid cycles.

3 Parameterised Complexity of Model Checking in PDL

In this section, we study the MC question under various parameterisations. Table 2 contains a complete list of the results.

Proposition 13 ([14, Thm. 3.2]). *MC is* **NP**-*complete*.

Theorem 14. *MC parameterised by* formula-tw *is* **paraNP**-*complete*.

Proof. The upper bounds follows from Proposition 13. For lower bound we prove that 1-slice of the problem is **NP**-hard by reducing from 3SAT. The reduction provided by Ebbing and Lohmann (Proposition 13) uses Kripke semantics (as they aim for a modal logic related result). We slightly modify it to fit our presentation (the correctness proof is the same). Let $\Phi := C_1 \wedge \dots C_m$ be a 3CNF over the variables $\{x_1, \dots, x_n\}$. We form an instance $\langle T, \Psi \rangle$ of $\mathcal{PDL}$-MC such that $\mathrm{VAR}(\Psi) = \{p_1, \dots, p_n, r_1 \dots, r_n\}$. The team $T = \{s_1, \dots, s_m\}$ contains m assignments where each assignment $s_i \colon \mathrm{VAR}(\Psi) \to \{0, 1\}$ is defined as follows,

$$
\begin{aligned}
s_i(p_j) = s_i(r_j) = 1 && \text{if } x_j \in C_i, \\
s_i(p_j) = 0, s_i(r_j) = 1 && \text{if } \neg x_j \in C_i, \\
s_i(p_j) = s_i(r_j) = 0 && \text{otherwise.}
\end{aligned}
$$

Finally, let $\Psi := \bigvee\limits_{j=1}^{n} (r_j \wedge \mathsf{dep}(; p_j))$. The proof of $T \models \Psi \iff \Phi \in \mathrm{SAT}$ follows from [14, Thm. 3.2]. Notice that none other parameter except formula-tw is fixed in advance. The syntax circuit of Ψ yields a tree, as a consequence, formula-tw $= 1$. This completes the proof. $\qquad\square$

Notice that the formula in the reduction from 3SAT has fixed arity for any dependence atom (that is, dep-arity $= 0$). As a consequence, we obtain the following corollary.

Corollary 15. *MC parameterised by* dep-arity *is* **paraNP**-*complete*.

The main source of difficulty in the model checking problem seems to be the split-junction operator. For a team of size k and a formula with only one split-junction there are 2^k many candidates for the correct split and each can

---

Algorithm 1. check(T, Φ), recursive bottom-up algorithm solving the MC parameterised by team-size.

Input : A $\mathcal{PDL}$-formula Φ and a team T
Output: true if $T \models \Phi$, otherwise false

1 **foreach** *non-root node v in the syntax tree* **do** $L_v = \{\emptyset\}$
2 **foreach** *leaf ℓ of the syntax tree* **do** // find all possible sub-teams for ℓ
3    $L_\ell = \{\emptyset\}$
4   **foreach** $P \subseteq T$ **do**
5   **if** $\ell = X$ *and* $\forall s \in P : s(X) = 1$ **then** $L_\ell \leftarrow L_\ell \cup \{P\}$
6   **else if** $\ell = \neg X$ *and* $\forall s \in P : s(X) = 0$ **then** $L_\ell \leftarrow L_\ell \cup \{P\}$
7   **else if** $\ell = \mathsf{dep}(Q; r)$ *and* $\forall s_i \forall s_j \bigwedge_{q \in Q} s_i(q) = s_j(q) \Rightarrow s_i(r) = s_j(r)$ **then**
8          $\lfloor L_\ell \leftarrow L_\ell \cup \{P\}$

9 **foreach** α_1, α_2 *with* $\alpha = \alpha_1 \circ \alpha_2$ *and* $L_{\alpha_i} \neq \emptyset$ *for $i = 1, 2$* **do**
10   **foreach** $P \in L_{\alpha_1}, Q \in L_{\alpha_2}$ **do**
11   **if** $\circ = \wedge$ *and* $P = Q$ **then** $L_\alpha \leftarrow L_\alpha \cup \{P\}$
12   **else if** $\circ = \vee$ **then** $L_\alpha \leftarrow L_\alpha \cup \{P \cup Q\}$

13 **if** $T \in L_\Phi$ **then return** true **else return** false

---

be verified in polynomial time. As a result, an exponential runtime in the input length seems necessary. However, if the team size (k) is considered as a parameter then the problem can be solved in polynomial time with respect to the input size and exponentially in the parameter. We consider both parameters (team-size and #splits) in turn.

Theorem 16. MC *parameterised by team-size is in* **FPT**.

Proof. We claim that Algorithm 1 solves the task in fpt-time. The correctness follows from the fact that the procedure is simply a recursive definition of truth evaluation of $\mathcal{PDL}$-formulas in bottom-up fashion.

Recall that the input formula Φ is a binary tree. The procedure starts at the leaf level by checking whether for each subformula α and each subteam $P \subseteq T$, $P \models \alpha$. Then recursively, if $P \models \alpha_i$ for $i = 1, 2$ and there is a subformula α such that $\alpha = \alpha_1 \wedge \alpha_2$ then it answers that $P \models \alpha$. Moreover, if $P_i \models \alpha_i$ for $i = 1, 2$ and there is a subformula α such that $\alpha = \alpha_1 \vee \alpha_2$ then it answers that $P \models \alpha$ where $P = P_1 \cup P_2$.

The first loop runs in $O^\star(2^k)$ steps for each leaf node and there are $|\Phi|$ many iterations, which gives a running time of $|\Phi| \cdot O^\star(2^k)$, where team-size $= k$. At each inner node, there are at most 2^k candidates for P and Q and as a consequence, at most 2^{2k} pairs that need to be checked. This implies that the loop for each inner node can be implemented in $O^\star(2^{2k})$ steps. Furthermore, the loop runs once for each pair of subformulas α_1, α_2 such that $\alpha_1 \circ \alpha_2$ is also a subformula of Φ. This gives a running time of $|\Phi| \cdot O^\star(2^{2k})$ for this step. Finally, in the last step a set of size k needs to be checked against a collection containing 2^k such sets, this can be done in $k \cdot O(2^k)$ steps.

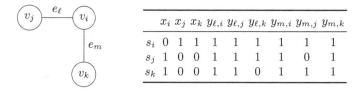

	x_i	x_j	x_k	$y_{\ell,i}$	$y_{\ell,j}$	$y_{\ell,k}$	$y_{m,i}$	$y_{m,j}$	$y_{m,k}$
s_i	0	1	1	1	1	1	1	1	1
s_j	1	0	0	1	1	1	1	0	1
s_k	1	0	0	1	1	0	1	1	1

Fig. 4. A graph $\mathcal{G} : \langle\{v_i, v_j, v_k\}, \{e_l, e_m\}\rangle$ and a corresponding team.

We conclude that the above procedure solves the MC problem in $p(|\Phi|) \cdot O(2^{2k})$ steps for some polynomial p. The fact that we do not get a blow-up in the number of subformulas is due to the reason that the formula tree is binary. The procedure operates on a pair of subformulas in each step and the label size $(|L_\alpha|)$ at the end of this step is again bounded by 2^k.   □

Regarding the parameter #splits, we show **paraNP**-completeness by reducing from the 3-colouring problem (3COL) and applying Proposition 3. The idea of the reduction from 3COL is to construct a team as shown in Fig. 4 in combination with the disjunction of three times the formula $\bigwedge_{e_k=\{v_i,v_j\}} \mathsf{dep}(y_k; x_i)$. Intuitively, vertices of the graph correspond to assignments in the team and the three splits then map to the three colours.

Theorem 17 ($\star$). MC *parameterised by* #splits *is* **paraNP**-*complete*.

The following cases then can be easily deduced.

Theorem 18 ($\star$). MC *under the parameterisations* formula-size, formula-depth *or* #variables *is* **FPT**.

Finally, the case for formula-team-tw follows due to Corollary 12 in conjunction with the **FPT** result for team-size (Lemma 16).

Corollary 19. MC *parameterised by* formula-team-tw *is* **FPT**.

4 Satisfiability

In this section, we study SAT under various parameterisations, so the question of whether there exists a team T for a given formula Φ such that $T \models \Phi$. Notice first, that the question is equivalent to finding a singleton team. This is since $\mathcal{PDL}$ is downwards closed. Consequently, if there is a satisfying team, then a singleton team satisfies the formula. As a result, team semantics coincides with the usual Tarskian semantics. This facilitates, for example, determining the truth value of disjunctions in the classical way. Accordingly, simplifying the notation a bit, for SAT we now look for an assignment rather than a singleton team that satisfies the formula.

Corollary 20. *The problem* SAT *under the parameterisations* formula-team-tw *and* formula-tw *is same*.

The following result is obtained by classical SAT being **NP**-complete [8, 26].

Corollary 21. SAT *under the parameterisations* team-size, *or* dep-arity *is* **paraNP**-*complete*.

Proof. The 1-slice regarding team-size (a singleton team is the same as an assignment), and the 0-slice regarding dep-arity (no dependence atoms at all) is **NP**-hard.   □

Turning towards treewidth, notice that classical propositional SAT is fixed-parameter tractable when parameterised by treewidth due to Samer and Szeider [34, Thm. 1]. However, we are unable to immediately utilise their result because Samer and Szeider study CNF-formulas and we have arbitrary formulas instead. Yet, Lück et al. [28, Cor. 4.7] studying temporal logics under the parameterised approach, classified, as a byproduct, the propositional satisfiability problem with respect to arbitrary formulas to be fixed-parameter tractable.

Corollary 22. SAT *parameterised by* formula-tw *is in* **FPT**.

Proof. As before, we need to find a singleton team. This implies that split-junctions have the same semantics as classical disjunctions and dependence atoms are always satisfied. So replacing every occurrence of a dependence atom $dep(P; Q)$ by $\top$ yields a propositional logic formula. This substitution does not increase the treewidth. Then the result follows by Lück et al. [28, Cor. 4.7].   □

Now, we turn towards the parameter #splits. We present a procedure that constructs a satisfying assignment s such that $s \models \Phi$ if there is one and otherwise it answers no. The idea is that this procedure needs to remember the positions where a modification in the assignment is possible. We show that the number of these positions is bounded by the parameter #splits.

Consider the syntax tree of Φ where, as before, multiple occurrences of subformulas are allowed. The procedure starts at the leaf level with satisfying assignment candidates (partial assignments, to be precise). Reaching the root it confirms whether it is possible to have a combined assignment or not. We assume that the leaves of the tree consist of literals or dependence atoms. Accordingly, the internal nodes of the tree are only conjunction and disjunction nodes. The procedure sets all the dependence atoms to be trivially *true* (as we satisfy them via every singleton team). Additionally, it sets all the literals satisfied by their respective assignment. Ascending the tree, it checks the relative conditions for conjunction and disjunction by joining the assignments and thereby giving rise to conflicts. A conflict arises (only at a conjunction node) when two assignments are joined with contradicting values for some variable. At this point, it sets this variable x to a conflict state c. At disjunction nodes the assignment stores that it has two options and keeps the assignments separately.

Joining a *true*-value from a dependence atom affects the assignment only at disjunction nodes. This corresponds to the intuition that a formula of the form $dep(P; Q) \vee \psi$ is true under any assignment. Whereas, at a conjunction node, when an assignment s joins with a *true*, the procedure returns the assignment s.

Since at a split the procedure returns both assignments, for k splits there could be $\leq 2^k$-many assignment choices. At the root node if at least one assignment is consistent then we have a satisfying assignment. Otherwise, if all the choices contain conflicts over some variables then there is no such satisfying singleton team.

Theorem 23 ($\star$). SAT *parameterised by* #*splits is* **FPT**. *Moreover, there is an algorithm that solves the problem in* $O(2^{\#splits(\Phi)} \cdot |\Phi|^{O(1)})$ *for any* $\Phi \in \mathcal{PDL}$.

The remaining cases follow easily.

Theorem 24 ($\star$). SAT *parameterised by* #variables, formula-size *or* formula-depth *is* **FPT**.

A satisfiability Variant
The shown results suggest that it might be interesting to study the following variant of SAT, in which we impose an additional input 1^m (unary encoding) with $m \geq 2$ and ask for a satisfying team of size m. Let us call the problem m-SAT. We wish to emphasise that m-SAT is not the same as the SAT parameterised by team-size.

Theorem 25 ($\star$). m-SAT *under the parameterisations* #variables, formula-size, *or* formula-depth *is* **FPT**.

Neither the arity of the dependence atoms nor the team-size alone are fruitful parameters which follows from Corollary 21.

Corollary 26. m-SAT *parameterised by* team-size, *or* dep-arity *is* **paraNP-complete.**

5   Conclusion

In this paper, we started a systematic study of the parameterised complexity of model checking and satisfiability in propositional dependence logic. For both problems, we exhibited a complexity dichotomy (see Table 2): depending on the parameter, the problem is either **FPT** or **paraNP**-complete. Interestingly, there exist parameters for which MC is easy, but SAT is hard (team-size) and *vice versa* (#splits).

In the end, we introduced a satisfiability question which also asks for a team of a given size (m-SAT). This has not been studied at all in the setting of team logics, yet. We pose it as an interesting problem to study. Here, we leave the cases for #splits and formula-tw open for further research.

As future work, we want to study combinations of the studied parameters, e.g., #splits + dep-arity. This parameter is quite interesting, as dep-arity alone is always hard for all three problems, whereas adding #splits allows for SAT to reach **FPT**. It is also interesting to observe that in both of our reductions for proving hardness of MC under the parametrisation #splits and dep-arity, if dep-arity is fixed then #splits is unbounded and vice versa.

Another important question for future research is to consider the parameterised version of validity and implication problem for $\mathcal{PDL}$. Finally, we aim, besides answering the open cases, to study further operators such as independence and inclusion atoms.

References

1. Abramsky, S., Kontinen, J., Väänänen, J., Vollmer, H. (eds.): Dependence Logic, Theory and Applications. Springer, Basel (2016). https://doi.org/10.1007/978-3-319-31803-5
2. Bläsius, T., Friedrich, T., Schirneck, M.: The parameterized complexity of dependency detection in relational databases. In: Guo, J., Hermelin, D. (eds.) 11th International Symposium on Parameterized and Exact Computation (IPEC 2016). Leibniz International Proceedings in Informatics (LIPIcs), Dagstuhl, Germany, vol. 63, pp. 6:1–6:13. Schloss Dagstuhl-Leibniz-Zentrum fuer Informatik (2017). http://drops.dagstuhl.de/opus/volltexte/2017/6920, https://doi.org/10.4230/LIPIcs.IPEC.2016.6
3. Chen, H., Mengel, S.: A trichotomy in the complexity of counting answers to conjunctive queries. In: Arenas, M., Ugarte, M. (eds.) 18th International Conference on Database Theory, ICDT 2015, Brussels, Belgium, 23–27 March 2015. LIPIcs, vol. 31. pp. 110–126. Schloss Dagstuhl - Leibniz-Zentrum fuer Informatik (2015). https://doi.org/10.4230/LIPIcs.ICDT.2015.110
4. Chen, H., Mengel, S.: Counting answers to existential positive queries: a complexity classification. In: Milo, T., Tan, W.-C. (eds.) Proceedings of the 35th ACM SIGMOD-SIGACT-SIGAI Symposium on Principles of Database Systems, PODS 2016, San Francisco, CA, USA, 26 June–01 July 2016, pp. 315–326. ACM (2016). https://doi.org/10.1145/2902251.2902279
5. Chen, H., Mengel, S.: The logic of counting query answers. In: 32nd Annual ACM/IEEE Symposium on Logic in Computer Science, LICS 2017, Reykjavik, Iceland, 20–23 June 2017, pp. 1–12. IEEE Computer Society (2017). https://doi.org/10.1109/LICS.2017.8005085
6. Chen, H., Müller, M.: The fine classification of conjunctive queries and parameterized logarithmic space. TOCT $7(2)$, 7:1–7:27 (2015). https://doi.org/10.1145/2751316
7. Ciardelli, I., Groenendijk, J., Roelofsen, F.: Towards a logic of information exchange - an inquisitive witness semantics. In: Bezhanishvili, G., Löbner, S., Marra, V., Richter, F. (eds.) TbiLLC 2011. LNCS, vol. 7758, pp. 51–72. Springer, Heidelberg (2013). https://doi.org/10.1007/978-3-642-36976-6_6
8. Cook, S.A.: The complexity of theorem-proving procedures. In: Harrison, M.A., Banerji, R.B., Ullman, J.D. (eds.) Proceedings of the 3rd Annual ACM Symposium on Theory of Computing, Shaker Heights, Ohio, USA, 3–5 May 1971, pp. 151–158. ACM (1971). https://doi.org/10.1145/800157.805047
9. Cygan, M., et al.: Parameterized Algorithms. Springer, Cham (2015). https://doi.org/10.1007/978-3-319-21275-3
10. Downey, R.G., Fellows, M.R.: Fundamentals of Parameterized Complexity. Texts in Computer Science. Springer, London (2013). https://doi.org/10.1007/978-1-4471-5559-1
11. Durand, A., Hannula, M., Kontinen, J., Meier, A., Virtema, J.: Approximation and dependence via multiteam semantics. Ann. Math. Artif. Intell. $83(3–4)$, 297–320 (2018). https://doi.org/10.1007/s10472-017-9568-4

12. Durand, A., Hannula, M., Kontinen, J., Meier, A., Virtema, J.: Probabilistic team semantics. In: Ferrarotti, F., Woltran, S. (eds.) FoIKS 2018. LNCS, vol. 10833, pp. 186–206. Springer, Cham (2018). https://doi.org/10.1007/978-3-319-90050-6_11
13. Durand, A., Mengel, S.: Structural tractability of counting of solutions to conjunctive queries. Theory Comput. Syst. **57**(4), 1202–1249 (2015). https://doi.org/10.1007/s00224-014-9543-y
14. Ebbing, J., Lohmann, P.: Complexity of model checking for modal dependence logic. In: Bieliková, M., Friedrich, G., Gottlob, G., Katzenbeisser, S., Turán, G. (eds.) SOFSEM 2012. LNCS, vol. 7147, pp. 226–237. Springer, Heidelberg (2012). https://doi.org/10.1007/978-3-642-27660-6_19
15. Flum, J., Grohe, M.: Parameterized Complexity Theory. Texts in Theoretical Computer Science. An EATCS Series. Springer, Heidelberg (2006). https://doi.org/10.1007/3-540-29953-X
16. Galliani, P.: Inclusion and exclusion dependencies in team semantics - on some logics of imperfect information. Ann. Pure Appl. Log. **163**(1), 68–84 (2012). https://doi.org/10.1016/j.apal.2011.08.005
17. Grädel, E.: Model-checking games for logics of imperfect information. Theor. Comput. Sci. **493**, 2–14 (2013). https://doi.org/10.1016/j.tcs.2012.10.033
18. Grädel, E., Kontinen, J., Väänänen, J., Vollmer, H.: Logics for dependence and independence (Dagstuhl seminar 15261). Dagstuhl Reports **5**(6), 70–85 (2016). http://drops.dagstuhl.de/opus/volltexte/2016/5508, https://doi.org/10.4230/DagRep.5.6.70
19. Grohe, M., Schwentick, T., Segoufin, L.: When is the evaluation of conjunctive queries tractable? In: Vitter, J.S., Spirakis, P.G., Yannakakis, M. (eds.) Proceedings on 33rd Annual ACM Symposium on Theory of Computing, Heraklion, Crete, Greece, 6–8 July 2001, pp. 657–666. ACM (2001). https://doi.org/10.1145/380752.380867
20. Hannula, M., Kontinen, J., Virtema, J.: Polyteam semantics. In: Artemov, S., Nerode, A. (eds.) LFCS 2018. LNCS, vol. 10703, pp. 190–210. Springer, Cham (2018). https://doi.org/10.1007/978-3-319-72056-2_12
21. Hannula, M., Kontinen, J., Virtema, J., Vollmer, H.: Complexity of propositional logics in team semantic. ACM Trans. Comput. Log. **19**(1), 2:1–2:14 (2018). https://doi.org/10.1145/3157054
22. Hella, L., Kuusisto, A., Meier, A., Virtema, J.: Model checking and validity in propositional and modal inclusion logics. J. Log. Comput. (2019). https://doi.org/10.1093/logcom/exz008
23. Kontinen, J., Müller, J.-S., Schnoor, H., Vollmer, H.: Modal independence logic. J. Log. Comput. **27**(5), 1333–1352 (2017). https://doi.org/10.1093/logcom/exw019
24. Krebs, A., Meier, A., Virtema, J.: A team based variant of CTL. In: Grandi, F., Lange, M., Lomuscio, A. (eds.) 22nd International Symposium on Temporal Representation and Reasoning, TIME 2015, Kassel, Germany, 23–25 September 2015, pp. 140–149. IEEE Computer Society (2015). https://doi.org/10.1109/TIME.2015.11
25. Krebs, A., Meier, A., Virtema, J., Zimmermann, M.: Team semantics for the specification and verification of hyperproperties. In: Potapov, I., Spirakis, P.G., Worrell, J. (eds.) 43rd International Symposium on Mathematical Foundations of Computer Science, MFCS 2018, Liverpool, UK, 27–31 August 2018. LIPIcs, vol. 117, pp. 10:1–10:16. Schloss Dagstuhl - Leibniz-Zentrum fuer Informatik (2018). https://doi.org/10.4230/LIPIcs.MFCS.2018.10
26. Levin, L.: Universal search problems. Probl. Inf. Transm. **9**(3), 115–116 (1973)

27. Lohmann, P., Vollmer, H.: Complexity results for modal dependence logic. Studia Logica **101**(2), 343–366 (2013). https://doi.org/10.1007/s11225-013-9483-6
28. Lück, M., Meier, A., Schindler, I.: Parameterised complexity of satisfiability in temporal logic. ACM Trans. Comput. Log. **18**(1), 1:1–1:32 (2017). https://doi.org/10.1145/3001835
29. Mahmood, Y., Meier, A.: Parameterised complexity of model checking and satisfiability in propositional dependence logic. CoRR, abs/1904.06107 (2019). http://arxiv.org/abs/1904.06107
30. Meier, A., Reinbold, C.: Enumeration complexity of poor man's propositional dependence logic. In: Ferrarotti, F., Woltran, S. (eds.) FoIKS 2018. LNCS, vol. 10833, pp. 303–321. Springer, Cham (2018). https://doi.org/10.1007/978-3-319-90050-6_17
31. Pippenger, N.: Theories of Computability. Cambridge University Press, Cambridge (1997)
32. Robertson, N., Seymour, P.D.: Graph minors. V. Excluding a planar graph. J. Comb. Theory Ser. B **41**(1), 92–114 (1986). https://doi.org/10.1016/0095-8956(86)90030-4
33. Samer, M., Szeider, S.: Fixed-parameter tractability. In: Biere, A., Heule, M., van Maaren, H., Walsh, T. (eds.) Handbook of Satisfiability. Frontiers in Artificial Intelligence and Applications, vol. 185, pp. 425–454. IOS Press (2009). https://doi.org/10.3233/978-1-58603-929-5-425
34. Samer, M., Szeider, S.: Algorithms for propositional model counting. J. Discrete Algorithms **8**(1), 50–64 (2010)
35. Väänänen, J.A.: Dependence Logic - A New Approach to Independence Friendly Logic. London Mathematical Society Student Texts, vol. 70. Cambridge University Press, Cambridge (2007). http://www.cambridge.org/de/knowledge/isbn/item1164246/?site_locale=de_DE
36. Väänänen, J.A.: Modal dependence logic. In: Apt, K., van Rooij, R. (eds.) New Perspectives on Games and Interaction. Amsterdam University Press, Amsterdam (2008)
37. Yang, F., Väänänen, J.: Propositional logics of dependence. Ann. Pure Appl. Log. **167**(7), 557–589 (2016). https://doi.org/10.1016/j.apal.2016.03.003

Utilizing Deep Learning and RDF to Predict Heart Transplantation Survival

Dennis Medved[1,2]([✉]), Johan Nilsson[2], and Pierre Nugues[1]

[1] Department of Computer Science, Lund University, Lund, Sweden
{dennis.medved,pierre.nugues}@cs.lth.se
[2] Department of Clinical Sciences, Cardiothoracic Surgery,
Lund University and Skåne University Hospital, Lund, Sweden
johan.nilsson@med.lu.se

Abstract. In this paper, we describe the conversion of three different heart transplantation data sets to a Resource Description Framework (RDF) representation and how it can be utilized to train deep learning models. These models were used to predict the outcome of patients both pre- and post-transplant and to calculate their survival time.

The International Society for Heart & Lung Transplantation (ISHLT) maintains a registry of heart transplantations that it gathers from grafts performed worldwide. The American organization United Network for Organ Sharing (UNOS) and the Scandinavian Scandiatransplant are contributors to this registry, although they use different data models.

We designed a unified graph representation covering these three data sets and we converted the databases into RDF triples. We used the resulting triplestore as input to several machine learning models trained to predict different aspects of heart transplantation patients.

Recipient and donor properties are essential to predict the outcome of heart transplantation patients. In contrast with the manual techniques we used to extract data from the tabulated files, the RDF triplestore together with SPARQL, enables us to experiment quickly and automatically with different combinations of features sets, to predict the survival, and simulate the effectiveness of organ allocation policies.

1 Introduction

Heart transplantations are life saving procedures that made it possible to extend the median survival time to 12 years for patients with end-stage heart diseases. Unfortunately, patients have to wait a relatively long time before being transplanted, because of a limited donor supply that forces the surgeons to prioritize the recipients.

The understanding of factors that predict mortality could help the doctors with the prioritization task and improve the post-operation care. With an

This research was supported by Heart Lung Foundation, The Swedish Research Council, and the eSSENCE program.

© Springer Nature Switzerland AG 2020
A. Herzig and J. Kontinen (Eds.): FoIKS 2020, LNCS 12012, pp. 175–190, 2020.
https://doi.org/10.1007/978-3-030-39951-1_11

improved outcome prediction, surgeons could be more confident in the transplantation performance. In addition, a better allocation of organs would make it possible to increase the survival as well as the number of organs that can be used. The availability of medical databases which have been created during the last two decades, and the application of machine-learning methods, such as deep learning, have led to the development of advanced models of survival prediction.

Patient and donor factors are essential to predict the mortality of heart transplantations [14,17]. [7] provides an eloquent advocacy of the importance of such factors, or features, in the success of machine-learning projects.

We wanted to mine the feature sets from the patient variables and integrate data from all our sources. We designed a unified, extendable, Resource Description Framework (RDF) representation of the variables. Our goal was to make the data extraction easier, using the different registries that were available to us, and simplify the feature engineering for the machine learning models.

The usage of RDF to store the patient data helped us streamline the development process of such survival models.

2   Related Works

There are several papers that detail the creation of RDF-based medical databases [1,3,6,9,16]. [4] used them in the cardiological field, but we have not found any previous work aiming specifically at heart transplant data.

3   Medical Registries

The International Society for Heart & Lung Transplantation (ISHLT) maintains a registry of heart transplantations it collects from national or regional organizations across the world. ISHLT aggregates the data submitted by the contributing organizations. The American organization United Network for Organ Sharing (UNOS) and the Scandinavian Scandiatransplant are two such contributing institutions. In total, ISHLT contains about 100,000 recorded heart transplantations.

ISHLT could be seen as a superset of all the included databases, in regards to the patients recorded, but it only contains a subset of the variables that are contained within the different registries. Overall, ISHLT is restricted to variables that are frequently recorded by the different regional registries.

The three data sources we considered: ISHLT, UNOS, and Scandiatransplant, have different structures, a different number of variables, use different variable names, and may use different units or encoding of the data.

The variables contained in these databases pertain to both recipient, donor, and the operation itself. It can for example be the age, weight, gender, or blood group of the patients.

UNOS contains the largest number of variables, about 500. ISHLT, for example, does not feature the variable *crossmatch_done*, a patient compatibility test, that is available in UNOS.

4   Resource Description Framework

RDF is a model and language for representing information about resources of different types. It is particularly useful for storing metadata about resources. RDF is based on the idea of making statements about resources in expressions of the form subject-predicate-object, known as triples. The subject denotes the resource, and the predicate denotes traits or aspects of the resource, and expresses a relationship between the subject and the object [18].

For example, one way to represent the statement "The Patient is 55 years old" in RDF is as the triple: a subject denoting "the patient", a predicate denoting "has the age", and an object denoting "55 years".

RDF is intended for situations in which this information needs to be processed by applications, rather than being only displayed to people. It provides a common framework for expressing this information so it can be exchanged between applications without loss of meaning.

RDF is an abstract model with several serialization formats so the particular encoding for resources or triples varies from format to format.

5   Representing the Data in RDF

The ISHLT, UNOS, and Scandiatransplant data sets are normally distributed to the researchers as SAS or CSV files. We started from the CSV files and we converted them to an RDF format.

The CSV files represent the transplants as rows, where each column is a variable for the transplant. In the RDF conversion, we mapped each row to a head node and we created leaf nodes for the selected variables.

The data sets use different names to denote the same variables. For example, the most recent blood creatinine value for the recipient patient is *Most rec. Creat.* in Scandiatransplant, *creat* in ISHLT, and *creat_trr* in UNOS. See Fig. 1.

We created unified names for about 140 of the variables, such as *aaot:creatinine* for the creatinine value, where the *aaot* prefix stands for *Algorithms and Applications for Organ Transplantation.*

We had to encode the data in a unified way between the databases, for example binary variables were both recorded as Y/N and 1/0, and categorical variables often used different codes to encode the data between the registries.

As previously mentioned, UNOS has more variables than ISHLT and Scandiatransplant. We used the UNOS variable names, when they had no counterpart in the other two registries.

We also added metadata about the variables containing the original variable name, as well the new one, the description of the variable, the source form of the data, the unit where it is applicable, as well as comments, and start and possibly end date of the recording of the variable.

We cleaned the data, created consistent encodings between the variables and built the RDF skeleton using the OpenRefine application [8] (Fig. 2).

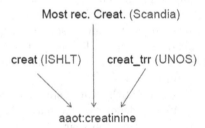

Fig. 1. A unification of the variable representing the most recent creatinine level of the recipient.

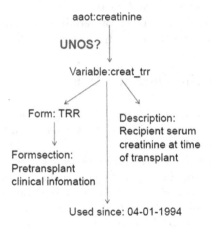

Fig. 2. *aaot:creatinine* metadata for the UNOS part of the database.

6   Querying the Database

We created a SPARQL endpoint to be able to query the data. Compared with the tedious copy-and-paste techniques we used to previously create the data sets and to test our survival prediction programs, SPARQL offers an easier way to extract relevant data samples.

Although the RDF database can be used to do statistics and exploratory analysis, our major use of the database is as input to the machine learning algorithms. We want to answer questions such as:

What variables are important for heart transplantation survival and how do they affect the outcome?

7   Deep Learning Models

Artificial neural networks are models inspired by the human brain that approximate functions used in machine learning, such as classification or regression. They consist of networks of neurons that emulate the properties of their real counterparts.

The neurons propagate signals depending on the weight of their connections. These connection strengths are tuned during the training step from observations when the network learns what it should output for a certain set of inputs.

A feed forward network consists of three or more layers. The first layer is called the input layer, where the features are used as the initial input. The middle layers which can be one or more, are called hidden layers. Finally, the last layer, the output layer, which has as many nodes as the wanted amount of outputs from the model. A neural network with two or more hidden layers is usually referred to as a deep learning model (Fig. 3).

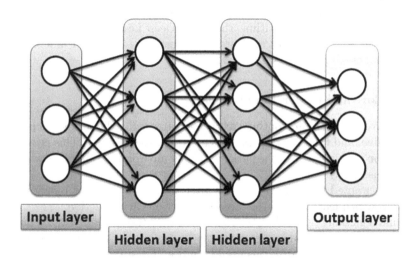

Fig. 3. The topology of a fully connected neural network with three input nodes, two hidden layers with four hidden nodes, and three output nodes.

These models have many practical applications, for example in computer vision, spam filtering, or medicine [2]. They have shown a superior predictive ability over more conventional models such as risk scores created by classical statistical methods [5].

8   Applications

We have used the RDF representation of the heart transplant patients to generate the input to different survival models regarding heart transplantation.

1. We first carried out an analysis on the features that had the largest impact on the post-transplant survival of the patient and to find locally optimal feature sets for different survival time periods [10].
2. Patients enter a waiting queue before they are transplanted and they may die in this queue if no appropriate organ becomes available for transplant. We designed a model to predict the outcome of patients awaiting heart transplant and we explored which features were the most predictive in assessing the result for the patients [11].
3. After a patient is being transplanted, the registries record his/her survival time. We trained a post-transplant model based on neural networks and we evaluated its performance against a more simple, point based model [13, 15]. We used data from UNOS instead of ISHLT in this experiment. This model is available via a web application (ihtsa.cs.lth.se), where a user can input a patient's data and the server returns the predicted survival. The application shows the survival prediction as a probability curve depending on the years after transplant.
4. And finally, we trained a pre- and post-transplant algorithm and we used it together with a discrete simulation model, to simulate a queue system for heart transplantation. This algorithm, the Lund Deep Learning Transplant Algorithm (LuDeLTA), enables analysts to evaluate the impact of different allocation policies on patient survival [12].

8.1 Explore and Understand the Data

Problem. We wanted to find optimal feature sets to predict the survival of patients after one, five, and ten year time periods after heart transplant. This is a binary classification problem, where the patient is either alive or dead after the respective time periods.

We also wanted to rank the features after their importance, for each period, to find out which features had the largest impact on the prediction.

To find a globally optimal feature set requires 2^n tests, where n is the number of features. This is infeasible even for a moderate number of features. Using the 482 features we had available, this would require $2^{482} \approx 1.25 \times 10^{145}$ tests.

Method. We applied a greedy forward selection and a greedy backward elimination that enabled us to find a locally optimal subset. Determining such a feature set is computationally feasible as opposed to a global optimal.

Table 1. The best validation set AUROC values for 1, 5, and 10 years, found using a search with 482 possible features [10].

Years	AUROC
1	0.6990
5	0.6892
10	0.7509

The greedy forward selection starts from a subset of the features, which can be empty, and adds one feature from the remaining set to the current subset. The selection procedure uses the new subset to produce the classification probabilities. These probabilities are then used to calculate an evaluation metric. The feature which improves the performance the most is then added to the current feature set for the next generation. The procedure is repeated if it improves the score of the preceding subset over a certain threshold Δ. If there is no improvement, we use the current feature set for a backward elimination instead.

The backward elimination removes the features one by one from the starting set and the resulting feature set is used to produce the classification probabilities. If the score improves on the preceding generation, then the process is repeated with the resulting feature set.

If two following forward selections and backward eliminations do not improve the score, the process is stopped and the resulting feature set corresponds to a local optimum.

Starting from the empty feature set and doing a full forward search of the 482 features, meaning that every feature is added, would result in about 100,000 models being tested. A number somewhat smaller than 1.25×10^{145}, but it is only locally optimal though.

We used logistic regression as the machine learning model for the search procedure, mainly because it has a short training time for each model, and few hyperparameters to tune.

For each generation in a forward and backward search, there is no dependence between the models. This makes it quite easy to parallelize. We parallelized the search using the Spark framework, to distribute the workload on a local cluster.

To assess the importance of the variables, we did a forward search from an empty feature set and recorded the order in which they were added. This roughly corresponds to the most important features for each time period.

Results. We found locally optimal feature sets, utilizing the available 482 features, for each time period, using our logistic regression model. Table 1 shows the best results for these feature sets, using area under the receiver operating characteristic curve (AUROC) as the metric.

The validation AUROC scores that are about the same for 1 and 5 years, but approximately 8 % points higher for 10 years. This is somewhat nonintuitive and we tried without success to find confounding factors to explain these results. A possible explanation is that there are much more positive examples, that is, dead patients, for 10 years compared with 1 and 5 years. Another bias is that many patients of this cohort are censored compared to 1 and 5 years: About 50% of the patients are censored after 10 year time period.

We listed the most important features, found using a forward search, in Table 2.

Table 2. The ten first features added for a forward search for the 1, 5, and 10 year time periods. Features in **bold** decreases the survival probability and features in non-bold increases the probability. Modified and updated from [10]

Rank	1 year	5 years	10 years
1	**Anti viral**	**Ethnicity: white**	**Days in status: 1**
2	**Creatinine**	Creatinine clearance	Days in status: 2
3	**Height**	**Func. status: very sick**	Days in status: 1b
4	**Donor age**	**Donor age**	**Donor angiogram: no**
5	**Ventricular assist**	**Ventricular assist**	**Func. status: very sick**
6	Ventricular assist: none	**Donor ischemic time**	Research immunosuppressive
7	**Serum bilirubin**	Func. status: cares for self	Func. status: cares for self
8	**Donor ischemic time**	Func. status: some assistance	**Diabetes**
9	Other therapies	Func. status: normal activity	**Anti viral**
10	**Dialysis**	Func. status: assistance	Func. status: assistance

8.2 Predict the Survival Before Transplant

Problem. Estimating the probability of dying in the waiting list given a waiting time could support the decision of surgeons on the priority of a transplantation. In addition, knowing the probability for a patient to be transplanted within a certain time frame would help plan operation resources and inform the patient. Extending the models to predict the amount of days a patient may survive in the queue could be used in a queue simulation system.

We carried out the prediction at three different time points: 180 days, 365 days, and 730 days, and we categorized the patient status with three possible outcomes: still waiting, transplanted, or dead in the waiting list.

We chose to use these time periods, because a patient should have a survival time of less than a year, predicted by a physician, to be placed in the waiting list. Although a small fraction of the patients may survive several years in the wait list.

There are other outcomes for patient standing in a heart transplantation queue, such as being too sick to be operated, but most patients are either transplanted or die while waiting for an organ.

Method. We created a neural network with two hidden layers and 128 nodes in each layer. The hidden layers used the rectified linear unit as activation function and the final output layer used a softmax activation. We used categorical cross entropy as the loss function and adamax as the optimizer. Dropout was used as a regularization technique, to reduce potential overfitting. We used the Keras framework to represent this model.

In our model, we included 87 variables as input, describing the patients in the queue that were available at the time of listing. Example of such features are age, sex, weight, and blood group.

We wanted to know which features contributed the most to the result of the classification. We utilized backward elimination to find these features.

Table 3. The F1 values for 180, 365, and 730 days obtained on the test set [11].

Days	F1 (micro)	F1 (macro)
180	0.750	0.675
365	0.760	0.680
730	0.888	0.680

Table 4. The ten most contributing features for each time period in order of importance, found using an ablation study [11]

Rank	180 days	365 days	730 days
1	Urgency status 2	BMI	BMI
2	Weight	Weight	Weight
3	BMI	Height	Height
4	Height	Urgency status 2	Urgency status 2
5	Inotropes	Creatine clearance	Creatinine
6	Blood group: AB	Inotropes	Functional status
7	Life support	Blood group: A	Pulmonary Vascular Resistance
8	Blood group: B	Life support	Educational level: none
9	Inotropic support	Blood group: AB	Ventricular assist type: LVAD + RVAD
10	Ethnicity: black	Blood group: B	Educational level: grade school

Results. Table 3 shows the best obtained F1 values for 180, 365, and 730 days, respectively. Because there is more than two classes, the F1 score needs to be averaged. It was calculated using both micro and macro averaging.

The macro average takes the average of the precision and recall of the system on the different classes. When the examples are unevenly distributed across the classes, the macro average method is less biased toward the largest class.

The ten most contributing features were found through a complete backward elimination, also known as an ablation study, for each time period, and is presented in Table 4. Using only the ten most important features resulted in a decrease of only about 2% (absolute difference) from the F1 macro score with all the features. This means that most of the predictive power from the ANN comes from a few features.

The features shared by all of the three sets are: urgency status 2, weight, height and BMI. BMI can be considered a feature transformation of weight and height as BMI = weight × height^2, but it provided extra predictive information over the constituent variables. A sufficiently complex neural network could probably approximate this transformation and therefore BMI would probably not be needed.

8.3 Predict the Survival After Transplant

Problem. One of the most limiting factors of the number of heart transplants performed is the lack of donor organs and a conservative allocation policy that results in the loss of about half of the organs being offered. An improved prediction of the outcome would augment the confidence in the post-transplantation performance and make it possible to optimize the allocation of organs. Furthermore, it would enable practitioners to determine the risk of early and late graft dysfunction more accurately and improve donor and recipient management.

Although there exist several survival models within cardiac surgery, currently there is no accepted tool for estimating the outcome after heart transplantation. In recent years, some risk score algorithms designed to predict post-transplantation performance have been developed. One of the most notable was the Index for Mortality Prediction After Cardiac Transplantation (IMPACT).

IMPACT was created with a data set of heart transplant patients between 1997 to 2008 that were collected from the UNOS database. IMPACT only utilizes recipient variables. By apportioning points according to the relative importance of the variables for the one-year mortality, a risk index was created. The points are after that converted to a predicted probability of one-year mortality by a formula derived from logistic regression.

The International Heart Transplantation Survival Algorithm (IHTSA) was developed on the ISHLT registry, with patients who were transplanted between 1994 and 2010. IHTSA utilizes both recipient and donor variables. The survival model consists of a flexible nonlinear generalization of the standard Cox proportional hazard model. Instead of using a single prediction model, this model integrates ensembles of artificial neural networks. In addition, its prediction capability is not limited to one year.

We wanted to determine the most suitable risk stratification model for heart transplantation by comparing the IMPACT and IHTSA algorithms.

Method. We included all the adult heart transplant patients (>17 years) from January 1997 to December 2011, from the UNOS database. The data set was divided into two temporal cohorts: transplantation done before 2009 (derivation cohort) and after or during 2009 (test cohort). These time periods were chosen because both IMPACT and IHTSA were developed on patients between 1997–2008 and we wanted disjoint sets (derivation and test) to evaluate the prediction performance.

We used the cohorts as input to both algorithms and then evaluated the performance for both methods.

The discriminatory power for one-year mortality was assessed by calculating the AUROC. We compared the statistical significance of the difference between the AUROC of the two models using the non-parametric DeLong's test. To evaluate the discrimination for long-term survival of the patients, we utilized the Harrell's concordance index (C-index). We used a z-score test to compare the C-indexes.

Table 5. The AUROC values for one-year mortality with the different cohorts using IMPACT and IHTSA respectively.

Time period	AUROC (95% CI)		
	IMPACT	IHTSA	P-Value
1997–2008	0.61 (0.59–0.62)	0.69 (0.68–0.70)	0.001
2009–2011	0.61 (0.58–0.63)	0.65 (0.63–0.68)	0.001

Table 6. The Harrells C-index for survival for the different cohorts using IMPACT and IHTSA respectively [13].

Time period	C-index (95% CI)		
	IMPACT	IHTSA	P-Value
1997–2008	0.56 (0.56–0.56)	0.62 (0.61–0.62)	0.001
2009–2011	0.58 (0.56–0.61)	0.63 (0.61–0.65)	0.001

Results. As shown in Table 5, the IHTSA model has a significantly higher discrimination compared with the IMPACT model for one-year mortality, P = 0.001, corresponding to an error reduction of 11.7%. Harrell's C-index for the recalibrated IHTSA compared with IMPACT was substantially larger, as shown in Table 6, with about a 4% absolute difference for the later time era.

The calibration plot, Fig. 4, shows that the predictive mortality compared with actual mortality was more consistent over all deciles for the ITHSA model, compared with the IMPACT model.

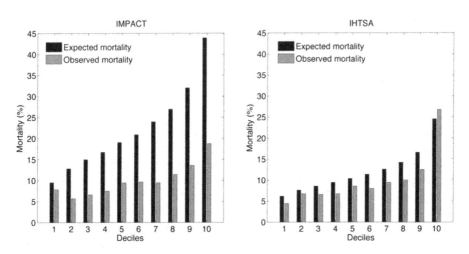

Fig. 4. The observed (gray bars) and expected mortality (black bars), in percent, for each decile, for the IMPACT and IHTSA models, in the test cohort (2009–2011). The patients are divided into deciles according to their expected mortality, and the observed mortality was derived for each decile.

We have shown that a flexible nonlinear artificial neural network model (IHTSA), utilizing deep learning techniques, exhibits better discrimination and accuracy than a more traditional risk score model (IMPACT) for predicting one-year mortality. We made public the results of the IHTSA model in the form of a web-based batch calculator, that could be used as a virtual recipient-donor matching tool.

8.4   Simulating the Impact of Allocation Policies

Problem. Allocation policies in heart transplantation are used to decide how patients awaiting transplant will be paired with hearts from donors. There is a trade-off between medical justice, giving everyone an equal chance for a transplant, and medical utility, which aims at making the best use of a scarce resource.

Predictions models are, most of the time, optimized for the prediction of a single patient, and not applicable to a larger group of patients. This is the reason why the simulation of the whole queue system in an organ allocation process better fits the goal of selecting a policy that maximizes the benefit over all the patients.

Simulating a transplantation queue requires the creation of a model of the queue. This model can thereafter be used to simulate the impact of different policies, on several possible metrics. Examples of potential metrics are the number of deaths in the waiting list, the mean survival time after transplant, and the end size of the waiting list.

The selection of the best allocation policy can be seen as an optimization problem, where you try to maximize predefined metrics by selecting an appropriate policy.

Method. We used a discrete event model to simulate the allocation process.

We chose a Poisson process to simulate the arrival of recipients and donors. This is achieved by selecting patients, without replacement, from the all of the real patients from that specific year.

We created two prediction models; one to simulate the removal of patients from the wait list, mainly caused by death, and the other to predict the survival after heart transplant. A similar model architecture is used. The main difference is the input features. The pre-transplant prediction uses 87 features, while the post-transplant utilizes 267 features. We have called this model: Lund Deep Learning Transplant Algorithm (LuDeLTA).

In addition to our own model LuDeLTA, we also used the IHTSA model to predict the post-graft survival of the patients. We evaluated the different allocation methods with both models.

We selected four allocation policies we wanted to evaluate. The policies were the following: wait time, clinical rules, and neural networks in two different versions. Wait time prioritized the patients with the longest wait time in the queue. Clinical rules ranked the patients based on simple rules based on weight, gender, age, and blood group. Allocation based on neural networks ordered the patients after predicted survival time and chose the patients with the longest predicted survival time for each donor, using either the IHTSA or LuDeLTA as the survival model.

Table 7. Performance metrics of the LuDeLTA models [12].

Metric	Pre-transplant	Post-transplant
AUROC 1 year	0.89	0.66
C-index	0.80	0.61

Table 8. Mean survival of patients results from simulating heart allocation policies. The results in bold should be taken with some care, because they use the same algorithm to both allocate and evaluate the patients. This makes the allocation method heavily biased and the numbers in bold should rather represent a possible maximum ceiling using the corresponding method [12].

Allocation policy	Mean survival IHTSA (days)	Mean survival LuDeLTA (days)
Wait time	4,285	4,309
Clinical rules	4,349	4,309
Neural network: IHTSA	**4,976**	4,719
Neural network: LuDeLTA	4,541	**5,668**

Results. We evaluated the LuDeLTA models using the AUROC for the one year mortality, and the long time survival using The Harrells C-index on the validation set. Results are shown in Table 7. The predicted mean survival on the wait list without transplant was 447 days using our pre-transplant survival model.

Table 9. Results from simulating heart allocation policies.

Allocation policy	Number transplanted	Number dead wait list	Number alive wait list	Mean wait time (days)
Wait time	9,469	5,485	444	139
Clinical rules	9,345	5,481	572	150
Neural network: IHTSA	9,469	4,801	1128	150
Neural network: LuDeLTA	9,469	4,993	936	110

The results for the different allocation policies can be found in the Tables 8 and 9. In Table 8, two of the values use the same algorithm to both allocate and evaluate the patients. This makes the allocation method heavily biased and the numbers in bold should rather represent a possible maximum ceiling using the corresponding method. The predicted mean survival using LuDeLTA, for allocating according to wait time was about 4,300 days, clinical rules 4,300 days, and using IHTSA 4,700 days.

The transplant policies based on the neural network models or wait time utilize all of the available organs, while using clinical rules lead to a discard of 124 hearts.

We have shown that an organ transplant queue can be simulated by utilizing neural networks to predict survival, both pre- and post-transplant. Additionally we have shown that using neural networks as the allocation policy, could possibly result in longer survival post-transplant for the patients.

9   Conclusion

The creation of the RDF representation has simplified the use of the three registries. It enabled us to utilize a unified interface to query the data using SPARQL, which made it easier to handle the patient variables.

We have successfully created several deep learning models using this patient data. Prediction using these models have produced results that were comparable to state-of-the-art systems.

10   Future Work

The models produced by this work could then be introduced as components in a tool that the doctors, who are involved in the transplant process can use. Such a tool could consist of web page, where the physicians would enter potential donors, where the tool would predict the survival for each patient in the waiting list and each recipient-donor pair after transplant, and may combine these two metrics and rank the patients after predicting survival. This could be used to improve the doctors decision process and help with the allocation of a potential heart donor.

We are also interested in including journals and operation descriptions to our database, to see if we could improve the outcome prediction of the transplants based on the written text in these documents.

Another interesting problem to explore would be to use machine learning on genetic data of patients to see if the outcome could be predicted depending on their DNA sequence.

Acknowledgements. This work is based on OPTN data as of October 1, 2013 and was supported in part by the Health Resources and Services Administration contract 234-2005-370011C. The content is the responsibility of the authors alone and does not necessarily reflect the views or policies of the Department of Health and Human Services, nor does mention of trade names, commercial products, or organizations imply endorsement by the U.S. Government. This research was supported by Heart Lung Foundation, The Swedish Research Council, and the eSSENCE program.

References

1. Barzdins, G., Rikacovs, S., Veilande, M., Zviedris, M.: Ontological re-engineering of medical databases. In: Proceedings of the Latvian Academy of Sciences, vol. 63, p. 156. De Gruyter Open Sp. z oo (2009)
2. Baxt, W.G.: Application of artificial neural networks to clinical medicine. Lancet **346**(8983), 1135–1138 (1995)

3. Boulos, M.N.K., Roudsari, A.V., Carson, E.R.: Towards a semantic medical web: Healthcybermap's tool for building an RDF metadata base of health information resources based on the qualified Dublin core metadata set. Med. Sci. Monit. **8**(7), MT124–MT126 (2002)
4. Bratsas, C., Quaresma, P., Pangalos, G., Maglaveras, N.: Using ontologies to build a knowledge base of cardiology problems and algorithms. In: Computers in Cardiology, 2004, pp. 609–612. IEEE (2004)
5. Cucchetti, A., et al.: Artificial neural network is superior to meld in predicting mortality of patients with end-stage liver disease. Gut **56**(2), 253–258 (2007)
6. Dieng-Kuntz, R., Minier, D., Ruzicka, M., Corby, F., Corby, O., Alamarguy, L.: Building and using a medical ontology for knowledge management and cooperative work in a health care network. Comput. Biol. Med. **36**(7–8), 871–892 (2006)
7. Domingos, P.: A few useful things to know about machine learning. Commun. ACM **55**(10), 78–87 (2012). https://doi.org/10.1145/2347736.2347755. http://doi.acm.org/10.1145/2347736.2347755
8. Ham, K.: Openrefine (version 2.5). http://openrefine.org. Free, open-source tool for cleaning and transforming data. J. Med. Libr. Assoc. JMLA **101**(3), 233 (2013)
9. Lindemann, G., Schmidt, D., Schrader, T., Keune, D.: The resource description framework (RDF) as a modern structure for medical data. Int. J. Biol. Med. Sci. **4**(2) (2009)
10. Medved, D., Nugues, P., Nilsson, J.: Selection of an optimal feature set to predict heart transplantation outcomes. In: 2016 38th Annual International Conference of the IEEE Engineering in Medicine and Biology Society (EMBC), pp. 3290–3293, August 2016
11. Medved, D., Nugues, P., Nilsson, J.: Predicting the outcome for patients in a heart transplantation queue using deep learning. In: 2017 39th Annual International Conference of the IEEE Engineering in Medicine and Biology Society (EMBC), pp. 74–77. IEEE (2017)
12. Medved, D., Nugues, P., Nilsson, J.: Simulating the outcome of heart allocation policies using deep neural networks. In: 2018 40th Annual International Conference of the IEEE Engineering in Medicine and Biology Society (EMBC). IEEE (2018, to appear)
13. Medved, D., Ohlsson, M., Höglund, P., Andersson, B., Nugues, P., Nilsson, J.: Improving prediction of heart transplantation outcome using deep learning techniques. Sci. Rep. **8**(1), 3613 (2018)
14. Nilsson, J., Ohlsson, M., Höglund, P., Ekmehag, B., Koul, B., Andersson, B.: Artificial neural networks - relative importance of different recipient-donor characteristic combinations on survival after heart transplantation. J. Hear. Lung Transplant. **30**, S68 (2011)
15. Nilsson, J., Ohlsson, M., Höglund, P., Ekmehag, B., Koul, B., Andersson, B.: The international heart transplant survival algorithm (IHTSA): a new model to improve organ sharing and survival. PLoS ONE **10**(3), e0118644 (2015)
16. Riaño, D., Real, F., López-Vallverdú, J.A., Campana, F., Ercolani, S., Mecocci, P., Annicchiarico, R., Caltagirone, C.: An ontology-based personalization of healthcare knowledge to support clinical decisions for chronically ill patients. J. Biomed. Inform. **45**(3), 429–446 (2012)

17. Weiss, E.S., et al.: Creation of a quantitative recipient risk index for mortality prediction after cardiac transplantation (IMPACT). Ann. Thorac. Surg. **92**(3), 914–922 (2011). https://doi.org/10.1016/j.athoracsur.2011.04. 030. http://www.sciencedirect.com/science/article/pii/S0003497511009350
18. WWWC: RDF 1.1 concepts and abstract syntax (2014). https://www.w3.org/TR/rdf11-concepts/

Game Description Logic with Integers: A GDL Numerical Extension

Munyque Mittelmann[(✉)] and Laurent Perrussel

Université de Toulouse - IRIT, Toulouse, France
{munyque.mittelmann,laurent.perrussel}@irit.fr

Abstract. Many problems can be viewed as games, where one or more agents try to ensure that certain objectives hold no matter the behavior from the environment and other agents. In recent years, a number of logical formalisms have been proposed for specifying games among which the Game Description Language (GDL) was established as the official language for General Game Playing. Although numbers are recurring in games, the description of games with numerical features in GDL requires the enumeration from all possible numeric values and the relation among them. Thereby, in this paper, we introduce the Game Description Logic with Integers (GDLZ) to describe games with numerical variables, numerical parameters, as well as to perform numerical comparisons. We compare our approach with GDL and show that when describing the same game, GDLZ is more compact.

Keywords: Game Description Language · Knowledge representation · General Game Playing

1 Introduction

Many problems, as multiagent planning or process synchronization, can be viewed as games, where one or more agents try to ensure that certain objectives hold no matter the behavior from the environment and other agents [4]. Thereby, a number of logical formalisms have been proposed for specifying game structures and its properties, such as the Game Logic [10,11], the Dynamic Game Logic for sequential [16] and simultaneous games [17], the GameGolog language [4] and so on. Among this formalisms, the Game Description Language (GDL) [1,7] has been established as the official language for the General Game Playing (GGP) Competition. Due to the GDL limitations, such as its restriction to deterministic games with complete state information, several works investigate GDL extensions to improve its expressiveness. Zhang and Thielscher (2014) [18] provide a GDL extension using a modality for linear time and state transition structures. They also propose two dual connectives to express preferences in strategies.

Another extension is called GDL with Incomplete Information (GDL-II) and it was proposed to describe nondeterministic games with randomness and incomplete state knowledge [12,13]. A different approach to deal with this problem is

© Springer Nature Switzerland AG 2020
A. Herzig and J. Kontinen (Eds.): FoIKS 2020, LNCS 12012, pp. 191–210, 2020.
https://doi.org/10.1007/978-3-030-39951-1_12

the Epistemic GDL, that allows to represent imperfect information games and provides a semantical model that can be used for reasoning about game information and players's epistemic status [6]. GDL with Imperfect Information and Introspection (GDL-III) is an extension of GDL-II to include epistemic games, which are characterized by rules that depend on the knowledge of players [14,15]. In order to model how agents can cooperate to achieve a desirable goal, Jiang et al. (2014) present a framework to combine GDL with the coalition operators from Alternating-time Temporal Logic and prioritized strategy connectives [5].

Although numbers are recurring in game descriptions (e.g. Monopoly, Nim game), neither GDL or its extensions incorporate numerical features. In these approaches, numbers can be designed as index in propositions or actions but not directly used as state variables. Thereby, describing games with numerical features can lead to an exhaustive enumeration of all possible numeric values and the relation between them. In the context of planning problems, numerical features have been introduced in Planning Domain Description Language (PDDL) by its first versions [3,9] and improved by PDDL 2.1 [2,8]. In PDDL 2.1, a world state contains an assignment of values to a set of numerical variables. These variables can be modified by action effects and used in expressions to describe actions' preconditions and planning goals.

Similarly to the approach of PDDL 2.1, in this paper, we introduce the GDL extension Game Description Logic with Integers (GDLZ) that incorporates numerical variables, parameters and comparisons. Regarding that board games are mainly described with discrete values, our approach only considers the integer set. We compare our approach with GDL and show that a game description in GDLZ is more compact than the corresponding description in GDL.

This paper is organized as follows. In Sect. 2, we introduce the framework by means of state transition structures and we present the language syntax and semantics. In Sect. 3, we define the translation between GDLZ and GDL and we compare both languages. Section 4 concludes the paper, bringing final considerations[1].

2   Game Description Logic with Integers

In this section, we introduce a logical framework for game specification with integer numbers. The framework is an extension from the GDL state transition model and language [18], such that it defines numerical variables and parameters. We call the framework Game Description Logic with Integers, denoted GDLZ.

To describe a game, we first define a game signature, that specifies who are the players (the agents), what are the possible actions for each player and what are the aspects that describe each state in the game (the propositions and numerical variables). We define a game signature as follows:

---

[1] Due to the space limitation, we omitted most of the Propositions and Theorems proofs. All the proofs are available at https://arxiv.org/abs/1912.01876.

Definition 1. *A game signature $\mathcal{S}$ is a tuple $(N, \mathcal{A}, \Phi, X)$, where:*

- $N = \{r_1, r_2, \cdots, r_k\}$ *is a nonempty finite set of agents;*
- $\mathcal{A} = \bigcup_{r \in N} A^r$ *where* $A^r = \{a_1^r(\bar{z}_1), \cdots, a_m^r(\bar{z}_m)\}$ *consists of a nonempty set of actions performed by agent* $r \in N$, *where* $\bar{z}_i \in \mathbb{Z}^l$ *is a possibly empty tuple of l integer values representing the parameters for the action* a_i^r, *$i \leq m$ and $l \in \mathbb{N}$. For convenience, we occasionally write a_i^r for denoting an action* $a_i^r(\bar{z}_i) \in \mathcal{A}$;
- $\Phi = \{p, q, \cdots\}$ *is a finite set of atomic propositions for specifying individual features of a game state;*
- $X = \langle x_1, x_2, \cdots, x_n \rangle$ *is a tuple of numerical variables for specifying numerical features of a game state.*

Given a game signature, we define a state transition model, that allows us to represent the key aspects of a game, such as the winning states for each agent, the legal actions in each state and the transitions between game states.

Definition 2. *Given a game signature $\mathcal{S} = (N, \mathcal{A}, \Phi, X)$, a state transition ST model M is a tuple $(W, \bar{w}, T, L, U, g, \pi_\Phi, \pi_\mathbb{Z})$, where:*

- W *is a nonempty set of states;*
- $\bar{w} \in W$ *is the initial state;*
- $T \subseteq W$ *is a set of terminal states;*
- $L \subseteq W \times \mathcal{A}$ *is a legality relation, describing the legal actions at each state;*
- $U : W \times D \to W$ *is an update function, where* $D = \prod_{r \in N} A^r$ *denote the set of joint actions, specifying the transitions for each joint state;*
- $g : N \to 2^W$ *is a goal function, specifying the winning states for each agent;*
- $\pi_\Phi : W \to 2^\Phi$ *is the valuation function for the state propositions;*
- $\pi_\mathbb{Z} : W \to \mathbb{Z}^n$ *is the valuation function for the state numerical variables, such that $\pi_\mathbb{Z}(w)$ is a tuple of integer values assigned to the variables X at state $w \in W$. Let $\pi_\mathbb{Z}^i(w)$ denote the i-th value of $\pi_\mathbb{Z}(w)$.*

Given $d \in D$, let $d(r)$ be the individual action for agent r in the joint action d. Let $L(w) = \{a \in \mathcal{A} \mid (w, a) \in L\}$ be the set of all legal actions at state w.

Definition 3. *Given an ST-model $M = (W, \bar{w}, T, L, U, g, \pi_\Phi, \pi_\mathbb{Z})$, a path is a finite sequence of states $\bar{w} \xrightarrow{d_1} w_1 \xrightarrow{d_2} \cdots \xrightarrow{d_e} w_e$ such that $e \geq 0$ and for any $j \in \{1, \cdots, e\}$: (i) $\{w_0, \cdots, w_{e-1}\} \cap T = \emptyset$, where $w_0 = \bar{w}$; (ii) $d_j(r) \in L(w_{j-1})$ for any $r \in N$; and (iii) $w_j = U(w_{j-1}, d_j)$.*

A path δ is *complete* if $w_e \in T$. Given $\delta \in \mathcal{P}$, let $\delta[j]$ denotes the j-th reachable state of δ, $\theta(\delta, j)$ denotes the joint action taken at stage j of δ; and $\theta_r(\delta, j)$ denotes the action of agent r taken at stage j of δ. Finally, the length of a path λ, written $|\lambda|$, is defined as the number of joint actions.

Describing a game with the ST-model is not practical, especially when modeling large games. Hereby, given a game signature $\mathcal{S} = (N, \mathcal{A}, \Phi, X)$, we introduce a variant of the language for GDL ($\mathcal{L}_{GDL}$ for short) to describe a GDLZ game in a more compact way by encoding its rules.

2.1   Syntax

The *language* is denoted by $\mathcal{L}_{GDLZ}$ and a *formula* φ in $\mathcal{L}_{GDLZ}$ is defined by the following Backus-Naur Form (BNF) grammar:

$$\varphi ::= p \mid initial \mid terminal \mid legal(a^r(\bar{z})) \mid wins(r) \mid does(a^r(\bar{z})) \mid \neg\varphi \mid \varphi \wedge \varphi \mid$$
$$\bigcirc\varphi \mid z > z \mid z < z \mid z = z \mid \langle\bar{z}\rangle$$

where, $p \in \Phi, r \in N, a^r \in A^r, \bar{z}$ is a number list and z is a numerical term.

Let ε denote the empty word. A number list $\bar{z}$ is defined as:

$$\bar{z} ::= z \mid z, \bar{z} \mid \varepsilon.$$

Finally, a numerical term z is defined by $\mathcal{L}_z$, which is generated by the following BNF:

$$z ::= z' \mid x' \mid add(z, z) \mid sub(z, z) \mid min(z, z) \mid max(z, z)$$

where $z' \in \mathbb{Z}$ and $x' \in X$.

Other connectives $\vee, \rightarrow, \leftrightarrow, \top$ and $\bot$ are defined by $\neg$ and $\wedge$ in the standard way. The comparison operators $\leq, \geq$ and $\neq$ are defined by $\vee, >, <$ and $=$, respectively, as follows: (i) $z_1 < z_2 \vee z_1 = z_2$, (ii) $z_1 > z_2 \vee z_1 = z_2$ and (iii) $z_1 > z_2 \vee z_1 < z_2$.

Intuitively, *initial* and *terminal* specify the initial state and the terminal state, respectively; $does(a^r(\bar{z}))$ asserts that agent r takes action a with the parameters $\bar{z}$ at the current state; $legal(a^r(\bar{z}))$ asserts that agent r is allowed to take action a with the parameters $\bar{z}$ at the current state; and $wins(r)$ asserts that agent r wins at the current state. The formula $\bigcirc\varphi$ means "φ holds at the next state". The formulas $z_1 > z_2, z_1 < z_2, z_1 = z_2$ means that a numerical term z_1 is greater, less and equal to a numerical term z_2, respectively. Finally, $\langle\bar{z}\rangle$ asserts the current values for the numerical variables, i.e. the i-th variable in X has the i-th value in $\bar{z}$, for $0 \leq i \leq |X|$. Notice that $\langle\bar{z}\rangle$ could be represented by a conjunction over each $x_i \in X$ of formulas $x_i = z_i$, where $z_i \in \mathcal{L}_z$ is the current value of the variable x_i. However, $\langle\bar{z}\rangle$ provides a short cut and it is more meaningful, in the sense that it is strictly related to the valuation of the numerical variables in a given state.

For numerical terms, $add(z_1, z_2)$ and $sub(z_1, z_2)$ specify the value obtained by adding and subtracting z_2 from z_1, respectively. The formulas $min(z_1, z_2)$ and $max(z_1, z_2)$ specify the minimum and maximum value between z_1 and z_2, respectively. The extension of the comparison operators $>, <, =, \leq, \geq$ and $\neq$ to multiple arguments is straightforward.

If φ is not in the form $\neg\varphi', \bigcirc\varphi'$ or $\varphi' \wedge \varphi''$, for any $\varphi', \varphi'' \in \mathcal{L}_{GLDZ}$, then φ is called an atomic formula. We say that a numerical variable occurs in an atomic formula φ if (i) φ is either in the form $legal(a^r(\bar{z})), does(a^r(\bar{z}))$ or $\langle\bar{z}\rangle$ and there is a $x \in X$ in the numerical list $\bar{z}$; (ii) φ is either in the form $z_1 < z_2$, $z_1 > z_2$ or $z_1 = z_2$ and $z_1 \in X$ or $z_2 \in X$.

2.2   Semantics

The semantics for the GDLZ language is given in two steps. First, we define function v to assign the meaning of numerical terms $z \in \mathcal{L}_z$ in a specified state (Definition 4). Next, a formula $\varphi \in \mathcal{L}_{GDLZ}$ is interpreted with respect to a stage in a path (Definition 5).

Definition 4. *Given an ST-model M, a state w and the functions minimum and maximum[2] let us define function $v : W \times \mathcal{L}_z \to \mathbb{Z}$, associating any $z_i \in \mathcal{L}_z$ in a state $w \in W$ to a number in $\mathbb{Z}$:*

$$v(z_i, w) = \begin{cases} z_i & \text{if } z_i \in \mathbb{Z} \\ \pi_{\mathbb{Z}}^i(w) & \text{if } z_i \in X \\ \pi_{\mathbb{Z}}^i(w) & \text{if } z_i = x_i \text{ \& } x_i \in X \\ v(z_i', w) + v(z_i'', w) & \text{if } z_i = add(z_i', z_i'') \\ v(z_i', w) - v(z_i'', w) & \text{if } z_i = sub(z_i', z_i'') \\ minimum(v(z_i', w), v(z_i'', w)) & \text{if } z_i = min(z_i', z_i'') \\ maximum(v(z_i', w), v(z_i'', w)) & \text{if } z_i = min(z_i', z_i'') \end{cases}$$

Definition 5. *Let M be an ST-Model. Given a complete path δ of M, a stage j on δ, a formula $\varphi \in \mathcal{L}_{GDLZ}$ and function v, we say φ is true (or satisfied) at j of δ under M, denoted by $M, \delta, j \models \varphi$, according with the following definition:*

$M, \delta, j \models p$	iff	$p \in \pi_\Phi(\delta[j])$
$M, \delta, j \models \neg\varphi$	iff	$M, \delta, j \not\models \varphi$
$M, \delta, j \models \varphi_1 \wedge \varphi_2$	iff	$M, \delta, j \models \varphi_1$ and $M, \delta, j \models \varphi_2$
$M, \delta, j \models initial$	iff	$\delta[j] = \bar{w}$
$M, \delta, j \models terminal$	iff	$\delta[j] \in T$
$M, \delta, j \models wins(r)$	iff	$\delta[j] \in g(r)$
$M, \delta, j \models legal(a^r(\bar{z}))$	iff	$a^r(v(z, \delta[j]) : z \in \bar{z}) \in L(\delta[j])$
$M, \delta, j \models does(a^r(\bar{z}))$	iff	$\theta_r(\delta, j) = a^r(v(z, \delta[j]) : z \in \bar{z})$
$M, \delta, j \models \bigcirc\varphi$	iff	if $j < |\delta|$, then $M, \delta, j + 1 \models \varphi$
$M, \delta, j \models z_1 > z_2$	iff	$v(z_1, \delta[j]) > v(z_2, \delta[j])$
$M, \delta, j \models z_1 < z_2$	iff	$v(z_1, \delta[j]) < v(z_2, \delta[j])$
$M, \delta, j \models z_1 = z_2$	iff	$v(z_1, \delta[j]) = v(z_2, \delta[j])$
$M, \delta, j \models \langle\bar{z}\rangle$	iff	$\langle v(z, \delta[j]) : z \in \bar{z}\rangle = \pi_{\mathbb{Z}}(\delta[j])$

A formula φ is globally true through δ, denoted by $M, \delta \models \varphi$, if $M, \delta, j \models \varphi$ for any stage j of δ. A formula φ is *globally true* in an ST-Model M, written $M \models \varphi$, if $M, \delta \models \varphi$ for all complete paths δ in M, that is, φ is true at every reachable state. A formula φ is *valid*, denoted by $\models \varphi$, if it is globally true in every ST-model of an appropriate signature. Finally, let Σ be a set of formulas in $\mathcal{L}_{GDLZ}$, then M is a *model* of Σ if $M \models \varphi$ for all $\varphi \in \Sigma$.

---

[2] Through the rest of this paper, the functions $minimum(a, b)$ and $maximum(a, b)$ respectively return the minimum and maximum value between $a, b \in \mathbb{Z}$.

Whenever $j \geq |\delta|$, the validity of $M, \delta, j \models \bigcirc \varphi$ is irrelevant, since $\delta[j]$ is the last state reachable in δ. A formula $\langle \bar{z} \rangle$ is valid at a stage j in a path δ under M only when it corresponds to the valuation of the numerical variables at $\delta[j]$.

The following propositions show that if a player does an action at a stage in a path, then (i) he does not any other action in the same stage and (ii) the action taken is legal.

Proposition 1. $\models does(a^r(\bar{z})) \rightarrow \bigwedge_{b^r \neq a^r \in A^r} \bigwedge_{\bar{z}' \neq \bar{z} \in \mathbb{Z}^n} \neg does(b^r(\bar{z}'))$.

Proposition 2. $\models does(a^r(\bar{z})) \rightarrow legal(a^r(\bar{z}))$.

Next, we illustrate the representation of a game with numerical features in GDLZ. First, we define the game signature and the game description in $\mathcal{L}_{GDLZ}$. Next, we define the ST-model by which it is possible to evaluate the $\mathcal{L}_{GDLZ}$ semantics. Finally, we illustrate a path in the game.

*Example 1. ($\langle \gamma_1, \cdots, \gamma_k \rangle$-**Nim Game**) A $\langle \gamma_1, \cdots, \gamma_k \rangle$-Nim Game consists in k heaps. Each heap starts with γ_i sticks, where $1 \leq i \leq k$. Two players take turns in removing sticks from one heap. The game ends when all heaps are empty. A player wins if it is not his turn when the game ends.*

To represent a $\langle \gamma_1, \cdots, \gamma_k \rangle$-Nim Game in terms in GDLZ, we first specify the agents, the actions, the propositions, and the numerical variables involved in the game. Thus, the game signature, written $\mathcal{S}_{k\text{-nim}}$, is described as follows:

- $N_{k\text{-nim}} = \{Player_1, Player_2\}$;
- $A^r_{k\text{-nim}} = \{reduce^r(m, s) \mid s \in \mathbb{N}, 1 \leq m \leq k\} \cup \{noop^r\}$, where $reduce^r(m, s)$ denotes the action that player r removes s sticks from the m-th heap and $noop^r$ denotes that player r does action noop;
- $\Phi_{k\text{-nim}} = \{turn(r) \mid r \in \{Player_1, Player_2\}\}$, where $turn(r)$ says that it is player r's turn now;
- $X_{k\text{-nim}} = \langle heap_i \mid 1 \leq i \leq k \rangle$, where $heaps_i$ represents the amount of sticks in the i-th heap.

Given a player $r \in N_{k-min}$, we denote $-r$ as the opponent of r, i.e. $-r = Player_1$ if $r = Player_2$ and $-r = Player_2$ otherwise. The rules of the $\langle \gamma_1, \cdots, \gamma_k \rangle$-Nim Game can be expressed by GDLZ- formulas as shown Fig. 1.

Statement 1 says that the $Player_1$ has the first turn and that the k heaps starts with $\gamma_1, \cdots, \gamma_k$ sticks, respectively. Statement 2 and 3 specify the winning states for each player and the terminal states of the game, respectively. The player who has not the turn when all the heaps become empty wins the game, and the game ends if all the heaps are empty. Statements 4 and 5 specify the preconditions of each action (legality). The player who has the turn can reduce s sticks from the m-th heap if $1 \leq s \leq heap_m$. The other player can only do noop. Statements 6 and 7 define what is true at the next state: the m-th heap will be subtracted by s if a player takes the action of reducing the m-th heap by s, otherwise it will keep its current value. Finally, Statement 8 specifies the turn-taking. Let $\Sigma_{k\text{-nim}}$ be the set of rules 1-8.

1. $initial \leftrightarrow turn(Player_1) \wedge \neg turn(Player_2) \wedge \langle \gamma_1, \cdots, \gamma_k \rangle$
2. $\bigwedge_{r \in N} wins(r) \leftrightarrow \neg turn(r) \wedge turn(-r) \wedge \langle 0, \cdots, 0 \rangle$
3. $terminal \leftrightarrow \langle 0, \cdots, 0 \rangle$
4. $\bigwedge_{r \in N} \bigwedge_{m \in \{1 \ldots k\}} \bigwedge_{s \in \{1 \ldots \gamma_m\}} legal(reduce^r(m, s)) \leftrightarrow 1 \leq s \leq heap_m$
   $\wedge turn(r)$
5. $\bigwedge_{r \in N} legal(noop^r) \leftrightarrow \neg turn(r)$
6. $\bigwedge_{i \in \{1 \ldots k\}} \bigwedge_{h_i \in \{1 \ldots \gamma_i\}} terminal \wedge \langle h_1, \cdots, h_k \rangle \rightarrow \bigcirc \langle h_1, \cdots, h_k \rangle$
7. $\bigwedge_{i \in \{1 \ldots k\}} \bigwedge_{h_i \in \{1 \ldots \gamma_i\}} \neg terminal \wedge \langle h_1, \cdots, h_k \rangle$
   $\wedge (\bigvee_{r \in N} \bigvee_{m \in \{1 \ldots k\}} \bigvee_{s \in \{1 \ldots \gamma_m\}} does(reduce^r(m, s))) \rightarrow$
   $\bigcirc \langle h_1, \cdots, sub(h_m, s), \cdots, h_k \rangle$
8. $\bigwedge_{r \in N} turn(r) \rightarrow \bigcirc \neg turn(r) \wedge \bigcirc turn(-r)$

Fig. 1. $\langle \gamma_1, \cdots, \gamma_k \rangle$-Nim Game represented by $\Sigma_{k\text{-nim}}$

Since the semantics for the language is based on the state transition model, we next specify the ST-model for this game, written $M_{k\text{-nim}}$, as follows:

- $W_{k\text{-nim}} = \{\langle t_1, t_2, \langle x_1, \cdots, x_k \rangle \rangle : t_1 \in \{turn(Player_1), \neg turn(Player_1)\} \& t_2 \in \{turn(Player_2), \neg turn(Player_2)\} \& x_i \in \mathbb{N}, \text{ for } 1 \leq i \leq k\}$ is the set of states, where t_1, t_2 specify the turn taking and x_i represents the amount of sticks in the i-th heap, i.e. the integer value assigned to $heap_i$;
- $\bar{w}_{k\text{-nim}} = \langle turn(Player_1), \neg turn(Player_2), \langle \gamma_1, \cdots, \gamma_k \rangle \rangle$;
- $T_{k\text{-nim}} = \{\langle turn(Player_1), \neg turn(Player_2), \langle 0, \cdots, 0 \rangle \rangle, \langle \neg turn(Player_1), turn(Player_2), \langle 0, \cdots, 0 \rangle \rangle\}$, i.e. all heaps are empty;
- $L_{k\text{-nim}} = \{(\langle t_1, t_2, \langle x_1, \cdots, x_k \rangle \rangle, reduce^r(m, s)) : t_r = turn(r) \& 1 \leq s \leq x_m\} \cup \{(\langle t_1, t_2, \langle x_1, \cdots, x_k \rangle \rangle, noop^r) : t_r = \neg turn(r)\}$, for all $\langle t_1, t_2, \langle x_1, \cdots, x_k \rangle \rangle \in W_{k\text{-nim}}$ and $r \in N_{k\text{-nim}}$;
- $U_{k\text{-nim}} : W_{k\text{-nim}} \times D_{k\text{-nim}} \rightarrow W_{k\text{-nim}}$ is defined as follows: for all $\langle t_1, t_2, \langle x_1, \cdots, x_k \rangle \rangle \in W_{k\text{-nim}}$ and all $(reduce^r(m, s), noop^{-r}) \in D_{k\text{-nim}}$, let $U_{k\text{-nim}}(\langle t_1, t_2, \langle x_1, \cdots, x_k \rangle \rangle, (reduce^r(m, s), noop^{-r})) = \langle t'_1, t'_2, \langle x'_1, \cdots, x'_k \rangle \rangle$, such that $\langle t'_1, t'_2, \langle x'_1, \cdots, x'_k \rangle \rangle$ are the same as $\langle t_1, t_2, \langle x_1, \cdots, x_k \rangle \rangle$, except by its components t'_1, t'_2 and x'_i which are updated as follows: $t'_1 = turn(Player_1)$ iff $t_2 = turn(Player_2)$, otherwise $t'_1 = \neg turn(Player_1)$; $t'_2 = turn(Player_2)$ iff $t_1 = turn(Payer_1)$, otherwise, $t'_2 = \neg turn(Player_2)$; and for $1 \leq i \leq k$:

$$x'_i = \begin{cases} x_i - s & \text{if } reduce^r(i, s) \text{ and } 1 \leq s \leq x_i \\ x_i & \text{otherwise} \end{cases}$$

For all $\langle t_1, t_2, \langle x_1, \cdots, x_k \rangle \rangle \in W_{k\text{-nim}}$ and all $(a^r, a^{-r}) \neq (reduce^r(m, s), noop^{-r}) \in D_{k\text{-nim}}$, let $U_{k\text{-nim}}(\langle t_1, t_2, \langle x_1, \cdots, x_k \rangle \rangle, (a^r, a^{-r})) = \langle t_1, t_2, \langle x_1, \cdots, x_k \rangle \rangle$.

- $g_{k\text{-nim}}(r) = \{\langle t_1, t_2, \langle 0, \cdots, 0 \rangle \rangle\}$, where $t_r = \neg turn(r)$ and $t_{-r} = turn(r)$.

Finally, for each state $w = \langle t_1, t_2, \langle x_1, \cdots, x_k \rangle \rangle \in W_{k\text{-nim}}$, let

- $\pi_{\Phi, k\text{-nim}}(w) = \{turn(r) : t_r = turn(r)\}$;
- $\pi_{\mathbb{Z}, k\text{-nim}}(w) = \langle x_1, \cdots, x_k \rangle$.

Let $M_{k\text{-nim}} = (W_{k\text{-nim}}, \bar{w}_{k\text{-nim}}, T_{k\text{-nim}}, L_{k\text{-nim}}, U_{k\text{-nim}}, g_{k\text{-nim}}, \pi_{\Phi, k\text{-nim}}, \pi_{\mathbb{Z}, k\text{-nim}})$ be the ST-model for the k-Nim Game.

Consider, for instance, $k = 2$ and $\langle \gamma_1, \gamma_2 \rangle = \langle 5, 3 \rangle$, i.e. there are only two heaps and their starting values are 5 and 3, respectively. Figure 2 illustrates a path in $M_{k\text{-nim}}$. The state w_0 represents the initial state. In w_0, it is the turn of $Player_1$ and he removes 5 sticks from the first heap. In the state w_1, the first heap is empty and players can only remove sticks from the second heap. It is now $Player_2$'s turn and he reduces 2 sticks from the second heap. In the state w_2, $Player_1$ removes the last stick from the second heap. Finally, in the state w_3, there is no stick remaining in any heap, thereby it is a terminal state. Since it is $Player_2$'s turn, $Player_1$ wins the game.

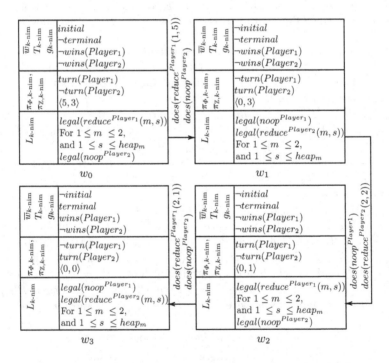

Fig. 2. A Path in $M_{k\text{-nim}}$, where $k = 2$ and $\langle \gamma_1, \gamma_2 \rangle = \langle 5, 3 \rangle$

The next proposition shows that soundness does hold, i.e. the framework provides a sound description for the k-Nim Game. Notice that as $M_{k\text{-nim}}$ is not the unique model for $\Sigma_{k\text{-nim}}$, thereby, the completeness does not hold.

Proposition 3. $M_{k\text{-}nim}$ *is an ST-model and it is a model of* $\Sigma_{k\text{-}nim}$.

In the next section, we show that the model checking for GDLZ is decidable in polynomial-time deterministic Turing machines (denoted PTIME), which is the same complexity then the model checking for GDL. In other words, the addition of numerical features in GDL does not increase the complexity at verifying the validity of a formula at a stage of a path in a model.

2.3   Model Checking

The *model checking problem* for GDLZ is the following: Given a GDLZ-formula φ, an ST-model M, a path δ of M and a stage j on δ, determine whether $M, \delta, j \models \varphi$ or not.

Let $Sub(\varphi)$ be the set of all subformulas[3] of φ. Algorithm 1 works in the following way: first it gets all subformulas of φ and orders them in S by its ascending length. Thus, $S(|\varphi|) = \varphi$, i.e. the position $|\varphi|$ in the vector S corresponds to the formula φ itself, and if ϕ_i is a subformula of ϕ_j, then $i < j$. An induction on S label each subformula ϕ_i depending on whether or not ϕ_i is true in M at $\delta[j]$. If ϕ_i does not have any subformula, its truth value is obtained directly from the semantics. Since S is ordered by the formulas length, if ϕ_i is either in the form $\phi' \wedge \phi''$ or $\neg\phi'$ the algorithm labels ϕ_i according to the label assigned to ϕ' and/or ϕ''. If ϕ_i is in the form $\bigcirc\phi'$, its label will be recursively defined according to ϕ' truth value in $\delta[j+1]$. As Algorithm 1 visits each node at most once, and the number of nodes in the tree is not greater than the size of φ, it can be clearly implemented in a polynomial-time deterministic Turing machine with PTIME.

Algorithm 1. $isTrue(M, \delta, j, \varphi)$

    Input: an ST-model M, a path δ of M, a stage j and a formula $\varphi \in \mathcal{L}_{GDLZ}$.
    Output: true if $M, \delta, j \models \varphi$, and **false** otherwise
1:   $S \leftarrow Sub(\varphi)$ ordered by ascending length
2:   Let $reg[1 \cdots size(S)]$ be a boolean array
3:   **for** $i \leftarrow 1$ to $size(S)$ **do**
4:      $\phi \leftarrow S[i]$
5:      **if** $(\phi = \phi' \wedge \phi'')$ **then**
6:         $reg[i] \leftarrow reg[getIndex(S, \phi')] \wedge reg[getIndex(S, \phi'')]$
7:      **else if** $(\phi = \bigcirc\phi')$ **then**
8:         $reg[i] \leftarrow isTrue(M, \delta, j+1, \phi')$
9:      **else if** $(\phi = \neg\phi')$ **then**
10:      $reg[i] \leftarrow \neg reg[getIndex(S, \phi')]$
11:      **else** $reg[i] \leftarrow M, \delta, j \models \phi$
       return $reg[size(S)]$

---

[3] We say that ψ is a subformula of $\varphi \in \mathcal{L}_{GDLZ}$ if either (i) $\psi = \varphi$; (ii) φ is of the form $\neg\varphi'$ or $\bigcirc\varphi'$ and ψ is a subformula of φ'; or (iii) φ is of the form $\varphi' \wedge \varphi''$ and ψ is a subformula of either φ' or φ''.

In Sect. 3.3 we show that $\mathcal{L}_{GDL} \subseteq \mathcal{L}_{GDLZ}$, i.e. any formula in GDL is also a formula in GDLZ. Thereby, Algorithm 1 can also be used in the model checking problem for GDL.

3   Translation Between GDLZ and GDL

In this section, we investigate translation maps among GDLZ and GDL models and descriptions. We first consider the general case where the GDLZ ST-model can have infinite components. Next, we restrict to the case where a GDLZ ST-model is finite. Finally, we compare both languages in order to show the succinctness of GDLZ descriptions over GDL descriptions.

Given a GDLZ ST-model M, a complete path δ in M and a formula $\varphi \in \mathcal{L}_{GDLZ}$, in the Sects. 3.1 and 3.2 our goal is to construct a GDL ST-model M', a path δ' in M' and a formula $\varphi' \in \mathcal{L}_{GDL}$ such that, for any stage j on δ, if $M, \delta, j \models \varphi$ then $M', \delta', j \models \varphi'$.

3.1   From GDLZ Paths and Models to GDL Models

In a GDL ST-model, the sets of states, actions and atomic propositions are finite. Since it does not hold for GDLZ ST-models, it is not possible to define a complete translation from every GDLZ model to a GDL model. However, since any GDLZ path is a finite sequence of states and joint actions, we can define a partial translation from GDLZ ST-models to GDL ST-models based on the reached states and joint actions performed in a complete path. In other words, we can translate a run in a GDLZ model into a GDL model. Let us formally describe the translation.

Through the rest of this section, we fix the GDLZ ST-model $M = (W, \bar{w}, T, L, U, g, \pi_\Phi, \pi_Z)$ with a game signature $\mathcal{S} = (N, \mathcal{A}, X, \Phi)$ and the complete path $\delta = \bar{w} \xrightarrow{d_1} w_1 \xrightarrow{d_2} \cdots \xrightarrow{d_e} w_e$ in M.

Given the path δ in M, we next define a shortcut to refer to the smallest and biggest integers occurring in δ and the set of all actions performed in δ.

Definition 6. *Given M and δ, we denote δ_{min} and δ_{max} as the smallest and biggest integer, respectively, occurring in any parameter list z from any action $a \in \{d_1, d_2, \cdots, d_e\}$ and in any $\pi_Z(w)$, for $w \in \{\bar{w}, w_1, \cdots, d_e\}$.*

Definition 7. *Given M and δ, let $\mathcal{A}^\delta = \{d_j(r) : r \in N \ \& \ 1 \leq j \leq e\}$ denote the set of all actions performed in δ.*

Since we are aware of the path numerical range, we are able to construct a partial model translation. The translation is restricted to the states and actions involved in a given path.

Definition 8. *Given a GDLZ ST-model M and δ, we construct an associated GDL ST-model $M' = (W', \bar{w}, T', L', U', g', \pi')$ with a game signature $\mathcal{S}' = (N, \mathcal{A}', \Phi')$. The components $\bar{w}$ and $N = \{r_1, \cdots, r_k\}$ are the same for M and M'.*

The propositional set Φ' is constructed over both Φ and X as follows: $\Phi' = \{p, smaller(z_1, z_2), bigger(z_1, z_2), equal(z_1, z_2), succ(z_1, z_2), prec(z_1, z_2), x(q) : p \in \Phi, x \in X, \delta_{min} \leq q, z_1, z_2 \leq \delta_{max}\}$. The notation $x(q)$ represents the proposition "variable x has the value q".

For integrating the GDLZ comparison operators $<, >$ and $=$ in GDL, we need to define the order between the numerical terms in the translated model. Let $\pi_z \subset \Phi'$ denote a set of propositions describing the numerical order, such as: $\pi_z = \{succ(z, z+1), prec(z+1, z), equal(z_1, z_1) : \delta_{min} \leq z < \delta_{max} \& \delta_{min} \leq z_1 \leq \delta_{max}\} \cup \{smaller(z_1, z_2) : \delta_{min} \leq z_1, z_2 \leq \delta_{max} \& z_1 < z_2\} \cup \{bigger(z_1, z_2) : \delta_{min} \leq z_1, z_2 \leq \delta_{max} \& z_1 > z_2\}$.

For any $a^r(z_1, \cdots, z_l) \in \mathcal{A}^\delta$, $a^r_{z_1, \cdots, z_l} \in \mathcal{A}'$. We define an action translation $Tr^a : \mathcal{A}^\delta \to \mathcal{A}'$ associating every action in $\mathcal{A}^\delta$ with an action in $\mathcal{A}' : Tr^a(a^r(z_1, \cdots, z_l)) = a^r_{z_1, \cdots, z_l}$, where $a^r(z_1, \cdots, z_l) \in \mathcal{A}^\delta$.

The M' components W', T', L', U', g' and π' are defined as follows:

- $W' = \{\bar{w}, w_1, \cdots, w_e\}$
- $T' = \{w_e\}$;
- $L' = \{(w_{j-1}, Tr^a(d_j(r)) : r \in N \ \& \ 1 \leq j \leq e\}$;
- $U'(w_{j-1}, (Tr^a(d_j(r_1)), \cdots, Tr^a(d_j(r_k)))) = w_j$, for $1 \leq j \leq e\}$;
- $g'(r) = \{\{w_e\}\}$ if $w_e \in g(r)$, otherwise $g'(r) = \emptyset$, for $r \in N$;
- $\pi'(w) = \{\pi_\Phi(w)\} \cup \{\pi_z\} \cup \{x(q) : q \in \pi_{\mathbb{Z}}(w), x \in X\}$, for $w \in W'$.

We say that $M' = (W', \bar{w}', T', L', U', g', \pi')$ with the signature $\mathcal{S}' = (N, \mathcal{A}', \Phi')$ is the ST-model translation of M restricted over δ and write $Tr^m(M, \delta)$.

The path translation assigns each action appearing on it to the appropriated GDL action through Tr^a, i.e. the action translation.

Definition 9. Given the agent set $N = \{r_1, \cdots, r_k\}$, define a path translation $Tr^\lambda : \delta \to \delta'$ associating a path $\delta = \bar{w} \xrightarrow{d_1} w_1 \xrightarrow{d_2} \cdots \xrightarrow{d_e} w_e$ in M with a path δ' in $Tr^m(M, \delta)$: $Tr^\lambda(\delta) = \bar{w} \xrightarrow{d'_1} w_1 \xrightarrow{d'_2} \cdots \xrightarrow{d'_e} w_e$, where $d'_i = (Tr^a(d_i(r_1)), \cdots, Tr^a(d_i(r_k)))$, for $1 \leq i \leq e$.

As shown next propositions, given a path in a GDLZ model, the translation of the GDLZ model is a GDL model. Moreover, the translation of a path in a GDLZ model is a path in the translation of the GDLZ model.

Proposition 4. If M is a GDLZ model and δ a complete path in M, then $Tr^m(M, \delta)$ is a GDL ST-model.

Proposition 5. If δ is a path in a GDLZ model M then $Tr^\lambda(\delta)$ is a path in $Tr^m(M, \delta)$.

Next, we show how to translate GDLZ formulas to GDL. Likewise to the model translation, the translation is restricted to a path.

From GDLZ Paths and Formulas to GDL Formulas. Let us briefly recall GDL grammar. Given a GDL game signature $S' = (N, \mathcal{A}', \Phi')$, a formula $\varphi' \in \mathcal{L}_{GDL}$ is defined by the following BNF:

$$\varphi' ::= p \mid initial \mid terminal \mid legal(a^r) \mid wins(r) \mid does(a^r) \mid \neg\varphi \mid \varphi \wedge \varphi \mid \bigcirc\varphi$$

where $p \in \Phi'$, $r \in N$ and $a^r \in \mathcal{A}'$.

Given a path δ in a GDLZ ST-model M, we next define a translation for formulas in $\mathcal{L}_{GDLZ}$ to $\mathcal{L}_{GDL}$. Each numerical term $z \in \mathcal{L}_z$ occurring in a formula $\varphi \in \mathcal{L}_{GDLZ}$ is translated by its semantic interpretation through function v (see Definition 4).

Definition 10. *Given a GDLZ ST-model M with $\mathcal{S} = (N, \mathcal{A}, X, \Phi)$, a path δ in M, a stage j in δ and function v (see Definition 4). A translation Tr^φ from a formula $\varphi \in \mathcal{L}_{GDLZ}$ in a state $\delta[j]$ to a formula $\varphi' \in \mathcal{L}_{GDL}$ is defined as follows:*

- $Tr^\varphi(\varphi, \delta[j]) = \varphi$ *for all* $\varphi \in \Phi \cup \{initial, terminal, wins(r)\}$;
- $Tr^\varphi(\neg\varphi, \delta[j]) = \neg Tr^\varphi(\varphi, \delta[j])$;
- $Tr^\varphi(\varphi_1 \wedge \varphi_2, \delta[j]) = Tr^\varphi(\varphi_1, \delta[j]) \wedge Tr^\varphi(\varphi_2, \delta[j])$;
- $Tr^\varphi(\bigcirc\varphi, \delta[j]) = \bigcirc Tr^\varphi(\varphi, \delta[j+1])$;
- $Tr^\varphi(legal(a^r(\bar{z})), \delta[j]) = legal(Tr^a(a^r(v(z) : z \in \bar{z})))$ *iff* $legal(a^r(v(z, \delta[j]) : z \in \bar{z}) = \theta_r(\delta, j)$; *otherwise* $Tr^\varphi(legal(a^r(\bar{z})), \delta[j]) = \neg legal(Tr^a(a^r(v(z) : z \in \bar{z})))$;
- $Tr^\varphi(does(a^r(\bar{z})), \delta[j]) = does(Tr^a(a^r(v(z) : z \in \bar{z})))$;
- $Tr^\varphi(\langle\bar{z}\rangle, \delta[j]) = \bigwedge_{i=1}^{|\bar{z}|} x_i(v(q_i, \delta[j]))$;
- $Tr^\varphi(z_1 < z_2, \delta[j]) = smaller(v(z_1, \delta[j]), v(z_2, \delta[j]))$;
- $Tr^\varphi(z_1 > z_2, \delta[j]) = bigger(v(z_1, \delta[j]), v(z_2, \delta[j]))$;
- $Tr^\varphi(z_1 = z_2, \delta[j]) = equal(v(z_1, \delta[j]), v(z_2, \delta[Tr^\varphi(\varphi, \delta[t])]))$.

Where $r \in N$, $x_i \in X$, q_i is the i-th value in $\bar{z}$ and $0 \le i \le |\bar{z}|$.

Given a path in a GDLZ model, we show that the translation of a GDLZ formula is a GDL formula. Furthermore, if the GDLZ formula is valid at a stage in the path, its translation will be valid at the same stage in the translated path in the translated model.

Proposition 6. *Given a GDLZ ST-model M, a path δ in M, a stage j in δ and function v, if φ is a formula in $\mathcal{L}_{GDLZ}$ then $Tr^\varphi(\varphi, \delta[j])$ is a formula in $\mathcal{L}_{GDL}$.*

Theorem 1. *If $M, \delta, j \models \varphi$ then $Tr^m(M, \delta), Tr^\lambda(\delta), j \models Tr^\varphi(\varphi, \delta[j])$.*

Proof. Given a GDLZ model $M = (W, \bar{w}, T, L, U, g, \pi_\Phi, \pi_Z)$, with the game signature $\mathcal{S} = (N, \mathcal{A}, X, \Phi)$, a complete path δ, a stage j on δ, a formula $\varphi \in \mathcal{L}_{GDLZ}$ and the function v. Let $M' = (W', \bar{w}, T', L', U', g', \pi')$, with $\mathcal{S}' = (N, \mathcal{A}', \Phi')$, be the GDL translation of M, i.e. $M' = Tr^m(M, \delta)$, $\delta' = Tr^\lambda(\delta)$ and δ_{min}, $\delta_{max} \in \mathbb{Z}$ denote the integer bounds in δ.

For any integers $\delta_{min} \le z_1, z_2 < \delta_{max}$, $\pi_z \subseteq \pi'(\delta[j])$ enumerates its predecessor and successor and define all the cases were $bigger(z_1, z_2)$, $smaller(z_1, z_2)$ and $equal(z_1, z_2)$ are true. Let $\varphi' = Tr^\varphi(\varphi, \delta[j])$. We assume that $M, \delta, j \models \varphi$ and show that then we have $M', \delta', j \models \varphi'$ for every φ.

- If φ is on the form $p \in \Phi$, we have $Tr^\varphi(p, \delta[j]) = p$. By $\mathcal{L}_{GDLZ}$ semantics, we know that $p \in \pi_\Phi(\delta[j])$. In the ST-model translation, we have the valuation function constructed such that $\pi'(\delta[j]) = \{\pi_\Phi(\delta[j])\} \cup \{\pi_z\} \cup \{x(q) : q \in \pi_\mathbb{Z}(\delta[j]), x \in X\}$. Then, $p \in \pi'(\delta[j]')$ and $M', \delta', j \models p$;
- If φ is either on the form $\neg\psi$, $\varphi_1 \wedge \varphi_2$, $initial$, $terminal$, $wins(r)$, $legal(a^r(\bar{z}))$, $does(a^r(\bar{z}))$, or $\bigcirc\psi$, since Tr^a and Tr^φ assigns each GDLZ action and formula to an unique GDL state, action and formula, respectively, due to both languages semantics it is easy to see that $M', \delta', j \models Tr^\varphi(\varphi, \delta[j])$, whenever $M, \delta, j \models \varphi$;
- If φ is on the form $z_1 > z_2$, we have $Tr^\varphi(z_1 > z_2, \delta[j]) = bigger(v(z_1, \delta[j]), v(z_2, \delta[j]))$. By $\mathcal{L}_{GDLZ}$ semantics, we know that $v(z_1, \delta[j]) > v(z_2, \delta[j])$, i.e. $v(z_1, \delta[j])$ is bigger then $v(z_2, \delta[j])$, then $bigger(v(z_1, \delta[j]), v(z_2, \delta[j])) \in \pi_z$. $\pi_z \subseteq \pi'(\delta[j])$ defines $bigger(v(z_1, \delta[j]), v(z_2, \delta[j]))$ such that it is true, iff $v(z_1, \delta[j]) > v(z_2, \delta[j])$. Thus, $M', \delta', j \models bigger(v(z_1, \delta[j]), v(z_2, \delta[j]))$;
- If φ is either on the form $z_1 < z_2$ or $z_1 = z_2$, the proof proceeds as in the previous case;
- If φ is on the form $\langle\bar{z}\rangle$, $Tr^\varphi(\bar{z}, \delta[j]) = \bigwedge_{i=1}^{|z|} x_i(v(q_i, w))$, where $x_i \in X$ and q_i is the i-th value of $\bar{z}$. By $\mathcal{L}_{GDLZ}$ semantics, we know that $\bar{z} = \pi_\mathbb{Z}(\delta[j])$. Since, by the ST-model translation each $x_i(q_i) \in \Phi'$ and $\pi'(\delta[j]) = \{\pi_\Phi(\delta[j])\} \cup \{\pi_z\} \cup \{x(q) : q \in \pi_\mathbb{Z}(\delta[j]), x \in X\}$, we have that $M', \delta', j \models x_1(q_1)$, $M', \delta', j \models x_2(q_2)$ and so on, thus $M', \delta', j \models \bigwedge_{i=1}^{|z|} x_i(v(q_i, \delta[t]))$.

Because it is a partial translation based on a path, the legal actions are restricted to the ones performed in the path. To overcome this issue, in the next section we show how to define complete translations over GDLZ models and formulas. The following complete translation is limited to the finite GDLZ models.

3.2   From Finite GDLZ Model to GDL Model

Let us consider the case where the GDLZ ST-model has finite components. In this case, we are able to define a complete model translation, instead of partial based on a path. In other words, all possible runs over the finite GDLZ ST-model can be translated. Next, we characterize a finite GDLZ ST-model.

Definition 11. *Given two arbitrary bounds $z_{min} \leq z_{max} \in \mathbb{Z}$, a finite GDLZ ST-model $M_f = (W_f, \bar{w}_f, T_f, L_f, U_f, g_f, \pi_{\Phi f}, \pi_{\mathbb{Z} f})$, with the game signature $\mathcal{S}_f = (N_f, \mathcal{A}_f, X_f, \Phi_f)$ is a subset of GDLZ ST-models that have the following aspects: (i) $z_{min} \leq z_i \leq z_{max}$, for any $a^r(z_1, \cdots, z_l) \in \mathcal{A}_f$, $1 \leq i \leq o$ and $r \in N_f$; (ii) W_f and $\mathcal{A}_f$ are finite sets; and (iii) $z_{min} \leq q_i \leq z_{max}$, for any $\langle q_1 \cdots q_n \rangle = \pi_\mathbb{Z}(w)$, $1 \leq i \leq n$ and $w \in W_f$.*

Through the rest of this section, we fix the bounds z_{min} and z_{max} as well as the finite GDLZ ST-model $M_f = (W_f, \bar{w}_f, T_f, L_f, U_f, g_f, \pi_{\Phi f}, \pi_{\mathbb{Z} f})$ with a game signature $\mathcal{S}_f = (N_f, \mathcal{A}_f, X_f, \Phi_f)$ and $N_f = \{r_1, \cdots, r_k\}$. Let us show how any finite GDLZ ST-model can be translated into a GDL ST-model.

Definition 12. *Given the finite GDLZ ST-model M_f and its signature S_f, we define the GDL ST-model $M'_f = (W_f, \bar{w}_f, T_f, L'_f, U'_f, g_f, \pi'_f)$ with a game signature $S'_f = (N_f, A'_f, \Phi'_f)$. The components $W_f, \bar{w}_f, T_f, g_f$ and N_f are the same for M_f and M'_f.*

We construct Φ'_f over both Φ_f, X_f and its values. Although X_f is a finite set, each one of its components has an integer value in each state $w \in W_f$. As Φ'_f is finite, we construct it with the bounds z_{min} and $z_{max} \in \mathbb{Z}$. Since $\mathbb{Z}$ is a countable set, for any z_{min} and z_{max}, we can define a finite enumeration of integer values.

The set of atomic propositions is defined as follows: $\Phi'_f = \{p, smaller(z_1, z_2),$ $bigger(z_1, z_2), equal(z_1, z_2), succ(z_1, z_2), prec(z_1, z_2), x(q) : p \in \Phi, x \in X_f, z_{min} \leq q, z_1, z_2 \leq z_{max}\}$.

We define an action translation $Tr_f^a : A_f \rightarrow A'_f$ associating every action in A_f with an action in A'_f as follows: $Tr_f^a(a^r(z_1, \cdots, z_l)) = a^r_{z_1, \cdots, z_l}$, where $a^r(z_1, \cdots, z_l) \in A_f, z_{min} \leq z_i \leq z_{max}$ and $0 \leq i \leq l\}$.

Note that Tr^a is an injective function. Thereby, we can define the GDL components A'_f and L'_f based on Tr^a, as follows: (i) $A'_f = \{Tr_f^a(a^r(z_1, \cdots, z_l)) : a^r(z_1, \cdots, z_l) \in A_f\}$; and (ii) $L'_f = \{(w, Tr_f^a(a)) : (w, a) \in L_f\}$.

For each $w \in W_f$, each $r \in N_f$ and each joint action $(a^{r_1}, \cdots, a^{r_k}) \in \prod_{r \in N_f} A^r_f$, where $A^r_f \in A_f$, the update function is defined as: $U'_f(w, (Tr_f^a(a^{r_1}), \cdots, Tr_f^a(a^{r_k}))) = U_f(w, (a^{r_1}, \cdots, a^{r_k}))$.

Let $\pi_{zf} \subset \Phi_f$ denote a set of propositions describing the numerical order, such that $\pi_{zf} = \{succ(z, z+1), prec(z+1, z), equal(z_1, z_1) : z_{min} \leq z < z_{max} \& z_{min} \leq z_1 \leq z_{max}\} \cup \{smaller(z_1, z_2) : z_{min} \leq z_1, z_2 \leq z_{max} \& z_1 < z_2\} \cup \{bigger(z_1, z_2) : z_{min} \leq z_1, z_2 \leq z_{max} \& z_1 > z_2\}$.

Finally, for all $w \in W_f$, we construct the valuation π'_f as follows: $\pi'_f(w) = \{\pi_{\Phi f}(w) \cup \pi_{zf} \cup \{x(q) : q \in \pi_{\mathbb{Z}f}(w), x \in X_f\}\}$.

We say that M'_f is a bounded ST-model translation of M_f and write $Tr_f^m(M_f)$.

The path translation consists at assigning each action appearing on it to the appropriated GDL action through Tr_f^a.

Definition 13. *Define a path translation $Tr_f^\lambda : \delta_f \rightarrow \delta'_f$ associating every path $\delta_f = \bar{w}_f \xrightarrow{d_1} w_1 \xrightarrow{d_2} \cdots \xrightarrow{d_e} w_e$ in M_f with a path δ'_f in M'_f: $Tr_f^\lambda(\delta_f) = \bar{w}_f \xrightarrow{d'_1} w_1 \xrightarrow{d'_2} \cdots \xrightarrow{d'_e} w_e$, where $d_i = (a^{r_1}, \cdots, a^{r_k}) \in D_f$, $D_f = \prod_{r \in N_f} A^r_f$, $A^r_f \in A_f$, $w_i \in W_f$, $d'_i = (Tr_f^a(a^{r_1}), \cdots, Tr_f^a(a^{r_k}))$ and $1 \leq i \leq e$.*

It follows that the translations of a finite GDLZ model and a path in a finite GDLZ model are a model and a path in GDL, respectively.

Proposition 7. *If M_f is a finite GDLZ model then $Tr_f^m(M_f)$ is a GDL model.*

Proposition 8. *If δ_f is a path in a finite GDLZ model M_f then $Tr_f^\lambda(\delta_f)$ is a path in $Tr_f^m(M_f)$.*

Next, we show a complete translate from GDLZ formulas to GDL formulas. Likewise to the model translation, we use arbitrary bounds to restrict the numerical range in the formulas.

From Bounded GDLZ Formulas to GDL Formulas. Assuming a GDLZ game signature $\mathcal{S}_f = (N_f, \mathcal{A}_f, \varPhi_f, X_f)$, the semantics of a numerical variable $x \in X_f$ in a $\mathcal{L}_{GDLZ}$ formula is evaluated depending on the current game state.

To translate the meaning of a numerical variable $x \in X_f$ occurring in an atomic formula $\varphi \in \mathcal{L}_{GDLZ}$ in the form $legal(a^r(\bar{z}))$, $does(a^r(\bar{z}))$, $\langle \bar{z} \rangle$, $z_1 < z_2$, $z_1 > z_2$ or $z_1 = z_2$, Algorithm 2, denoted $removeVar(\varphi)$, defines an intermediate formula φ_x as the disjunction from all possible values $z_{min} \le q \le z_{max}$ for x in φ and $x(q)$. Algorithm 2 stops when there is no more occurrence of numerical variables in the resulting formula.

Algorithm 2. $removeVar(\varphi)$

Input: a formula $\varphi \in \mathcal{L}_{GDLZ}$. Assume the variable set X_f and $z_{min} \le z_{max}$.
Output: a partially translated formula.

1: $I \leftarrow \{z_{min}, \cdots, z_{max}\}$
2: **if** $(\varphi = \text{``}legal(a^r(z_1, \cdots, z_m))\text{''})$ **then**
3:     **for each** $z_i \in (z_1, \cdots, z_m)$ **do**
4:         **if** $z_i \in X_f$ **then return** $\bigvee_{q_i \in I}(removeVar(legal(a^r(z_1, \cdots, q_i, \cdots, z_m)) \wedge z_i(q_i))$
5: **else if** $(\varphi = \text{``}does(a^r(z_1, \cdots, z_m))\text{''})$ **then** Proceeds as the previous case.
6: **else if** $(\varphi = \text{``}\langle z_1, \cdots, z_m \rangle\text{''})$ **then**
7:     **for each** $z_i \in \langle z_1, \cdots, z_m \rangle$ **do**
8:         **if** $z_i \in X_f$ **then return** $\bigvee_{q_i \in I}(removeVar(\langle z_1, \cdots, q_i, \cdots, z_m \rangle) \wedge z_i(q_i))$
9: **else if** $(\varphi = \text{``}z_1 < z_2\text{''})$ **then**
10:     **if** $z_1 \in X_f$ **then return** $\bigvee_{q_1 \in I}(removeVar(q_1 < z_2) \wedge z_1(q_1))$
11:     **if** $z_2 \in X_f$ **then return** $\bigvee_{q_2 \in I}(removeVar(z_1 < q_2) \wedge z_2(q_2))$
12: **else if** $(\varphi = \text{``}z_1 > z_2\text{''}$ or $\varphi = \text{``}z_1 = z_2\text{''})$ **then** Proceeds as the previous case.
    return φ

A numerical simple term z_f is defined by $\mathcal{L}_{z_f}$, which is generated by the following BNF: $z_f ::= z' \mid add(z_f, z_f) \mid sub(z_f, z_f) \mid min(z_f, z_f) \mid max(z_f, z_f)$, where $z' \in \mathbb{Z}$. Note that $\mathcal{L}_{z_f} \subseteq \mathcal{L}_z$. Each numerical term $z_f \in \mathcal{L}_{z_f}$ occurring in a formula $\varphi \in \mathcal{L}_{GDLZ}$ is translated by its semantic interpretation through function v_f, defined in a similar way to Definition 4:

Definition 14. *Let us define function $v_f : \mathcal{L}_{z_f} \to \mathbb{Z}$, associating any $z_f \in \mathcal{L}_{z_f}$ to a number in $\mathbb{Z}$:*

$$v_f(z_f) = \begin{cases} z_i & if\ z_f \in \mathbb{Z} \\ v_f(z_f') + v_f(z_f'') & if\ z_f = add(z_f', z_f'') \\ v_f(z_f') - v_f(z_f'') & if\ z_f = sub(z_f', z_f'') \\ minimum(v_f(z_f'), v_f(z_f'')) & if\ z_f = min(z_f', z_f'') \\ maximum(v_f(z_f'), v_f(z_f'')) & if\ z_f = min(z_f', z_f'') \end{cases}$$

The complete formula translation is restricted to bounded formulas, which are are defined as follows:

Definition 15. $\varphi \in \mathcal{L}_{GDLZ}$ *is a bounded formula if, for any numerical term* z_f *occurring in* φ, *we have* $z_f \in \mathcal{L}_{z_f}$ *and* $z_{min} \leq v_f(z) \leq z_{max}$ *or if there is no occurrence of numerical terms in* φ.

We next define a translation map for bounded formulas in $\mathcal{L}_{GDLZ}$ to formulas in $\mathcal{L}_{GDL}$. Each numerical simple term $z_f \in \mathcal{L}_{z_f}$ occurring in a formula $\varphi \in \mathcal{L}_{GDLZ}$ is translated by its semantic interpretation through function v_f (see Definition 14).

Definition 16. *Given the GDLZ game signature* $\mathcal{S}_f = (N_f, \mathcal{A}_f, X_f, \Phi_f)$ *and function* v_f, *a translation* Tr_f^φ *from a bounded formula* $\varphi \in \mathcal{L}_{GDLZ}$ *to a formula* $\varphi' \in \mathcal{L}_{GDL}$ *is defined as* $Tr_f^\varphi = Tr_f^z(removeVar(\varphi))$, *where* Tr_f^z *is specified as follows:*

- $Tr_f^z(\varphi) = \varphi$ *for all* $\varphi \in \Phi_f \cup \{initial, terminal, wins(r)\} \cup \{x(q) : x \in X_f, z_{min} \leq q \leq z_{max}\}$;
- $Tr_f^z(\neg\varphi) = \neg Tr_f^z(removeVar(\varphi)))$;
- $Tr_f^z(\varphi_1 \wedge \varphi_2) = Tr_f^z(removeVar(\varphi_1))) \wedge Tr_f^z(removeVar(\varphi_2)))$;
- $Tr_f^z(\bigcirc\varphi) = \bigcirc Tr_f^z(removeVar(\varphi)))$;
- $Tr_f^z(legal(a^r(\bar{z}))) = legal(Tr_f^a(a^r(v_f(z) : z \in \bar{z})))$;
- $Tr_f^z(does(a^r(\bar{z}))) = does(Tr_f^a(a^r(v_f(z) : z \in \bar{z})))$;
- $Tr_f^z(\langle\bar{z}\rangle) = \bigwedge_{i=1}^{|\bar{z}|} x_i(v_f(q_i))$;
- $Tr_f^z(z_1 < z_2,) = smaller(v_f(z_1), v_f(z_2))$;
- $Tr_f^z(z_1 > z_2) = bigger(v_f(z_1), v_f(z_2))$;
- $Tr_f^z(z_1 = z_2) = equal(v_f(z_1), v_f(z_2))$.

Where $r \in N_f$, $x_i \in X_f$, q_i *is the* i-*th value in* $\bar{z}$ *and* $0 \leq i \leq |\bar{z}|$.

Let us illustrate the translation of GDLZ formulas into GDL using Tr_f^φ.

Example 2. Let $I = \{z_{min}, \cdots, z_{max}\}$ and $\varphi_1 = does(reduce^r(heap_1, add(1,2)))$, where $heap_1 \in X_f$, then $Tr_f^\varphi(\varphi_1) = \bigvee_{h_1 \in \{z_{min}, \cdots, z_{max}\}}(does(reduce^r(h_1, 3)) \wedge heap_1(h_1))$.

The translation of a GDLZ formula is a GDL formula. Furthermore, if the GDLZ formula is valid at a stage in the path in a finite GDLZ model, then its translation will be valid at the same stage in the translated path in the translated model.

Proposition 9. *If* $\varphi \in \mathcal{L}_{GDLZ}$ *then* $Tr_f^\varphi(\varphi) \in \mathcal{L}_{GDL}$.

Theorem 2. *If* M_f *is a finite GDLZ ST-model,* $\varphi \in \mathcal{L}_{GDLZ}$ *is a bounded formula and* $M_f, \delta_f, j \models \varphi$ *then* $Tr_f^m(M_f), Tr_f^\lambda(\delta_f), j \models Tr_f^\varphi(\varphi)$.

In the next section, we briefly describe how to translate GDL ST-models into GDLZ ST-models. Besides that, we show that GDL is a sublanguage of GDLZ.

3.3 From GDL to GDLZ

Conversely, we show that any GDL ST-model can be transformed into a GDLZ ST-model. Given a GDL ST-model $M' = (W, \bar{w}, T, L, U, g, \pi')$ with a game signature $\mathcal{S}' = (N, \mathcal{A}, \Phi)$, we define an associated GDLZ ST-model $M = (W, \bar{w}, T, L, U, g, \pi_\Phi, \pi_Z)$ with the game signature $\mathcal{S} = (N, \mathcal{A}, X, \Phi)$, such that all elements are the same, except by π_Φ, π_Z and X and X. These GDLZ components are defined as follows: (i) $\pi_\Phi(w) = \pi'(w)$; (ii) $\pi_Z(w) = \emptyset$; and (iii) $X = \emptyset$.

It follows that any formula $\varphi \in \mathcal{L}_{GDL}$ is also a formula in GDLZ, i.e. $\varphi \in \mathcal{L}_{GDLZ}$.

Proposition 10. *If $\mathcal{S}' = (N, \mathcal{A}, \Phi')$ and $\mathcal{S} = (N, \mathcal{A}, X, \Phi)$ are GDL and GDLZ game signatures, respectively, and $\Phi' \subseteq \Phi$, then $\mathcal{L}_{GDL} \subseteq \mathcal{L}_{GDLZ}$.*

3.4 Succinctness

Next, we compare $\mathcal{L}_{GDLZ}$ and $\mathcal{L}_{GDL}$ in order to show the succinctness of $\mathcal{L}_{GDLZ}$ in describing the same game. The following definition specifies when two sets of formulas in GDLZ and GDL describe the same game.

Definition 17. *Two sets of formulas $\Sigma_{GDLZ} \subseteq \mathcal{L}_{GDLZ}$ and $\Sigma_{GDL} \subseteq \mathcal{L}_{GDL}$ describe the same game either (i) if $\Sigma_{GDLZ} = \{\varphi : \varphi \in \Sigma_{GDL}\}$ and $\mathcal{L}_{GDL}$ and $\mathcal{L}_{GDL}$ have, respectively, $\mathcal{S}' = (N, \mathcal{A}, \Phi)$ and $\mathcal{S} = (N, \mathcal{A}, \emptyset, \Phi)$; (ii) if $\Sigma_{GDL} = \{Tr^\varphi(\varphi, \delta[j]) : \varphi \in \Sigma_{GDLZ}\}$, given a GDLZ ST-model M, a path δ in M and a stage j in δ or (iii) if $\Sigma_{GDL} = \{Tr_f^\varphi(\varphi) : \varphi \in \Sigma_{GDLZ}\}$, where every $\varphi \in \Sigma_{GDLZ}$ is a bounded formula.*

The following theorem show that (i) a GDLZ description has less subformulas and (ii) if we compare with the path translation, the growth is linear, if we compare with the complete translation, the growth is exponential.

Theorem 3. *If Σ_{GDLZ} and Σ_{GDL} are two sets of formulas in $\mathcal{L}_{GDLZ}$ and $\mathcal{L}_{GDL}$, respec., describing the same game, then $|Sub(\Sigma_{GDLZ})| \leq |Sub(\Sigma_{GDL})|$.*

Proof. Assume the GDL and GDLZ game signatures $\mathcal{S}' = (N, \mathcal{A}', \Phi')$ and $\mathcal{S} = (N, \mathcal{A}, X, \Phi)$, respectively. Since Σ_{GDLZ} and Σ_{GDL} describe the same game, by Definition 17, we have either: (i) $\Sigma_{GDLZ} = \{\varphi : \varphi \in \Sigma_{GDL}\}$, $\mathcal{S}' = (N, \mathcal{A}, \Phi)$ and $\mathcal{S} = (N, \mathcal{A}, \emptyset, \Phi)$; (ii) $\Sigma_{GDL} = \{Tr^\varphi(\varphi, \delta[j]) : \varphi \in \Sigma_{GDLZ}\}$, for a GDLZ ST-model M, a path δ in M and a stage j in δ, or (iii) if $\Sigma_{GDL} = \{Tr_f^\varphi(\varphi) : \varphi \in \Sigma_{GDLZ}\}$, where every $\varphi \in \Sigma_{GDLZ}$ is a bounded formula. In the first case, $\mathcal{A}' = \mathcal{A}$, $\Phi' = \Phi$, $X = \emptyset$ and $\Sigma_{GDLZ} = \{\varphi : \varphi \in \Sigma_{GDL}\}$, we clearly have $|\Sigma_{GDLZ}| = |\Sigma_{GDL}|$ and $|Sub(\Sigma_{GDLZ})| = |Sub(\Sigma_{GDL})|$.

Given a path δ in a GDLZ ST-model M and a stage j, let us now consider the case (ii) where $\Sigma_{GDL} = \{Tr^\varphi(\varphi, \delta[j]) : \varphi \in \Sigma_{GDLZ}\}$. From $Tr^\varphi(\varphi, \delta[j])$, we have that any translation assigns φ to a corresponding φ' where $|Sub(\varphi)| = |Sub(\varphi')|$, except in the case where φ is of the form $\langle \bar{z} \rangle$. If φ is of the form $\langle \bar{z} \rangle$, then φ' will be constructed as $\bigwedge_{i=1}^{|\bar{z}|} x_i(v(q_i, w))$, where $x_i \in X$ and q_i is the i-th value of $\bar{z}$. Thus, $|Sub(\varphi')| = |\bar{z}||Sub(\varphi)|$. Since $|Sub(\varphi)| = 1$, then $|Sub(\varphi')| = |\bar{z}|$. Denote $\Sigma_l = \Sigma_{GDLZ} - \{\langle\bar{z}\rangle : \langle\bar{z}\rangle \in \mathcal{L}_{GDLZ}\}$, i.e. Σ_l is the subset of Σ_{GDLZ} without any formula $\langle\bar{z}\rangle$. Thereby $|Sub(\Sigma_l)| = |Sub(\{Tr^\varphi(\varphi, \delta[j]) : \varphi \in \Sigma_l)\}|)$. Assuming k as the amount of formulas in the form $\langle\bar{z}\rangle \in \Sigma_{GDLZ}$, we have $|Sub(\Sigma_{GDL})| = |Sub(\Sigma_l)| + |\bar{z}|k$. Thereby, in the second case, we have $|Sub(\Sigma_{GDLZ})| \leq |Sub(\Sigma_{GDL})|$.

Let us consider case (iii), where $\Sigma_{GDL} = \{Tr^\varphi_f(\varphi) : \varphi \in \Sigma_{GDLZ}\}$ and every $\varphi \in \Sigma_{GDLZ}$ is a bounded formula. Let $\mu = z_{max} - z_{min}$. The proof for case (iii) proceeds in the same way that for case (ii), except in the situation where there are numerical variables occurring in any $\varphi \in \Sigma_{GDLZ}$. If we have at least one numerical variable occurring in φ, we know that φ is either in the form $legal(a^r(\bar{z}))$, $does(a^r(\bar{z}))$, $\langle\bar{z}\rangle$, $z_1 < z_2$, $z_1 > z_2$ or $z_1 = z_2$. Thereby, $|\varphi| = 1$ and $|Tr^\varphi_f(removeVar(\varphi))| = 2\mu^\eta \times |\varphi|$, where η is the amount of numerical variables occurring in φ. Thereby, $|\varphi| < |Tr^\varphi_f(removeVar(\varphi))|$ and $|Sub(\Sigma_{GDLZ})| \leq |Sub(\Sigma_{GDL})|$. Denote $\Sigma'_l = \Sigma_{GDLZ} - \{\langle\bar{z}\rangle : \langle\bar{z}\rangle \in \mathcal{L}_{GDLZ}\} - \{\varphi \in \mathcal{L}_{GDLZ} : \text{there}$ is at least one numerical variable in $\varphi\}$. Assuming k as the amount of formulas in the form $\langle\bar{z}\rangle \in \Sigma_{GDLZ}$ and κ as the amount of formulas where occurs η numerical variables, we have $|Sub(\Sigma_{GDL})| = |Sub(\Sigma'_l)| + 2\mu^\eta\kappa + |\bar{z}|k$.

Theorem 4. *Given $\Sigma_{GDLZ} \subseteq \mathcal{L}_{GDLZ}$, a GDLZ ST-model M with the game signature $\mathcal{S} = (N, \mathcal{A}, \Phi, X)$:*

1. *If $\Sigma_{GDL} = \{Tr^\varphi(\varphi, \delta[j]) : \varphi \in \Sigma_{GDLZ}\}$, given a path δ in M and a stage j in δ, then $|Sub(\Sigma_{GDL})|$ grows in the order $\mathcal{O}(n)$, where $n = |Sub(\Sigma_l)| + |X|k$, the value k represents the amount of formulas in the form $\langle\bar{z}\rangle$ in Σ_{GDLZ} and $\Sigma_l = \Sigma_{GDLZ} - \{\langle\bar{z}\rangle : \langle\bar{z}\rangle \in \mathcal{L}_{GDLZ}\}$, i.e. Σ_l is the subset of Σ_{GDLZ} without any formula $\langle\bar{z}\rangle$;*

2. *If $\Sigma_{GDL} = \{Tr^\varphi_f(\varphi) : \varphi \in \Sigma_{GDLZ}\}$, where every $\varphi \in \Sigma_{GDLZ}$ is a bounded formula, then $|Sub(\Sigma_{GDL})|$ grows in the order $\mathcal{O}(n + \kappa\mu^\eta)$, where $n = |Sub(\Sigma_l)| + |X|k$, the value k represents the amount of formulas in the form $\langle\bar{z}\rangle$ in Σ_{GDLZ}, $\Sigma_l = \Sigma_{GDLZ} - \{\langle\bar{z}\rangle : \langle\bar{z}\rangle \in \mathcal{L}_{GDLZ}\}$ and η is the amount of numerical variables occurring in κ variables.*

The partial translation Tr^φ only concerns a fragment of the GDLZ model, that is the part of the model involved in a specific path. The size of a formula translated through Tr^φ has a linear growth over the number of numerical variables in X and the number of formulas in the form $\langle\bar{z}\rangle$. Conversely, Tr^φ_f is a complete translation over finite GDLZ models. To represent a GDLZ formula in a GDL formula regardless of a specific path, we should remove the occurrence of numerical variables as numerical terms (see Algorithm 2). This procedure exponentially increases the size of the translated formula, depending mainly on the occurrence of numerical variables in the original GDLZ formula.

4   Conclusion

In this paper, we have introduced a GDL extension to describe games with numerical aspects, called GDLZ. In GDLZ, states are evaluated with propositions and an assignment of integer values to numerical variables. This allows us to define the terminal and goal states in terms of the numerical conditions. Furthermore, we define actions with numerical parameters, such that these parameters can influence over the action legality and over the state update. The language was extended mainly to include the representation of numerical variables and integer values as well as to allow numerical comparison.

We defined translations between GDLZ and GDL game models and formulas. Since GDL models have finite components, we can not define a complete model translation for any GDLZ model. We first defined a partial translation from any GDLZ model restricted to a specified path, i.e. only a run in the game is represented. Second, we defined a complete translation from GDLZ models with finite components and bounded formulas. We show that, in both cases, a translated GDLZ model, path or formula is a GDL model, path or formula, respectively. Furthermore, we prove that if a formula is satisfied at a stage in a path under a GDLZ model, its translation will also be satisfied at the same stage in the translated path under the translated model.

Finally, we show that, if we have a GDLZ and a GDL description for the same (finite) game, the GDLZ description is more succinct or equal, in terms of the quantity of subformulas in the description. More precisely, if the GDL game description is based on the partial translation from a GDLZ description restricted to one path, it is linearly larger then the GDLZ description. When we consider the complete model translation, the GDL description is exponentially larger than the GDLZ description.

Future work may extend GDLZ to define numerical rewards to players, stating their achievement when the game ends. It means that numerical variables may not have values assigned in some state of the model. Our aim is to investigate this new kind of numerical models. In our framework, it is possible to define both concurrent and sequential games. However, the legality of an agent's action is independent from the actions of other agents. Thereby, it may be inappropriate to describe concurrent games where the actions of two agents change the same numerical variable. To overcome this limitation, future work may explore the definition of the legality function over joint actions.

Acknowledgments. Munyque Mitttelmann and Laurent Perrussel acknowledge the support of the ANR project AGAPE ANR-18-CE23-0013.

References

1. Genesereth, M., Love, N., Pell, B.: General game playing: overview of the AAAI competition. AI Mag. **26**(1), 1–16 (2005). http://www.aaai.org/ojs/index.php/aimagazine/article/viewArticle/1813
2. Gerevini, A.E., Saetti, A., Serina, I.: An approach to efficient planning with numerical fluents and multi-criteria plan quality. Artif. Intell. **172**(8–9), 899–944 (2008). https://doi.org/10.1016/j.artint.2008.01.002
3. Ghallab, M., et al.: PDDL - The Planning Domain Definition Language. Technical report, AIPS-98 Planning Competition Committee (1998). http://citeseerx.ist.psu.edu/viewdoc/summary?doi=10.1.1.37.212

4. Giacomo, G.D., Lesp, Y., Pearce, A.R.: Situation calculus-based programs for representing and reasoning about game structures. In: Proceedings of the Twelfth International Conference on the Principles of Knowledge Representation and Reasoning (KR 2010), pp. 445–455 (2010)
5. Jiang, G., Zhang, D., Perrussel, L.: GDL meets ATL: a logic for game description and strategic reasoning. In: Pham, D.N., Park, S.B. (eds.) PRICAI 2014: Trends in Artificial Intelligence, pp. 733–746. Springer, Cham (2014). https://doi.org/10.1007/978-3-319-13560-1_58
6. Jiang, G., Zhang, D., Perrussel, L., Zhang, H.: Epistemic GDL: a logic for representing and reasoning about imperfect information games. In: IJCAI International Joint Conference on Artificial Intelligence, January 2016, pp. 1138–1144 (2016)
7. Love, N., Genesereth, M., Hinrichs, T.: General Game Playing: Game Description Language Specification. Technical report LG-2006-01, Stanford University, Stanford, CA (2006). http://logic.stanford.edu/reports/LG-2006-01.pdf
8. Fox, M., Long, D.: PDDL2.1: An extension to PDDL for expressing temporal planning domains. J. Artif. Intell. Res. 20, 1–48 (2003). http://citeseerx.ist.psu.edu/viewdoc/summary?doi=10.1.1.68.1957
9. McDermott, D.M.: The 1998 AI planning systems competition. AI Mag. 21(2), 35 (2000). https://doi.org/10.1609/AIMAG.V21I2.1506
10. Parikh, R.: The logic of games and its applications. North-Holland Math. Stud. 102, 111–139 (1985). https://doi.org/10.1016/S0304-0208(08)73078-0
11. Pauly, M., Parikh, R.: Game logic - an overview. Studia Logica 75(2), 165–182 (2003). https://doi.org/10.1023/A:1027354826364
12. Schiffel, S., Thielscher, M.: Representing and reasoning about the rules of general games with imperfect information. J. Artif. Intell. Res. 49, 171–206 (2014)
13. Thielscher, M.: A general game description language for incomplete information games. In: Proceedings of the Twenty-Fourth AAAI Conference on Artificial Intelligence (AAAI-2010), pp. 994–999 (2010). https://www.aaai.org/ocs/index.php/AAAI/AAAI10/paper/view/1727
14. Thielscher, M.: GDL-III: a proposal to extend the game description language to general epistemic games. In: Proceedings of the European Conference on Artificial Intelligence (ECAI), vol. 285, pp. 1630–1631. Hague (2016). https://doi.org/10.3233/978-1-61499-672-9-1630
15. Thielscher, M.: GDL-III: a description language for epistemic general game playing. IJCAI International Joint Conference on Artificial Intelligence, pp. 1276–1282 (2017)
16. Van Benthem, J.: Games in dynamic-epistemic logic. Bull. Econ. Res. 53(4), 219–248 (2001). https://doi.org/10.1111/1467-8586.00133
17. van Benthem, J., Ghosh, S., Liu, F.: Modelling simultaneous games in dynamic logic. Synthese 165(2), 247–268 (2008). https://doi.org/10.1007/s11229-008-9390-y
18. Zhang, D., Thielscher, M.: Representing and reasoning about game strategies. J. Philos. Logic 44(2), 203–236 (2014). https://doi.org/10.1007/s10992-014-9334-6

Craig Interpolation of Epistemic Logics with Distributed Knowledge

Ryo Murai[1(✉)] and Katsuhiko Sano[2]

[1] Graduate School of Humanities and Human Sciences,
Hokkaido University, Sapporo, Japan
rmurai270@gmail.com
[2] Faculty of Humanities and Human Sciences,
Hokkaido University, Sapporo, Japan
v-sano@let.hokudai.ac.jp

Abstract. Distributed Knowledge among agents is an important topic in multi-agent systems. While semantic studies of distributed knowledge have been done by several authors in the context of epistemic logic, there are a few proof-theoretic studies. This paper provides cut-free Gentzen-style sequent calculi for epistemic logics with distributed knowledge and establishes Craig Interpolation Theorem for the logics by a constructive method, i.e., Maehara method.

Keywords: Epistemic logic · Distributed knowledge · Sequent calculus · Craig Interpolation Theorem

1 Introduction

"Distributed knowledge" is a notion developed in the community of multi-agent epistemic logic [3,10]. As Ågotnes et al. [1, Section 1] say, the notion of "a group G has a distributed knowledge of φ" is understood as: *It follows from the combination of knowledge of members of G that φ were true before any communication or other events took place.*[1] For example, a group consisting of a and b has distributed knowledge of a fact q, when a knows $p \to q$ and b knows p. Due to its aggregative nature, the notion of distributed knowledge may be applicable to the field of cooperative problem solving, knowledge base merging, and judgement aggregation, as [8] suggests.

In the context of epistemic logic, distributed knowledge is expressed as a modal operator D_G, parameterized by a group of agents and the satisfaction of $D_G\varphi$ at a state w is defined as: φ holds at all states v such that v can be reached

---

[1] Fagin et al. [3, p. 3] state as intuitive description for distributed knowledge "a group has distributed knowledge of a fact φ if the knowledge of φ is distributed among its members, so that by pooling their knowledge together the members of the group can deduce φ". This seems clearer, at first sight, than the explanation we give here. Ågotnes et al. [1] states, however, that the above intuitive description is inappropriate by an illustrative example given in [1, Section 1].

© Springer Nature Switzerland AG 2020
A. Herzig and J. Kontinen (Eds.): FoIKS 2020, LNCS 12012, pp. 211–221, 2020.
https://doi.org/10.1007/978-3-030-39951-1_13

in a single step from w for all agents in G, i.e., wR_av for all agents $a \in G$, where R_a is a binary relation on the states.

The study of distributed knowledge so far is mainly model-theoretic [1,6,16, 19] and proof-theoretic study has been not so active. As far as we know, existing sequent calculi for logic with distributed knowledge are presented only in [7,8,13]. [8] contains a natural G3-style (without structural rules) formalization, in which each formula has a label, [13] contains a Gentzen-style sequent calculus for **S4** distributed knowledge logic which is simpler than the one we are interested in, in that the operator is not parameterized by group G, and [7] contains Gentzen-style and Kanger-style sequent calculi for **S5** distributed knowledge logic with the same type of operator as the second one.

We propose Gentzen-style sequent calculi (without label) for five kinds of multi-agent epistemic propositional logics with distributed knowledge operators, parameterized by groups, which are reasonable generalization of sequent calculi for basic modal logic, and we prove the cut elimination theorem for four of them. Using a method described in [9], Craig interpolation theorem is also established for the four systems, in which not only condition of propositional variables but also that of agents is taken into account. This is a new result for logic for distributed knowledge, as far as we know. Craig interpolation theorem does not hold for some expansions of basic modal logic [2], so the result suggests the logics with distributed knowledge are "good" expansions of basic modal logic in this sense.

We proceed as follows. Section 2 provides the necessary preliminaries of distributed epistemic logics. We fix our language and give semantic definition of a distributed knowledge operator D_G. We also introduce the known Hilbert-style axiomatization of epistemic logics with distributed knowledge. In Sect. 3, we propose Gentzen-style sequent calculi for the logics defined in Sect. 2. We also establish the equipollence results between the sequent calculi and the Hilbert-style axiom systems introduced in Sect. 2 (Theorem 1). In Sect. 4, we prove our main technical results, Cut Elimination Theorem (Theorem 2) and Craig Interpolation Theorem (Theorem 3). In Sect. 5, we state possible future work.

2   Overview of Epistemic Logics with Distributed Knowledge Operators

2.1   Language

We denote a finite set of agents by Agt. We call a *nonempty* subset of Agt "group" and denote it by G, H, etc. Let Prop be a countable set of propositional variables and Form be the set of formulas defined inductively by the following clauses:

$$\text{Form} \ni \varphi ::= p \in \text{Prop} \mid \bot \mid \top \mid \neg\varphi \mid \varphi \to \varphi \mid D_G\varphi.$$

It is noted that $\wedge$ and $\vee$ are defined in the same way as in the classical propositional logic. We also define the epistemic operator $K_a\varphi$ (read "agent a knows that φ") as $D_{\{a\}}\varphi$. As noted above, an expression of the form $D_\emptyset\varphi$ is *not* a well-formed formula, since we have excluded $\emptyset$ from our definition of groups.

2.2 Kripke Semantics

We introduce the ordinary Kripke semantics for multi-agent epistemic logic here. Let W be a possibly countable set of states, $(R_a)_{a \in \mathsf{Agt}}$ be a family of binary relations on W, indexed by agents, and V be a valuation function $\mathsf{Prop} \to \mathcal{P}(W)$. We call a pair $F = (W, (R_a)_{a \in \mathsf{Agt}})$ a *frame* and a tuple $M = (W, (R_a)_{a \in \mathsf{Agt}}, V)$ a *model*. For a model $M = (W, (R_a)_{a \in \mathsf{Agt}}, V)$ and a state $w \in W$, a pair (M, w) is called a *pointed model*. *Satisfaction relation* $M, w \models \varphi$ on pointed models and formulas is defined recursively as follows:

$$
\begin{array}{ll}
M, w \models p & \text{iff } w \in V(p), \\
M, w \models \bot & \text{Never}, \\
M, w \models \top & \text{Always}, \\
M, w \models \neg\varphi & \text{iff } M, w \not\models \varphi, \\
M, w \models \varphi \to \psi & \text{iff } M, w \not\models \varphi \text{ or } M, w \models \psi, \\
M, w \models D_G\varphi & \text{iff for all } v \in W, \text{ if } (w, v) \in \bigcap_{a \in G} R_a \text{ then } M, v \models \varphi.
\end{array}
$$

It is noted from our definition of $K_a\varphi := D_{\{a\}}\varphi$ that the satisfaction of $K_a\varphi$ at a state w of a model M is given as follows:

$$
M, w \models K_a\varphi \text{ iff for all } v \in W, \text{ if } (w, v) \in R_a \text{ then } M, v \models \varphi.
$$

Given a frame $F = (W, (R_a)_{a \in \mathsf{Agt}})$, we say that a formula φ is *valid* in F (notation: $F \models \varphi$) if $(F, V), w \models \varphi$ for every valuation function V and every $w \in W$. Moreover, a formula φ is valid in a class $\mathbb{F}$ of frames if $F \models \varphi$ for every $F \in \mathbb{F}$. Let us say that a set Γ of formulas *defines* a class $\mathbb{F}$ of frames if, for every frame F, $F \in \mathbb{F}$ is equivalent to: $F \models \varphi$ for all $\varphi \in \Gamma$.

2.3 Hilbert Systems

We review the known Hilbert system for epistemic logics with D_G operators (cf. [3]). Hilbert system $\mathsf{H}(\mathbf{K}_D)$ is defined as in the following table.

Hilbert System $\mathsf{H}(\mathbf{K}_D)$
(Taut) all instantiations of propositional tautologies
(Incl) $D_G\varphi \to D_H\varphi \ (G \subseteq H)$
(K) $D_G(\varphi \to \psi) \to (D_G\varphi \to D_G\psi)$
(MP) From $\varphi \to \psi$ and φ infer ψ
(Nec) From φ infer $D_G\varphi$
Additional Axiom Schemes
(T) $D_G\varphi \to \varphi$
(4) $D_G\varphi \to D_G D_G\varphi$
(5) $\neg D_G\varphi \to D_G \neg D_G\varphi$

For additional axioms schemes, we note that (T), (4) and (5) define the class of reflexive, transitive and Euclidean frames, respectively (here, e.g., a "reflexive" frame means that R_a is reflexive for all agents $a \in \mathsf{Agt}$). Hilbert systems

H($\mathbf{KT}_D$), H($\mathbf{K4}_D$), H($\mathbf{S4}_D$), and H($\mathbf{S5}_D$) are defined as axiomatic expansions of H($\mathbf{K}_D$) with (T), (4), (T) and (4), and (T) and (5), respectively). Given any Hilbert system $\mathbf{X}$ above, the notion of provability is defined as usual.

For Hilbert systems H($\mathbf{K}_D$), H($\mathbf{KT}_D$), H($\mathbf{S4}_D$) and H($\mathbf{S5}_D$), the following soundness and completeness results are known [3] (we cannot find any explicit reference on H($\mathbf{K4}_D$), private communication by Thomas Ågotnes).

Fact 1. *Each of Hilbert systems* H($\mathbf{K}_D$), H($\mathbf{KT}_D$), H($\mathbf{S4}_D$) *and* H($\mathbf{S5}_D$) *is sound and complete with regard to the class of frames defined by additional axiom schemes.*

3 Sequent Calculi of Epistemic Logics with Distributed Knowledge

A *sequent* is a pair of finite multi-sets of formulas Γ and Δ denoted by "$\Gamma \Rightarrow \Delta$", whose reading is "if all formulas in Γ hold then some formulas in Δ hold." We now propose our sequent calculi for the logics for distributed knowledge as in Table 1. Axioms, Structural Rules and Propositional Logical Rules are common to $\mathbf{LK}$ [4,5] and the rest are new. We note that when $n = 0$, e.g., in the rule (D) of Table 1, the multi-set is regarded as the empty multi-set and thus $\bigcup_{i=1}^{n} G_i$ is regarded as $\emptyset$. A sequent $\Gamma \Rightarrow \Delta$ is *derivable* in each calculus G($\mathbf{X}$) if there exists a finite tree of sequents, whose root is $\Gamma \Rightarrow \Delta$ and each node of which is inferred by some rule in G($\mathbf{X}$). We write it as $\vdash_{G(\mathbf{X})} \Gamma \Rightarrow \Delta$. We introduce a notion of "principal formula" for a proof described later. A principal formula is defined for each inference rule, except for the axioms and (Cut) rule and is informally expressed as "a formula, on which the inference rule acts". A principal formula of the structural rules, the rules for $\rightarrow$ and the rule $(D \Rightarrow)$ is a formula appearing in the lower sequent, which is not contained in Γ or Δ. A principal formula of the rules for D_G operator other than $(D \Rightarrow)$ is every formula in the lower sequent.

Remark 1. The idea underlying the rule (D) is similar to that of an inference rule called "$R12$" described in [15, section 4]. Sequent calculi G($\mathbf{KT}_D$),G($\mathbf{K4}_D$), G($\mathbf{S4}_D$), and G($\mathbf{S5}_D$) are constructed based on the known sequent calculi for $\mathbf{KT}, \mathbf{K4}, \mathbf{S4}$, and $\mathbf{S5}$, respectively (surveyed in [14,18]). For example, inference rules for the modal operator $\square$ of the sequent calculus for $\mathbf{S4}$ is as follows:

$$\frac{\square\Gamma \Rightarrow \varphi}{\square\Gamma \Rightarrow \square\varphi} \ (\Rightarrow \square) \qquad \frac{\varphi, \Gamma \Rightarrow \Delta}{\square\varphi, \Gamma \Rightarrow \Delta} \ (\square \Rightarrow).$$

We note that for any epistemic logic $\mathbf{X}$ with distributed knowledge under consideration, H($\mathbf{X}$) and G($\mathbf{X}$) are equipollent in the following sense, and hence that each system G($\mathbf{X}$) deserves its own name.

Theorem 1 (Equipollence). *Let* $\mathbf{X}$ *be any of* $\mathbf{K}_D$, $\mathbf{KT}_D$, $\mathbf{K4}_D$, $\mathbf{S4}_D$, *and* $\mathbf{S5}_D$. *Then, the following hold.*

Table 1. Sequent Calculi for $\mathbf{K}_D$, $\mathbf{KT}_D$, $\mathbf{K4}_D$, $\mathbf{S4}_D$, and $\mathbf{S5}_D$

Axioms

$$\dfrac{}{\varphi \Rightarrow \varphi}\ (Id) \qquad \dfrac{}{\bot \Rightarrow}\ (\bot) \qquad \dfrac{}{\Rightarrow \top}\ (\top)$$

Structural Rules

$$\dfrac{\Gamma \Rightarrow \Delta}{\Gamma \Rightarrow \Delta, \varphi}\ (\Rightarrow w) \qquad \dfrac{\Gamma \Rightarrow \Delta}{\varphi, \Gamma \Rightarrow \Delta}\ (w \Rightarrow) \qquad \dfrac{\Gamma \Rightarrow \Delta, \varphi, \varphi}{\Gamma \Rightarrow \Delta, \varphi}\ (\Rightarrow c) \qquad \dfrac{\varphi, \varphi, \Gamma \Rightarrow \Delta}{\varphi, \Gamma \Rightarrow \Delta}\ (c \Rightarrow)$$

$$\dfrac{\Gamma \Rightarrow \Delta, \varphi \quad \varphi, \Pi \Rightarrow \Sigma}{\Gamma, \Pi \Rightarrow \Delta, \Sigma}\ (Cut)$$

Propositional Logical Rules

$$\dfrac{\varphi, \Gamma \Rightarrow \Delta}{\Gamma \Rightarrow \Delta, \neg\varphi}\ (\Rightarrow \neg) \qquad \dfrac{\Gamma \Rightarrow \Delta, \varphi}{\neg\varphi, \Gamma \Rightarrow \Delta}\ (\neg \Rightarrow)$$

$$\dfrac{\varphi, \Gamma \Rightarrow \Delta, \psi}{\Gamma \Rightarrow \Delta, \varphi \to \psi}\ (\Rightarrow \to) \qquad \dfrac{\Gamma \Rightarrow \Delta, \varphi \quad \psi, \Gamma \Rightarrow \Delta}{\varphi \to \psi, \Gamma \Rightarrow \Delta}\ (\to \Rightarrow)$$

Logical Rules for D_G of $\mathbf{K}_D$

$$\dfrac{\varphi_1, \ldots, \varphi_n \Rightarrow \psi \quad (\bigcup_{i=1}^n G_i \subseteq G)}{D_{G_1}\varphi_1, \ldots, D_{G_n}\varphi_n \Rightarrow D_G\psi}\ (D)$$

Logical Rules for D_G of $\mathbf{KT}_D$

$$\dfrac{\varphi_1, \ldots, \varphi_n \Rightarrow \psi \quad (\bigcup_{i=1}^n G_i \subseteq G)}{D_{G_1}\varphi_1, \ldots, D_{G_n}\varphi_n \Rightarrow D_G\psi}\ (D) \qquad \dfrac{\varphi, \Gamma \Rightarrow \Delta}{D_G\varphi, \Gamma \Rightarrow \Delta}\ (D \Rightarrow)$$

Logical Rules for D_G of $\mathbf{K4}_D$

$$\dfrac{\varphi_1, \ldots, \varphi_n, D_{G_1}\varphi_1, \ldots, D_{G_n}\varphi_n \Rightarrow \psi \quad (\bigcup_{i=1}^n G_i \subseteq G)}{D_{G_1}\varphi_1, \ldots, D_{G_n}\varphi_n \Rightarrow D_G\psi}\ (\Rightarrow D^{\mathbf{K4}_D})$$

Logical Rules for D_G of $\mathbf{S4}_D$

$$\dfrac{D_{G_1}\varphi_1, \ldots, D_{G_n}\varphi_n \Rightarrow \psi \quad (\bigcup_{i=1}^n G_i \subseteq G)}{D_{G_1}\varphi_1, \ldots, D_{G_n}\varphi_n \Rightarrow D_G\psi}\ (\Rightarrow D^{\mathbf{S4}_D}) \qquad \dfrac{\varphi, \Gamma \Rightarrow \Delta}{D_G\varphi, \Gamma \Rightarrow \Delta}\ (D \Rightarrow)$$

Logical Rules for D_G of $\mathbf{S5}_D$

$$\dfrac{D_{G_1}\varphi_1, \ldots, D_{G_n}\varphi_n \Rightarrow D_{H_1}\psi_1, \ldots, D_{H_m}\psi_m, \chi \quad (\bigcup_{i=1}^n G_i \cup \bigcup_{j=1}^m H_j \subseteq G)}{D_{G_1}\varphi_1, \ldots, D_{G_n}\varphi_n \Rightarrow D_{H_1}\psi_1, \ldots, D_{H_m}\psi_m, D_G\chi}\ (\Rightarrow D^{\mathbf{S5}_D})$$

$$\dfrac{\varphi, \Gamma \Rightarrow \Delta}{D_G\varphi, \Gamma \Rightarrow \Delta}\ (D \Rightarrow)$$

1. *If $\vdash_{\mathsf{H(X)}} \varphi$, then $\vdash_{\mathsf{G(X)}} \Rightarrow \varphi$.*
2. *If $\vdash_{\mathsf{G(X)}} \Gamma \Rightarrow \Delta$, then $\vdash_{\mathsf{H(X)}} \bigwedge \Gamma \to \bigvee \Delta$, where $\bigwedge \emptyset := \top$ and $\bigvee \emptyset := \bot$.*

Proof. We show the case of $\mathbf{K}_D$. The idea for proof is common to the rest. Here we focus on item 2 alone. We show item 2 by induction on the structure of the derivation for the sequent $\Gamma \Rightarrow \Delta$. We deal with the case for the rule (D) only. Suppose we have a derivation

$$\frac{\begin{array}{c} \mathcal{D} \\ \varphi_1, \ldots, \varphi_n \Rightarrow \psi \end{array} \quad (\bigcup_{i=1}^n G_i \subseteq G)}{D_{G_1}\varphi_1, \ldots, D_{G_n}\varphi_n \Rightarrow D_G \psi} \ (D).$$

We show $\vdash_{\mathsf{H(K}_D)} \bigwedge_{i=1}^n D_{G_i}\varphi_i \to D_G\psi$. We have $\vdash_{\mathsf{H(K}_D)} \bigwedge_{i=1}^n \varphi_i \to \psi$ as the induction hypothesis for the derivation $\mathcal{D}$. From this, we can infer by necessitation $\vdash_{\mathsf{H(K}_D)} D_G(\bigwedge_{i=1}^n \varphi_i \to \psi)$. By this and axiom (K), we have $\vdash_{\mathsf{H(K}_D)} D_G(\bigwedge_{i=1}^n \varphi_i) \to D_G\psi$, which is equivalent to $\vdash_{\mathsf{H(K}_D)} \bigwedge_{i=1}^n D_G\varphi_i \to D_G\psi$. Therefore, it suffices to show that $\vdash_{\mathsf{H(K}_D)} \bigwedge_{i=1}^n D_{G_i}\varphi_i \to \bigwedge_{i=1}^n D_G\varphi_i$, which is equivalent to $\vdash_{\mathsf{H(K}_D)} \bigwedge_{i=1}^n D_{G_i}\varphi_i \to D_G\varphi_i$ for any $i \in \{1, \ldots, n\}$. This is evident because we have a propositional tautology $\vdash_{\mathsf{H(K}_D)} \bigwedge_{i=1}^n D_{G_i}\varphi_i \to D_{G_i}\varphi_i$ and the axiom (Incl) $\vdash_{\mathsf{H(K}_D)} D_{G_i}\varphi_i \to D_G\varphi_i$. □

4 Main Proof-Theoretic Results

4.1 Cut Elimination

The cut elimination theorem does not hold for $\mathsf{G}(\mathbf{S5}_D)$, because the application of (Cut) rule in the following derivation cannot be eliminated [11].

$$\frac{\dfrac{\dfrac{\dfrac{\overline{p \Rightarrow p} \ (Id)}{\neg p, p \Rightarrow} \ (\neg \Rightarrow)}{D_{\{a\}}\neg p, p \Rightarrow} \ (D \Rightarrow)}{p \Rightarrow \neg D_{\{a\}}\neg p} \ (\Rightarrow \neg) \qquad \dfrac{\dfrac{\dfrac{\dfrac{\overline{D_{\{a\}}\neg p \Rightarrow D_{\{a\}}\neg p} \ (Id)}{\Rightarrow \neg D_{\{a\}}\neg p, D_{\{a\}}\neg p} \ (\Rightarrow \neg)}{\Rightarrow D_{\{a\}}\neg D_{\{a\}}\neg p, D_{\{a\}}\neg p} \ (\Rightarrow D^{\mathbf{S5}_D})}{\neg D_{\{a\}}\neg p \Rightarrow D_{\{a\}}\neg D_{\{a\}}\neg p} \ (\neg \Rightarrow)}{\dfrac{p \Rightarrow D_{\{a\}}\neg D_{\{a\}}\neg p}{\Rightarrow p \to D_{\{a\}}\neg D_{\{a\}}\neg p} \ (\Rightarrow \to)}} \ (Cut)$$

Therefore, we establish the cut elimination theorem for our sequent calculi except for $\mathsf{G}(\mathbf{S5}_D)$.

Theorem 2 (Cut Elimination). *Let $\mathbf{X}$ be any of $\mathbf{K}_D$, $\mathbf{KT}_D$, $\mathbf{K4}_D$, and $\mathbf{S4}_D$. Then, the following holds: If $\vdash_{\mathsf{G(X)}} \Gamma \Rightarrow \Delta$, then $\vdash_{\mathsf{G^-(X)}} \Gamma \Rightarrow \Delta$, where $\mathsf{G^-(X)}$ denotes a system "$\mathsf{G(X)}$ minus the cut rule".*

To prove the theorem, we consider a system $\mathsf{G^*(X)}$, in which the cut rule is replaced by a "extended" cut rule defined as:

$$\frac{\Gamma \Rightarrow \Delta, \varphi^n \quad \varphi^m, \Sigma \Rightarrow \Theta}{\Gamma, \Sigma \Rightarrow \Delta, \Theta} \ (ECut),$$

where φ^n denotes the multi-set of n-copies of φ and $n, m \geq 0$. Since $(ECut)$ is the same as (Cut) when we set $n = m = 1$, it is obvious that if $\vdash_{\mathsf{G(X)}} \Gamma \Rightarrow \Delta$, then $\vdash_{\mathsf{G^*(X)}} \Gamma \Rightarrow \Delta$, so it suffices to show the following.

Lemma 1. *Let* **X** *be any of* $\mathbf{K}_D$, $\mathbf{KT}_D$, $\mathbf{K4}_D$, *and* $\mathbf{S4}_D$. *Then, the following holds: If* $\vdash_{\mathsf{G^*(X)}} \Gamma \Rightarrow \Delta$, *then* $\vdash_{\mathsf{G^-(X)}} \Gamma \Rightarrow \Delta$.

Proof. Let **X** be any of $\mathbf{K}_D$, $\mathbf{KT}_D$, $\mathbf{K4}_D$, and $\mathbf{S4}_D$. Suppose $\vdash_{\mathsf{G^*(X)}} \Gamma \Rightarrow \Delta$ and fix one derivation for the sequent. To obtain an $(ECut)$-free derivation of $\Gamma \Rightarrow \Delta$, it is enough to concentrate on a derivation whose root is derived by $(ECut)$ and which has no other application of $(ECut)$. In what follows, we let **X** be $\mathbf{K}_D$. Let us suppose that $\mathcal{D}$ has the following structure:

$$\cfrac{\cfrac{\mathcal{L}}{\Gamma \Rightarrow \Delta, \varphi^n} \;(\mathbf{rule}_\mathcal{L}) \quad \cfrac{\mathcal{R}}{\varphi^m, \Sigma \Rightarrow \Theta} \;(\mathbf{rule}_\mathcal{R})}{\Gamma, \Sigma \Rightarrow \Delta, \Theta} \;(ECut) \quad ,$$

where the derivations $\mathcal{L}$ and $\mathcal{R}$ has no application of $(ECut)$ and $\mathbf{rule}_\mathcal{L}$ and $\mathbf{rule}_\mathcal{R}$ are meta-variables for the name of rule applied there. Let the number of logical symbols (including D_G) appearing in φ be $c(\mathcal{D})$ and the number of sequents in $\mathcal{L}$ and $\mathcal{R}$ be $w(\mathcal{D})$. We show the lemma by double induction on $(c(\mathcal{D}), w(\mathcal{D}))$. If $n = 0$ or $m = 0$, we can derive the root sequent of $\mathcal{D}$ without using $(ECut)$ by weakening rules. So, in what follows we assume $n, m > 0$.

Then, it is sufficient to consider the following four cases:[2]

1. $\mathbf{rule}_\mathcal{L}$ or $\mathbf{rule}_\mathcal{R}$ is an axiom.
2. $\mathbf{rule}_\mathcal{L}$ or $\mathbf{rule}_\mathcal{R}$ is a structural rule.
3. $\mathbf{rule}_\mathcal{L}$ or $\mathbf{rule}_\mathcal{R}$ is a logical rule and a cut formula φ is not principal (in the sense we have specified in Sect. 3) for that rule.
4. $\mathbf{rule}_\mathcal{L}$ and $\mathbf{rule}_\mathcal{R}$ are both logical rules (including (D)) for the same logical symbol and a cut formula φ is principal for each rule.

We concentrate on a rule (D) and the case involving the rule (D) is case 4 only, so we only comment on case 4 where both $\mathbf{rule}_\mathcal{L}$ and $\mathbf{rule}_\mathcal{R}$ are rules (D). In that case, the given derivation $\mathcal{D}$ has the following structure.

$$\cfrac{\cfrac{\mathcal{L}'}{\varphi_1, \ldots, \varphi_n \Rightarrow \psi} \;(\bigcup_{i=1}^n G_i \subseteq G) \quad \cfrac{\mathcal{R}'}{\psi^m, \psi_1, \ldots, \psi_m \Rightarrow \chi} \;(G \cup \bigcup_{j=1}^m H_j \subseteq H)}{D_{G_1}\varphi_1, \ldots, D_{G_n}\varphi_n, D_{H_1}\psi_1, \ldots, D_{H_m}\psi_n \Rightarrow D_H\chi} \;(ECut)$$

where the left premise derives $D_{G_1}\varphi_1, \ldots, D_{G_n}\varphi_n \Rightarrow D_G\psi$ by (D) and the right premise derives $(D_G\psi)^m, D_{H_1}\psi_1, \ldots, D_{H_m}\psi_m \Rightarrow D_H\chi$ by (D).

---

[2] In case 4, we assume the condition for both rule applications, because if the one of the two rule applications does not satisfy the condition, the whole derivation should be categorized into one of the rest cases.

The derivation $\mathcal{D}$ can be transformed into the following derivation $\mathcal{E}$:[3]

$$
\cfrac{
\cfrac{\mathcal{L}'}{\varphi_1,\ldots,\varphi_n \Rightarrow \psi} \quad \cfrac{\mathcal{R}'}{\psi^m, \psi_1,\ldots,\psi_m \Rightarrow \chi}
}{
\cfrac{\varphi_1,\ldots,\varphi_n,\psi_1,\ldots,\psi_m \Rightarrow \chi}{D_{G_1}\varphi_1,\ldots,D_{G_n}\varphi_n, D_{H_1}\psi_1,\ldots,D_{H_m}\psi_m \Rightarrow D_H\chi} \ (D)
} \ (ECut) \quad (\bigcup_{i=1}^n G_i \cup \bigcup_{j=1}^m H_j \subseteq H)
$$

.

We call $\mathcal{E}'$ its subderivation whose root sequent is $\varphi_1,\ldots,\varphi_n,\psi_1,\ldots,\psi_m \Rightarrow \chi$. The derivation $\mathcal{E}'$ have no application of $(ECut)$ and $c(\mathcal{E}') < c(\mathcal{D})$. Hence, by induction hypothesis, there exists an $(ECut)$-free derivation $\tilde{\mathcal{E}}'$ having the same root sequent. Replacing the derivation $\mathcal{E}'$ by $\tilde{\mathcal{E}}'$ in $\mathcal{E}$, we obtain an $(ECut)$-free derivation for the sequent $D_{G_1}\varphi_1,\ldots,D_{G_n}\varphi_n, D_{H_1}\psi_1,\ldots,D_{H_m}\psi_m \Rightarrow D_H\chi$ as required.   □

4.2   Craig Interpolation Theorem

As an application of the cut elimination theorem, Craig interpolation theorem can be derived, using a Maehara method described in [9] (application of the method to basic modal logic can be also found in [12]). To state a main lemma for proving Craig Interpolation Theorem, some definitions are needed.

Definition 1 (Partition). *A partition for a sequent $\Gamma \Rightarrow \Delta$ is defined as a tuple $\langle(\Gamma_1 : \Delta_1);(\Gamma_2 : \Delta_2)\rangle$, such that the multi-set union of Γ_1 and Γ_2 (Δ_1 and Δ_2) is equal to Γ (Δ, respectively).*

Definition 2. *For a formula φ, $\mathsf{Prop}(\varphi)$ is defined as the set of propositional variables appearing in φ. For a multi-set of formulas Γ, $\mathsf{Prop}(\Gamma)$ is defined as $\bigcup_{\varphi \in \Gamma} \mathsf{Prop}(\varphi)$. Similarly, $\mathsf{Agt}(\varphi)$ is defined as the set of agents appearing in φ and $\mathsf{Agt}(\Gamma)$ as $\bigcup_{\varphi \in \Gamma} \mathsf{Agt}(\varphi)$*

The following is a key lemma for Craig Interpolation Theorem.

Lemma 2. *Let $\mathbf{X}$ be any of $\mathbf{K}_D$, $\mathbf{KT}_D$, $\mathbf{K4}_D$, and $\mathbf{S4}_D$. Suppose $\vdash_{\mathsf{G}(\mathbf{X})} \Gamma \Rightarrow \Delta$. Then, for any partition $\langle(\Gamma_1 : \Delta_1);(\Gamma_2 : \Delta_2)\rangle$ for the sequent $\Gamma \Rightarrow \Delta$, there exists a formula φ called "interpolant", satisfying the following:*

1. $\vdash_{\mathsf{G}(\mathbf{X})} \Gamma_1 \Rightarrow \Delta_1, \varphi$ and $\vdash_{\mathsf{G}(\mathbf{X})} \varphi, \Gamma_2 \Rightarrow \Delta_2$.
2. $\mathsf{Prop}(\varphi) \subseteq \mathsf{Prop}(\Gamma_1, \Delta_1) \cap \mathsf{Prop}(\Gamma_2, \Delta_2)$.
3. $\mathsf{Agt}(\varphi) \subseteq \mathsf{Agt}(\Gamma_1, \Delta_1) \cap \mathsf{Agt}(\Gamma_2, \Delta_2)$.

Proof. We prove the case of $\mathbf{K}_D$ by induction on the structure of a derivation for $\Gamma \Rightarrow \Delta$. Fix the derivation and name it $\mathcal{D}$. By the cut-elimination theorem (Theorem 2), we can assume that $\mathcal{D}$ is cut-free. We treat only the case of (D) below (for other cases, the reader is referred to [12]). Suppose $\mathcal{D}$ is of the form

---

[3] Note that the condition $\bigcup_{i=1}^n G_i \cup \bigcup_{j=1}^m H_j \subseteq H$ in $\mathcal{E}$ can be obtained by the conditions $\bigcup_{i=1}^n G_i \subseteq G$ and $G \cup \bigcup_{j=1}^m H_j \subseteq H$ in $\mathcal{D}$ through "cutting" G.

$$\frac{\dfrac{\mathcal{E}}{\varphi_1,\ldots,\varphi_n \Rightarrow \psi} \quad (\bigcup_{i=1}^{n} G_i \subseteq G)}{D_{G_1}\varphi_1,\ldots,D_{G_n}\varphi_n \Rightarrow D_G\psi} (D).$$

There are the following two partitions of $D_{G_1}\varphi_1,\ldots,D_{G_n}\varphi_n \Rightarrow D_G\psi$:

(a) a partition $\langle (D_{G_1}\varphi_1,\ldots,D_{G_k}\varphi_k : \emptyset); (D_{G_{k+1}}\varphi_{k+1},\ldots,D_{G_n}\varphi_n : D_G\psi)\rangle$.
(b) a partition $\langle (D_{G_1}\varphi_1,\ldots,D_{G_k}\varphi_k : D_G\psi); (D_{G_{k+1}}\varphi_{k+1},\ldots,D_{G_n}\varphi_n : \emptyset)\rangle$.

Since the space is limited, we focus on case (b) alone. By induction hypothesis on $\mathcal{E}$ for a partition $\langle (\varphi_1,\ldots,\varphi_k : \psi); (\varphi_{k+1},\ldots,\varphi_n : \emptyset)\rangle$, we have derivations for $\varphi_1,\ldots,\varphi_k \Rightarrow \psi, \chi$ and $\chi, \varphi_{k+1},\ldots,\varphi_n \Rightarrow$ for some formula χ. If $k < n$, we can choose $\neg D_{\bigcup_{i=k+1}^{n} G_i}\neg\chi$ as a required interpolant, because we have following derivations:

$$\frac{\dfrac{\dfrac{\text{I.H.}}{\varphi_1,\ldots,\varphi_k \Rightarrow \psi, \chi}}{\neg\chi, \varphi_1,\ldots,\varphi_k \Rightarrow \psi}(\neg\Rightarrow) \quad (\bigcup_{i=k+1}^{n} G_i \cup \bigcup_{i=1}^{k} G_i = \bigcup_{i=1}^{n} G_i \subseteq G)}{\dfrac{D_{\bigcup_{i=k+1}^{n} G_i}\neg\chi, D_{G_1}\varphi_1,\ldots,D_{G_k}\varphi_k \Rightarrow D_G\psi}{D_{G_1}\varphi_1,\ldots,D_{G_k}\varphi_k \Rightarrow D_G\psi, \neg D_{\bigcup_{i=k+1}^{n} G_i}\neg\chi}(\Rightarrow\neg)}(D)$$

$$\frac{\dfrac{\dfrac{\text{I.H.}}{\chi, \varphi_{k+1},\ldots,\varphi_n \Rightarrow}}{\varphi_{k+1},\ldots,\varphi_n \Rightarrow \neg\chi}(\Rightarrow\neg) \quad (\bigcup_{i=k+1}^{n} G_i \subseteq \bigcup_{i=k+1}^{n} G_i)}{\dfrac{D_{G_{k+1}}\varphi_{k+1},\ldots,D_{G_n}\varphi_n \Rightarrow D_{\bigcup_{i=k+1}^{n} G_i}\neg\chi}{\neg D_{\bigcup_{i=k+1}^{n} G_i}\neg\chi, D_{G_{k+1}}\varphi_{k+1},\ldots,D_{G_n}\varphi_n \Rightarrow}(\neg\Rightarrow)}(D)$$

Furthermore, the interpolant enjoys the condition 2 and 3 by the induction hypothesis and a simple calculation. If $k = n$, we can choose χ (equivalent to $\bot$) as an interpolant. $\qquad\square$

Theorem 3 (Craig Interpolation Theorem). *Let* $\mathbf{X}$ *be any of* $\mathbf{K}_D$, $\mathbf{KT}_D$, $\mathbf{K4}_D$, *and* $\mathbf{S4}_D$. *Given that* $\vdash_{\mathsf{G}(\mathbf{X})} \varphi \Rightarrow \psi$, *there exists a formula* χ *satisfying the following conditions:*

1. $\vdash_{\mathsf{G}(\mathbf{X})} \varphi \Rightarrow \chi$ *and* $\vdash_{\mathsf{G}(\mathbf{X})} \chi \Rightarrow \psi$.
2. $\mathsf{Prop}(\chi) \subseteq \mathsf{Prop}(\varphi) \cap \mathsf{Prop}(\psi)$.
3. $\mathsf{Agt}(\chi) \subseteq \mathsf{Agt}(\varphi) \cap \mathsf{Agt}(\psi)$.

We note that not only the condition for propositional variables but also the condition for agents can be satisfied.

Proof. When we set $\Gamma := \varphi$ and $\Delta := \psi$, and take a partition $\langle (\varphi : \emptyset); (\emptyset : \psi)\rangle$, Lemma 2 proves Craig Interpolation Theorem. $\qquad\square$

5   Concluding Remark

To conclude the paper, we would like to mention possible directions of further research. First, we may provide sequent calculi $\mathbf{KD}_D$, $\mathbf{KD4}_D$ or $\mathbf{KD45}_D$ to establish the cut-elimination and Craig interpolation theorems. Second, we may establish the subformula property for $\mathbf{S5}_D$ along the line of [17,18], which proves the property on a sequent calculus for $\mathbf{S5}$ though the calculus is not cut-free. We note that even if the sequent calculus for $\mathbf{S5}$ is not cut-free but still we can apply Maehara method to establish Craig interpolation theorem (cf [12]). Third, it is interesting to see if we can construct a cut-free sequent calculus for $\mathbf{S5}_D$ on the basis of one of the known cut-free calculi for $\mathbf{S5}$ (with label or with the notion of hypersequent). Fourth, we may check what follows from Craig interpolation theorem. It is known that Craig interpolation theorem entails Beth definability theorem or Robinson consistency theorem in many systems. It is interesting to see whether these hold for our logics. Fifth, we may work on computational aspects of the logics of distributed knowledge. Especially, a decision problem is worth pursuing, because there is a known method for proving decidability based on cut elimination theorem [4,5,12]. In general, sequent calculus is tractable proof system from a computational viewpoint, because a derivation in sequent calculus is a finite tree, which is common in computer science. Therefore, our work can be a basis for computational research about distributed knowledge. Finally, we may establish completeness results on epistemic logics with distributed knowledge other than the ones mentioned in Fact 1 (if it has not been done).

Acknowledgement. We thank three reviewers of FoIKS 2020 for their helpful comments. The work of both authors was partially supported by JSPS KAKENHI Grant-in-Aid for Scientific Research (C) Grant Number 19K12113 and JSPS Core-to-Core Program (A. Advanced Research Networks). The second author was also partially supported by JSPS KAKENHI Grant-in-Aid for Scientific Research (B) Grant Number 17H02258.

References

1. Ågotnes, T., Wáng, Y.N.: Resolving distributed knowledge. Artif. Intell. **252**, 1–21 (2017). https://doi.org/10.1016/j.artint.2017.07.002
2. ten Cate, B.: Interpolation for extended modal languages. J. Symbolic Log. **70**(1), 223–234 (2005). https://doi.org/10.2178/jsl/1107298517
3. Fagin, R., Halpern, J.Y., Vardi, M.Y., Moses, Y.: Reasoning About Knowledge. MIT Press, Cambridge (1995)
4. Gentzen, G.: Untersuchungen über das logische Schließen I. Mathematische Zeitschrift **39**(1), 176–210 (1935). https://doi.org/10.1007/BF01201353
5. Gentzen, G.: Untersuchungen über das logische Schließen II. Mathematische Zeitschrift **39**(1), 405–431 (1935). https://doi.org/10.1007/BF01201363
6. Gerbrandy, J.: Bisimulations on planet Kripke. Ph.D. thesis, University of Amsterdam (1999)

7. Giedra, H.: Cut free sequent calculus for logic $S5_n(ED)$. Lietuvos matematikos rinkinys **51**, 336–341 (2010)
8. Hakli, R., Negri, S.: Proof theory for distributed knowledge. In: Sadri, F., Satoh, K. (eds.) CLIMA 2007. LNCS (LNAI), vol. 5056, pp. 100–116. Springer, Heidelberg (2008). https://doi.org/10.1007/978-3-540-88833-8_6
9. Maehara, S.: Craig no interpolation theorem (Craig's interpolation theorem). Sugaku **12**(4), 235–237 (1961). https://doi.org/10.11429/sugaku1947.12.235
10. Meyer, J.J.C., Hoek, W.V.D.: Epistemic Logic for AI and Computer Science. Cambridge University Press, New York (1995)
11. Ohnishi, M., Matsumoto, K.: Gentzen method in modal calculi II. Osaka Math. J. **11**(2), 115–120 (1959)
12. Ono, H.: Proof-theoretic methods in nonclassical logic-an introduction. In: Takahashi, M., Okada, M., Dezani-Ciancaglini, M. (eds.) Theories of Types and Proofs, MSJ Memoirs, vol. 2, pp. 207–254. The Mathematical Society of Japan, Tokyo, Japan (1998). https://doi.org/10.2969/msjmemoirs/00201C060
13. Pliuskevicius, R., Pliuskeviciene, A.: Termination of derivations in a fragment of transitive distributed knowledge logic. Informatica Lith. Acad. Sci. **19**, 597–616 (2008)
14. Poggiolesi, F.: Gentzen Calculi for Modal Propositional Logic. Trends in Logic. Springer, Dordrecht (2010). https://doi.org/10.1007/978-90-481-9670-8
15. Punčochář, V., Sedlár, I.: Substructural logics for pooling information. In: Baltag, A., Seligman, J., Yamada, T. (eds.) LORI 2017. LNCS, vol. 10455, pp. 407–421. Springer, Heidelberg (2017). https://doi.org/10.1007/978-3-662-55665-8_28
16. Roelofsen, F.: Distributed knowledge. J. Appl. Non-Class. Log. **17**(2), 255–273 (2007)
17. Takano, M.: Subformula property as a substitute for cut-elimination in modal propositional logics. Math. Jpn. **37**, 1129–1145 (1992)
18. Takano, M.: A semantical analysis of cut-free calculi for modal logics. Rep. Math. Log. **53**, 43–65 (2018)
19. Van Der Hoek, W., Van Linder, B., Meyer, J.J.: Group knowledge is not always distributed (neither is it always implicit). Math. Soc. Sci. **38**(2), 215–240 (1999)

On the Dynamics of Structured Argumentation: Modeling Changes in Default Justification Logic

Stipe Pandžić[(✉)]

Department of Theoretical Philosophy, Bernoulli Institute for Mathematics, Computer Science and Artificial Intelligence, University of Groningen, Groningen, The Netherlands
s.pandzic@rug.nl

Abstract. This paper studies information changes in default justification logic with argumentation semantics. We introduce dynamic operators that combine belief revision and default theory tools to define both prioritized and non-prioritized operations of contraction, expansion and revision for justification logic-based default theories. This combination enriches both default logics and belief revision techniques. We argue that the kind of attack called "undermining" amounts to those operations that contract a knowledge base by an attacked formula.

Keywords: Justification logic · Default theory · Defeaters · Formal argumentation · Reasoning dynamics

1 Introduction

This paper investigates the dynamics of default theories with justification logic formulas. Our logic has been presented in [28] as a theory of structured arguments where justification formulas of the form $t : F$ are interpreted as arguments that can defeat other formulas by means of undercut or rebuttal. Technically, the workings of undercut and rebuttal rely on defining default theories with default rules based on justification logic formulas. In such rules, justification terms codify defeasible inferences in their structure. In this paper, we will take a step further to add reasoning about changes to default theories.

The existing work about dynamics in formal argumentation [9,11–13,33] almost entirely focuses on abstract argumentation frameworks in the style of [14]. The literature on the dynamics of structured argumentation is limited to the DeLP framework [2], where the dynamics is understood as adding or removing strict and defeasible rules, and ASPIC+ [26], where the dynamic component is meant to resolve symmetric attacks by updating preferences. The current

I wish to thank Allard Tamminga, Barteld Kooi and Rineke Verbrugge for their useful advice on this project. My research is supported by Ammodo KNAW award *Rational Dynamics and Reasoning*.

© Springer Nature Switzerland AG 2020
A. Herzig and J. Kontinen (Eds.): FoIKS 2020, LNCS 12012, pp. 222–241, 2020.
https://doi.org/10.1007/978-3-030-39951-1_14

paper advances this line of research by specifying a variety of dynamic operators for modeling changes of argument systems based on justification logic.

We will show that introducing dynamic operators for justification logic default theories enables us to model an additional kind of defeat: *undermining*. According to [16], an argument is undermined if its premises or assumptions are attacked. Defeating an argument by attacking its premise or its assumption is not new to structured argumentation. In assumption-based argumentation (ABA) [15], all attacks are reduced to this type of attacks and in ASPIC+ [30], ordinary premises of an argument are susceptible to undermining. However, these systems do not provide an insight into the logical workings of undermining, because they neither specify a concrete logical language nor inference rules.[1]

In our default theories, undermining can be given a precise logical interpretation. While undercut and rebuttal rely on the uncertainty of defeasible arguments, undermining changes the context from which agents make further inferences. For a specific default theory, this context is determined by the set of starting premises, sometimes also called "axioms". Our idea is that, since undermining targets axioms that are considered as a given, it should be modeled as a result of non-inferential information inputs that require contracting the set of axioms of a default theory. This means that we will define undermining by "climbing up" the definitions of more fundamental operations of default theory changes. To elicit the reasoning process behind undermining, we specify four different logical operations that model undermining: prioritized and non-prioritized contraction and prioritized and non-prioritized revision.

The paper is organized as follows. Section 2 introduces the basics of the logic of default justifications first introduced in [27,29] and developed as a theory of structured arguments in [28]. This logic already models non-monotonic behavior with the use of undercut and rebuttal. Section 3 is the main technical contribution of this paper, where we define dynamic operations for default theories with justification formulas. The operations we introduce combine base revision operations as defined in, e.g., [20] with a specific kind of standard Reiter default rules. Our approach to defining the dynamic operators for default theory revision has most in common with the approach of [4], which deals with the dynamics of Reiter's default theories. It turns out by the end of the paper that undermining attacks on premises correspond to those dynamic operations that involve either contraction or a variant of non-prioritized contraction defined in Sect. 3.

2   Logic of Default Reasons

The logic of default justifications was first defined in [27] and our definitions in this section follow those given in [29]. We start by outlining the underlying justification logic on which we base default theories. Before we formally introduce

---

[1] In fact, ABA does not distinguish between different kinds of attacks and models each attack as that on premises or what we call here undermining. In ASPIC+, undermining is taken as a primitive notion of attack, which is different from rebuttal or undercut only by virtue of targeting "ordinary" premises of an argument.

syntax, several preliminary remarks on how to build justification assertions of the type $t : F$ are in order.

One of the basic operations of standard justification logics is *Application* ('·') exemplified by the following formula $u : (F \to G) \to (t : F \to (u \cdot t) : G)$. By focusing only on the propositional content of the formula, one may notice the familiar structure of a *modus ponens* inference from $F \to G$ and F to G. The added machinery of reason terms syntactically captures the dependence of the consequent G on the propositions $F \to G$ and F. This dependence is witnessed by the structure of the reason term $(u \cdot t)$ which shows that the reason u has been applied to the reason t.

In addition to application, all standard justification logics include the operation known as *Sum* '+'. Sum admits of merging two reason terms into a single term as done, for example, in the formula $t : F \to (t + u) : F$. The intuition behind Sum is that, if evidence t justifies some formula F, then adding new evidence u to t will not invalidate t being a reason for F. Thus, standard justifications are inherently non-defeasible.

The underlying logic for our default theories is a standard justification logic with non-defeasible and truth-inducing reasons.[2] We assume both Application and Sum operations for this logic. Moreover, to ensure that all reasons in this logic are truth-inducing, the logic contains the axiom $t : F \to F$ called *Factivity*. This axiom corresponds to the modal logic truth axiom: $\Box F \to F$ read as "If F is known, then F". Our choice of basing default theories on a logic of non-defeasible and truth-inducing reasons is in line with the strategy of standard default logics [3], where defeasible conclusions are based on certain information.

2.1   Logic of Factive Reasons JT$_{CS}$

Syntax, Axioms and Rules. The basic format of justification assertions is "$t : F$", where a justification term "t" is informally interpreted as a reason or justification for "F". The set Tm consists of all justification terms, constructed from variables $x_1, \ldots, x_n, \ldots$ and proof constants $c_1, \ldots, c_n, \ldots$ by means of operations $\cdot$ and $+$. The following BNF gives the grammar of justification terms:

$$t ::= x \mid c \mid (t \cdot t) \mid (t + t)$$

where x is a variable denoting an unspecified justification and c is a proof constant, taken as atomic within the system. A set of subterms $Sub(t)$ is defined by induction on the construction of a term t. The set Fm consists of exactly all JT$_{CS}$ formulas based on the countable set of propositional atoms $\mathcal{P}$ and defined by the following BNF:

$$F ::= \top \mid P \mid (F \to F) \mid (F \vee F) \mid (F \wedge F) \mid \neg F \mid t : F$$

---

[2] The first variant of justification logic, the logic of proofs (LP), was developed in [5]. The logic of non-defeasible and factive reasons that we use here was first defined in [10]. For more basic information on its relation to other justification logics see [17]. For recent overviews of justification logic systems, see [6] and [22].

where $P \in \mathcal{P}$ and $t \in Tm$.

We can now define the logic $\mathbf{JT_{CS}}$, which is the weakest logic with non-defeasible reasons containing axiom schemes for the two basic operations $\cdot$ and $+$. These are the axioms and rules of $\mathbf{JT_{CS}}$:

A0. *All the instances of propositional logic tautologies from Fm*
A1. $t : (F \rightarrow G) \rightarrow (u : F \rightarrow (t \cdot u) : G)$ (Application)
A2. $t : F \rightarrow (t + u) : F$; $u : F \rightarrow (t + u) : F$ (Sum)
A3. $t : F \rightarrow F$ (Factivity)
R0. *From F and $F \rightarrow G$ infer G* (Modus ponens)
R1. *If F is an axiom instance of A0–A3 and $c_n, c_{n-1}, \ldots, c_1$ proof constants, then infer*
$c_n : c_{n-1} : \cdots : c_1 : F$ (Iterated axiom necessitation)

The set of all justified formulas produced by the use of R1 is called a *Constant Specification* ($\mathcal{CS}$) set:

Definition 1 (Constant specification). $\mathcal{CS} = \{c_n : c_{n-1} : \cdots : c_1 : F \mid$ F *is an instance of A0-A3, $c_n, c_{n-1}, \ldots, c_1$ are proof constants and $n \in \mathbb{N}$*}.

The intuition behind the set of formulas $\mathcal{CS}$ in R1 is that the basic logical axioms are taken to be justified by proof constants and so are the formulas likewise produced. In the set of formulas $\mathcal{CS}$ from Definition 1, any axiom instance can be labelled with any sequence of proof constants. A constant specification of this type is thus appropriately called *Total Constant Specification* ($\mathcal{TCS}$) and the logic we defined with the use of $\mathcal{TCS}$ is called $\mathbf{JT_{TCS}}$.

Following the idea that each logical axiom is justified by exactly one constant, we require that the use of proof constants respects the following two conditions and we say that $\mathcal{CS}$ is:[3]

- Axiomatically appropriate: for each axiom instance A, there is a constant c such that $c : A \in \mathcal{CS}$ and for each formula $c_n : c_{n-1} : \cdots : c_1 : A \in \mathcal{CS}$ such that $n \geq 1$, there is a constant c_{n+1} such that $c_{n+1} : c_n : c_{n-1} : \cdots : c_1 : A \in \mathcal{CS}$;
- Injective: each proof constant c justifies at most one formula.

According to the restrictions on an axiomatically appropriate and injective $\mathcal{CS}$, we replace R1 of $\mathbf{JT_{TCS}}$ with the following rule to obtain the logic $\mathbf{JT_{CS}}$:

R1* *If F is an axiom instance of A0–A3 and $c_n, c_{n-1}, \ldots, c_1$ proof constants such that $c_n : c_{n-1} : \cdots : c_1 : F \in \mathcal{CS}$, then infer $c_n : c_{n-1} : \cdots : c_1 : F$*

We say that the formula F is $\mathbf{JT_{CS}}$-provable ($\mathbf{JT_{CS}} \vdash F$) if F can be derived using the axioms A0–A3 and rules R0 and R1.

---

[3] For example, one such constant specification set could be generated by assigning a Gödel number to each axiom instance and to each instance of R1.

Semantics. The semantics for $\mathbf{JT_{CS}}$ is an adapted version of the semantics for the logic of proofs (**LP**) given by [25].

Definition 2 ($\mathbf{JT_{CS}}$ model). *We define a function* reason assignment *based on* $\mathcal{CS}$, $*(\cdot) : Tm \to 2^{Fm}$, *a function mapping each term to a set of formulas from Fm. We assume that it satisfies the following conditions:*

(1) If $F \to G \in *(t)$ and $F \in *(u)$, then $G \in *(t \cdot u)$,
(2) $*(t) \cup *(u) \subseteq *(t + u)$ and
(3) If $c : F \in \mathcal{CS}$, then $F \in *(c)$.

A truth assignment $v : \mathcal{P} \to \{True, False\}$ *is a function assigning truth values to propositional atoms in $\mathcal{P}$. We define the interpretation $\mathcal{I}$ as a pair $(v, *)$. For an interpretation $\mathcal{I}$, $\models$ is a truth relation on the set of formulas of $\mathbf{JT_{CS}}$. We say that, for any formula $t : F \in Fm$, $\mathcal{I} \models t : F$ iff $F \in *(t)$. Truth conditions for atomic propositions, $\neg$, $\to$, $\wedge$ and $\vee$ are defined as usual. An interpretation $\mathcal{I}$ is* reflexive *iff the truth relation for $\mathcal{I}$ fulfills the following condition:*

– For any term t and any formula F, if $F \in *(t)$, then $\mathcal{I} \models F$.

The consequence relation of the logic of factive reasons $\mathbf{JT_{CS}}$ is defined on reflexive interpretations:

Definition 3 ($\mathbf{JT_{CS}}$ consequence relation). $\Sigma \models F$ *iff for all reflexive interpretations $\mathcal{I}$, if $\mathcal{I} \models B$ for all $B \in \Sigma$, then $\mathcal{I} \models F$.*

For a set of formulas $\Gamma \subseteq Fm$ and the JT_{CS} consequence relation $\models$ defined above, a $\mathbf{JT_{CS}}$ closure of Γ is given by $Th^{JT_{CS}}(\Gamma) = \{F | \Gamma \models F\}$. For a closure $Th^{JT_{CS}}(\Gamma)$, it holds that $\mathcal{CS} \subseteq Th^{JT_{CS}}(\Gamma)$.

2.2 Logic of Default Reasons

Building on the $\mathbf{JT_{CS}}$ syntax, we introduce the definition of the *default theory*:

Definition 4 (Default Theory). *A* default theory T *is a pair (W, D), where W is a finite set of $\mathbf{JT_{CS}}$ formulas and D is a countable set of default rules.*

Each default rule is of the following form:

$$\delta = \frac{t : F :: (u \cdot t) : G}{(u \cdot t) : G}.$$

The informal reading of the default δ is: "If t is a reason justifying F, and it is consistent to assume that $(u \cdot t)$ is a reason justifying G, then $(u \cdot t)$ is a defeasible reason justifying G". The default rule δ introduces a unique reason term u, which means that, for a default theory T, the following three conditions are required:

(1) For any formula $v : H$ such that $v : H \in Th^{JT_{CS}}(W)$, it holds that $u \neq v$;
(2) For any formula $H \in W$, $u : (F \to G)$ is not a subformula of H and

(3) For any default rule $\delta' \in D$ such that $\delta' = \frac{t':F' \,::\, (u' \cdot t'):G'}{(u' \cdot t'):G'}$, if $u = u'$, then $F = F'$ and $G = G'$.

In the rule δ, the formula $t : F$ is called the *prerequisite*, $(u \cdot t) : G$ above the line is called the *consistency requirement* and $(u \cdot t) : G$ below the line is called the *consequent* of δ. We refer to each of these formulas by $pre(\delta)$, $req(\delta)$ and $cons(\delta)$ respectively.

Every default rule produces a reason term whose structure codifies an application operation step. Notice that in δ above, in contrast to axiom A1, we do not require the formula $u : (F \rightarrow G)$ to be a part of the knowledge base. Instead, $u : (F \rightarrow G)$ is the underlying assumption of δ on the basis of which we are able to extend an incomplete knowledge base. The propositions of this kind are important in the system since they function as rules allowing for default steps, but they are also specific $\mathbf{JT_{CS}}$ formulas. They will be referred to as "warrants", because their twofold role mirrors that of Toulmin's argument warrants ([34], p. 91). Warrants extend the interpretation of the application operation "$\cdot$" and each warrant is made explicit by means of a function *warrant assignment*: $\#(\cdot) : D \rightarrow Fm$. The function maps each default rule to a specific justified conditional as follows: $\#(\delta_i) = u : (F \rightarrow G)$, where $\delta_i \in D$ and $\delta_i = \frac{t:F \,::\, (u \cdot t):G}{(u \cdot t):G}$, for some reason term t, a unique reason term u and some formulas F and G. It is important that uniqueness of u does not prevent two default rules to share a warrant formula. This reflects also the informal idea of warrants as general rules that are, in principle, applicable to different starting data.[4]

A set of all such underlying warrants of default rules is called *Warrant Specification* ($\mathcal{WS}$) set.

Definition 5 (Warrant specification). *For a default theory $T = (W, D)$, justified defeasible conditionals are given by the Warrant Specification set:*

$$\mathcal{WS}^T = \#[D] = \{u : (F \rightarrow G) \mid \#(\delta) = u : (F \rightarrow G) \text{ and } \delta \in D\}.$$

The basis of operational semantics for a default theory $T = (W, D)$ is the procedure of collecting new, reason-based information from the available defaults. This procedure is defined following Antoniou's [3] operational semantics for Reiter's default theories. A *sequence* of default rules $\Pi = (\delta_0, \delta_1, \dots)$ is a possible order in which a list of default rules without multiple occurrences from D is applied (Π is possibly empty). Applicability of defaults is determined in the following way: for a set of $\mathbf{JT_{CS}}$-closed formulas Γ we say that a default rule $\delta = \frac{t:F \,::\, (u \cdot t):G}{(u \cdot t):G}$ is applicable to Γ iff $t : F \in \Gamma$ and $\neg(u \cdot t) : G \notin \Gamma$. Default consequents are brought together in the set of $\mathbf{JT_{CS}}$ formulas that represents the current evidence base: $In(\Pi) = Th^{JT_{CS}}(W \cup \{cons(\delta) \mid \delta \text{ occurs in } \Pi\})$. The set $In(\Pi)$ pools reason-based formulas whose acceptability depends on the acceptability of other available (counter-)reasons.

---

[4] Formally, we also do not require that $t = t'$ holds in the antecedent of condition (3) for the general definition of defaults above. This reflects the independence of the warrant $(u \cdot t) : G$ from the data $t : F$ to which we apply the warrant.

We need to further specify those sequences of defaults that are significant for a theory T namely, default processes. For a sequence Π, the initial segment of the sequence is denoted as $\Pi[k]$, where k indicates the number of elements contained in that segment of the sequence and a minimal number of defaults for the sequence Π. Any segment $\Pi[k]$ is also a sequence. We can now define default processes:

Definition 6 (Process). *A sequence of default rules Π is a process of a default theory $T = (W, D)$ iff every k such that $\delta_k \in \Pi$ is applicable to the set $In(\Pi[k])$, where $\Pi[k] = (\delta_0, \ldots \delta_{k-1})$.*

The kind of process that we are focusing on is called *closed* process: a process Π is said to be closed iff every $\delta \in D$ that is applicable to $In(\Pi)$ is already in Π.

The possibility to refer to warrants within the language enables us to model *undercutting* defeaters. They defeat other reasons by denying that their warrant provides support for the conclusion in an undercutting circumstance.

Definition 7 (Undercut). *A reason u undercuts reason t being a reason for F in a set of $\mathbf{JT_{CS}}$ formulas $\Gamma \subseteq In(\Pi[k])$ iff $\bigvee_{(v) \in Sub(t)} u : \neg[v : (G \to H)] \in Th^{JT_{CS}}(\Gamma)$ and $v : (G \to H) \in \mathcal{WS}^{\Pi'}$ for a process Π' of T.*

We say that a set $\Gamma \subseteq In(\Pi[k])$ undercuts reason t being a reason for F iff, for a subterm v of t, $\neg[v : (G \to H)] \in Th^{JT_{CS}}(\Gamma)$ and $v : (G \to H) \in \mathcal{WS}^{\Pi'}$ for a process Π' of T.

For any default theory $T = (W, D)$, an agent always considers potential extension sets of $\mathbf{JT_{CS}}$ formulas that meet the following conditions:

1. $W \subseteq \Gamma$ and
2. $\Gamma \subseteq \{W \cup cons(\Pi) \mid \Pi$ is some process of $T\}$.

Besides undercut, reasons may conflict each other due to the fact that they cannot be both added to a same consistent process. For example, if a knowledge base contains a reason for a statement F, then any default introducing a reason for $\neg F$ will be inapplicable. This enables us to obtain a formal representation of *rebuttal* among reasons for free. Formally, every rebuttal will imply an undercut for formulas that cannot possibly extend a knowledge base together. Thus, we exclusively rely on the definition of undercut to characterize all the standard argumentation extension notions [14] in justification logic. For any potentially acceptable set Γ we say that Γ is conflict-free if its closure is undercut-free and we define the notion of acceptability of a justified formula $t : F$:

Definition 8 (Acceptability). *For a process Π of a default theory $T = (W, D)$, a formula $t : F \in cons(\Pi)$ is acceptable w.r.t. a set of $\mathbf{JT_{CS}}$ formulas $\Gamma \subseteq In(\Pi)$ iff for each undercutting reason u for t being a reason for F such that $u : G \in In(\Pi)$, $Th^{JT_{CS}}(\Gamma)$ undercuts u being a reason for G.*

Definition 9 (JT$_{\mathbf{CS}}$ Extensions). *We define multiple argumentation theory extensions for any default theory $T = (W, D)$ based on the $\mathbf{JT_{CS}}$ language:*

JT$_{\mathbf{CS}}$ Admissible Extension. *A potential extension set of* **JT$_{\mathbf{CS}}$** *formulas* $\Gamma \subset In(\Pi)$ *is a* **JT$_{\mathbf{CS}}$** *admissible extension of a default theory* $T = (W, D)$ *iff* $Th^{JT_{CS}}(\Gamma)$ *is conflict-free, each formula* $t : F \in \Gamma$ *is acceptable w.r.t.* Γ *and* Π *is closed.*

JT$_{\mathbf{CS}}$ Preferred Extension. *A closure* $Th^{JT_{CS}}(\Gamma)$ *of a* **JT$_{\mathbf{CS}}$** *admissible extension* Γ *is a* **JT$_{\mathbf{CS}}$** *preferred extension of* T *iff for any other* **JT$_{\mathbf{CS}}$** *admissible extension* Γ', $\Gamma \not\subset \Gamma'$.

JT$_{\mathbf{CS}}$ Complete Extension. *A closure* $Th^{JT_{CS}}(\Gamma)$ *of a* **JT$_{\mathbf{CS}}$** *admissible extension* Γ *is a* **JT$_{\mathbf{CS}}$** *complete extension of* T *iff for each closed process* Π *for which* $\Gamma \subset In(\Pi)$ *is* **JT$_{\mathbf{CS}}$** *admissible it holds that, if a formula* $t : F \in cons(\Pi)$ *is acceptable w.r.t., then* $t : F$ *belongs to* Γ.

JT$_{\mathbf{CS}}$ Grounded Extension. *A* **JT$_{\mathbf{CS}}$** *complete extension* $Th^{JT_{CS}}(\Gamma)$ *is the unique* **JT$_{\mathbf{CS}}$** *grounded extension of* T *if* Γ *is the smallest potential extension with respect to set inclusion such that* $Th^{JT_{CS}}(\Gamma)$ *is a* **JT$_{\mathbf{CS}}$** *complete extension of* T.

JT$_{\mathbf{CS}}$ Stable Extension. *A conflict-free closure* $Th^{JT_{CS}}(\Gamma)$ *of a potential extension* Γ *is a* **JT$_{\mathbf{CS}}$** *stable extension iff* Γ *undercuts all the formulas* $t : F \in cons(\Pi)$ *outside* $Th^{JT_{CS}}(\Gamma)$, *for any process* Π *of* T.

It turns out that a large subclass of Dung's frameworks, excluding only some kinds of attack cycles, can be proven to be a special case of our logic via establishing the correspondences of extensions. This result is shown in [28].

The above presented account of default reasons suffices to represent reasoning from an incomplete knowledge base, but it does not represent reasoning with information changes that alter the facts from which an agent starts to reason. Still, the basic account can already model one type of non-monotonic behavior induced by the definition of undercut. We will refer to this way of revising as "inferential" revision. The workings of undercut can be illustrated with Toulmin's famous example ([34], p. 92) of arguing for the claim that Harry is a British subject. This claim "can be defended by appeal to the information that he was born in Bermuda, for this datum lends support to our conclusion on account of the warrants implicit in the British Nationality Acts...". The example is translated into a justification logic default as follows. Given the fact that Harry was born in Bermuda (B), an agent can conclude that Harry is a British subject (S):

$$\delta_1 = \frac{r : B :: (s \cdot r) : S}{(s \cdot r) : S}.$$

The default can be read as follows: "If r is a reason justifying that Harry was born in Bermuda and it is consistent to assume that $(s \cdot r)$ is a reason justifying that Harry is a British subject, then $(s \cdot r)$ is a defeasible reason justifying that Harry is a British subject".

However, if the agent were in possession of the additional information saying that both Harry's parents are aliens (P), the "general authority of the warrant" $s : (B \rightarrow S)$ for the claim S would have to be set aside. This is modeled with the following rule that introduces an undercutting reason:

$$\delta_2 = \frac{t : P :: (u \cdot t) : \neg[s : (B \to S)]}{(u \cdot t) : \neg[s : (B \to S)]}.$$

The consequent can be read as follows: "$(u \cdot t)$ is a defeasible reason denying that the reason s justifies that if Harry was born in Bermuda, then Harry is a British subject". This is a classical argumentation theory example of a defeater that leads to the suspension of the conclusion supported by the reason $(s \cdot r)$. For a default theory $T_1 = (W, D)$ with $W = \{r : B, t : P\}$ and $D = \{\delta_1, \delta_2\}$, the process (δ_1, δ_2) corresponds to such course of reasoning with revised **JT**$_{CS}$ extensions. Notice that the warrant underlying δ_2 can also be questioned in a further course of reasoning. For example, one could find out that one of Harry's parents was settled in Bermuda at the time when he was born, thus reinstating the authority of the warrant used in δ_1.

Notice that in the logic above, the argument $(s \cdot r) : S$ is susceptible to attack due to the fallibility of inference δ_1 that is characteristic for *defeasible* reasoning. For the argument $(s \cdot r) : S$ to be undermined, we consider a wider Toulminian interpretation of the argument that includes the formula $r : B$ as the data for the argument. Since $r : B$ is in the set W, the only possibility to attack $r : B$ is to remove it from W and to thereby undermine $(s \cdot r) : S$. This kind of attack on arguments is studied under the paradigm of *plausible* reasoning (see [31], pp. 59–61 for details on this distinction). In this paradigm, arguments are taken to be susceptible to attack due to the uncertainty of their premises. The aim of the current work is to unify the two paradigms in a single logical formalism.

3   Dynamic Operations for Default Theories: Introducing Undermining Attack

As mentioned above, undermining can be interpreted as an attack on the formulas that are considered to be facts. In our view, undermining is essentially non-inferential because introducing conflicting information that undermines facts cannot be done with the use of warrants.[5] For a default theory, these facts are represented by the set of justification logic formulas W and, in constructing a defeasible argument, such formulas can be prerequisites of default rules. A plausible interpretation of undermining defeaters would be that they propose alternative states of facts which ground further reasoning steps. To be able to incorporate factual changes, we need methods based on belief revision methods. Our selection of the belief-revision operations follows the way in which default theories are defined—since the set of facts W is typically finite, it is natural to use operators for sets that do not require closure. Therefore, our choice is to make use of base revision operators [20] instead of the AGM operators [1].

---

[5] The non-inferential view of information change is also relevant for human interaction. As Hlobil [21] argues, we can believe by accepting testimonies, but we cannot make inferences by merely accepting testimony. Two testimonies that contradict each other are to be, *ceteris paribus*, equally treated and the acceptance of new information is not the same process as inferentially extending the existing (incomplete) information.

To model changes to default theories, we will use the capacity of default logic to represent two levels of information certainty. The top-level of information certainty is represented by formulas that are included in all extensions. Typical examples of such formulas are those contained in a set of facts W. The lower-level of information certainty is represented by formulas whose status is contingent on whether it becomes defeated by other available information. Such formulas are typically consequents of default rules. Our goal is to use the two levels and to define dynamic operators that can bring about the changes that fix whether a formula is included in or excluded from all extensions, but also to define *non-prioritizing* operations that leave the status of a formula undecided.

To be able to model the dynamics at the two levels of information certainty, we extend the above defined default theories with defaults without warrants, which correspond to Reiter's supernormal defaults, but (possibly) containing justification assertions:

$$\delta = \frac{\top :: F}{F}.$$

Standard default rules with justification assertions encode inferential steps supported by warrants. In contrast to inferential steps, supernormal defaults will be used to represent non-inferential, information-changing actions in which an agent accepts that a formula can be included in (at least) one extension. We will extend sets of defaults with supernormal defaults whenever we represent introducing uncertain information to a theory T or relegate information from W to the status of uncertain information.

Why would we want to make changes only to the lower-level of information certainty or alter a default theory at the level of some, instead of all extensions? Sometimes, an agent has doubts with respect to whether it is safe to include some information or not and, analogously, whether it is safe to remove some information or not. In the standard base revision approach to modeling information change, incoming information is always prioritized over the existing one, which is ensured by the success postulate. Consider again the example of the agent reasoning about Harry's eligibility for British nationality. It is possible that, according to the census record, Harry was born in Bermuda and, according to the military record, he was born outside Bermuda. The fact that the agent first collected the census record data and then collected the military record data cannot justify the prioritization of the newly acquired information. If the agent does not know which information source is reliable, the order of data input is irrelevant. In these cases, default logic can avoid the "naive" priority ascription by the use of multiple extensions. The rest of this section gives a solution to the problem of non-prioritized change of default theories, along with the more standard prioritized change. In a case of non-prioritized change, the corresponding dynamic operator uses supernormal defaults with an aim to alter the lower-level of information certainty of a default theory. On our interpretation of undermining attacks, whether undermining fully or partially realizes its defeating potential depends on whether the new information is prioritized or not.

3.1   Default Theory Expansion

The first kind of change we want to consider corresponds to learning new information. For example, adding a formula F to a set of facts W can be based on the information provided by some information channel. The formal operation that naively adds new information without checking the joint consistency of the resulting set of beliefs is called "expansion":

Definition 10 (Expansion). *For a default theory $T = (W, D)$ and a formula F, $T_F^+ = (W_F^+, D)$ is the expansion of the default theory T, where W_F^+ is the base expansion of the set W such that $W_F^+ = W \cup \{F\}$.*

If the added information results in an inconsistent set W_F^+, any definable **JT$_{CS}$** extension will be inconsistent. Notice that default theory expansion can already cause non-monotonic behaviour on the level of default theory extensions. For example, if the added formula is a prerequisite for a default rule with an undercutter for some other default consequent, the new information can result in removing elements from **JT$_{CS}$** extensions of T.

An agent can approach accepting incoming information more cautiously. If the agent accepts new information as a plausible premise, but hesitates to consider it a fact, the change is made to the set of default rules:

Definition 11 (Conservative Expansion). *For a default theory $T = (W, D)$ and a formula F, $T_F^\times = (W, D \cup \{\delta_F\})$ is the conservative expansion of the default theory T with F, where $\delta_F = \frac{T \,::\, F}{F}$.*

Notice that the operation $\times$ opens up a possibility that the formula F is included in all extensions, but it can also be excluded from all extensions. For example, if $\neg F$ is contained in $Th^{JT_{CS}}(W)$, then δ_F is not applicable. The following statements characterize the introduced operators.[6]

Proposition 12. *For a default theory $T = (W, D)$ with unwarranted default rules and a **JT$_{CS}$** formula F it holds that*

(a) *If F is not a contradiction, then F is contained in each **JT$_{CS}$** extension of the theory T_F^+.*
(b) *If F is not a contradiction and if $\neg F$ is not contained in any **JT$_{CS}$** extension of T, then F is contained in each **JT$_{CS}$** extension of the theory $T_F^\times$.*
(c) *If W is not inconsistent and if $\neg F$ is contained in $Th^{JT_{CS}}(W)$, then F is not contained in any **JT$_{CS}$** extension of the theory $T_F^\times$.*

3.2   Default Theory Contraction

How does an agent give up on some information, if the information is proven to be unreliable? We will again differentiate between two strategies of giving up on information or *contracting* default theories: one of them aims to remove a

---

[6] Proving Proposition 12 is straightforward. Details are omitted due to space limitations.

formula from all extensions and another leaves the possibility that extensions still contain the formula. One problem we face in removing a formula from all theory extensions is that the base contraction of a set of facts is necessary, but not sufficient to secure that the formula will not be reintroduced by the application of a default rule. To illustrate the need for such operation, consider that changes in information may cause that a certain source of justification t is denied its reliability as a reason for some formula F.

To deal with this problem, we propose to put constraints on the application of default rules. The aim of constraints on application is to prevent an unwanted formula F to become a part of any default theory extension. Such "application-constrained" default theories can be defined for any default theory for which the closure of a set of facts W does not entail F:

Definition 13 (Application-Constrained Default Theory). *For a finite set of* **JT$_{\mathbf{CS}}$** *formulas W such that $\neg F \notin Th^{JTcs}(W)$, a countable set of default rules D and a finite set of* **JT$_{\mathbf{CS}}$** *formulas $C = W \cup \{F\}$, an application-constrained default theory $_{[F]}T$ is defined as a triple (W, D, C) such that:*

$$In(\Pi) = Th^{JTcs}(C \cup \{cons(\delta) \mid \delta \ occurs \ in \ \Pi\}).$$

For any application-constrained default theory, the expansion operation $(_{[F]}T)_G^+$ and the conservative expansion operation $(_{[F]}T)_G^\times$ are both defined analogously to the corresponding default theory operations, with the following two convention in effect. Firstly, if a formula $F \in Th^{JTcs}(W \cup \{G\})$, then $(_{[F]}T)_G^+ = (_{[\emptyset]}T)_G^+ = T_G^+$, where $(_{[\emptyset]}T)_G^+$ is an application-constrained default theory for which $C = W$. Secondly, if a formula $\neg F \in Th^{JTcs}(W \cup \{G\})$, then $(_{[F]}T)_G^+ = T_G^+$.

We first define a contraction operation that aims at removing a formula at the level of a whole default theory. The operation corresponds to the action of removing information when an agent is confident that the information is not reliable. To achieve this in a default theory, a formula has to be removed from the set of facts by a base contraction and its reintroduction should be prevented. In the definition of contraction, *remainder sets* will be used: for any set of **JT$_{\mathbf{CS}}$** formulas Γ and a formula F, the remainder set $\Gamma \perp F$ is defined as the set of maximal subsets of Γ that do not entail F.

Definition 14 (Contraction). *For a default theory $T = (W, D)$ and a formula F, the application-constrained theory $_{[\neg F]}T_F^- = (W_F^-, D \cup D_{!F}, C)$ is the contraction of the default theory T by F, where*

1. *W_F^- is the (full) meet contraction of the set W such that $W_F^- = \bigcap(W \perp F)$,*
2. *$D_{!F} = \{\delta_G \mid \delta_G = \frac{\top :: G}{G} \ for \ every \ G \in W \setminus \bigcap(W \perp F)\}$ and*
3. *$C = W_F^- \cup \{\neg F\}$.*

Notice that an application-constrained default theory $_{[\neg F]}T_F^-$ is definable for any theory T since, due to condition 1, the formula F cannot be an element of the set $Th^{JTcs}(W_F^-)$.

The combination of the constraint set C and the set of default rules $D_{!F}$ provides a balanced solution for avoiding extremely cautious and extremely incautious behavior. Since the set of formulas $W \perp F$ usually contains many elements, theory contraction operations need to include a procedure of selecting the formulas that can be kept after contracting by F. It is difficult to define such procedures in a principled and intuitively plausible way. In default theory contraction, we do not need to force selection by a function. Instead, the choice of formulas selected upon contraction depends on the type of extension that is being computed. For example, a $\mathbf{JT_{CS}}$ preferred extension corresponds to the idea of maxichoice contraction, while $\mathbf{JT_{CS}}$ extension corresponds to the idea of full meet contraction ([20], pp. 12–13).

Using again the two-leveled perspective on changing default theories, we can define a more conservative way of giving up a belief. In conservative contraction, agents are reluctant to entirely give up on some information, but the information is no longer considered to be a fact. To relegate the status of a formula in such a way within a default theory, the formula is removed from the set of facts and then reintroduced through application of a supernormal default rule.

Definition 15 (Conservative Contraction). *For a default theory* $T = (W, D)$ *and a formula* F, $T_F^{\dot{-}} = (W_F^{-}, D \cup D_{!F})$ *is the conservative contraction of the default theory* T *by* F, *where*

1. W_F^{-} *is the (full) meet contraction of the set* W *such that* $W_F^{-} = \bigcap(W \perp F)$ *and*
2. $D_{!F} = \{\delta_G \mid \delta_G = \frac{\top :: G}{G} \text{ for every } G \in W \setminus \bigcap(W \perp F)\}.$

Clearly, $_{[\neg F]}T_F^{\dot{-}}$ and $_{[\neg F]}T_F^{-}$ are defining the same application-constrained default theory.

Conservative contraction is an open-ended operation in the sense that it does not preclude the possibility of reintroducing a formula F in an extension through a default rule application. Furthermore, it leaves open the possibility that F occurs in all extensions of the resulting default theory.[7] In ([4], p. 1149), a different approach has been taken. Namely, that of securing that there is at least one extension added that does not contain the formula removed from the set of facts. In our view, it is unnecessary to have such an operation. If some formula is not regarded to be a fact, but it is still plausible that the formula is true, accepting it as the only available information might be the only reasonable action. Instead of "forcing" an extension without the formula, conservative contraction enables the possibility of an extension without the formula. If there is no support for the contrary statement whatsoever, an agent might still need to hold on to the only available information. The following statement immediately follows from conditions 1 and 3 of Definition 14.

---

[7] Analogously, conservative expansion might not guarantee that there will be any extension containing a formula F, after a default theory has been conservatively expanded with F.

Proposition 16. *For a default theory $T = (W, D)$ with unwarranted default rules and a non-tautological $\mathbf{JT_{CS}}$ formula F, it holds that F is not contained in any $\mathbf{JT_{CS}}$ extension of the theory $_{[\neg F]}T_F^-$.*

3.3 Default Theory Revision

The task of adding new information to the set of facts by the expansion operation (Definition 10) can lead to an inconsistent set of facts. A more realistic dynamic operator for adding information needs to specify a process by which an agent adds information inconsistent with W without being committed to an inconsistent set of facts. One possible way is to only add information via the conservative expansion operation (Definition 11), but this comes with an obvious flaw: an agent is not able to confidently replace an old, unreliable piece of information with a new, reliable one. This is one of the motivations to define a default theory revision operator that not only adds a formula, but also removes inconsistent formulas at one of the two levels of the default theory.

A revision operation can be defined from a combination of the expansion and contraction operations.[8] In our approach, we will follow the traditional arrangement of the operations ([20], p. 203), namely, removing formulas will precede adding a formula. This kind of revision operations in which contraction is followed by expansion are called "internal revision" operators and we define the following four internal revision operators for each combination of the operations.

Definition 17 (Revision Operators). *For a default theory $T = (W, D)$ and a formula F, (internal) revision operators for T are defined as follows:*

1. $T_F^{\mp} = (_{[F]}T_{\neg F}^-)_F^+$
2. $T_F^{\stackrel{-}{\times}} = (_{[F]}T_{\neg F}^-)_F^{\times}$
3. $T_F^{\dotplus} = (T_{\neg F}^{\div})_F^+$
4. $T_F^{*} = (T_{\neg F}^{\div})_F^{\times}$

The variety of possible revision operators raises the question about what kinds of revision strategies they represent.[9]

---

[8] If we were to exhaust all possible combinations, eight revision operators could be defined. Note that the revision operation symbols we use below reflect the composition of the introduced revision operations that are defined in terms of contraction and expansion variants. The symbols are not intended to be in continuity with the standard usage of revision operation symbols.

[9] Note that the second output theory $(_{[F]}T_{\neg F}^-)_F^{\times}$ of Definition 17 is an application-constrained default theory $((_{[F]}T_{\neg F}^-)_F^+$ is, by our convention, a default theory after F has been added to the set of facts). The fact that $(_{[F]}T_{\neg F}^-)_F^{\times}$ is an application-constrained theory might cause problems if we want to make our operators *global*, rather than *local*, and enable iterated revision. A solution to this problem would be to allow for iterated contraction and generalize the contraction operation to application-constrained theories. This could be done if we allow that an application-constrained theory $_{[F]}T$ can be further constrained by a formula G. We leave the details of developing iterated variants of the present operators for the future work.

We can show that the four operations amount to two strategies. Again, as in the cases of expansion and contraction, one strategy is meant to revise confidently and the other strategy more conservatively. The key to show this is to prove that the operations $\mp$, $\divideontimes$ and $\dotplus$ give equivalent extensions in revising a default theory with some formula F. This is the result stated in the (a) clause of Proposition 18.

Proposition 18. *For a default theory $T = (W, D)$ with unwarranted default rules and a* $\mathbf{JT_{CS}}$ *formula F, it holds that*

(a) *If F is not a contradiction, then F is contained in all $\mathbf{JT_{CS}}$ extensions of the theories $T_F^{\mp}, T_F^{\divideontimes}$ and $T_F^{\dotplus}$.*

(b) *If F is not a contradiction, then F is contained in at least one $\mathbf{JT_{CS}}$ extension of the theory $T_F^{\divideontimes}$.*

(c) *If F is not a contradiction, then $\neg F$ is not contained in all $\mathbf{JT_{CS}}$ extensions of $T_F^{\divideontimes}$.*

Proof. To prove that (a) holds, consider the three revision operators $\mp$, $\divideontimes$ and $\dotplus$ and the resulting theories $T_F^{\mp}$, $T_F^{\divideontimes}$ and $T_F^{\dotplus}$. For the case of the default theory $T_F^{\mp}$, it follows from Proposition 16 that $\neg F$ is not contained in any $\mathbf{JT_{CS}}$ extension of $_{[F]}T_{\neg F}^{-}$. By Proposition 12 (a), F is contained in each $\mathbf{JT_{CS}}$ extensions of $(_{[F]}T_{\neg F}^{-})_F^{\dotplus}$.

For the case of the default theory $T_F^{\divideontimes}$, it follows from Proposition 16 that $\neg F$ is not contained in any $\mathbf{JT_{CS}}$ extension of $_{[F]}T_{\neg F}^{-}$. Moreover, the constraint set C contains the formula F, which means that the default rule $\frac{\top : : F}{F}$ is applicable to any $\mathbf{JT_{CS}}$ extension of the conservative expansion $(_{[F]}T_{\neg F}^{-})_F^{\divideontimes}$ of the theory $_{[F]}T_{\neg F}^{-}$. Therefore, F is contained in each $\mathbf{JT_{CS}}$ extensions of $(_{[F]}T_{\neg F}^{-})_F^{\times}$.

For the case of the default theory $T_F^{\dotplus}$, consider that the base contraction of W ensures that $\neg F$ cannot be contained in the set of facts $W_{\neg F}^{-}$ of the default theory $T_{\neg F}^{\div}$, but $\neg F$ can still be reintroduced by applying the defaults from $D_{!\neg F}$. However, after expanding the theory $T_{\neg F}^{\div}$ by F, the inclusion of the formula $\neg F$ into any $\mathbf{JT_{CS}}$ extension of the theory $T_F^{\dotplus}$ is blocked and, by Proposition 12(a), F is in contained each $\mathbf{JT_{CS}}$ extensions of $T_F^{\dotplus}$.

To prove that (b) holds, consider that the base contraction of W ensures that $\neg F$ cannot be contained in the set of facts $W_{\neg F}^{-}$ for the conservative contraction $T_{\neg F}^{\div}$. This means that, for the conservative expansion $(T_{\neg F}^{\div})_F^{\divideontimes}$, it holds that the default rule $\frac{\top : : F}{F}$ is applicable to $Th^{JT_{CS}}(W)$ and, therefore, contained in at least one $\mathbf{JT_{CS}}$ extension of $T_F^{\divideontimes}$.

To prove (c), consider that after the base contraction of W by $\neg F$, $\mathbf{JT_{CS}}$ extensions of $T_{\neg F}^{\div}$ are $\mathbf{JT_{CS}}$ consistent. Since we also know that, after the conservative expansion $(T_{\neg F}^{\div})_F^{\times}$, (b) holds, then (c) holds.

To show the equivalence of the operators $\mp$, $\divideontimes$ and $\dotplus$, we first say that for any σ-extension, where

$$\sigma \in \{\mathbf{JT_{CS}} \text{ admissible}, \mathbf{JT_{CS}} \text{ complete}, \mathbf{JT_{CS}} \text{ grounded}, \mathbf{JT_{CS}} \text{ preferred}, \mathbf{JT_{CS}} \text{ stable}\},$$

$\sigma(T)$ is the set of all σ-extensions for a theory T. Then we prove that for any default theory T, the default theories $T_F^{\mp}, T_F^{\times}$ and $T_F^{\dotplus}$ realize the same set of extensions under any $\mathbf{JT_{CS}}$ extension-based semantics for default theories. The following result is obtainable from Proposition 18(a) together with the fact that none of the three operators $\mp$, $\divideontimes$ and $\dotplus$ change the status of formulas that do not take part in F-implying sets:

Theorem 19. *For any default theory $T = (W, D)$, a $\mathbf{JT_{CS}}$ formula F and the (internal) revision operators $\mp$, $\divideontimes$ and $\dotplus$, it holds that $\sigma(T_F^{\mp}) = \sigma(T_F^{\times}) = \sigma(T_F^{\dotplus})$.*

Intuitively, the three operations represent a type of revision in which an agent confidently includes new and possibly inconsistent information into all $\mathbf{JT_{CS}}$ extensions. Another option specified by the operator $\divideontimes$ is to accept the new information in some extensions while maintaining the old information in other extensions. The revision operators comply to the two-leveled view of default semantics: the first three revision operators of Definition 17 fix the status of a revision at the level of a default theory as a whole, while the last revision operator targets at modifying only some extensions. Any of the three operations $T_F^{\mp}, T_F^{\times}$ and $T_F^{\dotplus}$ will be referred to as the *Revision* of T with F and the operation $T_F^{\divideontimes}$ will be referred to as the *Conservative Revision* of T with F.

3.4   The Notion of Undermining

Finally, we are now able to say in what way the dynamic operations connect to the notion of undermining defeat. It was mentioned in the Introduction that by undermining we understand the attack whereby argument premises are being questioned. This intuition can now be cashed out by using those dynamic operators for default theories that involve contracting a default theory.

Definition 20 (Undermining). *For a default theory $T = (W, D)$ and a $\mathbf{JT_{CS}}$ formula F such that $F \in W$ and $F = pre(\delta)$ for some $\delta \in D, F$ is undermined iff W is contracted by F by applying any of the following operations to T:*

1. $_{[\neg F]}T_F^-$ *(Contraction)*
2. $T_F^{\dotdiv}$ *(Conservative Contraction)*
3. $T_G^{\mp}, T_G^{\times}$ *or* $T_G^{\dotplus}$ *for $\mathbf{JT_{CS}}$ inconsistent formulas F and G (Revision)*
4. $T_G^{\divideontimes}$ *for $\mathbf{JT_{CS}}$ inconsistent formulas F and G (Conservative Revision).*

Notice that there is no requirement on the structure of F. However, each meaningful undermining targets justification assertions because W cannot be successfully contracted by a tautology and justification assertions are the only other type of formula occurring as a default prerequisite. Not every attack on the premises results in confidently revising the set of facts. It is possible that undermining leaves an agent undecided as to whether newly acquired information or older information should be prioritized.

Starting from the theory T_1 defined in Sect. 2, we can give a formalized undermining example from the beginning of this section to show the difference between inferential and non-inferential ways of information acquisition. Recall that the agent started to reason from the information that Harry was born in Bermuda. This piece of information is represented in the set of facts W with the formula $r : B$, where r can now be taken to reflect the source of information as, e.g., data from census records. However, if the information based on military records says that Harry was born outside Bermuda, and having no means to resolve this conflict of information, the theory T_1 needs to be conservatively revised. The theory $T^*_{1_{v:\neg B}}$ is the revision of T_1 with the formula $v : \neg B$, where v reflects the new source of information for the claim that Harry was not born in Bermuda.

To see this revision in more detail, recall that the theory $T_1 = (W, D)$ consisted of the set of facts $W = \{r : B, t : P\}$ and the set of defaults $D = \{\delta_1, \delta_2\}$. The first dynamic operation in revising with $v : \neg B$ is contracting the theory by $\neg v : \neg B$. The resulting theory $T^{\div}_{1_{\neg v : \neg B}} = (W^-_{\neg v : \neg B}, D \cup D_{!\neg v : \neg B})$ consists of the set of facts $W^-_{\neg v : \neg B} = \{t : P\}$ and the set of defaults D extended with the default $\delta_{r:B} = \frac{\top :: r : B}{r : B}$.

Finally, the agent conservatively expands the theory $T^{\div}_{1_{\neg v : \neg B}}$ with the information that Harry was not born in Bermuda. The new default theory is defined as $T^*_{1_{v:\neg B}} = (T^{\div}_{1_{\neg v:\neg B}})^{\times}_{v:\neg B}$. The change of the theory after conservative expansion with $v : \neg B$ amounts to adding the new default rule $\delta_{v:\neg B} = \frac{\top :: v : \neg B}{v : \neg B}$, which means that the new set of defaults is $D \cup D_{!\neg v:\neg B} \cup \{\delta_{v:\neg B}\}$. The revised theory $T^*_{1_{v:\neg B}}$ changes the default processes in which the agent reasons about Harry's nationality and, eventually, changes the structure of acceptable reasons by changing the way in which $\mathbf{JT_{CS}}$ extensions are computed.

4   Related Work and Conclusions

As mentioned in the Introduction, our approach to structured argumentation dynamics builds on similar ideas as the approach to the dynamics of standard default theories from [4]. Antoniou's approach significantly differs from ours in the way he treats those changes that add or remove a formula at the level of some, but not necessarily all extensions. Unlike our conservative expansion and conservative contraction, none of Antoniou's operations leaves the inclusion status of a formula undecided. For instance, to secure that a formula is not contained in at least one extension, Antoniou ([4], p. 1149) adds a new extension where introducing the formula is blocked by adding either a new atom or its negation to any default, dependent on whether they are allowed to be in a same extension or not.

Some approaches to default reasoning such as [23] and [24] represent the idea of defaults in dynamic epistemic logic. The main focus of [23] is to embed supernormal defaults in a multi-agent modal logic with knowledge, belief and update modalities. The authors show that Reiter's extensions can be represented as a result of consecutive jump actions to default conclusions, but they do not focus on how such extensions are revised due to information changes. In [24], a preference modality is introduced to differ between known and (provisionally) preferred information. A non-monotonic belief revision component consists in changing preferences as a result of obtaining knowledge.

Baltag, Renne and Smets [7,8] and Renne [32] define extensions of justification logic in which agents may acquire new information that defeats the reasons they accepted. The logics combine belief revision and dynamic epistemic logic techniques to model a kind of defeat that seems to correspond to undermining. However, each of the logics assumes prioritizing new information and none of these logics is able to model undercut and rebuttal. Even so, approaches based on dynamic epistemic logic are attractive because they open up a possibility of developing a multi-agent justification logic with defeaters.

We indicated in the Introduction that the work in the area of the dynamics of argumentation frameworks without argument structures is already well-developed. Among the approaches, it is worth mentioning those that follow the belief revision methods applied to Dung's frameworks such as [9] and [12]. In [9], the authors start from a labelling approach to Dung's frameworks and constraints on a framework's output. Their focus is on finding the best way to recover a rational output given a framework and a constraint on its output. For this, they use ordering of conflict-free labellings in a way that the most rational conflict-free labelling is chosen when none of complete labellings respects the constraint. In the work by [12], we find two kinds of revision operators. One of them revises a Dung framework by taking a propositional formula as a means to represent the new information, while the other operation revises an input framework by information in the form of another framework. Both operations give a single output framework respecting a particular type of rankings on extensions.

Finally, our paper contributes to the study of non-prioritized belief revision operations, that is, such operations for which the new information has no special priority due to its novelty [19]. The way in which our operators are defined meaningfully combines resources from both belief revision and default logic. The relation between belief revision and non-monotonic reasoning has long been a matter of discussion [18] among AI researchers. Although it was not our aim to discuss the relation between modeling reasoning with *incomplete information* in default theories and modeling reasoning with *changing information* in belief revision, we showed that our justification logic creates a useful junction for the two approaches.

As a result of connecting the two reasoning paradigms, the logic presented here, we can model both plausible and defeasible reasoning. According to [31], "argumentation models of plausible reasoning locate all fallibility of an argument in its premises, while argumentation models of defeasible reasoning locate all fallibility in its defeasible inferences". To the best of our knowledge, the system presented here is the first logic to combine the two approaches by modeling all the standard notions of defeat in AI: rebuttal, undercut and undermining.

References

1. Alchourrón, C.E., Gärdenfors, P., Makinson, D.: On the logic of theory change: partial meet contraction and revision functions. J. Symbolic Log. **50**(2), 510–530 (1985)
2. Alfano, G., Greco, S., Parisi, F., Simari, G.I., Simari, G.R.: An incremental approach to structured argumentation over dynamic knowledge bases. In: Thielscher, M., Toni, F., Wolter, F. (eds.) Sixteenth International Conference on Principles of Knowledge Representation and Reasoning, KR 2018 (2018)
3. Antoniou, G.: Nonmonotonic Reasoning. MIT Press, Cambridge (1997)
4. Antoniou, G.: On the dynamics of default reasoning. Int. J. Intell. Syst. **17**(12), 1143–1155 (2002)
5. Artemov, S.N.: Explicit provability and constructive semantics. Bull. Symbolic Log. **7**, 1–36 (2001)
6. Artemov, S.N., Fitting, M.: Justification Logic: Reasoning with Reasons, Cambridge Tracts in Mathematics, vol. 216. Cambridge University Press, Cambridge (2019)
7. Baltag, A., Renne, B., Smets, S.: The logic of justified belief change, soft evidence and defeasible knowledge. In: Ong, L., de Queiroz, R. (eds.) WoLLIC 2012. LNCS, vol. 7456, pp. 168–190. Springer, Heidelberg (2012). https://doi.org/10.1007/978-3-642-32621-9_13
8. Baltag, A., Renne, B., Smets, S.: The logic of justified belief, explicit knowledge, and conclusive evidence. Ann. Pure Appl. Log. **165**(1), 49–81 (2014)
9. Booth, R., Kaci, S., Rienstra, T., van der Torre, L.: A logical theory about dynamics in abstract argumentation. In: Liu, W., Subrahmanian, V.S., Wijsen, J. (eds.) SUM 2013. LNCS (LNAI), vol. 8078, pp. 148–161. Springer, Heidelberg (2013). https://doi.org/10.1007/978-3-642-40381-1_12
10. Brezhnev, V.: On the logic of proofs. In: Striegnitz, K. (ed.) Proceedings of the Sixth ESSLLI Student Session, Helsinki, pp. 35–46 (2001)
11. Coste-Marquis, S., Konieczny, S., Mailly, J.G., Marquis, P.: On the revision of argumentation systems: minimal change of arguments statuses. In: Baral, C., De Giacomo, G., Eiter, T. (eds.) Fourteenth International Conference on the Principles of Knowledge Representation and Reasoning, KR 2014 (2014)
12. Diller, M., Haret, A., Linsbichler, T., Rümmele, S., Woltran, S.: An extension-based approach to belief revision in abstract argumentation. In: Yang, Q., Wooldridge, M. (eds.) Twenty-Fourth International Joint Conference on Artificial Intelligence, IJCAI 2015 (2015)
13. Doutre, S., Herzig, A., Perrussel, L.: A dynamic logic framework for abstract argumentation. In: Baral, C., De Giacomo, G., Eiter, T. (eds.) Fourteenth International Conference on the Principles of Knowledge Representation and Reasoning, KR 2014 (2014)
14. Dung, P.M.: On the acceptability of arguments and its fundamental role in nonmonotonic reasoning, logic programming and n-person games. Artif. Intell. **77**(2), 321–357 (1995)
15. Dung, P.M., Kowalski, R.A., Toni, F.: Assumption-based argumentation. In: Rahwan, I., Simari, G.R. (eds.) Argumentation in Artificial Intelligence, pp. 199–218. Springer, Boston (2009). https://doi.org/10.1007/978-0-387-98197-0_10
16. van Eemeren, F.H., Garssen, B., Krabbe, E.C.W., Henkemans, A.F.S., Verheij, H.B., Wagemans, J.H.M.: Argumentation and artificial intelligence. In: Handbook of Argumentation Theory, pp. 615–675. Springer, Boston (2014). https://doi.org/10.1007/978-0-387-98197-0

17. Fitting, M.: Justification logics, logics of knowledge, and conservativity. Ann. Math. Artif. Intell. **53**(1–4), 153–167 (2008)
18. Gärdenfors, P.: Belief revision and nonmonotonic logic: two sides of the same coin? In: van Eijck, J. (ed.) JELIA 1990. LNCS, vol. 478, pp. 52–54. Springer, Heidelberg (1991). https://doi.org/10.1007/BFb0018432
19. Hansson, S.O.: A survey of non-prioritized belief revision. Erkenntnis **50**(2–3), 413–427 (1999)
20. Hansson, S.O.: A Textbook of Belief Dynamics: Theory Change and Database Updating. Kluwer Academic Publishers, Dordrecht (1999)
21. Hlobil, U.: We cannot infer by accepting testimony. Philos. Stud. 1–10 (2018). https://doi.org/10.1007/s11098-018-1142-3
22. Kuznets, R., Studer, T.: Logics of Proofs and Justifications. College Publications, Wenham (2019)
23. van Linder, B., van der Hoek, W., Meyer, J.J.C.: The dynamics of default reasoning. Data Knowl. Eng. **3**(21), 317–346 (1997)
24. Meyer, J.-J.C., van der Hoek, W.: Non-monotonic reasoning by monotonic means. In: van Eijck, J. (ed.) JELIA 1990. LNCS, vol. 478, pp. 399–411. Springer, Heidelberg (1991). https://doi.org/10.1007/BFb0018455
25. Mkrtychev, A.: Models for the logic of proofs. In: Adian, S., Nerode, A. (eds.) LFCS 1997. LNCS, vol. 1234, pp. 266–275. Springer, Heidelberg (1997). https://doi.org/10.1007/3-540-63045-7_27
26. Modgil, S., Prakken, H.: Resolutions in structured argumentation. In: Verheij, B.H., Szeider, S., Woltran, S. (eds.) Computational Models of Argument: Proceedings of COMMA 2012, pp. 310–321. IOS Press (2012)
27. Pandžić, S.: A logic of default justifications. In: Fermé, E., Villata, S. (eds.) 17th International Workshop on Nonmonotonic Reasoning, NMR 2018, pp. 126–135 (2018)
28. Pandžić, S.: Logic of defeasible argumentation: constructing arguments in justification logic. Unpublished manuscript (2019)
29. Pandžić, S.: Reifying default reasons in justification logic. In: Beierle, C., Ragni, M., Stolzenburg, F., Thimm, M. (eds.) Proceedings of the KI 2019 Workshop on Formal and Cognitive Reasoning, DKB-KIK 2019, CEUR Workshop Proceedings, vol. 2445, pp. 59–70 (2019)
30. Prakken, H.: An abstract framework for argumentation with structured arguments. Argument Comput. **1**(2), 93–124 (2010)
31. Prakken, H.: Historical overview of formal argumentation. IfCoLog J. Log. Their Appl. **4**(8), 2183–2262 (2017)
32. Renne, B.: Multi-agent justification logic: communication and evidence elimination. Synthese **185**(1), 43–82 (2012)
33. de Saint-Cyr, F.D., Bisquert, P., Cayrol, C., Lagasquie-Schiex, M.C.: Argumentation update in YALLA (yet another logic language for argumentation). Int. J. Approximate Reasoning **75**, 57–92 (2016)
34. Toulmin, S.E.: The Uses of Argument. Cambridge University Press, Cambridge (2003)

Logic-Based Approach to Incremental Monitoring and Optimization on Strongly Distributed Data Streams

Elena V. Ravve[✉]

Ort Braude College, Karmiel, Israel
cselena@braude.ac.il

Abstract. In this paper, we systematically adopt logical reduction techniques to monitoring and optimization on distributed data streams. The first technique: Feferman-Vaught reductions, which describe how the queries over a disjoint union of data streams can be computed from queries over the components and queries over the index set. The second one: the syntactically defined translation schemes, which describe possible transformations of data. Combination of these two techniques allows us to consider not only monitoring and optimization on disjoin unions of data streams but rather on much richer compositions. We call them *strongly distributed data streams*. Our approach is applicable to both homogeneous and heterogeneous data streams. While, as a rule, the known approaches provide some approximation of the solution of the original problem, our method derives solutions over the components of a strongly distributed data stream, such that their further proceeding gives a result that is equivalent to the solution of the original problem on the given data stream.

Keywords: Data streams · Incremental · Parallel and distributed monitoring · Logical reductions · Syntactically defined translation schemes

1 Introduction

In our reality, data from all fields of business, industry, communication, academic and private lives are of high volume, distributed, and dynamic. Dynamic character of the data leads us to the situation of dealing with data streams rather than with static databases. As it was shown in particular in [28], the high volume and distributed character of the data make it unfeasible to collect the distinct data streams to a central node for processing.

That is why, stream monitoring and optimization on the distributed case is of great interest. Often, the monitoring problem consists of determining whether the value of a global function, defined on the combination of all streams, crossed a certain threshold; while, the optimization problem consists of optimization of a value of a global function. One wishes to reduce communication by transforming

© Springer Nature Switzerland AG 2020
A. Herzig and J. Kontinen (Eds.): FoIKS 2020, LNCS 12012, pp. 242–262, 2020.
https://doi.org/10.1007/978-3-030-39951-1_15

the global monitoring problem to the testing of local constraints, checked independently at the nodes, or computation of the global minimum or maximum from the local (not necessarily optimal) values, evaluated on the components.

According to Garofalakis, cf. [24]: *Large-scale stream processing applications rely on continuous, event-driven monitoring, that is, real-time tracking of measurements and events, rather than one-shot answers to sporadic queries... Furthermore, the vast majority of these applications are inherently distributed, with several remote monitor sites observing their local, high-speed data streams and exchanging information through a communication network. This distribution of the data naturally implies critical communication constraints that typically prohibit centralizing all the streaming data ... Finally, an important requirement of large-scale event monitoring is the effective support for tracking complex, holistic queries that provide a global view of the data by combining and correlating information across the collection if remote monitor sites.*

1.1 Monitoring and Optimization on Streams: Related Work

Monitoring and optimization on streams over distributed systems has a long history. Babcock and Olston, cf. [1], study a class of queries that continuously report the k largest values obtained from distributed data streams. Olston, Jiang and Widom, cf. [35], present users register continuous queries with precision requirements at the central stream processor, which installs filters at remote data sources. The filters adapt to changing conditions to minimize stream rates while guaranteeing that all continuous queries still receive the updates necessary to provide answers of adequate precision at all times. Already in [5], the authors describe two working stream processing systems, *Aurora** and *Medusa*. These systems were designed in order to explore complementary solutions to architectural challenges facing the design of large-scale distributed stream processing systems.

In [25], techniques for computing small space representations of massive data streams were introduced. These are inspired by traditional Wavelet-based approximations that consist of specific linear projections of the underlying data. General *"sketch"*-based methods for capturing various linear projections are presented. These methods are used in order to provide a pointwise and rangesum estimation of data streams. The key idea in such sketching techniques is to represent a streaming frequency vector v using a much smaller sketch vector that can be easily maintained as the updates incrementally rendering v are streaming by, and provide probabilistic guarantees for the quality of the data approximation. Algorithms of [6] rely on tracking general-purpose randomized sketch summaries of local streams at remote sites along with concise prediction models of local site behavior in order to produce highly communication- and space/time-efficient solutions.

An algorithm that determines whether the norm of the average vector in a distributed system is below a certain threshold, or more generally, whether it lies inside some convex set, is provided in [47]. However, it does not deal with general functions. Functional approximation in a distributed setting is considered in [7], but only lower and upper bounds for vector norm functions are obtained.

Geometric monitoring, cf. [31,45,46], proved useful for constructing local constraints for general functions. Tracking schemes of [23] rely on a novel combination of the geometric method with compact sketch summaries of local data streams, and maintain approximate answers with provable error guarantees, while optimizing space and processing costs at each remote site and communication cost across the network.

It can be easily seen that most of the proposed methods produce a powerful *approximate query* that incorporates several complex analysis queries. Extending the proposed ideas to general, scalable distributed architectures raises several theoretical and practical challenges From a more foundational perspective, there is a need for developing new models and theories for studying the complexity of such continuous distributed computations, cf. [24].

1.2   A Motivating Example: Alternation of Data Streams

Let us consider a toy example, inspired by [44], in order to illustrate our main idea. Assume that we are given two streams: the *First* stream and the *Second* stream. These streams are coded as ordered sequences of their elements, which may be labeled. The first element of a stream is labeled by star ($\star$). These streams are combined in a resulting stream by alternation. Finite constant number of elements of each stream are labeled by $\bowtie$ that is aimed to define the switch points of the streams. The combination of these streams is defined as follows.

Combination rules: The resulting stream is composed from a fragment of the first stream from its first element, labeled by $\star$ till its first element, labeled by $\bowtie$. This element, labeled by the $\bowtie$, is the last element of the first stream that is included into the resulting stream. Then, the resulting stream includes all the elements from the second stream from its first element, labeled by $\star$ until its first element, labeled by $\bowtie$. Then, we switch back to the first stream an so on, see Fig. 1.

Some elements of the streams may be also labeled by $\oplus$ and some elements of the streams are labeled by $\otimes$. Assume that we should compute and analyze a global monitoring or optimization quantitative property on the resulting stream. Consider, for example, a monitoring property that counts the number of pairs, for which the first element is labeled by $\oplus$ and the second element is labeled by $\otimes$. Locally, one monitors one component and transfers locally observed results to the centralized final proceeding.

In order to propagate the computation to the components (two original streams), we observe, that a pair of such elements of the resulting stream is labeled by $\oplus\otimes$ in two cases.

Case 1: Both elements belong to the same local stream. The first condition is translated to computation of the number of $\oplus\otimes$ sequences in the original streams. There are two such pairs on Fig. 1: one in Fragment 1 and one more in Fragment 4. They are marked by two vertical ovals.

Second stream ◇   □ First stream

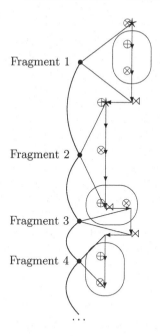

Fragment 1

Fragment 2

Fragment 3

Fragment 4

Fig. 1. Alternation of two data streams.

Case 2: These elements are coming from different local streams. The second condition is translated to computation of the number of pairs of elements x and y from different streams, which satisfy the following properties:

- **Last element of a fragment is labeled by** $\oplus$: x is labeled by both $\oplus$ and $\bowtie$;
- **First element of a fragment is labeled by** $\otimes$: y is labeled by $\otimes$ and either it is also labeled by $\star$ or its immediate predecessor in its local stream is labeled by $\bowtie$.
- **The fragments are succeeding:** x belongs to a fragment that is an immediate predecessor of the fragment of y in the resulting stream.

There exists one[1] pair of elements on Fig. 1, which satisfies all the conditions of the second case: the last element of a fragment is labeled by $\oplus$, the first element of a fragment is labeled by $\otimes$, the fragments are succeeding. The pair is found on the border between Fragment 2 and Fragment 3. This pair is marked by the horizontal oval.

The final calculation sums all the local results: three ovals on Fig. 1.

---

[1] The first element of the *First* stream satisfy condition: **First element of a fragment is labeled by** $\otimes$. However, its fragment does not satisfy condition: **The fragments are succeeding**.

From the consideration, we observe:

1. The propagation of the evaluation of the global monitoring property to the evaluations of the properties on the local fragments and the order of the fragments is exact: the final calculation on the results of the evaluation of the derived local properties is equivalent to the evaluation of the global one.
2. The derived properties do not depend upon particular streams but rather upon the combination rules. For all particular streams, which are combined, according to the predefined rules, the same derived properties must be checked.
3. The global and the derived properties may coincide (**Case 1**).
4. The global and the derived properties might not coincide (**Case 2**).
5. The derived properties may be checked on the original local data streams: **Last element of a fragment is labeled by** $\oplus$ or **First element of a fragment is labeled by** $\otimes$.
6. The derived properties may be checked on the index structure of the fragments: **The fragments are succeeding**.

Fortunately, the above properties are logically expressible in *Weighted Monadic Second Order Logic* (*WMSOL*), introduced by Droste and Gastin in [11,12] and further developed in [14]. More surprisingly, such reductions of computation of a global property on a resulting data stream to the derived properties on components may be sometimes converted to an algorithm. In this contribution, we prove that such algorithm exists for any *strongly distributed data stream*.

1.3   Novelty of the Contribution: Logical Tools

In this contribution, we propose a logic-based approach to incremental monitoring and optimization on data streams that is new in the context. The approach allows unification of the distributed and parallel computation and communication as well as significant reduction of the communication load. We do not use any approximation, our results are exact. Our approach is applicable to monitoring and optimization of any function (property) expressible in lots of extensions of *First Order Logic* (*FOL*). We systematically exploit logical reduction techniques to handling of big distributed data streams.

Logical reduction techniques come in two flavours. The first one: Feferman-Vaught reductions, which are applied in situations of distributed data. The reduction describes how the evaluation of properties over a distributed data can be computed from evaluation of properties over the components and evaluation of properties over the index set. Feferman-Vaught, cf. [21], reductions were first introduced in model theory. Their use in computer science was seemingly first suggested in [39] in the context of formal verification and model checking, and in [8] in the context of graph algorithms for graphs of bounded clique width. Incremental verification and coverage analysis of strongly distributed software systems is discussed in Analytic Methods in Systems and Software Testing [41]. For the algorithmic uses of the Feferman-Vaught Theorem see also [32].

The reductions are somehow analogue to *local violations* of [28]. One of the most recent contribution in the field is due to [17].

The second one: the syntactically defined translation schemes, known also in model theory as interpretations, cf. [27] for the case of finite models. They describe transformations of data and evaluation of properties. They give rise to two induced maps, translations and transductions. Transductions describe the induced transformation of data instances and the translations describe the induced transformations of evaluation of properties. The fundamental property of translation schemes describes how to compute transformed evaluation of properties in the same way Leibniz' Theorem describes how to compute transformed integrals. The fundamental property has a long history, but was first properly stated by Rabin, cf. [36]. One of the recent uses of interpretations in the field of database theory may be found in [26]. However, for our best knowledge, the technique has not be used yet in the context of monitoring and optimization of distributed data streams.

1.4   Incremental Monitoring and Optimization on Strongly Distributed Data Streams

Combination and adaptation of these techniques allow us to introduce the notion of *strongly distributed data streams* in the context of monitoring and optimization. For such data streams, we extend and generalize the known techniques of incremental monitoring and optimization. For the strongly distributed data streams, we derive the following main steps of monitoring and optimization:

1. Computation of the derived properties on the components;
2. Computation of the derived properties on the index;
3. Final computation, based on the computations above.

Our use of the logical reduction techniques in the field of the incremental monitoring and optimization on data streams is new. The notion of *strongly distributed systems* was introduced in [10] in the context of information systems. Then, it was adopted to the database theory in [40], and to the incremental reasoning and multi-agent systems in [37, 43]. Complexity analysis of the strongly distributed systems goes back to [38].

Our approach shows how evaluation of a global monitoring and optimization property, expressed as formula ϕ on the distributed data streams, may be syntactically reduced to incremental evaluations of algorithmically derived properties on components, the index set I and some post-proceeding. From our main Theorem 3, we derive a method for evaluation of ϕ on strongly distributed data streams, which proceeds as follows:

Preprocessing: Given ϕ and translation scheme Φ that describes combination of the local streams, but not the distributed data streams themselves; we construct a sequence of formulae $\psi_{i,j}$ and an evaluation function $F_{\Phi,\phi}$.

Incremental Computation: We compute the local values $\psi_{i,j}$ on each local component.

Final Solution: Theorem 3 now states that evaluation of ϕ of strongly distributed data streams may be algorithmically computed from $\psi_{i,j}$, using $F_{\Phi,\phi}$.

We emphasize the following:

Communication load: Note that the only values transferred between different computational components are the values of $\psi_{i,j}$, that significantly reduces the communication load.

Confidentiality: All meaningful information is still stored in the corresponding locations in the secure way and is not transferred. The transferred values $\psi_{i,j}$, as a rule, are meaningless without the knowledge about the final proceeding.

1.5   Structure of the Contribution

The paper is structured in the following way. Section 2 provides general logical background and shows how quantitative queries may be expressible as logical formulae. Section 3 introduces the notion of abstract translation schemes. Section 4 is the core section of the paper, where we define the notion of strongly distributed data streams and prove our main theorems. Using these theorems, we propose the general approach for incremental monitoring and optimization on the strongly distributed data streams. Section 5 summarizes the paper. We assume that the reader has general logical background, cf. [19]. The precise definitions of *Weighted Finite Automata (WFA)* and *Weighted Monadic Second Order Logic (WMSOL)* are rather provided in [14].

2   Quantitative Querying

Classical *Second Order Logic (SOL)* is like *First Order Logic*, but allows also variables and quantification over relation variables of various but fixed arities. The sublogic of *SOL*, where relation variables are restricted to be unary, is called *Monadic SOL(MSOL)*. The meaning function of formulae is explained for arbitrary τ–structures, where τ is the vocabulary, i.e., a finite set of relation and constant symbols.

Droste and Gastin in [11,12] generalized the fundamental theorems of Büchi's, cf. [4], and Elgot's, cf. [20], to some kinds of quantitative settings. Moreover, a weighted version of *MSOL* was introduced and it was proven that, for commutative semirings, the behaviours of weighted automata are precisely the formal power series definable with the weighted logic. In this contribution, we restrict ourselves to monitoring and optimization of quantitative properties of data streams, which are modeled as weighted words. The quantitative properties are expressed as formulae of *WMSOL* and evaluation of the properties are executed using *WFA*.

2.1 Weighted Monadic Second Order Logic

In this section, we introduce *WMSOL* on words. Extensions of *MSOL* to the case of investigation of quantitative properties of trees may be found in [13] that is based on [2].

Given semiring $\mathcal{K}$ and alphabet A. For each $a \in A$, P_a denotes a unary predicate symbol. The syntax of *WMSOL* over $\mathcal{K}$ is defined by basis

$$\beta ::= P_a(x) \mid \neg P_a(x) \mid x \leq y \mid \neg(x \leq y) \mid x \in X$$

and closure, where β_1 and β_2 are boolean formulae:

$$\beta ::= \neg(x \in X) \mid \beta_1 \wedge \beta_2 \mid \forall x.\beta_1 \mid \forall X.\beta_1.$$

Moreover,

$$\varphi ::= k \mid \beta \mid \phi \wedge \psi \mid \phi \vee \psi \mid \exists x.\phi \mid \exists X.\phi \mid \forall x.\phi \mid \forall X.\phi,$$

where $k \in K$ and $a \in A$. $WMSO(\mathcal{K}, A)$ denotes the collection of all such weighted formulae φ. We call β's *boolean formulae*. We call φ's *weighted formulae*. The syntax of *WMSOL* over $\mathcal{K}$ and A may be easily extended to the case of multiple $\leq$ relations.

Let $Free(\varphi)$ denote the set of all free variables of φ. Let $w = a_1 a_2 \ldots a_n \in A^\star$ with $a_i \in A$ and $w(i) = a_i (1 \leq i \leq n)$. The length of w is $|w| = n$. Let $\mathcal{V}$ be a finite set of first-order or second-order variables. A $(\mathcal{V}, w)$-assignment σ is a function mapping first-order variables in $\mathcal{V}$ to elements of $\{1, \ldots, |w|\}$ and second-order variables in $\mathcal{V}$ to subsets of $\{1, \ldots, |w|\}$. $\sigma[x \to i]$ is the $(\mathcal{V} \cup \{x\}, w)$ assignment that assigns x to i and acts like σ on all other variables. Similarly, $\sigma[X \to I]$ is defined for $I \subseteq \{1, \ldots, |w|\}$.

Let $\varphi \in WMSOL(\mathcal{K}, A)$ and $\mathcal{V}$ be a finite set of variables containing $Free(\varphi)$. The semantics of φ is a formal power series $[[\varphi]]_\mathcal{V} \in K\langle\langle A_\mathcal{V}^\star \rangle\rangle$. Let $(w, \sigma) \in A_\mathcal{V}^\star$. If σ is not a valid $\mathcal{V}$-assignment, then we put $[[\varphi]]_\mathcal{V}(w, \sigma) = 0$. Otherwise, we define $[[\varphi]]_\mathcal{V}(w, \sigma) \in K$ as follows:

- $[[k]]_\mathcal{V}(w, \sigma) = k$
- $[[P_a(x)]]_\mathcal{V}(w, \sigma) = \begin{cases} 1 \text{ if } w(\sigma(x)) = a \\ 0 \text{ otherwise} \end{cases}$
- $[[x \leq y]]_\mathcal{V}(w, \sigma) = \begin{cases} 1 \text{ if } \sigma(x) \leq \sigma(y) \\ 0 \text{ otherwise} \end{cases}$
- $[[x \in X]]_\mathcal{V}(w, \sigma) = \begin{cases} 1 \text{ if } \sigma(x) \in \sigma(X) \\ 0 \text{ otherwise} \end{cases}$
- If φ is of the form $P_a(x), (x \leq y)$ or $(x \in X)$ then
  $[[\neg\varphi]]_\mathcal{V}(w, \sigma) = \begin{cases} 1 \text{ if } [[\neg\varphi]]_\mathcal{V}(w, \sigma) = 0 \\ 0 \text{ if } [[\neg\varphi]]_\mathcal{V}(w, \sigma) = 1 \end{cases}$
- $[[\varphi \wedge \psi]]_\mathcal{V}(w, \sigma) = [[\varphi]]_\mathcal{V}(w, \sigma) \cdot [[\psi]]_\mathcal{V}(w, \sigma)$
- $[[\varphi \vee \psi]]_\mathcal{V}(w, \sigma) = [[\varphi]]_\mathcal{V}(w, \sigma) + [[\psi]]_\mathcal{V}(w, \sigma)$
- $[[\exists x.\varphi]]_\mathcal{V}(w, \sigma) = \Sigma_{1 \leq i \leq |w|}[[\varphi]]_{\mathcal{V} \cup \{x\}}(w, \sigma[x \to i])$

- $[[\exists X.\varphi]]_V(w,\sigma) =$
  $\Sigma_{I\in\{1,\dots,|w|\}}[[\varphi]]_{V\cup\{x\}}(w,\sigma[X \to I])$
- $[[\forall x.\varphi]]_V(w,\sigma) = \Pi_{1\leq i\leq|w|}[[\varphi]]_{V\cup\{x\}}(w,\sigma[x \to i])$
- We fix some order on the power set of $\{1,\dots,|w|\}$ so that the following product is defined even if $\mathcal{K}$ is not commutative: $[[\forall X.\varphi]]_V(w,\sigma) =$
  $\Pi_{I\in\{1,\dots,|w|\}}[[\varphi]]_{V\cup\{x\}}(w,\sigma[X \to I])$

Note that if φ is a sentence, i.e. has no free variables, then $[[\varphi]] \in K\langle\langle A^\star\rangle\rangle$.

2.2   Expressive Power of *WMSOL*

WMSOL over semirings and its fragments have considerable expressive power. As it was shouwn in [2,12,15,18,34], lots of optimization problems and counting problems are expressible in *WMSOL*. Here we quote use of only few commutative semirings from [12].

Example 1. $\mathcal{K} = (\mathbf{N},+,\cdot,0,1)$. Let $P_a(x)$ denote a unary predicate symbol. Formula $\varphi^{State} = \exists x P_a(x)$ counts how often a occurs in the word. Each choice of the unary predicate, affects the meaning of *"how often"* and generates lots of variations of the coverage properties. It means that the formula is a good candidate for monitoring of different coverage properties of data streams.

Example 2. $\mathcal{K}_{\mathbf{R}_{avg}} = (\mathbf{R}\cup\{-\infty\}, sup, avg, +, -\infty, 0)$.
Let P_{Send} stands for a send event, $P_{Receive}$ for a receive event and

$$\varphi = \forall x((P_{Send}(x) \to 1) \wedge (P_{Receive}(x) \to -1)),$$

where $(P_{Send}(x) \to 1)$ stands for $(P_{Send}(x) \wedge 1) \vee (\neg P_{Send}(x) \wedge 0)$ and similarly for $(P_{Send}(x) \to -1)$. Thus, the meaning of φ equals the average difference of *Send* events P_{Send} and *Receive* events $P_{Receive}$ in word w, cf. [15].

Example 3. $\mathcal{K} = ([0,1], max, \cdot, 0, 1)$ is the probability semiring.
Let $A = \{a_1,\dots,a_n\}$. Assume that each letter a_i has a reliability k_i. Then the series assigning to a word its reliability can be given by the first order formula $\forall x \bigvee_{1\leq i\leq n}(P_{a_i}(x) \wedge k_i)$. Note that the *damped window model*, which associates weights with the data in the stream, also falls in this formalism.

Results of [11,12] are extended in different papers. Mandrali and Rahonis, cf. [34], use *WFA* with discounting over commutative semirings. Droste and Meinecke, cf. [15], investigate also automata operations like average, limit superior, limit inferior, limit average, or discounting. In fact, there exists a more general cost model, where the weight of a run may be determined by a global valuation function, cf. [13]. One may also consider a weighted finite transition systems with weights from naturally ordered semirings, cf. [16]. Such semirings comprise distributive lattices as well as the natural numbers with ordinary addition and multiplication, and the max-plus-semiring.

3   Weighted Translation Schemes

The *syntactically defined translation schemes* describe transformations of logical structures. The notion of abstract translation schemes comes back to Rabin, [36]. They give rise to two induced maps, translations and transductions. Transductions describe the induced transformation of logical structures and the translations describe the induced transformations of logical formulae. Syntactically defined translation schemes for *WMSOL* over weighted trees were already introduced and investigated in [42]. In this contribution, we adopt the machinery to data streams modeled by weighted words.

Given commutative semiring $\mathcal{K}$ and finite alphabets Γ and Δ. For each $\gamma \in \Gamma$, P_γ denotes a unary predicate symbol (label). For each $\delta \in \Delta$, P_δ also denotes a unary predicate symbol. Let $R_\leq$ be a binary relation that defines $(x \leq y)$. Formal definition of the syntax and semantics of *WMSOL* over $\mathcal{K}$ may be found in [14]. Let $\tau = \langle R_\leq^\tau, P_\gamma^\tau | \gamma \in \Gamma \rangle$ and $\sigma = \langle R_\leq^\sigma, P_\delta^\sigma | \delta \in \Delta \rangle$ be the corresponding vocabularies. We model data streams as two-sorted logical structures. The first sort W corresponds to the universe of (positions in) words $w = a_1 a_2 \ldots a_n$ with $w(i) = a_i (1 \leq i \leq n)$. The meaning of $P(w(i))$ is that position i in a word is labeled by P. The second sort K is coming from the corresponding commutative semiring $\mathcal{K} = (K, +, \cdot, 0, 1)$. This sort is used in computations over $\mathcal{K}$.

Definition 1 (Translation Schemes $\Phi_\mathcal{K}$).
Let τ and σ be two vocabularies of weighted labeled words. Let $\phi, \psi_\leq, \psi_\delta | \delta \in \Delta$ be boolean WMSOL formulae. Let $\Phi = \langle \phi, \phi_K; \psi_\leq, \psi_\delta | \delta \in \Delta \rangle$. We say that $\Phi_\mathcal{K}$ is feasible for σ over τ if

- *ϕ has exactly 1 distinct free first order variable,*
- *ϕ_K is a tautology that has exactly 1 distinct free first order variable,*
- *$\psi_\leq$ has exactly 2 distinct free first order variables that re-defines $\leq$ relation,*
- *each ψ_δ has exactly 1 distinct free first order variable.*

Such a $\Phi_\mathcal{K}$[2] we also call a τ–σ–translation scheme or, shortly, a translation scheme, if the parameters are clear in the context.

With a translation scheme $\Phi_\mathcal{K}$, we can naturally associate a (partial) function $\Phi_\mathcal{K}^*$ from τ–streams to σ–streams.

Definition 2 (The induced map $\Phi_\mathcal{K}^*$).
Let $\mathcal{W}^\tau$ be a τ-stream and $\Phi_\mathcal{K}$ be feasible for σ over τ. The structure $\mathcal{W}^\sigma{}_{\Phi_\mathcal{K}}$ is defined as follows:

- **The two-sorted universe U, K contains sets:**

  1. *$U_{\Phi_\mathcal{K}} = \{a \in W^\tau : \mathcal{W}^\tau \models \phi(a)\}$;*
  2. *$K_{\Phi_\mathcal{K}} = K$.*

---

[2]   The introduced notion of $\Phi_\mathcal{K}$ can be naturally extended to the case of multiple $\leq$ relations in each of the involved vocabularies.

- *Relations:*

    1. *The interpretation of $R_{\leq}^{\sigma}$ is the set of pairs*

$$\mathcal{W}^{\sigma}{}_{\Phi_{\mathcal{K}}}(R_{\leq}) = \{(a_1, a_2) \in W^{\tau 2} : \mathcal{W}^{\tau} \models \psi_{\leq}(a_1, a_2)\};$$

    2. *The interpretation of each $\mathcal{W}^{\sigma}{}_{\Phi_{\mathcal{K}}}(P_{\delta})$ is the set*

$$\mathcal{W}^{\sigma}{}_{\Phi_{\mathcal{K}}}(P_{\delta}) = \{a \in W^{\tau} : \mathcal{W}^{\tau} \models \psi_{\delta}(a)\};$$

- $\Phi_{\mathcal{K}}{}^*$: *The partial function $\Phi_{\mathcal{K}}{}^*$ from weighted labeled words WLW to WLW $\Phi_{\mathcal{K}}{}^* : WLW(\tau) \to WLW(\sigma)$ is defined by $\Phi_{\mathcal{K}}{}^*(\mathcal{W}^{\tau}) = \mathcal{W}^{\sigma}{}_{\Phi_{\mathcal{K}}}.$*

Note that $\Phi_{\mathcal{K}}{}^*(\mathcal{W}^{\tau})$ is defined iff $\mathcal{W}^{\tau} \models \phi(a).$

Usually used *landmark data model* considers the data in a data stream from the beginning of the observation until now. In such a case, a *sliding window model* in data stream investigation, which considers the data from now up to a certain range in the past, is a special case of $\phi(a)$.

With a translation scheme $\Phi_{\mathcal{K}}$ we can also naturally associate a function $\Phi_{\mathcal{K}}{}^{\#}$ from $WMSOL(\sigma)$–formulae to $WMSOL(\tau)$–formulae.

Definition 3 (The induced map $\Phi_{\mathcal{K}}{}^{\#}$).
Let ζ be a σ–formula and $\Phi_{\mathcal{K}}$ be feasible for σ over τ. The formula $\zeta_{\Phi_{\mathcal{K}}}$ is defined inductively as follows:

1. *for $R_{\leq}^{\sigma}$ we put $\zeta_{\Phi_{\mathcal{K}}} = \psi_{\leq}(a_1, a_2)$;*
2. *for each P_{δ}^{σ} we put $\zeta_{\Phi_{\mathcal{K}}} = \psi_{\delta}(a)$;*
3. *for k we do nothing;*
4. *for the boolean connectives the translation distributes, i.e., if $\zeta = (\zeta_1 \wedge \zeta_2)$, then $\zeta_{\Phi_{\mathcal{K}}} = (\zeta_{1\Phi_{\mathcal{K}}} \wedge \zeta_{2\Phi_{\Theta}})$ and if $\zeta = \neg\zeta_1$, then $\zeta_{\Phi_{\mathcal{K}}} = \neg\zeta_{1\Phi_{\mathcal{K}}}.$*
5. *for the existential quantifier, we use relativization, i.e. if $\zeta = \exists v \zeta_1$ then we put $\zeta_{\Phi_{\mathcal{K}}} = \exists u \phi(u) \wedge (\zeta_{1\Phi_{\mathcal{K}}})$;*
6. *for second order variables X and a first order variable x or a constant we translate $X(x)$ by treating x like a relation symbol and put*

$$\zeta_{\Phi_{\mathcal{K}}} = \exists X (\forall x (X(x) \to (\phi(x) \wedge (\zeta_1)_{\Phi_{\mathcal{K}}}))).$$

- *Function $\Phi_{\mathcal{K}}{}^{\#} : WMSOL(\sigma) \to WMSOL(\tau)$ is defined by $\Phi_{\mathcal{K}}{}^{\#}(\zeta) = \zeta_{\Phi_{\mathcal{K}}}.$*

The following facts may be proved by induction.

Proposition 1.
Let $\Phi_{\mathcal{K}}$ be a τ–σ–translation scheme and ζ be a σ–formula. Then,

$$\mathcal{W}^{\tau} \models \Phi_{\mathcal{K}}{}^{\#}(\zeta) \text{ iff } \Phi^*(\mathcal{W}^{\tau}) \models \zeta.$$

Definition 4 (Composition of $\Phi_\mathcal{K}$'s).
Let $\tau = \langle R_\leq^\tau, P_\gamma^\tau | \gamma \in \Gamma \rangle$, $\sigma = \langle R_\leq^\sigma, P_\delta^\sigma | \delta \in \Delta \rangle$ and $\mu = \langle R_\leq^\mu, P_\lambda^\mu | \lambda \in \Lambda \rangle$ be vocabularies. Let $\mathcal{K}$ be a commutative semiring. Let $\Phi_\mathcal{K}^{\tau\sigma} = \langle \phi^{\tau\sigma}, \bar\phi_K; \psi_\leq^{\tau\sigma}, \psi_\delta^{\tau\sigma} | \delta \in \Delta \rangle$ be feasible for σ over τ and $\Phi_\mathcal{K}^{\sigma\mu} = \langle \phi^{\sigma\mu}, \phi_K; \psi_\leq^{\sigma\mu}, \psi_\lambda^{\sigma\mu} | \lambda \in \Lambda \rangle$ be feasible for μ over σ. Then, we denote by $\Phi_\mathcal{K}^{\tau\sigma} \circ \Phi_\mathcal{K}^{\sigma\mu}$ the translation scheme μ over τ given by $\langle \Phi_\mathcal{K}^{\tau\sigma\#}(\psi^{\sigma\mu}), \phi_K; \Phi_\mathcal{K}^{\tau\sigma\#}(\psi_\lambda^{\sigma\mu}) | \lambda \in \Lambda \rangle$. $\Phi_\mathcal{K}^{\tau\sigma}(\Phi_\mathcal{K}^{\sigma\mu})$ is called the composition of $\Phi_\mathcal{K}^{\sigma\mu}$ with $\Phi_\mathcal{K}^{\tau\sigma}$.

One can easily check that the syntactically defined composition of translation schemes has the following semantic property: $\Phi_\mathcal{K}^{\tau\sigma} \circ \Phi_\mathcal{K}^{\sigma\mu}(\mathcal{W}^\tau) = \Phi_\mathcal{K}^{\sigma\mu}(\Phi_\mathcal{K}^{\tau\sigma}(\mathcal{W}^\tau))$.

4   Incremental Analysis of Quantitative Properties of Strongly Distributed Streams

In this section, we discuss a way of obtaining weighted labeled words from components. We use the extended version of *WMSOL* and weighted labeled words with multiple $\leq$ relations.

4.1   Disjoint Union of Weighted Labeled Words

The *Disjoint Union* of a family of weighted labeled words is the simplest example of juxtaposing weighted labeled words, where none of the components are linked to each other.

Definition 5 (Finite Disjoint Union).
Let $\tau_\imath = \langle R_{\leq_\imath}^\tau, P_{\gamma_\imath}^\tau \rangle$, be a vocabulary of a weighted labeled word $\mathcal{W}_\imath$ over $\mathcal{K}$ and Γ. In the general case, the word over $\tau \cup I$ is

$$\mathcal{W} = \bigsqcup_{\imath \in I} \mathcal{W}_\imath = \langle \bigsqcup_{\imath \in I} \mathcal{W}_\imath, K; R_{\leq_\imath}^\tau(\imath \in I), P_{\gamma_\imath}^\tau(\imath \in I), P_\imath(\imath \in I) \rangle$$

for all $\imath \in I$, where $P_\imath(a)$ is true iff a came from $\mathcal{W}_\imath$, I is the finite index set.

Let us consider an example of disjoint union of two weighted labeled words:

Example 4 (Disjoint union of words). Assume we are given two τ_{alt}-words from example of Sect. 1.2, see Fig. 2. The precise definition of the corresponding vocabulary will be $\tau_{alt} = \langle R_\leq, P_\star, P_\oplus, P_\otimes, P_\Join^1, P_\Join^2 \rangle$ for up to four alternations. If more, say κ, alternations are allowed then up to $log(\kappa)$ unary predicates $P_\Join^\ell$ should be added. $\mathcal{K} = (\mathbf{N}, +, \cdot, 0, 1)$. The disjoint union of such words is presented as a logical structure in the following way:

$$\mathcal{W}^\cup = \langle \mathcal{W}_1 \dot\cup \mathcal{W}_2, \mathbf{N}; R_{\leq 1}, R_{\leq 2}, P_{\star 1}, P_{\star 2}, P_{\oplus 1}, P_{\oplus 2}, P_{\otimes 1}, P_{\otimes 1},$$

$$P_{\Join 1}^1, P_{\Join 1}^2, P_{\Join 2}^1, P_{\Join 2}^2, P_1, P_2 \rangle.$$

Note that the interpretation of $P_\Join^2$ in the second component is empty.

Second stream◇   □First stream

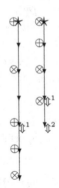

Fig. 2. Disjoint union of two data streams.

Now, the following theorem can be stated, cf. [42]:

Theorem 1.
Let I be a finite index set with ℓ elements. There exist commutative semirings $\mathcal{K}$ and weighted labeled words $\mathcal{W} = \bigsqcup_{i \in I} \mathcal{W}_i$ over $\mathcal{K}$ such that for every $\varphi \in$ WMSOL over $\mathcal{K}$ there are:

– a computation over WMSOL formulae

$$F_\varphi(\varpi_{1,1}, \ldots, \varpi_{1,j_1}, \ldots, \varpi_{\ell,1}, \ldots, \varpi_{\ell,j_\ell}),$$

– WMSOL–formulae $\psi_{1,1}, \ldots, \psi_{1,j_1}, \ldots, \psi_{\ell,1}, \ldots, \psi_{\ell,j_\ell}$

such that for every $\mathcal{W}_i$ and I as above with $\varpi_{i,j} = \varrho_{i,j}$ iff $[\psi_{i,j}] = \varrho_{i,j}$ we have,

$$[\varphi] = \varrho \quad \text{iff} \quad F_\varphi(\varpi_{1,1}, \ldots, \varpi_{1,j_1}, \ldots, \varpi_{\ell,1}, \ldots, \varpi_{\ell,j_\ell}) = \varrho.$$

Moreover, F_φ and the $\psi_{i,j}$ are computable from φ, ℓ and vocabularies alone, but are a tower of exponents in the quantifier depth of φ.

We list some options of the commutative semirings to choose:

Theorem 2. *The following semirings satisfy Theorem 1.*

- **Subset Semi-ring:** $(\mathcal{P}(A), \cap, \cup, \emptyset, A)$;
- **Boolean Semi-ring:** $(\{0,1\}, \vee, \wedge, 0, 1)$;
- **Fuzzy Semi-ring:** $([0,1], \vee, \wedge, 0, 1)$;
- **Extended natural number:** $(\mathbf{N} \cup \{\infty\}, +, \cdot, 0, 1)$;
- **Arctic Semi-ring:** $(\mathbf{R}_+ \cup \{-\infty\}, \max, +, -\infty, 0)$;
- **Tropical Semi-ring:** $(\mathbf{R}_+ \cup \{+\infty\}, \min, +, +\infty, 0)$.

Proof:

- **Subset Semi-ring:** By analyzing and extension of the proof in [21];
- **Boolean Semi-ring:** cf. [33];
- **Fuzzy Semi-ring:** By analyzing and extension of the proof in [33];
- **Extended natural number:** By analyzing and extension of the proof in [33];
- **Tropical Semi-ring:** cf. [42];
- **Arctic Semi-ring:** By analyzing and extension of the proof in [42].

4.2 Strongly Distributed Systems

Disjoint union as such is not very interesting. However, combining it with translation schemes gives us a rich repertoire of patching techniques. Let τ_1, τ be finite vocabularies, τ_1-weighted labeled words are pairwise disjoint for simplicity $\mathcal{W}_i (i \in I)$ and a τ-weighted labeled word $\mathcal{W}$ is the disjoint union of $\langle \mathcal{W}_i : i \in I \rangle$ with $\mathcal{W} = \bigsqcup_{i \in I} \mathcal{W}_i$.

Now, we generalize the notion of the disjoint union of weighted labeled words to *Strongly Distributed Streams* in the following way.

Definition 6.
Let I be a finite index set. Let $\mathcal{W} = \bigsqcup_{i \in I} \mathcal{W}_i$ be a τ-weighted labeled word. Furthermore let $\Phi_{\mathcal{K}}$ be a τ-σ WMSOL–translation scheme. The $\Phi_{\mathcal{K}}$-weighted sum of $\mathcal{W}_1, \ldots, \mathcal{W}_\ell$ over I is the weighted labeled word $\Phi^*_{\mathcal{K}}(\mathcal{W})$, or rather any weighted labeled word isomorphic to it. We denote the $\Phi_{\mathcal{K}}$-sum of $\mathcal{W}_1, \ldots, \mathcal{W}_\ell$ over I by $\bigsqcup_{i \in I}^{\Phi_{\mathcal{K}}} \mathcal{W}_i$.
A strongly distributed weighted stream over I is a stream that is isomorphic to a $\Phi_{\mathcal{K}}$-sum of $\mathcal{W}_1, \ldots, \mathcal{W}_\ell$.

Now, we come back to our Example 4, where we considered the disjoint union of two weighted labeled words. Using our technique of weighted translation schemes, we will proceed as follows:

Example 5 (Alternation of data streams).
Recall that we are given two τ-words and their disjoint union:

$$\mathcal{W}^\cup = \langle \mathcal{W}_1 \dot\cup \mathcal{W}_2, \mathbf{N}; R_{\leq_1}, R_{\leq_2}, P_{\star 1}, P_{\star 2}, P_{\oplus 1}, P_{\oplus 2}, P_{\otimes 1}, P_{\otimes 1},$$

$$P^1_{\bowtie 1}, P^2_{\bowtie 1}, P^1_{\bowtie 2}, P^2_{\bowtie 2}, P_1, P_2 \rangle.$$

We recall that we want to combine the stream in the following way: The resulting stream is composed from the fragment of the first stream from its first element, labeled by $\star$ till its first element, labeled by $\bowtie^1$. This element, labeled by the $\bowtie^1$, is the last element of the first stream that is included in the resulting stream. Then the combination stream includes all the elements from the second stream from its first element, labeled by $\star$ until its first element, labeled by $\bowtie^1$. Then we switch back to the first stream an so on, see Fig. 1. There exists finite constant number κ of switch points (up to two in each stream in our particular example). The corresponding translation scheme

$$\Phi_{\mathcal{K}}^{alt} = \langle \phi, \phi_{\mathcal{K}}; \psi_{\leq}, \psi_{\star}, \psi_{\oplus}, \psi_{\otimes} \rangle$$

is defined in the following way:

- $\phi, \phi_{\mathcal{K}}$ are any tautologies with exactly one free variable;
- $\psi_{\leq}(a_1, a_2) = \psi_{\leq}^{ori} \vee (\psi_{\leq}^{first} \vee \psi_{\leq}^{second} \vee \psi_{\leq}^{third} \vee \psi_{\leq}^{fourth})$, where
  - $\psi_{\leq}^{ori}(a_1, a_2) = (P_1(a_1) \wedge P_1(a_2) \wedge R_{\leq_1}(a_1, a_2)) \vee (P_2(a_1) \wedge P_2(a_2) \wedge R_{\leq_2}(a_1, a_2))$;
  - $\psi_{\leq}^{first}(a_1, a_2) = P_1(a_1) \wedge P_2(a_2) \wedge (P_{\bowtie_1}^1(a_1) \vee (\forall a_3(P_1(a_3) \wedge (P_{\bowtie_1}^1(a_3))) \rightarrow R_{\leq_1}(a_1, a_3)))$
    tells that any element of the first fragment of the first stream is smaller than any element of the second stream, see Fig. 1;
  - $\psi_{\leq}^{second}, \psi_{\leq}^{third}, \psi_{\leq}^{fourth}$ are similar to $\psi_{\leq}^{first}$;
- $\psi_{\star}(a) = P_1(a) \wedge P_{\star 1}(a)$;
- $\psi_{\oplus}(a) = (P_1(a) \wedge P_{\oplus 1}(a)) \vee (P_2(a) \wedge P_{\oplus 2}(a))$;
- $\psi_{\otimes}$ is similar to $\psi_{\oplus}$.

The desired global property counts the number of pairs, for which the first element is labeled by $\oplus$ and the second element is labeled by $\otimes$. The property is expressed in the following way:

$$\varphi = \exists a_1 \exists a_2 (\forall a_3 (R_{\leq}(a_1, a_3) \rightarrow R_{\leq}(a_2, a_3))) \bigwedge$$
$$(P_{\oplus}(a_1) \wedge P_{\otimes}(a_2) \wedge R_{\leq}(a_1, a_2) \wedge \neg R_{\leq}(a_2, a_1))).$$

Let us consider only a part of φ:

$$\eta(a_1, a_2) = P_{\oplus}(a_1) \wedge P_{\otimes}(a_2) \wedge R_{\leq}(a_1, a_2).$$

We substitute $\psi_{\oplus}(a_1), \psi_{\otimes}(a_2), \psi_{\leq}(a_1, a_2)$. We obtain:

$$\eta(a_1, a_2) = ((P_1(a_1) \wedge P_{\oplus 1}(a_1)) \bigvee (P_2(a_1) \wedge P_{\oplus 2}(a_1))) \bigwedge$$

$$((P_1(a_2) \wedge P_{\otimes 1}(a_2)) \bigvee (P_2(a_2) \wedge P_{\otimes 2}(a_2))) \bigwedge$$

$$((((P_1(a_1) \wedge P_1(a_2) \wedge R_{\leq_1}(a_1, a_2)) \vee (P_2(a_1) \wedge P_2(a_2) \wedge R_{\leq_2}(a_1, a_2))) \bigvee$$

$$(\psi_{\leq}^{first} \vee \psi_{\leq}^{second} \vee \psi_{\leq}^{third} \vee \psi_{\leq}^{fourth})).$$

Now, when we start to distribute the boolean connectives, we start to detect the cases, observed in Sect. 1.2. In fact, we easily locate:

Case 1: Both elements belong to the same original stream:

$$(P_1(a_1) \wedge P_1(a_2) \wedge P_{\oplus 1}(a_1) \wedge P_{\otimes 1}(a_2) \wedge R_{\leq_1}(a_1, a_2)) \wedge \dots) \bigvee$$
$$(P_2(a_1) \wedge P_2(a_2) \wedge P_{\oplus 2}(a_1) \wedge P_{\otimes 2}(a_2) \wedge R_{\leq_2}(a_1, a_2)) \wedge \dots).$$

Case 2: These elements are coming from different original streams requires more patience and persistence but definitely doable.

However, Theorem 3 guarantees that it is a general case rather than a nice coincidence.

Theorem 3.
Let I be a finite index set and let $\mathcal{W}$ be the Φ–sum of $\mathcal{W}_1,\ldots,\mathcal{W}_\ell$ over I. For every $\varphi \in WMSOL(\tau)$ that satisfies Theorem 1 there are:

- *a computation over weighted formulae*

$$F_{\Phi,\varphi}(\varpi_{1,1},\ldots,\varpi_{1,j_1},\ldots,\varpi_{\ell,1},\ldots,\varpi_{\ell,j_\ell}), \text{ and}$$

- *WMSOL–formulae $\psi_{1,1},\ldots,\psi_{1,j_1},\ldots,\psi_{\ell,1},\ldots,\psi_{\ell,j_\ell}$*

such that for every $\mathcal{W}_\imath$ and I as above with $\varpi_{\imath,\jmath} = \varrho_{\imath,\jmath}$ iff $[\psi_{\imath,\jmath}] = \varrho_{\imath,\jmath}$ we have

$$[\varphi] = \varrho \text{ iff } F_{\Phi,\varphi}(\varpi_{1,1},\ldots,\varpi_{1,j_1},\ldots,\varpi_{\ell,1},\ldots,\varpi_{\ell,j_\ell}) = \varrho.$$

Moreover, $F_{\Phi,\varphi}$ and the $\psi_{\imath,\jmath}$ are computable from $\Phi^{\#}$ and φ, but are a tower of exponents in the quantifier depth of φ.

Proof: By analyzing the proof of Theorem 1 in [42] and using Prop. 1.

Using composition of translated schemes, introduced in Definition 4, the application of Theorem 3 may be iterated in order to propagate evaluation of the computation of $\varphi \in WMSOL(\tau)$ to the components in more complicated combinations.

4.3 The General Approach

Assume we are given a distributed data stream and we are interested to monitor or optimize a quantitative property $\mathcal{P}_{Property}$ on the stream. In order to apply Theorem 3, we want to find commutative semiring $\mathcal{K}$ and vocabulary τ, such that we are able to formulate $\mathcal{P}_{Property}$ as $\varphi \in WMSOL(\tau)$. Moreover, if we manage to find such τ and semiring $\mathcal{K}$ and φ is restricted, then (according to the main theorem of [11,12]) there may be effectively constructed WFA $\mathcal{M}^{\varphi Property}$ that computes the value of $\varphi_{Property}$ on the data stream.

Moreover, assume the data stream may be obtained as a composition of translation schemes on the components of the data stream, such that Theorem 3 holds for the corresponding semiring $\mathcal{K}$. In such a case, the question whether $\mathcal{W}^\tau \models \varphi_{Property}$ can be reduced, according to Theorem 3, to the incremental computation on weighted formulae $F_{\Phi,\varphi_{Property}}(\varpi_{1,1},\ldots,\varpi_{\ell,j_\ell})$, such that

$$[\varphi_{Property}] = \varrho_{Property} \text{ iff } F_{\Phi,\varphi_{Property}}(\varpi_{1,1},\ldots,\varpi_{\ell,j_\ell}) = \varrho_{Property}.$$

In other words, given $\varphi_{Property}$ and $\Phi_{\mathcal{K}}$ of Definition 6, but not a particular $\mathcal{W}^\tau$, we construct a sequence of formulae $\psi_{\imath,\jmath}$ and evaluation function $F_{\Phi,\varphi_{Property}}$. Now, for any given $\mathcal{W}^\tau$, we compute the local values $\varpi_{\imath,\jmath}$ for each component. Then, we compute $F_{\Phi,\varphi_{Property}}$ such that

$$[\varphi_{Property}] = F_{\Phi,\varphi_{Property}}(\varpi_{1,1},\ldots,\varpi_{\ell,j_\ell}).$$

We emphasize that F_φ and the $\psi_{i,j}$ are computable from φ, ℓ and vocabularies alone. It means that from our main Theorem 3, we effectively derive an <u>algorithm</u> for computing $\varphi_{Property}$ on any strongly distributed data streams. On the other hand, the number of the derived formulae is a tower of exponents in the quantifier depth of φ. However, in practical applications, as a rule, the quantifier depth of formulae leads to simply exponential complexity. More detailed investigation of the complexity gains may be found in [38].

The algorithm proceeds as follows:

Preprocessing: Given $\varphi_{Property}$ and Φ, but not the strongly distributed data streams, we construct a sequence of formulae ψ_{ℓ,j_ℓ} and an evaluation function $F_{\Phi,\varphi_{Property}}$.

Incremental Computation: Given a particular strongly distributed data stream. We compute the local values ϖ_{ℓ,j_ℓ} on each local component.

Final Solution: Theorem 3 now states that $\varphi_{Property}$ on the strongly distributed data stream may be effectively computed from ϖ_{ℓ,j_ℓ}, using $F_{\Phi,\varphi_{Property}}$.

On the other hand, the choice of τ of $\mathcal{W}$ as well as $\mathcal{K}$, such that $\mathcal{P}_{Property}$ may be expressed as $WMSOL$-formula $\varphi_{Property}$, still should be done manually. One should be lucky to find them, as we are not able to guarantee their existence.

5   Conclusion and Discussion

In this contribution, we systematically adopt two logical reduction techniques to handling of distributed data streams. For our best knowledge, this is the first attempt to use logical machinery in the field of incremental monitoring and optimization on distributed streams and especially in its quantitative aspect. We used Weighted Monadic Second Order Logic as our quantitative query language.

The first technique are Feferman-Vaught reductions, which describe how the queries over a disjoint union of data streams can be computed from queries over the components and queries over the index set. The second one are the syntactically defined translation schemes, which describe possible transformations of data. We adopted the general framework of the syntactically defined translation schemes to the particular case of Weighted Monadic Second Order Logic on words. Combination of these two techniques allows us to consider not only disjoint unions of data streams but rather much richer compositions. We call them *strongly distributed data streams*. For such data streams, we extend and generalize the known approaches of incremental monitoring.

Our logical approach allows us to improve the known results. It is applicable to both homogeneous and heterogeneous streams. Our method derives queries over the components and queries over the index set, such that their proceeding gives a result that is equivalent to the answer of the original query rather than is its approximation. The method allows unification of the distributed and parallel computation as well as significant reduction of the communication load. It is safe in the sense that the transferred locally evaluated values are mostly meaningless without knowledge about the final proceeding.

Weighted Monadic Second Order Logic is not a unique way to express quantitative properties. In [9], a formalism for graph parameters definable in *MSOL*, here called *MSOLEVAL* with values in a ring **R**, was introduced. Labai and Makowsky in [30] provided two proofs that *RMSOL* (a fragment of *WMSOL* that does not contain a universal first-order quantification) and *MSOLEVAL* have the same expressive power over words.

Recently, other options of augmentation of classical logics with quantitative features have been introduced. In fact, *Quantitative Monadic Second Order Logic* was introduced in [29], which allows a generic framework for adding quantitative properties to any logic capable of expressing Boolean properties of words. Moreover, a multioperator monoid $\mho$ is a commutative monoid with additional operations on its carrier set. A weighted tree automaton over $\mho$ is a finite state tree automaton of which each transition is equipped with an operation of $\mho$. M-expressions over $\mho$ is defined in [22] in the spirit of formulae of *WMSOL*. In addition, new classes of weighted automata on words were introduced in [3]. Equipped with pebbles, they go beyond the class of recognizable formal power series: they capture weighted first order logic enriched with a quantitative version of *transitive closure*.

In other words, *WMSOL*is not the only quatative query language and *WTA* is not the only evaluation tool to handle quantitative properties of distributed systems. Thus, this apparatus opens a very large window for further research.

References

1. Babcock, B., Olston, C.: Distributed top-k monitoring. In: Proceedings of the 2003 ACM SIGMOD International Conference on Management of Data, pp. 28–39. SIGMOD 2003, ACM, New York, NY, USA (2003)
2. Bollig, B., Gastin, P.: Weighted versus probabilistic logics. In: Diekert, V., Nowotka, D. (eds.) DLT 2009. LNCS, vol. 5583, pp. 18–38. Springer, Heidelberg (2009). https://doi.org/10.1007/978-3-642-02737-6_2
3. Bollig, B., Gastin, P., Monmege, B., Zeitoun, M.: Pebble weighted automata and weighted logics. ACM Trans. Comput. Log. **15**(2), 15 (2014). https://doi.org/10.1145/2579819
4. Büchi, J.: Weak second-order arithmetic and finite automata. Z. Math. Logik-Grundlagen Math. **6**, 66–92 (1960)
5. Cherniack, M., et al.: Scalable distributed stream processing. In: CIDR 2003 - First Biennial Conference on Innovative Data Systems Research, Asilomar, CA, January 2003
6. Cormode, G., Garofalakis, M.: Approximate continuous querying over distributed streams. ACM Trans. Database Syst. **33**(2), 9:1–9:39 (2008)
7. Cormode, G., Muthukrishnan, S., Yi, K.: Algorithms for distributed functional monitoring. ACM Trans. Algorithms **7**(2), 211–2120 (2011). https://doi.org/10.1145/1921659.1921667
8. Courcelle, B., Makowsky, J.A., Rotics, U.: Linear time solvable optimization problems on graphs of bounded clique width. In: Hromkovič, J., Sýkora, O. (eds.) WG 1998. LNCS, vol. 1517, pp. 1–16. Springer, Heidelberg (1998). https://doi.org/10.1007/10692760_1

9. Courcelle, B., Makowsky, J., Rotics, U.: On the fixed parameter complexity of graph enumeration problems definable in monadic second order logic. Discrete Appl. Math. **108**(1–2), 23–52 (2001)
10. Cvrček, D.: Authorization model for strongly distributed information systems. Ph.D. thesis, Faculty of Electrical Engineering and Computer Science, Brno University of Technology, Czech Republic (2000)
11. Droste, M., Gastin, P.: Weighted automata and weighted logics, research Report LSV-05-13, Labiratoire Spécification et Vérification, Ecole Normal Supérieure Cachan 61, avenue du Président Wilson 94235 Cachan Cedex France (2005)
12. Droste, M., Gastin, P.: Weighted automata and weighted logics. Theoret. Comput. Sci. **380**, 69–86 (2007)
13. Droste, M., Götze, D., Märcker, S., Meinecke, I.: Weighted tree automata over valuation monoids and their characterization by weighted logics. In: Kuich, W., Rahonis, G. (eds.) Algebraic Foundations in Computer Science. LNCS, vol. 7020, pp. 30–55. Springer, Heidelberg (2011). https://doi.org/10.1007/978-3-642-24897-9_2
14. Droste, M., Kuich, W., Vogler, H.: Handbook of Weighted Automata, 1st edn. Springer, Berlin (2009)
15. Droste, M., Meinecke, I.: Describing average- and longtime-behavior by weighted MSO logics. In: Hliněný, P., Kučera, A. (eds.) MFCS 2010. LNCS, vol. 6281, pp. 537–548. Springer, Heidelberg (2010). https://doi.org/10.1007/978-3-642-15155-2_47
16. Droste, M., Meinecke, I., Šešelja, B., Tepavčević, A.: A cascade decomposition of weighted finite transition systems. In: Mauri, G., Leporati, A. (eds.) DLT 2011. LNCS, vol. 6795, pp. 472–473. Springer, Heidelberg (2011). https://doi.org/10.1007/978-3-642-22321-1_43
17. Droste, M., Paul, E.: A Feferman-Vaught decomposition theorem for weighted MSO logic. In: Potapov, I., Spirakis, P., Worrell, J. (eds.) 43rd International Symposium on Mathematical Foundations of Computer Science, MFCS 2018. LIPIcs, vol. 117, pp. 76:1–76:15 (2018)
18. Droste, M., Vogler, H.: Weighted logics for unranked tree automata. Theory Comput. Syst. **48**(1), 23–47 (2009)
19. Ebbinghaus, H., Flum, J.: Finite Model Theory. Perspectives in Mathematical Logic. Springer, Berlin (1995). https://doi.org/10.1007/978-3-662-03182-7
20. Elgot, C.: Decision problems of finite automata design and related arithmetics. Trans. Amer. Math. Soc. **98**, 21–52 (1961)
21. Feferman, S., Vaught, R.: The first order properties of products of algebraic systems. Fundamenta Mathematicae **47**, 57–103 (1959)
22. Fülöp, Z., Stüber, T., Vogler, H.: A Büchi-like theorem for weighted tree automata over multioperator monoids. Theory Comput. Syst. **50**(2), 241–278 (2012)
23. Garofalakis, M., Keren, D., Samoladas, V.: Sketch-based geometric monitoring of distributed stream queries. Proc. VLDB Endow. **6**(10), 937–948 (2013)
24. Garofalakis, M.: Querying distributed data streams - (invited keynote talk). In: Proceedings of 18th East European Conference Advances in Databases and Information Systems, ADBIS 2014, Ohrid, Macedonia, 7–10 September 2014, pp. 1–10 (2014)
25. Gilbert, A., Kotidis, Y., Muthukrishnan, S., Strauss, M.J.: One-pass wavelet decompositions of data streams. IEEE TKDE **15**, 2003 (2003)
26. Grädel, E., Siebertz, S.: Dynamic definability. In: 15th International Conference on Database Theory, ICDT 2012, Berlin, Germany, 26–29 March 2012, pp. 236–248 (2012). https://doi.org/10.1145/2274576.2274601

27. Immerman, N.: Descriptive Complexity. Graduate texts in computer science. Springer, New York (1999). https://doi.org/10.1007/978-1-4612-0539-5

28. Keren, D., Sagy, G., Abboud, A., Ben-David, D., Schuster, A., Sharfman, I., Deligiannakis, A.: Geometric monitoring of heterogeneous streams. IEEE Trans. Knowl. Data Eng. **26**(8), 1890–1903 (2014)

29. Kreutzer, S., Riveros, C.: Quantitative monadic second-order logic. In: 28th Annual ACM/IEEE Symposium on Logic in Computer Science, LICS 2013, New Orleans, LA, USA, 25–28 June 2013, pp. 113–122 (2013)

30. Labai, N., Makowsky, J.: Weighted automata and monadic second order logic. In: Proceedings Fourth International Symposium on Games, Automata, Logics and Formal Verification, GandALF 2013, Borca di Cadore, Dolomites, Italy, 29–31th August 2013, pp. 122–135 (2013)

31. Lazerson, A., Sharfman, I., Keren, D., Schuster, A., Garofalakis, M., Samoladas, V.: Monitoring distributed streams using convex decompositions. PVLDB **8**(5), 545–556 (2015)

32. Makowsky, J.: Algorithmic uses of the Feferman-Vaught theorem. Ann. Pure Appl. Logic **126**, 159–213 (2004)

33. Makowsky, J.A., Ravve, E.V.: Incremental model checking for decomposable structures. In: Wiedermann, J., Hájek, P. (eds.) MFCS 1995. LNCS, vol. 969, pp. 540–551. Springer, Heidelberg (1995). https://doi.org/10.1007/3-540-60246-1_159

34. Mandrali, E., Rahonis, G.: Recognizable tree series with discounting. Acta Cybernetica **19**(2), 411–439 (2009)

35. Olston, C., Jiang, J., Widom, J.: Adaptive filters for continuous queries over distributed data streams. In: Proceedings of the 2003 ACM SIGMOD International Conference on Management of Data, pp. 563–574. SIGMOD 2003, ACM, New York, NY, USA (2003)

36. Rabin, M.: A simple method for undecidability proofs and some applications. In: Hillel, Y.B. (ed.) Logic, Methodology and Philosophy of Science II. Studies in Logic, pp. 58–68. North Holland, Amsterdam (1965)

37. Ravve, E.V., Volkovich, Z., Weber, G.W.: A uniform approach to incremental automated reasoning on strongly distributed structures. In: Gottlob, G., Sutcliffe, G., Voronkov, A. (eds.) GCAI 2015. Global Conference on Artificial Intelligence. EasyChair Proceedings in Computing, vol. 36, pp. 229–251. EasyChair (2015)

38. Ravve, E., Volkovich, Z.: Four scenarios of effective computations on sum-like graphs. In: Proceedings of the The 9th International Multi-Conference on Computing in the Global Informationin Technology, pp. 1–8 (2014)

39. Ravve, E.: Model Checking for various notions of products. Master's thesis, Thesis, Department of Computer Science, Technion-Israel Institute of Technology (1995)

40. Ravve, E.: Incremental computations over strongly distributed databases. Concurrency Comput. Pract. Experience **28**(11), 3061–3076 (2016)

41. Ravve, E., Volkovich, Z.: Incremental verification and coverage analysis of strongly distributed systems. In: Kenett, R., Ruggeri, G., Faltin, F. (eds.) Analytic Methods in Systems and Software Testing. Wiley, Hoboken (2017)

42. Ravve, E., Volkovich, Z., Weber, G.W.: Effective optimization with weighted automata on decomposable trees. Optimization Journal, Special Issue on Recent Advances in Continuous Optimization on the Occasion of the 25th European Conference on Operational Research (EURO XXV 2012), vol. 63, pp. 109–127 (2014)

43. Ravve, E., Volkovich, Z., Weber, G.W.: Reasoning on strongly distributed multi-agent systems. In: Proceedings of the 17th International Symposium on Symbolic and Numeric Algorithms for Scientific Computing, pp. 251–256 (2015)

44. Schubert, E., Weiler, M., Kriegel, H.P.: Signitrend: scalable detection of emerging topics in textual streams by hashed significance thresholds. In: Proceedings of the 20th ACM SIGKDD International Conference on Knowledge Discovery and Data Mining, pp. 871–880. KDD 2014, ACM, New York, NY, USA (2014)

45. Sharfman, I., Schuster, A., Keren, D.: A geometric approach to monitoring threshold functions over distributed data streams. ACM Trans. Database Syst. **32**(4), 23 (2007)

46. Sharfman, I., Schuster, A., Keren, D.: Shape sensitive geometric monitoring. In: Proceedings of the Twenty-seventh ACM SIGMOD-SIGACT-SIGART Symposium on Principles of Database Systems, pp. 301–310. PODS 2008, ACM, New York, NY, USA (2008)

47. Wolff, R., Bhaduri, K., Kargupta, H.: A generic local algorithm for mining data streams in large distributed systems. IEEE Trans. Knowl. Data Eng. **21**(4), 465–478 (2009). https://doi.org/10.1109/TKDE.2008.169

Realisability of Choreographies

Klaus-Dieter Schewe[1]([✉]), Yamine Aït-Ameur[2], and Sarah Benyagoub[2]

[1] UIUC Institute, Zhejiang University, Haining, China
kd.schewe@intl.zju.edu.cn, kdschewe@acm.org
[2] Université de Toulouse, IRIT/INPT-ENSEEIHT, Toulouse, France
{yamine,sarah.benyagoub}@enseeiht.fr

Abstract. Choreographies prescribe the rendez-vous synchronisation of messages in a system of communicating finite state machines. Such a system is called *realisable*, if the traces of the prescribed communication coincide with those of the asynchronous system of peers, where the communication channels either use FIFO queues or multiset mailboxes. In this paper we provide two necessary conditions for synchronisability and hence for realisability of communication choreographies. We show that both conditions together are sufficient. A simple consequence is that realisability in the presence of a choreography becomes decidable. The conditions permit realisable choreographies to be obtained by means of composition, and then choreographies can be further refined into concurrent systems of communicating machines.

Keywords: Communicating system · Choreography · Synchronisability · Peer-to-peer system · Realisability

1 Introduction

A peer-to-peer (P2P) system is an asynchronous system of independent peers communicating through messages. On a very high level of abstraction one may disregard completely the internal computations performed by the peers and consider only the sequences of messages sent and received. Then each peer can be described by a finite state machine (FSM), the P2P system becomes a system of communicating FSMs, and its semantics is defined by the traces of messages sent. In addition, stability conditions may be taken into account, i.e. only those traces are considered in which all sent messages also have been received.

Such a trace semantics can be defined in various ways, e.g. using a separate channel organised as a FIFO queue for each ordered pair of distinct peers (see e.g. [9,12]). In particular, messages on the same channel are received in the same order as they have been sent and no message is lost. Alternatives are the use of such FIFO queues with only a single channel for each receiver (as e.g. in [2]) or the organisation of the channels as multisets (see e.g. [11]), which corresponds to mailboxes, from which messages can be received in arbitrary order. Naturally, one may also consider the possibility of messages being lost (see e.g. [10]).

© Springer Nature Switzerland AG 2020
A. Herzig and J. Kontinen (Eds.): FoIKS 2020, LNCS 12012, pp. 263–280, 2020.
https://doi.org/10.1007/978-3-030-39951-1_16

A common question investigated for communicating FSMs is whether the traces remain the same, if a rendez-vous (or handshake) synchronisation of (sending and receiving of) messages is considered, in which case the P2P system itself is also modelled as a FSM. This *synchronisability* problem has been claimed to be decidable in various publications (see e.g. [2]), but it was finally proven to be undecidable in general [12], though counterexamples are rather tricky.

The picture changes slightly in the presence of *choreographies*, i.e. FSMs that prescribe the rendez-vous synchronisation [3]. In this case the peers are projections of a choreography, and synchronisability becomes *realisability* of the given choreography. The rendez-vous composition of the projected peers coincides with the choreography, whereas in general projections of a rendez-vous composition of arbitrary peers may not coincide with the given peers. Also the distinction between *language synchronisability* based only on the message traces, and *synchronisability* based in addition on the stable configurations reached becomes obsolete.

One may ask, whether there are sufficient or necessary conditions on the choreography for realisability. In this article we will investigate this question and characterise realisability by two simple conditions that are both necessary and together sufficient. Actually, a hint on the sufficiency of these conditions is already given by the compositional approach to choreographies and the associated proof of realisability [5]. This compositional approach may then be taken further, when the communicating FSMs are refined by communicating machines, i.e. the message exchange is coupled with state-changing behaviour of the peers.

Our Contribution. The main contribution of this article is a characterisation of *synchronisability* by two necessary conditions on a communication choreography, which together are sufficient. One condition refers to sequences of messages in the choreography and excludes independence. Another condition refers to non-deterministic choice. Both conditions appear already in a restricted form in the correct-by-construction proofs in [4]. We generalise the conditions and prove also their necessity. We embed this result into a discussion of realisability under different semantics. As the conditions are purely syntactical, a simple consequence is that realisability in the presence of a choreography becomes decidable.

Related Work. The abstract view of P2P systems as communicating FSMs has already a long tradition [9], and there has been a longer chain of results addressing the decidability of the (language) synchronisability problem. Decidability has been claimed by Basu et al. in [2] for systems with separate FIFO queues for P2P channels as well as for combined queues per receiver. For both cases Finkel showed that (language) synchronisability is in fact undecidable [12]. Assuming a mailbox semantics, i.e. multisets instead of queues, decidability can be obtained [11], and this remains so even if messages can get lost [10]. However, the examples in [12] showing that previous claims of decidability are incorrect give already a hint that if the peers are projections of their rendez-vous composition, the decidability should hold. This will be the case for prescribed choreographies.

These investigations apply to arbitrary systems of peers, for which an overarching FSM is composed, either using communication channels organised as queues or multisets or rendez-vous synchronisation. If the rendez-vous synchronisation is prescribed by a *choreography*, the picture changes, as the peers become projections of the choreography [3]. This adds a conformity problem for choreographies [1] and extends synchronisability to the realisability of choreographies [6]. As our results in this paper will show, choreographies simplify the theory, as a choreography can always be regained by rendez-vous composition of its projection peers, and the problem of messages being sent but never received disappears.

A constructive approach to develop realisable choreographies and consequently P2P systems was brought up in [6]. The general idea is to exploit construction operators, by means of which realisable choreographies can be built out of a primitive base. The composition operators can be specified using Event-B [14], so the correctness of the construction can be verified, e.g. using the RODIN tool [5]. This actually exploits the sufficiency of our characterisation under moderate restrictions, but cannot be used to show also necessity. On the other hand it gives already hints for choreography repair [4]. The results in this article further strengthen the theoretical underpinnings of this correct-by-construction approach to realisable choreographies and permits to remove unnecessary assumptions.

Naturally, using Event-B in this context provides an open invitation to a refinement-based approach taking P2P systems defined by choreographies to communicating concurrent machines. Here we can exploit the work on concurrent Event-B [13]. The proposal to support the development of concurrent systems by multiple Event-B machines with concurrent runs has been derived from concurrent ASMs [7], and the introduction of messaging (as in [8] for concurrent ASMs) is straightforward.

Organisation of the Article. The remainder of this article is organised as follows. Section 2 is dedicated to preliminaries, i.e. we introduce all the notions that are relevant for the work: P2P systems, rendez-vous, P2P and mailbox composition, synchronisability, choreographies, realisability, etc. We also show how projection and composition interact and how to deal with non-deterministic FSMs and ϵ-transitions in this context. In Sect. 3 we address sufficient and necessary conditions for realisability, which gives our main result. We then indicate how violations to these conditions can be repaired. Finally, Sect. 4 contains a brief summary and outlook.

2   P2P Communication Systems and Choreographies

In a P2P system we need at least peers and messages to be exchanged between them. Therefore, let M and P be finite, disjoint sets, elements of which are called *messages* and *peers*, respectively. Each message $m \in M$ has a unique *sender* $s(m) \in P$ and a unique *receiver* $r(m) \in P$ with $s(m) \neq r(m)$. We use the notation $i \xrightarrow{m} j$ for a message m with $s(m) = i$ and $r(m) = j$. We also use the notation $!m^{i \to j}$ and $?m^{i \to j}$ for the *event* of sending or receiving the message m, respectively. Write M_p^s and M_p^r for the sets of messages, for which the sender or the receiver is p, respectively.

Let $s(M)$ and $r(M)$ denote the sets of send and receive events defined by a set M of messages. A *P2P system* over M and P is a family $\{\mathcal{P}_p\}_{p \in P}$ of finite state machines (FSMs) $\mathcal{P}_p$ over an alphabet $\Sigma_p = s(M_p^s) \cup r(M_p^r)$. By abuse of terminology $\mathcal{P}_p$ is also called a *peer*. Note that the FSM $\mathcal{P}_p$ may be deterministic or non-deterministic.

We write $\mathcal{P}_p = (Q_p, \Sigma_p, q_{0,p}, F_p, \delta_p)$, where Q_p is the finite set of states of the FSM, $q_{0,p} \in Q_p$ is the start state, $F_p \subseteq Q_p$ is the set of final states, and δ_p is the transition relation (or function), i.e. $\delta_p : Q_p \times \Sigma_p \to Q_p$ in case $\mathcal{P}_p$ is deterministic and $\delta_p \subseteq Q_p \times \Sigma_p \times Q_p$ in case $\mathcal{P}_p$ is non-deterministic. In the latter case let $\hat{\mathcal{P}}_p$ denote an equivalent deterministic FSM. We allow the subscript p to be omitted, if it is given by the context.

For our purpose here we always assume that all states are accepting, i.e. $F_p = Q_p$. As P is finite, we also assume without loss of generality that $P = \{1, \dots, n\}$.

2.1 Composition of Peers

A composition of a P2P system over M and P will be another FSM, the alphabet of which will be either M or $s(M) \cup r(M)$. The different ways to define a composition depend on several choices:

1. The sending and receiving of messages may use a rendez-vous (or handshake) synchronisation, i.e. it is assumed that a sent message is immediately received. This will lead to an FSM with alphabet M.
2. If no rendez-vous synchronisation is assumed we have to assume channels for the communication between the peers. For these we have the choice between channels for each pair of different peers or a single channel per receiving peer. The former case is usually referred to as *P2P semantics*, the latter one as *mailbox semantics*.
3. For the channels (in both P2P and mailbox semantics) we may assume an organisation as a FIFO queue or as a multiset. The former case includes the assumption that messages (in the same channel) are received in the same order they have been sent, whereas in the latter case messages may pass each other. Note that if multisets are assumed, then it becomes obsolete to distinguish between the P2P and the mailbox semantics, as the receiver may choose any message from a single or multiple channels.
4. Furthermore, we may restrict the capacity of channels to a maximum size, say $s_{i,j}$ for the P2P semantics or s_j for the mailbox semantics.
5. We may also allow messages to get lost, i.e. for a send event there may be no corresponding receive event. However, in this article we will not consider this possibility.

Definition 1. The *rendez-vous composition* of a P2P system $\{\mathcal{P}_p\}_{1\leq p\leq n}$ with $\mathcal{P}_p = (Q_p, \Sigma_p, q_{0p}, Q_p, \delta_p)$ is the FSM $\mathcal{C}_{rv} = (Q, M, q_0, Q, \delta)$ with $Q = Q_1 \times \cdots \times Q_n$, $q_0 = (q_{01}, \ldots, q_{0n})$, and $\delta((q_1, \ldots, q_n), i \overset{m}{\to} j, (q'_1, \ldots, q'_n))$ holds if $(q_i, !m^{i\to j}, q'_i) \in \delta_i$ and $(q_j, ?m^{i\to j}, q'_j) \in \delta_j$ hold, and $q_x = q'_x$ for all $x \notin \{i,j\}$.

Example 1. We adopt and modify the example in [12, Fig. 1] with $n = 3$. Preserving the notational conventions from above the transition relations δ_i are defined as follows:

$$\delta_1 : \quad (q_{01}, !m_a^{1\to 2}) \mapsto q_{11}, (q_{11}, !m_a^{1\to 2}) \mapsto q_{21}, (q_{21}, !m_b^{1\to 3}) \mapsto q_{31}$$

$$\delta_2 : \quad (q_{02}, ?m_a^{1\to 2}) \mapsto q_{12}, (q_{12}, ?m_a^{1\to 2}) \mapsto q_{22}, (q_{22}, ?m_c^{3\to 2}) \mapsto q_{32},$$
$$(q_{02}, ?m_c^{3\to 2}) \mapsto q_{42}, (q_{42}, !m_d^{2\to 3}) \mapsto q_{52}$$

$$\delta_3 : \quad (q_{03}, ?m_b^{1\to 3}) \mapsto q_{13}, (q_{13}, !m_c^{3\to 2}) \mapsto q_{23}, (q_{23}, ?m_d^{2\to 3}) \mapsto q_{33}$$

Then the transition relation δ in the rendez-vous composition is defined as follows (omitting states that are unreachable):

$$((q_{01}, q_{02}, q_{03}), 1 \overset{m_a}{\to} 2) \mapsto (q_{11}, q_{12}, q_{03}), ((q_{11}, q_{12}, q_{03}), 1 \overset{m_a}{\to} 2) \mapsto (q_{21}, q_{22}, q_{03})$$

$$((q_{21}, q_{22}, q_{03}), 1 \overset{m_b}{\to} 3) \mapsto (q_{31}, q_{22}, q_{13}), ((q_{31}, q_{22}, q_{13}), 3 \overset{m_c}{\to} 2) \mapsto (q_{31}, q_{32}, q_{23})$$

For the other possible choices let us first consider the P2P semantics, which gives rise to two more compositions[1].

Definition 2. The *P2P composition* of a P2P system $\{\mathcal{P}_p\}_{1\leq p\leq n}$ with $\mathcal{P}_p = (Q_p, \Sigma_p, q_{0p}, Q_p, \delta_p)$ is the automaton $\mathcal{C}_{p2p} = (Q, \Sigma, q_0, Q, \delta)$ satisfying the following conditions:

- The set of states is $Q = Q_1 \times \cdots \times Q_n \times (c_{ij})_{1\leq i\neq j\leq n}$, where each c_{ij} is either a finite queue (*FIFO channel semantics*) or a finite multiset (*multiset semantics*) with elements in M and $|c_{ij}| \leq s_{ij}$ (provided a maximum channel size s_{ij} is defined).
- The alphabet is $\Sigma = s(M) \cup r(M)$.
- The initial state is $q_0 = (q_{01}, \ldots, q_{0n}, ([])_{1\leq i\neq j\leq n})$ (*FIFO channel semantics*) or $q_0 = (q_{01}, \ldots, q_{0n}, (\langle\rangle)_{1\leq i\neq j\leq n})$ (*multiset semantics*), i.e. initially all channels are empty.
- The transition relation δ is defined by $\delta((q_1, \ldots, q_n, (c_{ij})_{1\leq i\neq j\leq n}), e, (q'_1, \ldots, q'_n, (c'_{ij})_{1\leq i\neq j\leq n}))$ if there exists i such that
  - $(q_i, e, q'_i) \in \delta_i$ holds, $q_x = q'_x$ for all $x \neq i$, and
    * either $e = !m^{i\to j}$ for some j, $c'_{ij} = c_{ij} \frown [i \overset{m}{\to} j]$ (*FIFO channel semantics*) or $c'_{ij} = c_{ij} \uplus \langle i \overset{m}{\to} j \rangle$ (*multiset semantics*), and $c_{k\ell} = c'_{k\ell}$ for all $(k, \ell) \neq (i, j)$

---

[1] Note that the automata defined in the next two definitions are not FSMs, strictly speaking, as there may be infinitely many states. Nonetheless, languages accepted by these automata can be defined analogously to FSMs.

* or $e = ?m^{j \to i}$ for some j and $c_{ji} = [j \xrightarrow{m} i]^\frown c'_{ji}$ (*FIFO channel semantics*) or $c_{ji} = \langle j \xrightarrow{m} i \rangle \uplus c'_{ji}$ (*multiset semantics*), and $c_{k\ell} = c'_{k\ell}$ for all $(k, \ell) \neq (j, i)$.

For later use we call a state $(q_1, \ldots, q_n, (c_{ij})_{1 \leq i \neq j \leq n})$ *stable* if and only if all channels c_{ij} are empty.

Analogously, we can define composition for the mailbox semantics.

Definition 3. The *mailbox composition* of a P2P system $\{\mathcal{P}_p\}_{1 \leq p \leq n}$ with $\mathcal{P}_p = (Q_p, \Sigma_p, q_{0p}, Q_p, \delta_p)$ is the automaton $\mathcal{C}_m = (Q, \Sigma, q_0, Q, \delta)$ satisfying the following conditions:

- The set of states is $Q = Q_1 \times \cdots \times Q_n \times (c_j)_{1 \leq j \leq n}$, where each c_j is either a finite queue (*FIFO channel semantics*) or a finite multiset (*multiset semantics*) with elements in M and $|c_j| \leq s_j$ (provided a maximum channel size s_j is defined).
- The alphabet is $\Sigma = s(M) \cup r(M)$.
- The initial state is $q_0 = (q_{01}, \ldots, q_{0n}, ([\,])_{1 \leq j \leq n})$ (*FIFO channel semantics*) or $q_0 = (q_{01}, \ldots, q_{0n}, (\langle \rangle)_{1 \leq j \leq n})$ (*multiset semantics*), i.e. initially all channels are empty.
- The transition relation δ is defined by $(q_1, \ldots, q_n, (c_j)_{1 \leq j \leq n}), e, (q'_1, \ldots, q'_n, (c'_j)_{1 \leq j \leq n}) \in \delta$ if there exists i such that
  - $(q_i, e, q'_i) \in \delta_i$ holds, $q_x = q'_x$ for all $x \neq i$, and
    * either $e = !m^{i \to j}$ for some j, $c'_j = c_j ^\frown [i \xrightarrow{m} j]$ (*FIFO channel semantics*) or $c'_j = c_j \uplus \langle i \xrightarrow{m} j \rangle$ (*multiset semantics*), and $c_k = c'_k$ for all $k \neq j$
    * or $e = ?m^{j \to i}$ for some j and $c_i = [j \xrightarrow{m} i]^\frown c'_i$ (*FIFO channel semantics*) or $c_i = \langle j \xrightarrow{m} i \rangle \uplus c'_i$ (*multiset semantics*), and $c_k = c'_k$ for all $k \neq i$.

As above we call a state $(q_1, \ldots, q_n, (c_j)_{1 \leq j \leq n})$ *stable* if and only if all channels c_j are empty.

In the following we consider four different compositions. The first three are rendez-vous composition $\mathcal{C}_{rv}$ as in Definition 1, P2P composition $\mathcal{C}_{p2p}$ with FIFO queue semantics as in Definition 2, and mailbox composition $\mathcal{C}_m$ with multiset semantics as in Definition 3. The fourth composition is *queue composition*, which is mailbox composition with FIFO queue semantics as in Definition 3, but in order to avoid confusion it will be denoted as $\mathcal{C}_q$. The fifth possible choice, P2P composition with multiset semantics, can be neglected, as Proposition 3 will show. We will prove this proposition after defining the trace semantics in the next subsection.

Let us briefly look at non-determinism. As peers are defined by FSMs, we may replace them by equivalent non-deterministic FSMs. The following proposition shows that this is compatible with the compositions, so in the sequel we may always assume without loss of generality that the FSMs are deterministic. The proof is straightforward.

Proposition 1. *Let $\mathcal{P} = \{\mathcal{P}_p\}_{1 \leq p \leq n}$ be a P2P system. Then the following hold:*

(i) *If all peers $\mathcal{P}_p$ are defined by deterministic FSMs, then also their composition $\mathcal{C}_c$ (with $c \in \{rv, p2p, m, q\}$) is deterministic.*

(ii) *Any deterministic $\hat{\mathcal{C}}_c$ (for $c \in \{rv, p2p, m, q\}$) equivalent to the c-composition $\mathcal{C}_c$ of $\mathcal{P}$ is also equivalent to the c-composition of $\hat{\mathcal{P}} = \{\hat{\mathcal{P}}_p\}_{1 \leq p \leq n}$.*

Let us further look at an extension of peers and compositions of P2P systems and permit ϵ-transitions[2] in the defining FSMs, i.e. a transition relation is defined as $\delta \subseteq Q \times (\Sigma \cup \{\epsilon\}) \times Q$. This can be used to extend also the composition operations:

- In case of the rendez-vous composition we have in addition $((q_1, \ldots, q_n), \epsilon, (q'_1, \ldots, q'_n)) \in \delta$ if and only if $(q_i, \epsilon, q'_i) \in \delta_i$ holds for some i and $q_j = q'_j$ for all $j \neq i$.
- In the case of P2P composition we have $((q_1, \ldots, q_n, (c_{ij})_{1 \leq i \neq j \leq n}), \epsilon, (q'_1, \ldots, q'_n, (c_{ij})_{1 \leq i \neq j \leq n})) \in \delta$ if and only if $(q_i, \epsilon, q'_i) \in \delta_i$ holds for some i and $q_j = q'_j$ for all $j \neq i$.
- In the case of mailbox composition we have $((q_1, \ldots, q_n, (c_j)_{1 \leq j \leq n}), \epsilon, (q'_1, \ldots, q'_n, (c_j)_{1 \leq j \leq n})) \in \delta$ if and only if $(q_i, \epsilon, q'_i) \in \delta_i$ holds for some i and $q_j = q'_j$ for all $j \neq i$.

In an FSM ϵ-transitions can be eliminated as follows: First, whenever $\delta(q, \epsilon, q')$ and (q', x, q'') hold for state $q, q', q'' \in Q$ and $x \in \Sigma$, then add $\delta(q, x, q'')$. Second, if the first rule cannot be applied anymore, remove all ϵ-transitions. The resulting automaton is equivalent to the given one. It is again straightforward to show that this elimination of ϵ-transitions is also compatible with the compositions. Therefore, we may always assume to deal with peers that do not use ϵ-transitions.

Proposition 2. *Let $\mathcal{P} = \{\mathcal{P}_p\}_{1 \leq p \leq n}$ be a P2P system, and let $\bar{\mathcal{P}} = \{\bar{\mathcal{P}}_p\}_{1 \leq p \leq n}$ be the corresponding P2P system, where $\bar{\mathcal{P}}_p$ results from $\mathcal{P}_p$ by the elimination of ϵ-transitions. Then the composition $\mathcal{C}_c$ (for $c \in \{rv, p2p, m, q\}$) of $\mathcal{P}$ and the corresponding c-composition $\bar{\mathcal{C}}_c$ of $\bar{\mathcal{P}}$ are equivalent.*

2.2   Trace Semantics

Peers as well as any composition of a P2P system are defined by (finite) automata, so their semantics is well defined by the notion of language accepted by them. However, in the case of P2P systems the alphabets of these automata are made out of messages or corresponding send and receive events. It is common to consider just sequences of sending events, i.e. for a word $w \in M^*$ let $\sigma(w)$ denote its restriction to its sending events. Formally, we have $\sigma(\epsilon) = \epsilon$, $\sigma(i \xrightarrow{m} j) = !m^{i \to j}$, and $\sigma(w_1 \cdot w_2) = \sigma(w_1) \cdot \sigma(w_2)$, where $\cdot$ denotes concatenation. Analogously, for words in $(s(M) \cup r(M))^*$ we have $\sigma(\epsilon) = \sigma(?m^{i \to j}) = \epsilon$, $\sigma(!m^{i \to j}) = !m^{i \to j}$, and $\sigma(w_1 \cdot w_2) = \sigma(w_1) \cdot \sigma(w_2)$.

---

[2] As we will see, ϵ-transitions are not needed, but they come in handy in proofs.

If $\mathcal{L}$ is the language accepted by an FSM $\mathcal{A}$ with alphabet M or $\Sigma = s(M) \cup r(M)$, then $\mathcal{L}(\mathcal{A}) = \sigma(\mathcal{L})$ is the *trace language* of $\mathcal{A}$. This applies for the cases where $\mathcal{A}$ is a peer $\mathcal{P}_p$ or a composition $\mathcal{C}_{rv}$, $\mathcal{C}_{p2p}$, $\mathcal{C}_m$ or $\mathcal{C}_q$ of a P2P system. Analogous to [12] for a P2P system $\mathcal{P} = \{\mathcal{P}_p\}_{1 \leq p \leq n}$ we use the notation $\mathcal{L}_0(\mathcal{P}) = \mathcal{L}(\mathcal{C}_{rv})$, $\mathcal{L}_\omega(\mathcal{P}) = \mathcal{L}(\mathcal{C}_c)$, where c is one of $p2p$, m or q, and furthermore $\mathcal{L}_k(\mathcal{P}) = \mathcal{L}(\mathcal{C}_c)$ for $k > 0$, if we require $s_{ij} \leq k$ or $s_j \leq k$ for all i,j.

If we restrict final states to be stable, we obtain a different language $\hat{\mathcal{L}}(\mathcal{C}_c) \subseteq \mathcal{L}(\mathcal{C}_c)$ (for $c \in \{p2p, m, q\}$), which we call the *stable trace language* of $\mathcal{C}_c$. $\hat{\mathcal{L}}_k(\mathcal{P})$ for $k > 0$ is defined analogously.

With these trace semantics we are now able to state and prove Proposition 3, which we announced already above.

Proposition 3. *For a P2P system* $\{\mathcal{P}_p\}_{1 \leq p \leq n}$ *its P2P composition with multiset semantics is equivalent to its mailbox composition* $\mathcal{C}_m$ *with multiset semantics.*

Proof. Let δ_{p2p} and δ denote the transition relation for the P2P composition with multiset semantics (as in Definition 2), and for the mailbox composition with multiset semantics (as in Definition 3), respectively. Then $\delta_{p2p}((q_1, \ldots, q_n, (c_{ij})_{1 \leq i \neq j \leq n}), e, (q'_1, \ldots, q'_n, (c'_{ij})_{1 \leq i \neq j \leq n}))$ holds if and only if $\delta((q_1, \ldots, q_n, (c_j)_{1 \leq j \leq n}), e, (q'_1, \ldots, q'_n, (c'_j)_{1 \leq j \leq n}))$ holds with $c_j = c_{1j} \cup \cdots \cup c_{nj}$ for all j. This implies that the trace languages accepted by both composed automata are the same. $\qquad\qquad\square$

A key notion is *language synchronisability*, which is based on the trace languages defined above and relates the "synchronous" rendez-vous composition with any of the other "asynchronous" compositions. If in addition we want to ensure that all sent messages are also received, then we take stable states into consideration. This leads to another notion of *synchronisability*. Obviously, if a P2P system $\mathcal{P}$ is synchronisable, it is also language synchronisable.

Definition 4. A P2P system $\mathcal{P} = \{\mathcal{P}_p\}_{1 \leq p \leq n}$ is called *language-synchronisable* with respect to P2P, mailbox or queue composition if $\mathcal{L}_0(\mathcal{P}) = \mathcal{L}_\omega(\mathcal{P})$ holds. $\mathcal{P} = \{\mathcal{P}_p\}_{1 \leq p \leq n}$ is called *synchronisable* with respect to P2P, mailbox or queue composition if $\mathcal{L}_0(\mathcal{P}) = \mathcal{L}_\omega(\mathcal{P}) = \hat{\mathcal{L}}_\omega(\mathcal{P})$ holds.

2.3 Choreography-Defined P2P Systems

Let us now look into choreographies. We define a *choreography* by an FSM $\mathcal{C} = (Q, M, q_0, F, \delta)$, where M is again a set of messages. As before we ignore final states and assume $F = Q$. Then every rendez-vous composition of a P2P system $\mathcal{P} = \{\mathcal{P}_p\}_{1 \leq p \leq n}$ defines a choreography. We are interested in P2P systems that are defined by a choreography, for which we need the notion of a projection.

Definition 5. Let $\mathcal{C} = (Q, M, q_0, Q, \delta)$ be a choreography with messages M and peers P. For $p \in P$ the *projection* $\pi_p(\mathcal{C})$ is the FSM $(Q, \Sigma, q_0, Q, \delta_p)$ with

$\Sigma = s(M) \cup r(M)$ and $(q, e, q') \in \delta_p$ if $e = !m^{p \rightarrow j}$ for some j with $(q, p \overset{m}{\rightarrow} j, q') \in \delta$, $e = ?m^{i \rightarrow p}$ for some i with $(q, i \overset{m}{\rightarrow} p, q') \in \delta$ or $e = \epsilon$ for $(q, i \overset{m}{\rightarrow} j, q') \in \delta$ with $p \notin \{i, j\}$.

The *peer* $\mathcal{P}_p$ *defined by* $\mathcal{C}$ is the FSM without ϵ-transitions corresponding to $\pi_p(\mathcal{C})$. A P2P system $\mathcal{P} = \{\mathcal{P}_p\}_{1 \leq p \leq n}$ is *choreography-defined* if there exists a choreography with peers $\mathcal{P}_p$ for all p.

There is a close relationship between rendez-vous compositions and choreography-defined P2P systems, which we exploit in the next section.

Proposition 4. *Each choreography $\mathcal{C}$ coincides (up to isomorphism) with the rendez-vous composition of its peers.*

Proof. Let the choreography be $\mathcal{C} = (Q, M, q_0, Q, \delta)$. The peers of $\mathcal{C}$ result from the projection $\pi_i(\mathcal{C})$ by eliminating ϵ-transitions, and the state sets Q_p of these peers are subsets of Q.

Let $\mathcal{C}_{rv} = (Q', M, q'_0, Q', \delta_{rv})$. Assume $((q_1, \ldots, q_n), i \overset{m}{\rightarrow} j, (q'_1, \ldots, q'_n)) \in \delta_{rv}$. According to Definition 1 there exist $i \neq j$ with $(q_i, !m^{i \rightarrow j}, q'_i) \in \delta_i$, $(q_j, ?m^{i \rightarrow j}, q'_j) \in \delta_j$, and $q_x = q'_x$ for all $x \notin \{i, j\}$.

The δ_i-transition corresponds to a sequence of δ'_i-transitions defined by the projection $\pi_i(\mathcal{C})$. Let this be $(q_{i,x-1}, \epsilon, q_{ix})$ for $x = 1, \ldots, k - 1$ and $(q_{i,k-1}, !m^{i \rightarrow j}, q_{ik})$ with $q_{i0} = q_i$ and $q_{ik} = q'_i$. Analogously, the δ_j-transition corresponds to a sequence of δ'_j-transitions defined by the projection $\pi_j(\mathcal{C})$, say $(q_{j,x-1}, \epsilon, q_{jx})$ for $x = 1, \ldots, \ell - 1$ and $(q_{j,\ell-1}, ?m^{i \rightarrow j}, q_{j\ell})$ with $q_{j0} = q_j$ and $q_{j\ell} = q'_j$.

Now let $\mathcal{C}'_{rv}$ be the rendez-vous composition of the projections. For this we obtain a sequence of ϵ-transitions starting in a state with i'th component q_i and j'th component q_j, leading to a state with i'th component $q_{i,k-1}$ and j'th component $q_{j,\ell-1}$. Furthermore, on this last state we have a δ'_{rv}-transition consuming $i \overset{m}{\rightarrow} j$.

If in $\mathcal{C}'_{rv}$ we omit the ϵ-transitions, only states $(q, \ldots, q)$ with $q \in Q$ remain. This shows that $\mathcal{C}_{rv}$ is isomorphic to $\mathcal{C}$. □

Proposition 4 implies that not all P2P systems are choreography-defined[3]. In fact, if a P2P system is choreography-defined, then it must consist of the peers defined by its rendez-vous composition.

Example 2. Consider the P2P system $\mathcal{P} = \{\mathcal{P}_p\}_{1 \leq p \leq 3}$ from Example 1, which cannot be choreography-defined. If it were choreography-defined, say by a choreography $\mathcal{C}$, then the peers defined by $\mathcal{C}$ would be the peers $\mathcal{P}_p$, and their rendez-vous composition would give back $\mathcal{C}$.

We computed the rendez-vous composition of $\mathcal{P}$ in Example 1. If we take the peers of this composition we obtain a P2P system with the following transition

---

[3] In fact, all counter-examples in [12] to previously claimed decidability results are P2P systems that are not choreography-defined.

relations:

$$\delta_1: \quad (q_{01}, !m_a^{1\to 2}) \mapsto q_{11}, (q_{11}, !m_a^{1\to 2}) \mapsto q_{21}, (q_{21}, !m_b^{1\to 3}) \mapsto q_{31}$$

$$\delta_2: \quad (q_{02}, ?m_a^{1\to 2}) \mapsto q_{12}, (q_{12}, ?m_a^{1\to 2}) \mapsto q_{22}, (q_{22}, ?m_c^{3\to 2}) \mapsto q_{32}$$

$$\delta_3: \quad (q_{03}, ?m_b^{1\to 3}) \mapsto q_{13}, (q_{13}, !m_c^{3\to 2}) \mapsto q_{23}$$

This is not the given P2P system $\mathcal{P}$.   □

Next we show that for choreography-defined P2P systems the synchronisability problem is much simpler than in the general case.

Proposition 5. *Let the P2P system* $\mathcal{P} = \{\mathcal{P}_p\}_{1 \le p \le n}$ *be choreography-defined. Then* $\mathcal{P}$ *is synchronisable if and only if it is language-synchronisable.*

Proof. According to Definition 4 and the remark following it we only have to show $\mathcal{L}_\omega(\mathcal{P}) \subseteq \hat{\mathcal{L}}_\omega(\mathcal{P})$.

Let $w^! \in \mathcal{L}_\omega(\mathcal{P})$ be a sequence of send events: $w^! = !m_1^{i_1 \to j_1} \dots !m_k^{i_k \to j_k}$. Let $w^? = ?m_1^{i_1 \to j_1} \dots ?m_k^{i_k \to j_k}$ denote the sequence of corresponding receive events. Then there exists a word w with $\sigma(w) = w^!$ that is accepted by the composition $\mathcal{C}_c$ ($c \in \{p2p, m, q\}$), and w results from interleaving $w^!$ and a subsequence of $w^?$ such that for each x the sending event $!m_x^{i_x \to j_x}$ precedes the corresponding receiving event $?m_x^{i_x \to j_x}$ in w, provided the latter one appears in w at all.

Let $w = w_1' \dots w_\ell'$. Then there exists a sequence of states $\gamma_0 \dots \gamma_\ell$ of $\mathcal{C}_c$ with $\gamma_x = (q_{x1}, \dots, q_{xn}, C^x)$ and transitions $(\gamma_{x-1}, w_x', \gamma_x) \in \delta$ for $x = 1, \dots, \ell$. Here we use C^x as a shortcut for $(c_{ij}^x)_{1 \le i \ne j \le n}$ or $(c_j^x)_{1 \le j \le n}$ depending on the semantics of the composition. The channels c_{ij}^ℓ (or c_j^ℓ, respectively) in the final accepting state contain the messages $i_y \overset{m_y}{\to} j_y$, for which $?m_y^{i_y \to j_y}$ does not appear in w.

According to Definitions 2 and 3 each transition $(\gamma_{x-1}, w_x', \gamma_x) \in \delta$ in the sequence is defined by a transition $(q_{x-1,i}, w_x', q_{xi}) \in \delta_i$ for some i, and either $w_x' = !m_y^{i \to j}$ or $w_x' = ?m_y^{j \to i}$ holds for some j. By projection we obtain sequences of states for each peer $\mathcal{P}_i$ together with transitions between them associated with the send events $!m_y^{i \to j}$ and the receive events $?m_y^{j \to i}$.

As we have $\mathcal{L}_\omega(\mathcal{P}) = \mathcal{L}_0(\mathcal{P})$ the sequence of messages $i_1 \overset{m_1}{\to} j_1, \dots, i_k \overset{m_k}{\to} j_k$ is accepted by $\mathcal{C}_{rv}$. Furthermore, as each peer is a projection of $\mathcal{C}_{rv}$, the projected sequences can be extended by transitions for the missing receive events, i.e. $?m_y^{i_y \to j_y}$ does not appear in w.

We can use these extensions to also define an extension of the sequence of states $\gamma_\ell \dots \gamma_{2k}$ of $\mathcal{C}_c$ with $\gamma_x = (q_{x1}, \dots, q_{xn}, C^x)$ and transitions $(\gamma_{x-1}, w_x', \gamma_x) \in \delta$ for $x = \ell + 1, \dots, 2k$, where $w_x' = ?m_j^{j \to i}$ holds for some j. In particular, due to Definitions 2 and 3 we must have that the channels c_{ij}^ℓ (or c_j^ℓ, respectively) in the final accepting state γ_{2k} must be empty.

This shows that $w_{ext} = w_1' \dots w_{2k}'$ is also accepted by $\mathcal{C}_c$. As we have $\sigma(w_{ext}) = \sigma(w) = w^!$, we conclude $w^! \in \hat{\mathcal{L}}_\omega(\mathcal{P})$.   □

As a consequence of Proposition 5 we may focus only on language-synchronisability. If a trace is accepted, then it will be accepted in a stable configuration. Furthermore, Proposition 4 allows us to identify the rendez-vous composition with the given choreography. This gives rise to the notion of realisability.

Definition 6. A choreography $\mathcal{C}$ is *realisable* with respect to P2P, mailbox or queue composition if $\mathcal{L}_0(\mathcal{P}) = \mathcal{L}_\omega(\mathcal{P})$ holds for the P2P system $\mathcal{P}$ defined by the projections of $\mathcal{C}$.

3   Characterisation of Realisability

We now investigate realisability of choreographies, which due to Proposition 4 are equivalent to the rendez-vous composition of their peers resulting from projections. We will first derive two necessary conditions for realisability for P2P and queue composition and discuss them also for mailbox composition. Then we will show that these conditions together are also sufficient, a result that was already indicated by the constructive approach in [4]. Finally, we investigate minimal changes to choreographies in case these conditions are violated.

In the following we fix a choreography $\mathcal{C} = (Q, M, q_0, Q, \delta)$. Both conditions will establish constraints on δ for two messages $i \xrightarrow{m_1} j$ and $k \xrightarrow{m_2} \ell$, but in both cases we need to exclude that these two messages are *independent* in the sense that they may appear in any order[4], i.e. we request that if there are states q_1, q_2, q_3 with $\delta(q_1, i \xrightarrow{m_1} j, q_2)$ and $\delta(q_2, k \xrightarrow{m_2} \ell, q_3)$, then we cannot have both $\delta(q_1, k \xrightarrow{m_2} \ell, q_2)$ and $\delta(q_2, i \xrightarrow{m_1} j, q_3)$.

3.1   The Sequence Condition

We first investigate sequences of messages as prescribed by a choreography. The sequence condition expresses that if two messages appear in a sequence, the sender of the second message must coincide with either the sender or the receiver of the preceding message.

> **Sequence Condition.** Whenever there are states $q_1, q_2, q_3 \in Q$ with $\delta(q_1, i \xrightarrow{m_1} j, q_2)$ and $\delta(q_2, k \xrightarrow{m_2} \ell, q_3)$ for non-independent messages $i \xrightarrow{m_1} j$ and $k \xrightarrow{m_2} \ell$, we must have $k \in \{i, j\}$.

Proposition 6. *If $\mathcal{C}$ is a realisable choreography with respect to P2P, queue or mailbox composition, then it satisfies the sequence condition.*

Proof. Due to Propositions 1 and 2 we may assume without loss of generality that the choreography $\mathcal{C}$ is deterministic and does not contain ϵ-transitions. We may further assume that all states are reachable.

Assume that the sequence condition is violated, so we have states q_1, q_2 and q_3 with $\delta(q_1, i \xrightarrow{m_1} j) = q_2$ and $\delta(q_2, k \xrightarrow{m_2} \ell) = q_3$ with non-independent messages $i \xrightarrow{m_1} j$ and $k \xrightarrow{m_2} \ell$ and pairwise different peers i, j, k. Then we have states q_{1i}, q_{2i}

---

[4] As the theory is based on FSMs, there is no possibility to express parallelism.

with $\delta_i(q_{1i}, !m_1^{i \to j}) = q_{2i}$, states q_{1j}, q_{2j} with $\delta_j(q_{1j}, ?m_1^{i \to j}) = q_{2j}$, states q_{2k}, q_{3k} with $\delta_k(q_{2k}, !m_2^{k \to \ell}) = q_{3k}$ and states $q_{2\ell}, q_{3\ell}$ with $\delta_\ell(q_{2\ell}, ?m_2^{k \to \ell}) = q_{3\ell}$. We may have $j = \ell$, but this will be irrelevant in the following.

Then $\mathcal{L}_0(\mathcal{C})$ contains words of the form $w!m_1^{i \to j}!m_2^{k \to \ell}$ corresponding to a sequence of states $q_0 \dots q_1 q_2 q_3$. As the messages are not independent, we have $w!m_2^{k \to \ell}!m_1^{i \to j} \notin \mathcal{L}_0(\mathcal{C})$. For the projected peers i, j and k we get sequences of states $\underbrace{q_{0i} \dots q_{1i}}_{\text{for peer } i}$ $\underbrace{q_{0k} \dots q_{2k}}_{\text{for peer } k}$ $\underbrace{q_{0j} \dots q_{1j}}_{\text{for peer } ij}$ generating fractions of w. Combining them gives rise to a sequence of states in $\mathcal{C}_c$ (for $c \in \{p2p, q, m\}$)

$$(q_{01}, \dots, q_{0n}, C^0) \dots \dots (\dots, q_{1i}, \dots, q_{1j}, \dots, q_{2k}, \dots, C^y),$$

where the C^x stand for the collection of queues or multisets as defined in Definitions 2 and 3. Initially, in C^0 all these queues or multisets are empty. Without loss of generality we may choose the sequence of states in such a way that also the queues or multisets in C^y are empty.

This may be continued by a state $(\dots, q_{2i}, \dots, q_{1j}, \dots, q_{2k}, \dots, C^{y+1})$ using $\delta_i(q_{1i}, !m_1^{i \to j}) = q_{2i}$, in which case $i \xrightarrow{m_1} j$ is added to the queue or multiset c_{ij} or c_j in C^{y+1}. We can then use further $\delta_k(q_{2k}, !m_2^{k \to \ell}) = q_{3k}$ to reach a state $(\dots, q_{2i}, \dots, q_{1j}, \dots, q_{3k}, \dots, C^{y+2})$ adding $k \xrightarrow{m_2} \ell$ to $c_{k\ell}$ or c_ℓ in C^{y+2}. This implies $w!m_1^{i \to j}!m_2^{k \to \ell} \in \mathcal{L}_\omega(\mathcal{C})$.

In addition, we may continue the sequence of states above by a different state $(\dots, q_{1i}, \dots, q_{1j}, \dots, q_{3k}, \dots, \tilde{C}^{y+1})$ using $\delta_k(q_{2k}, !m_2^{k \to \ell}) = q_{3k}$ thereby adding $k \xrightarrow{m_2} \ell$ to $c_{k\ell}$ or c_ℓ in $\tilde{C}^{y+1}$. Then we further use $\delta_i(q_{1i}, !m_1^{i \to j}) = q_{2i}$ to create the state $(\dots, q_{2i}, \dots, q_{1j}, \dots, q_{3k}, \dots, C^{y+2})$ adding $i \xrightarrow{m_1} j$ to the queue or multiset c_{ij} or c_j. This continuation implies $w!m_2^{k \to \ell}!m_1^{i \to j} \in \mathcal{L}_\omega(\mathcal{C})$, hence $\mathcal{L}_\omega(\mathcal{C}) \neq \mathcal{L}_0(\mathcal{C})$.   □

3.2   The Choice Condition

Next we investigate branching in the choreography. The choice condition expresses that if there is a choice of continuation with two different messages, then these messages must have the same sender.

> **Choice Condition.** Whenever there are states $q_1, q_2, q_3 \in Q$ with $\delta(q_1, i \xrightarrow{m_1} j, q_2)$, $\delta(q_1, k \xrightarrow{m_2} \ell, q_3)$ and $q_2 \neq q_3$ for non-independent messages $i \xrightarrow{m_1} j$ and $k \xrightarrow{m_2} \ell$, we must have $k = i$.

Proposition 7. *If $\mathcal{C}$ is a realisable choreography with respect to P2P, queue composition or mailbox composition, then it satisfies the choice condition.*

Proof. As in the proof of Proposition 6 we may assume without loss of generality that the choreography $\mathcal{C}$ is deterministic, does not contain ϵ-transitions, and all states are reachable.

Assume that the choice condition is violated, so we have states q_1, q_2 and q_3 with $\delta(q_1, i \xrightarrow{m_1} j, q_2)$ and $\delta(q_1, k \xrightarrow{m_2} \ell, q_3)$ with non-independent messages $i \xrightarrow{m_1} j$

and $k \xrightarrow{m_2} \ell$ and peers $i \neq k$. Then we have states q_{1i}, q_{2i} with $\delta_i(q_{1i}, !m_1^{i \to j}, q_{2i})$, states q_{1j}, q_{2j} with $\delta_j(q_{1j}, ?m_1^{i \to j}, q_{2j})$, states q_{1k}, q_{3k} with $\delta_k(q_{1k}, !m_2^{k \to \ell}, q_{3k})$ and states $q_{1\ell}, q_{3\ell}$ with $\delta_\ell(q_{1\ell}, ?m_2^{k \to \ell}, q_{3\ell})$. We may have $j \in \{k, \ell\}$ or $\ell \in \{i, j\}$, but this will be irrelevant in the following.

Then $\mathcal{L}_0(\mathcal{C})$ contains words of the form $w_1 = w!m_1^{i \to j}$ and $w_2 = w!m_2^{k \to \ell}$ corresponding to sequences of states $q_0 \ldots q_1 q_2$ and $q_0 \ldots q_1 q_3$, respectively.

The sequence for w_1 gives rise to sequences of states $q_{0i} \ldots q_{1i} q_{2i}$ and $q_{0k} \ldots q_{1k}$ for the peers i and k, respectively. From this we build a sequence of states

$$(q_{01}, \ldots, q_{0n}, C^0) \ldots (\ldots, q_{1i}, \ldots, q_{1k}, \ldots, C^y) \, (\ldots, q_{2i}, \ldots, q_{1k}, \ldots, C^{y+1})$$

of $\mathcal{C}_c$ (with $c \in \{p2p, m, q\}$ and the C^x standing for the collection of queues or multisets as defined in Definitions 2 and 3). For the last step we use the transition $\delta_i(q_{1i}, !m_1^{i \to j}, q_{2i})$ thereby adding $i \xrightarrow{m_1} j$ to the queue or mailbox c_{ij} or c_j in C^{y+1}.

This sequence can be continued using $\delta_k(q_{1k}, !m_2^{k \to \ell}, q_{3k})$ to reach a state $(\ldots, q_{2i}, \ldots, q_{3k}, \ldots, C^{y+2})$ and adding $k \xrightarrow{m_2} \ell$ to the queue or mailbox $c_{k\ell}$ or c_ℓ in C^{y+2}. This implies $w_1' = w!m_1^{i \to j}!m_2^{k \to \ell} \in \mathcal{L}_\omega(\mathcal{C})$.

Analogously, the sequence for w_2 gives rise to sequences of states $q_{0i} \ldots q_{1i}$ and $q_{0k} \ldots q_{1k} q_{3k}$. From this we build a sequence of states

$$(q_{01}, \ldots, q_{0n}, C^0) \ldots (\ldots, q_{1i}, \ldots, q_{1k}, \ldots, C^y) \, (\ldots, q_{1i}, \ldots, q_{3k}, \ldots, C^{y+1})$$

of $\mathcal{C}_c$ (with $c \in \{p2p, m, q\}$ and the C^x standing for the collection of queues or multisets as defined in Definitions 2 and 3). For the last step we use the transition $\delta_k(q_{1k}, !m_2^{k \to \ell}, q_{3k})$ adding $k \xrightarrow{m_2} \ell$ to the queue or mailbox $c_{k\ell}$ or c_ℓ in C^{y+1}.

This sequence can be continued using $\delta_i(q_{1i}, !m_1^{i \to j}, q_{2i})$ to reach a state $(\ldots, q_{2i}, \ldots, q_{3k}, \ldots, C^{y+2})$ thereby adding $i \xrightarrow{m_1} j$ to the queue or mailbox c_{ij} or c_j in C^{y+2}. This implies $w_2' = w!m_2^{k \to \ell}!m_1^{i \to j} \in \mathcal{L}_\omega(\mathcal{C})$.

However, as the messages are dependent, we cannot have both w_1' and w_2' in $\mathcal{L}_0(\mathcal{C})$, hence either $\mathcal{C}$ is not realisable or the assumption that the choice condition is violated cannot be true. $\qquad \square$

3.3 Sufficient Conditions for Realisability

Now we are ready to show our main result, which states that the sequence and choice conditions together are also sufficient for realisability. This is in accordance with the result achieved in [4], but removes unnecessary assumptions and extends it to different semantics for the composition.

Theorem 1. *A choreography $\mathcal{C}$ is a realisable with respect to P2P, queue or mailbox composition if and only if it satisfies the sequence and choice conditions.*

Proof. The necessity of the sequence and choice conditions has been shown in Propositions 6 and 7. We now show their sufficiency, for which it suffices to show $\mathcal{L}_\omega(\mathcal{C}) \subseteq \mathcal{L}_0(\mathcal{C})$.

As all states are accepting, the languages under consideration are prefix-closed. So we can proceed by induction over the length of words. The induction base for the empty word ϵ is trivial.

Now take $w = w'!m^{i \rightarrow j} \in \mathcal{L}_\omega(\mathcal{C})$. Then $w' \in \mathcal{L}_\omega(\mathcal{C})$ and hence $w' \in \mathcal{L}_0(\mathcal{C})$ by induction. Consider a sequence of states $q_0 \ldots q_x$ of $\mathcal{C}$ with transitions that correspond to accepting $\bar{w}$ with $\sigma(\bar{w}) = w'$. We have to show $w \in \mathcal{L}_0(\mathcal{C})$, which means that we have to find a state q_{x+1} with $(q_x, i \xrightarrow{m} j, q_{x+1}) \in \delta$.

If $w' = \epsilon$ holds, we have $q_x = q_0$, and $!m^{i \rightarrow j} \in \mathcal{L}_\omega(\mathcal{C})$ is only possible, if there is a transition $(q_{0i}, !m^{i \rightarrow j}, q_{1i}) \in \delta_i$ of the peer i in its initial state q_{i0}. Such a transition is the result of projecting some $(q_0, i \xrightarrow{m} j, q_1) \in \delta$, hence $w \in \mathcal{L}_0(\mathcal{C})$.

Now assume that $w' \neq \epsilon$ holds, so let us write $w' = w''!m_2^{k \rightarrow \ell}$ with some $w'' \in \mathcal{L}_\omega(\mathcal{C})$. We distinguish two cases.

Case 1. Assume $w''!m^{i \rightarrow j} \in \mathcal{L}_\omega(\mathcal{C})$. By induction we also have $w''!m^{i \rightarrow j} \in \mathcal{L}_0(\mathcal{C})$. Then omitting the last transition the sequence of states $q_0 \ldots q_{x-1}$ corresponds to accepting w''. Hence we have a choice in the choreography $\mathcal{C}$ with transitions $(q_{x-1}, k \xrightarrow{m_2} \ell, q_x) \in \delta$ and $(q_{x-1}, i \xrightarrow{m} j, q'_x) \in \delta$. As $\mathcal{C}$ satisfies the choice condition, this implies that either the two messages $k \xrightarrow{m_2} \ell$ and $i \xrightarrow{m} j$ have the same sender, i.e. $k = i$, or they are independent.

In case they are independent by definition there exists a state q_{x+1} with $(q'_x, k \xrightarrow{m_2} \ell, q_{x+1}) \in \delta$ and $(q_x, i \xrightarrow{m} j, q_{x+1}) \in \delta$, hence $w \in \mathcal{L}_0(\mathcal{C})$ as desired.

In case $k = i$ we obtain a choice in the projected peer i with transitions $(q_{x-1,i}, !m^{i \rightarrow j}, q'_{xi}) \in \delta_i$ and $(q_{x-1,i}, !m_2^{i \rightarrow \ell}, q_{xi}) \in \delta_i$. Furthermore, we must have also $(q_{yi}, !m^{i \rightarrow j}, q'_{y+1,i}) \in \delta_i$ for $y \geq x$ and some state $q_{y+1,i}$.

If there were transitions between q_{xi} and q_{yi} for peer i, then these can only be associated with receive events $?m_z^{\ell \rightarrow i}$. The sender must be ℓ, because $\mathcal{C}$ satisfies the sequence condition. The corresponding send events would appear in peer ℓ after $(q_{x-1,\ell}, ?m_2^{i \rightarrow \ell}, q_{x\ell}) \in \delta_\ell$, which contradicts $w \in \mathcal{L}_\omega(\mathcal{C})$. This gives $y = x$, i.e. we have a transition $(q_{xi}, !m^{i \rightarrow j}, q'_{x+1,i}) \in \delta_i$ for some state $q_{x+1,i}$.

As peer i is defined by projection of $\mathcal{C}$, there must exist a state q_{x+1} of $\mathcal{C}$ and transitions $(q_{x-1}, i \xrightarrow{m_2} \ell, q_x) \in \delta$ and $(q_x, i \xrightarrow{m} j, q_{x+1}) \in \delta$, hence $w \in \mathcal{L}_0(\mathcal{C})$ as desired.

Case 2. Now assume $w''!m^{i \rightarrow j} \notin \mathcal{L}_\omega(\mathcal{C})$. This is only possible, if either $k = i$ and the send event $!m_2^{k \rightarrow \ell}$ precedes the send event $!m^{i \rightarrow j}$ or $\ell = i$ and the receive event $?m_2^{k \rightarrow \ell}$ precedes the send event $!m^{i \rightarrow j}$. Let us look at these two subcases.

In case $k = i$ we have $(q_{x-1,i}, !m_2^{i \rightarrow \ell}, q_{xi}) \in \delta_i$ as well as $(q_{yi}, !m^{i \rightarrow j}, q'_{y+1,i}) \in \delta_i$ for $y \geq x$ and some state $q_{y+1,i}$. We can argue as in Case 1 that there cannot be any transitions between q_{xi} and q_{yi} for peer i, i.e. $y = x$. Otherwise such transitions could only be associated with receive events $?m_z^{\ell \rightarrow i}$ due to the sequence condition. The corresponding send events would appear in peer ℓ after $(q_{x-1,\ell}, ?m_2^{i \rightarrow \ell}, q_{x\ell}) \in \delta_\ell$, which contradicts $w \in \mathcal{L}_\omega(\mathcal{C})$. Therefore, we have a transition $(q_{xi}, !m^{i \rightarrow j}, q'_{x+1,i}) \in \delta_i$ for some state $q_{x+1,i}$.

As the peer i is defined by a projection of $\mathcal{C}$, we obtain states q_{x-1}, q_x and q_{x+1} of $\mathcal{C}$ with transitions $(q_{x-1}, k \xrightarrow{m_2} \ell, q_x) \in \delta$ and $(q_x, i \xrightarrow{m} j, q_{x+1}) \in \delta$, hence $w \in \mathcal{L}_0(\mathcal{C})$.

In case $\ell = i$ we have transitions $(q_{xi}, ?m_2^{k \to i}, q_{x+1,i}) \in \delta_i$—as before we can argue that there cannot be any other intermediate (receive) events—and $(q_{x+1,i}, !m^{i \to j}, q_{x+2,i}) \in \delta_i$. As the peer i is defined by projection, we obtain states q_{x-1}, q_x and q_{x+2} of $\mathcal{C}$ with transitions $(q_{x-1}, k \xrightarrow{m_2} \ell, q_x) \in \delta$ and $(q_x, i \xrightarrow{m} j, q_{x+2}) \in \delta$, hence $w \in \mathcal{L}_0(\mathcal{C})$.   □

A simple consequence of Theorem 1 is that for checking realisability in the presence of a choreography it suffices to check the choice and sequence conditions. As this is a purely syntactical check, we have decidability.

Corollary 1. *Realisability in the presence of a choreography is decidable.*

3.4 Choreography Repair

Let us finally look into the repair of choreographies, i.e. we investigate what can be done in case a choreography turns out not to be realisable. As we will also see in Example 3, this usually refers to a design error.

Violation of the Sequence Condition. Suppose we have a non-realisable choreography $\mathcal{C}$ violating the sequence condition, i.e. we have states q_1, q_2, q_3 with $(q_1, m_1^{i \to j}, q_2) \in \delta$, $(q_2, m_2^{k \to \ell}, q_3) \in \delta$ and $k \notin \{i, j\}$. Then we have the following minimal choices to repair the choreography:

1. Make the messages $m_1^{i \to j}$ and $m_2^{k \to \ell}$ independent by adding the transitions $(q_1, m_2^{k \to \ell}, q_2)$ and $(q_2, m_1^{i \to j}, q_3)$ to δ.
2. Add an additional intermediate message, i.e. add $(q_2, m_3^{i \to k}, q_2')$ or $(q_2, m_3^{j \to k}, q_2')$ with a new state q_2' to δ, and replace $(q_2, m_2^{k \to \ell}, q_3) \in \delta$ by $(q_2', m_2^{k \to \ell}, q_3)$. In order to enforce the sequence condition the sender of the additional message must be either i or j, and k must be either the sender or receiver, which implies that $m_3^{i \to k}$ and $m_3^{j \to k}$ are the only possibilities.

We will illustrate these repairs in Example 3 below.

Violation of the Choice Condition. Suppose now that we have a non-realisable choreography $\mathcal{C}$ violating the choice condition, i.e. we have states q_1, q_2, q_3 with $(q_1, m_1^{i \to j}, q_2) \in \delta$, $(q_1, m_2^{k \to \ell}, q_3) \in \delta$, $q_2 \neq q_3$ and $k \neq i$. Then we have the following minimal choices to repair the choreography:

1. Make the messages $m_1^{i \to j}$ and $m_2^{k \to \ell}$ independent by adding the transitions $(q_2, m_2^{k \to \ell}, q_3)$ and $(q_3, m_1^{i \to j}, q_2)$ to δ.
2. Add an additional message preceding either $m_1^{i \to j}$ or $m_2^{k \to \ell}$, which gives rise to the following two possibilities:
   (a) Replace $(q_1, m_1^{i \to j}, q_2)$ in δ by $(q_2', m_1^{i \to j}, q_2)$ with a new state q_2', and add $(q_1, m_3^{k \to i}, q_2')$ to δ. In order not to repair the choice condition the sender of the additional message must be k. In order not to violate the sequence condition i must be either sender or receiver of the new message, so $k \neq i$ implies that i must be the receiver.

(b) Replace $(q_1, m_2^{k \to \ell}, q_3)$ in δ by $(q_3', m_2^{k \to \ell}, q_3)$ with a new state q_3', and add $(q_1, m_3^{i \to k}, q_3')$ to δ. In order not to repair the choice condition the sender of the additional message must be i. In order not to violate the sequence condition k must be either sender or receiver of the new message, so $k \neq i$ implies that k must be the receiver.

We now present a final example showing violations of the sequence and choice conditions and their repair.

Example 3. Take a choreography $\mathcal{C}$ with the following transitions in the relation δ:

$$(q_0, m_0^{3 \to 2}, q_1) \ (q_1, m_1^{1 \to 2}, q_2) \ (q_2, m_2^{1 \to 2}, q_3) \ (q_3, m_3^{2 \to 1}, q_4) \ (q_2, m_3^{2 \to 1}, q_5)$$

In an interpretation (given in [4]) the peer 1 stands for a client, 2 for a server, and 3 for a communication agent. The latter one initialises a communication between client and server informing the server about a file transfer using message m_0. Then the client informs the server about the intended file transfer using message m_1. The client may revoke the request and send a cancellation message m_2, and the server informs the client about the completion of the task, either the successful file transfer or its abortion, using message m_3.

It is easy to see that $\mathcal{L}_0(\mathcal{C}) \neq \mathcal{L}_\omega(\mathcal{C})$ holds, so the choreography is not realisable. Both the sequence and the choice condition are violated.

The former one can be repaired by adding $(q_1, m_4^{2 \to 1}, q_1')$ or $(q_1, m_4^{3 \to 1}, q_1')$ with a new state q_1', and replacing $(q_1, m_1^{1 \to 2}, q_2) \in \delta$ by $(q_1', m_1^{1 \to 2}, q_2)$. Using the interpretation above this corresponds to either the server 2 or the communication agent 3 informing the client 1 that the file transfer can be started. Both options make sense.

The violated choice condition can be repaired by replacing $(q_2, m_2^{1 \to 2}, q_3)$ by $(q_2', m_2^{1 \to 2}, q_3)$ and adding $(q_2, m_4^{2 \to 1}, q_2')$ with a new state q_2'. In the interpretation above this corresponds to the server informing the client that the file transfer is ongoing but not completed, so cancellation is possible. We dispense with a discussion of alternative repair options.

4   Conclusion

In this article we re-investigated the synchronisability problem for communication FSMs under the presence of a choreography that prescribes the rendez-vous synchronisation of the peers. We discovered two necessary conditions on realisable choreographies and proved that these conditions together guarantee realisability. A consequence is decidability of realisability in the presence of a choreography. Hints for the sufficiency of these conditions originate from previous research on an Event-B-based correct-by-construction approach to the construction of realisable choreographies [4]. Our new sufficient and necessary characterisation of realisability strengthens the theoretical underpinnings of this approach and removes unnecessary assumptions. In forthcoming work we extended the RODIN-based proofs to the necessity of the conditions.

The approach can be taken further towards concurrent communicating machines that should result from stepwise refinement, i.e. P2P systems that go beyond communicating FSMs that are just abstractions emphasising the flow of messages. First, the characterisation permits the detection of possible repairs in case a condition is violated, i.e. it allows us to find minimal changes to the choreography that restore realisability. Clearly, such repairs have to be validated by a designer. What is needed in addition is a systematic reification by a system of concurrent machines. So the natural next step is an investigation of refinements by communicating machines based on Event-B. In this context an analysis of the realisation of the messaging channels is due, for which we expect the most natural semantics using mailboxes to turn out at the same time to be the simplest to be realised. This refinement method provides an open invitation for the continuation of this research towards a verifiable method for the specification and refinement of correct P2P systems.

References

1. Basu, S., Bultan, T.: Choreography conformance via synchronizability. In: Srinivasan, S., et al. (eds.) Proceedings of the 20th International Conference on World Wide Web (WWW 2011), pp. 795–804. ACM (2011)
2. Basu, S., Bultan, T.: On deciding synchronizability for asynchronously communicating systems. Theor. Comput. Sci. **656**, 60–75 (2016)
3. Basu, S., Bultan, T., Ouederni, M.: Deciding choreography realizability. In: Field, J., Hicks, M. (eds.) Proceedings of the 39th ACM SIGPLAN-SIGACT Symposium on Principles of Programming Languages (POPL 2012), pp. 191–202. ACM (2012)
4. Benyagoub, S., Aït-Ameur, Y., Ouederni, M., Mashkoor, A., Medeghri, A.: Formal design of scalable conversation protocols using Event-B: validation, experiments and benchmarks. J. Softw. Evol. Process (2019, to appear)
5. Benyagoub, S., Ouederni, M., Aït-Ameur, Y., Mashkoor, A.: Incremental construction of realizable choreographies. In: Dutle, A., Muñoz, C., Narkawicz, A. (eds.) NFM 2018. LNCS, vol. 10811, pp. 1–19. Springer, Cham (2018). https://doi.org/10.1007/978-3-319-77935-5_1
6. Benyagoub, S., Ouederni, M., Singh, N.K., Ait-Ameur, Y.: Correct-by-construction evolution of realisable conversation protocols. In: Bellatreche, L., Pastor, Ó., Almendros Jiménez, J.M., Aït-Ameur, Y. (eds.) MEDI 2016. LNCS, vol. 9893, pp. 260–273. Springer, Cham (2016). https://doi.org/10.1007/978-3-319-45547-1_21
7. Börger, E., Schewe, K.-D.: Concurrent abstract state machines. Acta Informatica **53**(5), 469–492 (2016)
8. Börger, E., Schewe, K.-D.: Communication in abstract state machines. J. UCS **23**(2), 129–145 (2017)
9. Brand, D., Zafiropulo, P.: On communicating finite-state machines. J. ACM **30**(2), 323–342 (1983)
10. Chambart, P., Schnoebelen, P.: Mixing lossy and perfect fifo channels. In: van Breugel, F., Chechik, M. (eds.) CONCUR 2008. LNCS, vol. 5201, pp. 340–355. Springer, Heidelberg (2008). https://doi.org/10.1007/978-3-540-85361-9_28
11. Clemente, L., Herbreteau, F., Sutre, G.: Decidable topologies for communicating automata with FIFO and bag channels. In: Baldan, P., Gorla, D. (eds.) CONCUR 2014. LNCS, vol. 8704, pp. 281–296. Springer, Heidelberg (2014). https://doi.org/10.1007/978-3-662-44584-6_20

12. Finkel, A., Lozes, É.: Synchronizability of communicating finite state machines is not decidable. In: Chatzigiannakis, I., et al. (eds.) 44th International Colloquium on Automata, Languages, and Programming (ICALP 2017). LIPIcs, vol. 80, pp. 122:1–122:14. Schloss Dagstuhl - Leibniz-Zentrum für Informatik (2017)
13. Schewe, K.-D.: Extensions to hybrid Event-B to support concurrency in cyber-physical systems. In: Abdelwahed, E.H., Bellatreche, L., Golfarelli, M., Méry, D., Ordonez, C. (eds.) MEDI 2018. LNCS, vol. 11163, pp. 418–433. Springer, Cham (2018). https://doi.org/10.1007/978-3-030-00856-7_28
14. Zoubeyr, F., Aït-Ameur, Y., Ouederni, M., Tari, K.: A correct-by-construction model for asynchronously communicating systems. STTT **19**(4), 465–485 (2017)

Schema Optimisation Instead of (Local) Normalisation

Bernhard Thalheim$^{(\boxtimes)}$ (ORCID)

Department of Computer Science, Christian-Albrechts University at Kiel, 24098 Kiel, Germany
thalheim@is.informatik.uni-kiel.de
http://www.is.informatik.uni-kiel.de/~thalheim

Abstract. Classical normalisation theory has a number of lacunas although it is commonly and widely accepted and it is the basis for database theory since the 80ies. Most textbooks and monographs still follow this approach despite the good number of open problems. Today, modern object-relational DBMS offer far better capabilities than the systems that have been built in the past based on the strict relational paradigm. Constraint maintenance has been oriented on transformation of structures to structures that are free of functional dependencies beside key constraints. The maintenance of coherence constraints such as two-type inclusion constraints has been neglected although this maintenance might be the most expensive one. In reality normalisation is local optimisation that exclusively considers functional dependency maintenance.

We thus need a different normalisation approach. This paper develops an approach towards optimisation of schemata and global normalisation. This approach results in a denormalisation and object-relational database schemata.

Keywords: Normalisation · Schema optimisation · Denormalisation · Performance · Object-relational databases · Global normalisation

1 Normalisation - the Good, the Bad, the Ugly

Normalisation is considered to be one of the pearls of database theory. There is almost no database course that does not teach this part of a theory. The main results have been achieved during the 70ies, 80ies and early 90ies. Since then the theory is considered to be completed although new DBMS (database management systems) and new database paradigms have been developed since then. There are very few publications on object-relational structures. XML approaches have mainly be following this research paradigm.

1.1 Local Vertical Normalisation Based on Functional and Other Dependencies

Local database normalization aims at the derivation of database structures that can easily be supported by the DBMS. In the past, DBMS supported keys, domain constraints and referenced-key-based inclusion constraints (so-called foreign-key-constraint). Therefore, it was a goal to derive another equivalent schema to the given one which has a set of integrity constraints that can be supported by the DBMS used for implementation. This approach can be understood as a descriptive approach to optimisation of database structuring depending on the platform for implementation. Normalisation as a concept is typically too narrow and too much focussed on local vertical normalisation.

© Springer Nature Switzerland AG 2020
A. Herzig and J. Kontinen (Eds.): FoIKS 2020, LNCS 12012, pp. 281–300, 2020.
https://doi.org/10.1007/978-3-030-39951-1_17

Three Kinds of Normalisation. Normalisation is mainly considered to be *vertical normalisation* on the basis of projection sets as mappings and join as the restoration operation. *Horizontal normalisation* [23] is based on selection and union. *Deductive normalisation* [30] is based on reduction of classes according to tuple-generating constraints and extended completion using this set of tuple-generating constraints (see, for instance, [3]). It is the most storage effective and the best computational method for normalisation as long as the tuple-generating dependency used for decomposition is acyclic [30,31]. The latter two normalization methods have not yet got a proper support by the database systems vendors. A common treatment for these three kinds has not yet developed.

The consideration of horizontal normalisation together with vertical normalisation has a number of advantages. A type may be first decomposed horizontally what would enable application of another vertical normalisation to one of the horizontal subclasses[1]. Additionally, horizontal normalisation enables in separation of a class into data that are stable and will not be changed and volatile data that are still changed. The first data class can then be supported by a large set of indexes and improve performance.

Normalisation Theory is a Relict of the 80ies. The relational database design book by [41] is one of the most comprehensive surveys on kinds of normal forms. It essentially considers almost 30 kinds. Some of them have been superseded by other, some of them did not find practical solutions or their specific kinds of applications.

Normalisation starts with the requirement of the first normal form, i.e. all attributes are atomic and do not have an inner structure. Consider, however, the *ZIP* as an attribute of an *address*. It has an inner structure which is guilty for the non-BCNF normalisation of addresses [21]. Meanwhile DBMS support user-defined data structures (UDT's). These data structures have their specific functions and predicates. The *address* example is an academic one while in practice address data are checked against public address databases. A similar observation can be made on the second normal form and others. They can be neatly supported by modern DBMS. The support goes far beyond what non-first-normal-form (NF^2) research provides.

Beside the flaws of the synthesis algorithm for normalisation discussed below we should pay attention whether the *collection step* (step 3 in normalisation: collect all FD's with the same left side into a singleton new relation type) is really appropriate. Some of the FD's may have a different flavour. Consider, for instance, an *organisation unit* which is *chaired by somebody*, has its *postal address*, has a main *financial funding*, a secretariat or *office*, and a *contact*. Why it should be represented as a singleton type? It is far better to use a star kind structure since each of the associations has its meaning and its usage in the database application. Such structures are also causing evolution problems.

Normalisation approaches typically do not consider the different meanings of the dependencies [35] but treat the set of all constraints under consideration as a set of elements which are of equal importance. Moreover, normalisation is based on classes

---

[1] This mechanism has been already used for fragmentation techniques for distributed databases. A similar approach has been proposed by Date [10] for handling NULL-polluted classes by rigid horizontal normalisation into NULL-free fragments (see also [25,40]).

of constraints such as functional dependencies and multivalued dependencies. As we shall see below, normalisation should however been based on sets of constraints from different classes and on the meaning of such constraint sets.

Another obstacle of normalisation theory is the assumption that relation classes are sets. SQL allows multi-sets. The normalisation theory for multi-sets has a different setting [18].

1.2   Local Vertical Normalisation

Local vertical normalisation plays a prominent role in normalisation theory. Research started with the introduction of FD-based normal forms (first, second, third in a good variety of notions, Boyce-Codd) and led to a good body of knowledge[2] Later functional dependencies have been generalised to domain, multi-valued, and hierarchical dependencies and the generalisation of the last one to join dependencies. This research resulted in introduction of further normal forms, e.g. fourth and fifth. The sixth normal form (6NF) has already been introduced with the DBMS MIMER in the mid-70ies.

Local normalisation is a good approach as long as we restrict the consideration to strictly equality-generation dependencies such as functional dependencies and singleton relational schemata with atomic attributes. In this case the so-called third normal form is achievable. The Boyce/Codd Normal Form (BCNF) is not achievable in any case. [21] has shown however that all known counterexamples are based on ill-defined structures. It can be shown [31] that either hierarchical decomposition or refined granularity of attributes result in BCNF structures.

Reasons to Normalise. [31] surveys the main targets of normalisation: (1) avoiding inadequate behaviour such as anomalies, (2) elimination of unnecessary redundancy, (3) prohibiting inconsistent data, (4) stability of database schemata during application evolution, (5) optimising database performance, and (6) maintenance of abstraction levels within a schema. We refer to [31] for discussion of other problems encountered for normalisation such as adequate BCNF representation, adequacy of decomposed schemata, competing normalisations of the same schema, and inadequacy of multivalued dependencies and other tuple-generating dependencies within the relational database model.

We observe that the first three reasons are rather operational one whereas the last two are tactical ones. The fourth reason is a strategic one that is the main source for later modernisation, migration, and re-engineering.

Did We Achieve the Six Targets? The answer is no. We achieved the first target and partially achieved the second target. The third target can only be achieved if all potential constraints and their influence on consistency is handled properly. The fourth target has not yet found good research solutions. After evolution schemata suffer from mannerism and look similar to Gothic cathedrals or chaotic sandcastles. After normalisation,

---

[2] We restrict the citations to the most essential ones for this paper and restrain to give a full survey of the research.

database performance might be good for data manipulation operations. Normalisation might result in a far worse behaviour for database querying. The advent of data warehouses is a reaction on this problem. The sixth target is not supported by current languages that force us to stay on one abstraction level.

1.3 Inclusion Constraint Maintenance After Decomposition

Literature often neglects the set of additional inclusion constraints that must be maintained after a decomposition of a class. Given a multivalued dependency $X \twoheadrightarrow Y$ for a partition X, Y, Z of the set of attribute of a relation type R. The class R^C can be decomposed into $R_1^C = \pi_{X \cup Y}(R^C)$ and $R_2^C = \pi_{X \cup Z}(R^C)$ in such a way that $R^C = R_1^C \bowtie R_2^C$ if the multivalued dependency is valid in R^C. We note that this multivalued dependency is implied by a functional dependency $X \rightarrow Y$.

This vertical decomposition of R^C into R_1^C and R_2^C must be maintained by pairwise inclusion dependencies $\pi_X(R_1) \subseteq \pi_X(R_2)$ and $\pi_X(R_2) \subseteq \pi_X(R_1)$. In the relational DBMS setting the pairwise inclusion constraint should be maintained by foreign key constraints, i.e. X should be a key in both R_1^C and R_2^C.

Question 1.Is there any good approach to inclusion constraint maintenance after decomposition?
A solution to this foreign key requirement is proposed in the RM/V2 model [9] by introduction of a third relation type $R_0^C = \pi_X(R^C)$.

Observation 1. This approach results in a number of additional auxiliary relation types what limits the effect of normalisation.[3]
We need to add to the normalisation approach also an *extended dependency preservation rule* that is often neglected in the literature:

Principle 1. The decomposition based on vertical normalisation adds to the decomposed types pairwise inclusion dependencies on intersecting attributes. The decomposition based on horizontal normalisation adds to the decomposed types an exclusion constraint.

This principle has already been implicitly used for the universal relation assumption. We observe however that pairwise inclusion dependencies may cause severe performance problems.

The union constraint for horizontal decomposition is implicit and is the basis for defining a view that combines by UNION the decomposed components into the original type[4]. The deductive normalisation [31] is another option.

1.4 Constraint Sets Instead of Sets of Constraints

The classical approach of computer science introduces syntax first. And then semantics is defined on top of syntax. As again discussed in [35], this approach is nice for

---

[3] We forbear from postulating these observations as theorems. They are rather simple and easy to check statements.

[4] We note that a database schema is typically not a database model. The schema must be enhanced by views to become a database model [36]. Since we have to use anyway views then we should better extensively use horizontal decomposition beside vertical decomposition.

computational handling and for inductive and incremental construction but completely unnatural for normal languages. Syntax, semantics, and pragmatics form a unit. The *syntax-semantics-separation* principle finds its rigid continuation in the separation of integrity constraints into classes that have some uniformity in their definition structure. This separation principle has been found well-acceptable by programmers and logicians. It is however completely counterintuitive [35]. Natural language use a holistic semiotic approach and do not artificially separate units that form a semiotic holistic statement. Additionally, constraints might have their own meaning [28] such as syntactic functional dependencies compared to dependencies that represent semantical units.

The main deficiency is the constraint acquisition problem. Since we need a treatment for sets a more sophisticated reasoning theory is required. One good candidate is visual or graphical reasoning that goes far beyond logical reasoning [12].

Star and snowflake structures used in OLAP approaches are the basis for an approach that handles structures as a complex within its structure and its semantics in the complex. With the advent of object-oriented and XML languages we learned lessons on object identification [2] and the co-use of set-based classes with pointers. These approaches can be considered as a starting point.

Let us extend the open problem (TIC7) (Real-life constraint sets [33]):

Problem 1. Provide a reasoning facility for treatment of heterogeneous sets of constraints instead of constraints from a given constraint class. Classify 'real life' constraint sets which can be easily maintained and specified.

In [13] we realised that the classical Hilbert-type reasoning (premises allow to derive a conclusion) should be replaced by another schema: some premises which are supported by other constraints allow to derive a conclusion. This set-of-support reasoning can be based on graphical reasoning means or spreadsheet reasoning schemata.

1.5   The Storyline of the Paper

Several reasons can be observed why local normalisation may be inadequate. We shall discuss some of them in the next Section. We restrict the discussion to relational database technology and to conceptualisation through the extended entity-relationship model [31]. Some of the (88 [sic!]) pitfalls of object orientation [39] and of XML orientation have similar causes but are outside the scope of this paper. Normalisation theory is so far exclusively built as a theory of vertical local normalisation. We might ask whether we should consider global vertical normalisation. Or at least other kinds of local normalisation as well. The main target of normalisation of optimisation of the overall database for all six targets. Instead of poly-optimisation for some of the six criteria we might use a less strict form by optimisation of some of the database types and by denormalising others.

Since poly-optimisation is typically unsolvable we develop a number of corrections to normalisation approaches and a general approach to denormalisation as a kernel for a general theory of optimisation.

2 Solutions for Classical Normalisation Approaches

2.1 Refining Synthesis Algorithms

Rigidity of Classical Synthesis Algorithms. The third step of the classical synthesis algorithm typically groups all attributes that can be inferred from the same set of attributes by functional dependencies. This approach groups then attributes that are potentially conceptually completely independent into one group. An alternative approach could be rigid non-grouping, i.e. the left hand side of a functional dependency $X \longrightarrow Y$ is the basis of k new types with attributes $X \cup B_i$ for $Y = \{B_1, ..., B_k\}, 1 \le i \le k$. Both approaches are extreme positions. We may observe, however, that some separation must be maintained.

Let us consider a simple example [35] of a relational type R: given attributes $attr(R) = \{A, B, D, F, G, I\}$ and a set of functional dependencies
$$\Sigma_R = \{A \longrightarrow IG, D \longrightarrow FG, IAB \longrightarrow D, IF \longrightarrow AG\}.$$
This FD set can be represented by the graph on the left side of Fig. 1. This set can be reduced by deleting $IF \longrightarrow G$ from the graph since it is derivable through the edges representing $IF \longrightarrow A$ and $A \longrightarrow G$. Furthermore, the set ABI can be reduced since the edge representing $A \longrightarrow I$ already supports subset reduction. No other reduction can be applied to the graph. We use the calculus for graphical reasoning [12] that is complete and sound. We use dotted lines for the subset relationship among subsets of attributes and arrows for functional dependencies.

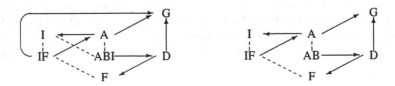

Fig. 1. The graph of the functional dependencies and the reduced cover of this set

We may directly derive a normalisation according to this graph reduction. Each constraint must be covered. We arrive with the synthesis algorithm to[5]:
$$R_1 = (\{A, G, I\}, \{A \longrightarrow GI, R_1[AI] \subseteq \supseteq R_2[AI]\}),$$
$$R_2 = (\{A, F, I\}, \{A \longrightarrow I, FI \longrightarrow A, R_2[F] \subseteq \supseteq R_4[F]\}),$$
$$R_3 = (\{A, B, D\}, \{AB \longrightarrow D, R_3[D] \subseteq \supseteq R_4[D], R_1[A] \subseteq \supseteq R_3[A]\}),$$
$$R_4 = (\{D, F, G\}, \{D \longrightarrow FG, R_1[G] \subseteq \supseteq R_4[G]\}).$$
The set $\{A, B\}$ is a key. We thus do not need an additional key type for the normalisation.

If we however take into account constraint maintenance and redundancy then we arrive at a smaller and better schema with the type:

---

[5] If we require that all inclusion dependencies are referential integrity constraints then we need 12 types for normalisation that results in a foreign-key-faithful decomposition: $R_1[\underline{A}], R_1[\underline{I}], R_1[\underline{A}, I], R_1[\underline{A}, G], R_2[\underline{F}], R_2[A, \underline{I}, F], R_3[\underline{B}], R_3[\underline{A}, \underline{B}], R_3[\underline{A}, \underline{B}, D],$ $R_4[\underline{D}], R_4[\underline{G}], R_4[\underline{D}, F, G]$ where the key of each new type $R_i[X] := \pi_X[R_i]$ is underlined.

$$R'_1 = (\{A, G\}, \{A \longrightarrow G, R_1[A] \subseteq \supseteq R_2[A]\})$$

due to the validity of the following derivation (reduction rule):

$$\frac{R_1[A, I] \subseteq \supseteq R_2[A, I], \; R_2 : A \to I, \; R_1 : A \to GI}{R_1 \bowtie R_2 = R'_1 \bowtie R_2, \; R'_1 : A \to G} \quad R'_1 = \pi_{A,G}(R_1).$$

This rule is based on general deduction rules and on equalities for the relational algebra:

Theorem 1 *(General deduction for closed Horn formulas). A sound and complete axiomatisation for closed Horn formulas $\forall...(\alpha \to \beta)$ consists of*

axioms

$$\overline{\alpha \to \beta} \qquad \text{for all facets of substructures } \beta \preceq \alpha,$$

augmentation rules for super-structures α^+ ***and sub-structures*** β^- $\quad \frac{\alpha \to \beta}{\alpha^+ \to \beta^-}$ *for either $\beta^- \preceq \beta$ and $\alpha \preceq \alpha^+$ or as well as $\alpha^+ = \alpha \sqcup \gamma$ and $\beta^- = \beta \sqcap \gamma$, and*

transitivity rules $\frac{\alpha \to \beta, \; \beta \to \gamma}{\alpha \to \gamma}$ *for all connecting pairs $(\alpha \to \beta, \beta \to \gamma)$.*

The proof of the theorem is based on the Boolean representation of the open first-order predicate calculus and on properties of implications. The completeness uses the same arguments as the classical FD completeness proof.

An alternative proof of the reduction rule is based on algebraic dependencies [23, 30].

We observe that graphical synthesis would result in a better behaviour. This structure is represented in Fig. 2. The hyper-network approach [24, 34] uses nodes as Venn diagrams of subsets of the set of attributes and directed edges between the nodes. The constraint set in Fig. 1 is given by the hyper-network representation in Figure 2. Compare this representation to the decomposition hypergraph for the classical synthesis algorithm [17]. The hyper-network representation is rather simple to read. Each of the edges must be represented by some new relation type. Moreover, the set $\{A, B\}$ is a key due to the graph node closure. Otherwise we may use a combination of nodes for the graph node closure. The second minimal key is $\{I, F, B\}$ which is not a node and thus would have been added to the decomposition if we would not have represented the first one. We notice that graphical reasoning is simpler for implications than Hilbert-type calculi, e.g. [3, 14, 40].

Questions One Might Ask for Normalisation Theory and Their Research Agenda. The classical normalisation theory is based on functional and multi-valued dependencies. Normalisation synthesis algorithms are deterministic. The result depends on the order of attributes and on the order of constraints considered in the algorithm. The minimal cover is not unique for a given set of constraints (even not polynomial according to the number of attributes in the worse case). We, thus, have a good number of opportunities for a normalisation.

Question 2. Which normalisation opportunity should be the best one?
The solution cannot be to consider only one of them. We might used a pragmatistic solution however: choose the most performing one and keep the knowledge on alternatives.

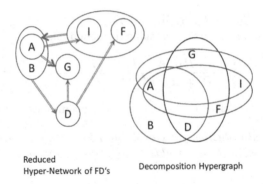

Reduced
Hyper-Network of FD's   Decomposition Hypergraph

Fig. 2. Graphical normalisation depicted as hyper-network and decomposition hypergraph.

The classical approach is based on hypergraphs. Instead we should use hyper-networks and meta hyper-networks [35] (see also Fig. 2).

Let us consider a very simple set Σ of functional dependencies for
$$R = (\{A, B, C, D, E, F, G, H\}, \Sigma) \text{ with}$$
$$\Sigma = \{A \to B \to C \to AD, D \to E \to F \to DG, G \to H \to G\}.$$
Σ has more than 50 minimal covers. Similar examples can be given for results of normalisation according to synthesis algorithms.

Observation 2. Attribute sets which are FD-equivalent can be given in an abstract form, i.e. we consider in a set of constraints the complex $[A, B]$ instead of $\{A \to B, B \to A\}$.

The attribute-wise consideration might be appropriate for first normal form definitions but it complicates reasoning on constraints.

We thus represent Σ by the rather simple FD set system $\{[A, B, C] \to [D, E, F] \to [G, H]\}$. It has 18 different BCNF normalisations (similar case in [16, 17].

Question 3. Why we should consider so many minimal covers and normal forms?

We note that multivalued dependencies are defined in the relational database theory in a mathematical manner. They are far better expressed by entity-relationship modelling languages [32] and far simpler to capture and to develop.

Question 4. Should we better develop a normalisation approach for entity-relationship schemata? Should we better consider a schema for normalisation instead of type-wise normalisation?

Good HERM Schemata are Typically the Best Normalisation. Folklore of ER modelling claims that the best normalisation is obtained if the main target of conceptual modelling is the specification of an ER schema. This claim is not valid in general since extended entity-relationship modelling languages such as HERM [31] are not cognitively complete. The ER approach provides however a far better solution to normalisation of relational schemata than normalisation theory, e.g. for multivalued and hierarchical dependencies. Moreover, the structure is more natural and thus better to comprehend.

A flaw of the first normalisation algorithms was corrected by a *key composition rule*, i.e. if the decomposed relational structure does not have a constructed type which contains some key of the old schema as a (sub-)structure then a new type is created for one of the minimal keys and added to this relational structure. This rule is nothing else as a decomposition for which a relationship type is added with a key that has the property that it is overwriting the key product of related decomposed types. The corresponding hyper-network has then nodes which are not incrementally layered and thus need a connecting element which can be then used as a key. The hyper-network approach also allows a generalised and less strict key composition rule.

2.2 Balancing Between Conceptualisation and Programming Adequacy

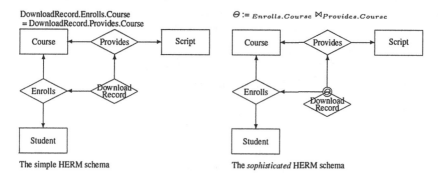

The representational conceptual schema

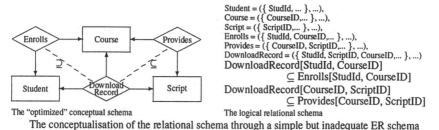

Student = ({ StudId, ... }, ...),
Course = ({ CourseID,... }, ...),
Script = ({ ScriptID,... }, ...),
Enrolls = ({ StudId, CourseID,... }, ...),
Provides = ({ CourseID, ScriptID,... }, ...),
DownloadRecord = ({ StudId, ScriptID, CourseID,... }, ...)
DownloadRecord[StudId, CourseID]
              $\subseteq$ Enrolls[StudId, CourseID]
DownloadRecord[CourseID, ScriptID]
              $\subseteq$ Provides[CourseID, ScriptID]

The "optimized" conceptual schema   The logical relational schema

The conceptualisation of the relational schema through a simple but inadequate ER schema

Fig. 3. The 'Janus' schema cluster for conceptual modelling

Database design and development is often based on the three-layer architecture. This architecture requires that the user can be supported by views defined on top of the conceptual schema. The conceptual schema is mapped to the logical and to the physical schemata. The last two are considered to be internal or implementation schemata. It is well known in database design [27,29] that the conceptual schema be represented by a 'Janus' schema, i.e. one schema (*representational conceptual schema*) that is used for conceptual representation, for view formulation and for conceptual programming and another schema (*optimized conceptual schema*) that is used for implementation issues including logical and physical programming. The two schemata are equivalent to each other and are tightly associated with each other by transformation mappings. A typical example of these two schemata is given in Figure 3. The example is taken from a script server project. Students enrolled in a course may download scripts that are provided by the course.

The optimised conceptual schema can be easily mapped to a structure that supports smooth operating of the database. We can deduct from this schema the internal representation, supporting structures such as indexes (in various kinds), internal concatenation or splitting of relations, introduction of generalisations, the navigational structure for access, the management of derived structures (derived attributes, views, macro-data on top of the micro-data of the database), criteria for the selection of internal storage and computational structures, the necessity for the introduction of artificial or surrogate keys, and for clustering or separation of records. These parameters are used for tuning and physical optimisation of the database. The sophisticated HERM schema uses the Θ-join for the correct building of the relationship type that records downloads. The optimised conceptual schema is equivalent to this schema due to the equivalence of the join decomposition and the inclusion constraints [31].

2.3   Accuracy of the Internal Database Structure

The internal database structure is ruled by the DBMS. The mappings from the conceptual schema to the internal schema must preserve a number of properties:

Preservation of content : The conceptual schema has been established by modelling the application domain and highlights data that are going to be stored in a database. The internal schema represents the same and only the same data that can be recorded through the conceptual schema and is based on the database modelling language of the platform assumed for implementation.

Preservation of access and modification : The access pathes and modification pathes that are implicitly or explicitly assumed for the conceptual schema are mapped to access and modification pathes that are entirely supported by the platform assumed for the implementation.

Preservation of maintenance efficiency : Integrity constraints are given in a declarative form in conceptual schemata. The efficiency of their maintenance is not considered. The (economic) value of constraints is neglected. Internal schemata must provide mechanisms for efficient integrity maintenance.

The first property is an element of any database course and well considered in most database books. The property can be treated on the basis of non-losslessness and dependency preservation. It might be enhanced by the requirement that each projection (or other range structure) should be needed for the reconstruction process. The second property is mainly solved by the professional experience of database professionals. It is typically not discussed in scientific publications and is an element of database operator education. The third property is often neglected in database research. Database operators have a very simple solution for this property: they switch off all those integrity preserving processes that become a bottleneck at database production time and switch on these constraints for a short time at the maintenance phase.

Preservation of access and modification is a fuzzy criterion since we often may not assume that any access and modification can be forecasted at design time. The co-design approach to database modelling [31] also takes into consideration functionality that can be envisioned. Optimisation of the internal schemata is based on profiles of functions, stored procedures and transactions (shortly processes) and on the cardinality profile of the relations. The first profile provides information on the kind of the operation, the frequency, the relations affected by the processes, the type of access (online, batch, prefetch or ad-hoc) and the changes applied to the databases. The cardinality profile of relations provides detailed information on the size of the relation (minimal, maximal and average), on the changes of the size over time and the associations among relations that must be maintained. Both profiles are compared with the modus of computation (bath, online, ad-hoc), with performance expectations (execution time, throughput, priority), with visibility of performance gaps (depending on operations frequency, organisation level of users, business process interaction) and with computation strategies for the operations (kind of operation, scheduling, auxiliary facilities, selection and storage alternatives, set sizes).

Decomposition approaches generate structures that easily support some of the constraints such as key constraints, domain constraints and key-based inclusion constraints. Typically, the maintenance complexity of such constraint sets is not taken into account. Moreover, decomposition algorithms may generate a large variety of decompositions that are semantically equivalent but pragmatically and technologically different. Typical normalisation algorithms are deterministic for a given set of functional dependencies, for an order of attributes and an order of the dependencies. Changes in the the last two orders result in different solutions of those algorithms.

2.4   Infomorphisms Among Schemata

We use the notion of infomorphisms as the general foundation for schema optimisation. Infomorphisms have been used for schema modernisation in [15, 38]. A typical example of an infomorphism is the association of a relational database schema and a sophisticated XML schema (with a good number of additional constraints since XML uses list (or tree) and reference types instead of values). The relational database schema that is obtained by the classical forgetful mapping from an entity-relationship schema is not an infomorphism since the ER structuring is richer than the relational structuring.

Let us consider two *database* **schemata** S_1 and S_2 consisting of *database types* of the form $T = (struc(T), \Sigma, \Sigma^*)$ with a structure definition, inner integrity constraints Σ defined on the type, and outer integrity constraints Σ^* that constrain the type by means of other types. Structure elements of types that are not defined by constructors are called *basic*.

Let us consider only *complete schemata*, i.e. those which types are complete relative to the outer constraints. Given furthermore, *basic domain types* B for the value foundation of the database. We use the abstract data type approach for basic domain types and presume for these types their value collections, their operations, and their predicates.

An *extended database schema* $\mathfrak{D} = (S, B, DOM)$ consists of a database schema and an assignment DOM of basic elements of its types to basic domain types.

The set of all $MOD((\mathcal{S}, \mathcal{B}, DOM))$ of all finite databases on $\mathfrak{M}$ consists of finite collections of classes for each type for which all constraints are valid, which values of objects in a class are given by DOM.

Let us now associate databases for different extended database schemata $\mathfrak{D}_1$ and $\mathfrak{D}_2$ by mappings mappings $\widehat{put}_{1,2}$ and $\widehat{put}_{2,1}$. These two mappings form an **infomorphism** of $MOD(\mathcal{D}_1)$ and $MOD(\mathcal{D}_2)$ if for i, j with $\{i, j\} = \{1, 2\}, i \neq j$ and for each database DB_i on $MOD(\mathcal{D}_i)$ there exists a database DB_j and on $MOD(\mathcal{D}_j)$ such that $\widehat{put}_{i,j}(DB_i) = DB_j$ from one side and $\widehat{put}_{j,i}(DB_j) = DB_i$ from the other side.

We may extend this notion also to views defined on each of the database schemata. The association among views can be based on the extract-transform-load (ETL) approach where extraction is based on a query language of the first schema, transformation is given by an infomorphism, and loading uses views which allow updates on the second database schema.

This notion is very general one. Infomorphisms are essentially transformations of one database to another one. These transformations are *information-invariant* in the sense that any database object collection can be associated with one and only one database object collection from the other extended database schema.

The infomorphism notion can be based on HERM schema operations ([31], Chapter 9.2.) in the case that we consider only classes with set semantics. The Σ^* dependence among types also includes inclusion constraints. Therefore, vertical normalisation can be directly expressed in this approach. In this case, we can represent transformations as graph-grammar rules which are defined on sub-schemata. Horizontal normalisation uses separating selection σ_{α_i} predicates which define a partition of singleton classes. Deductive normal forms use for the mapping a reduction operation from one side and a chase-like completion procedure for the second mapping.

Observation 3. Vertical normalisation, horizontal normalisation, and deductive normalisation are specific variants of infomorphisms.

We conclude now that a theory of infomorphisms can subsume the classical relational normalisation theory, especially vertical normalisation. It is, moreover, better since the pairwise inclusion constraints after decomposition must be integrated into the decomposed schema. Infomorphisms can be partially supported by schema construction rules for extended entity-relationship schemata. These rules follow the graph grammar approach.

2.5 Global and Local Vertical Normalisation

The synthesis algorithm is also based on *structure minimality*, i.e. the type structures form a Sperner set in the sense that $struct(T_1) \not\sqsubseteq struct(T_2)$ is not valid for any two types T_1 and T_2 of a schema $\mathcal{S}$. Structure minimality reduces the maintenance. It might, however, provide its advantages as we already illustrated for the RM/V2 approach. Moreover, additional structures such as overlapping and subtype indexes (as hedges of indexes) may support performance of computation and also input-output to a real essential extent.

Global normalisation concurrently and coherently considers all types of a database schema. The example in [31] allows to derive five different results of a normalisation of a small schema. Each of these schemata have their advantage.

We may define a result of a normalisation process that is applied to an entity-relationship schema as a type-wise transformation of the given schema by an info-morphism, i.e. the types of a schema are (vertically or horizontally) decomposed to a schema in which all types are in certain α-normal form ($\alpha \in \{1, 2, 3, 4, 5, 6, BCNF\}$). Since decomposed types may be a substructure of another one, we use these types only once. Entity, cluster, and relationship types are transformed by graph grammar rules [31]. The decomposition of a relationship type follows the procedure developed in [22].

A schema S is a *global α-normal form* schema if all its types are in α-normal form and if the schema is structure-minimal. Within a platform setting $\mathfrak{P}$, we add the requirement that all its integrity constraints can be supported by declarative means provided by the platform. Otherwise, the schema is called *(α,$\mathfrak{P}$)-unnormalised*.

Observation 4. *Global normalisation is based on an infomorphism.*

3   Denormalisation

... There are many database experts, particularly the more academic ones, who feel that any talk about denormalising a database is like a race car driving – a great way to mangle or kill yourself even if you know what your are doing. [8]

3.1   State-Of-the-Art for Denormalisation

We observe two camps where the first one is well acknowledged.

No denormalisation at all! Almost all[6] textbooks and monographs in the database area require strict normalisation. Local (vertical) normalisation of a singleton database type is well reflected in most database books (e.g. [1,5,19,41]) and publications, most database courses, and in actual database practice. It is considered as one of the pearls of database research and known to almost everybody who knows database technology. The provenance and acknowledgement is based on the facility it provides: keeping as much as possible locally and globally supporting only those processes that are inherently global. Both independence concepts of databases (conceptual independence and implementation independence) are based on localisation. [11] advocates lazy normalisation based on relevant and quickly to capture FD's, i.e. somehow *liberal normalisation*[7] for which not all functional dependencies that are valid in database schema are considered but only the really important ones[8]. Additionally, almost valid FD's might be more important than FD's that happens to be valid. The treatment of such dependencies would be based on the introduction

---

[6] We know so far only less than a handful books that do not require such.

[7] Many constraints can be omitted since integrity is also often managed through proper interfacing and exchange procedures without a chance for inconsistency as long as the data modification is exclusively based on interface or exchange view data. The development of a theory for this approach is one of the lacunas of database theory.

[8] We avoid the exponential size **trap** for sets of functional dependencies with this toleration of incompleteness of constraint sets. We consider only essential ones and completely or partially neglect others. This approach can be extended to a *theory of robust normalisation*.

of artificial identifiers, i.e. a heavy object identity pollution. A far better solution is horizontal decomposition with a class for which all identities are valid and an exception class in which the few exceptions are recorded. Horizontal decomposition can be combined with union views as long as the exceptions are disjoint from the normal case.

Liberal and controlled denormalisation whenever it is really necessary: Very few papers and books advocate or consider at least to some extent denormalisation (e.g. [6–8] or the discussion in [31]). The three central quality criteria for database installations are, however, performance, performance, and performance. The classical vertical local normalisation is useful as long as any casual user may query by any casual query at any time without considering performance. However, a database system contains of a (or a number of) DBMS with a number of databases on top of which a large massive of business procedures has been developed. These business procedures form the main part of the profile of the database. Casual queries are rather exceptions. The definition that is used for denormalisation is typically based on application of the natural join operator to relational types[9].

Our consulting experience and also observations on the why's for OLAP and data warehouse applications drives us to a completely different picture in many applications. The first setting of a database application is very often based on normalisation. This database becomes then step by step denormalised after the database is fully populated and operating. Already after one year of full operation, the database is partially normalised and also partially denormalised.

3.2   A Matter of Definition

Denormalisation has not yet been defined in the literature despite [4]. Essentially, we find two approaches (e.g. in [6, 8, 10] for simple forms):

Denormalisation as the inverse of normalisation: Given a schema with α-normalised types. Any non-trivial combination (typically by a join operation) of two or more types that defines an infomorphism is a denormalisation.

Denormalisation as the extension of a schema: Given a schema S. Any extension of the schema that is defined by an infomorphism is called denormalisation.

Typical extension operations are [8]: prejoined types for types with complex queries, reports added to the schema, mirrored types, type splitting by horizontal decomposition, partial combination of types, introduction of controlled redundancy for data, repeating groups, hierarchy tables, and overloading of types.

Given a $(\alpha, \mathfrak{P})$-normalised schema. Any infomorphism transformation of this schema to an unnormalised one is called ***denormalisation***. Index and other supporting means thus do not change the normalisation status of a schema.

We may consider at the same time normalised and denormalised schemata. A theory and techniques for denormalisation for physical schemata based on normalised conceptual (or logical) schemata have been developed in [29]. [20] lists some key effects of thoughtful denormalisation: definite improvement in query time, a potential increase in

---

[9] The validity of pairwise inclusion constraints is also neglected in this case.

update time or in storage space, a potential loss of data integrity due to certain deletions, the necessity for program transformations for all relevant queries and the overhead needed to reorganise some tables. Strict local normalisation may be inadequate. Denormalisation may result in a more complex maintenance complexity. It may also lead to complications for query formulation. It becomes easier to formulate incorrectly a query to a request meaning that the query does not correspond to the request. Often tricky view creation is used for denormalised tables. The denormalisation is considered a method for performance improvement despite discussed so far advantages of normalisation.

Our definition of denormalisation does not allow composition by equi-join since we have to avoid the NULL marker problem. NULL markers must be treated depending on their specific meaning, their occurrence, and their impact on computation. We however support co-existence of vertical and horizontal normalisation and denormalisation.

Therefore, it seems that normalisation is the best way for optimisation of database behaviour. A theory of denormalisation has not yet been proposed as far as we know.

Question 5. What are the denormalisation criteria? Is there any theory for it? Is there any 'playground' approach for consideration of (de)normalisation?

Instead, a number of heuristic rules for denormalisation are provided. These rules are based on observations for performance traps for some of the platforms and often use the 80/20% rule.

3.3   Denormalisation Driven by Optimisation

We base our approach on essentials of database performance forecasting, tuning techniques, and database distribution into fragments [4, 26]. Our approach has been implemented in an industrial setting and for performance improvement for a very large cluster of databases [4, 37].

Let us first define the performance portfolio of a database application and the profile of a DBMS. A *portfolio* consists of a set or collection of tasks. A *profile* of a DBMS specifies the services and the capability of a DBMS. The *extended database application schema* consists of the database schema, the business processes and the characterisation of the application demand by a *characterisation of the kind of computation* based on the description of the operations involved, the operation support, and the data volumina transferred for support of computation, the *visibility description of processes* for the business user that includes frequency of operations and their relation to business processes, the *description of the modes of computation* such as online, batch and interactive mode of computation or deferrable and immediate application of computation, the *performance properties and quality* based on the expected execution time for online etc. modes, based on the throughput expectation for queries, modifications and transactions, based on restrictions such as suitability or response time, and based on priority claims issued by the business user, the *criticality level* of the processes.

We derive now the measures for this application, a database schema, and a DBMS:

Data modification costs: Given a set M of modification operations m_i with their weight wm_i in the application, the frequency of application hm_i of each operation, their complexity of realisations m_i^* in the DBMS $\mathcal{P}$, and the complexity of integrity maintenance sm_i for the operation m_i.

The complexity m_i^* can be computed type-wise for types T from the schema $\mathcal{S}$, i.e. by m_{iT}^*.

The modification complexity $modify(\mathcal{S}, \mathcal{P}, M)$ is given by the formula:

$$\sum_{m_i \in M} \left(\left(\sum_T (m_{iT}^* + sm_i) \times wm_i \right) \times hm_i \right)$$

Query cost: Given a set Q of queries q_j with their weight wq_j in the application, the frequency of application hq_j of the query q_j, the complexity q_j^* of the realisation of q_j, and the complexity of integrity query imposed integrity maintenance sq_j for the query q_j.

The complexities can be computed type-wise for the types T from the schema $\mathcal{S}$, i.e. by q_{iT}^*.

The query complexity $querymodify(\mathcal{S}, \mathcal{P}, Q)$ is given by the formula:

$$\sum_{q_j \in Q} \left(\left(\sum_T (q_{jT}^* + sq_j) \times wq_j \right) \times hq_j \right)$$

The **schema complexity** $complexity(\mathcal{S}, \mathcal{P}, M, Q)$ is the sum of the modification complexity and of the query complexity.

An infomorphism can be now extended to the modification and to the query operations under consideration of the base types and the domain assignments $(\mathcal{B}, DOM)$.

We can now compare the complexity of the schema according to the modification and the query portfolio of a given application.

Given two extended database schemata $\mathfrak{D}_1 = (\mathcal{S}_1, \mathcal{B}_1, DOM_1)$ and $\mathfrak{D}_2 = (\mathcal{S}_2, \mathcal{B}_2, DOM_2)$ and an infomorphism $(\widehat{put}_{1,2}, \widehat{put}_{2,1})$ for these two schemata; further, given a platform $\mathcal{P}$, and a modification and query portfolio.

The extended database schema $\mathfrak{D}_1$ performs better than the database schema $\mathfrak{D}_2$ in a given setting $\mathcal{P}$ for a portfolio $M \cup Q$ if

$$complexity(\mathcal{S}_1, \mathcal{P}, M, Q) \; << \; complexity(\mathcal{S}_2, \mathcal{P}, M, Q) \qquad .$$

We use $<<$ as a denotation for an essential discrepancy of the two complexities.

We may thus derive the normalisation and denormalisation criterion for schema optimisation for given schemata where $\mathcal{S}_1$ is $(\alpha, \mathcal{P})$-normalised and $\mathcal{S}_2$ is $(\alpha, \mathcal{P})$-denormalised:

Use the normalised schema $\mathcal{S}_1$ if $\mathcal{S}_1$ performs better than $\mathcal{S}_2$ in the given setting.
Use the denormalised schema $\mathcal{S}_2$ if $\mathcal{S}_2$ performs better than $\mathcal{S}_1$ in the given setting.

We neglect within this approach the existence of casual queries and of casual data modification. This approach, however, supports typical applications where the retrieval and also the modification is well-defined at the development or at later maintenance time.

Observation 5. The optimisation approach to normalisation and denormalisation allows to coherently meet demands for the six reasons why we should normalise.

4 Conclusion

4.1 Summarising

Normalisation is considered to be one of the pearls of database theory and technology. We mainly consider, however, local vertical normalisation instead of global normalisation or horizontal normalisation. It seems that normalisation theory is a body of knowledge that is completely settled and well understood in most of its aspects. We discuss on the basis of simple examples that this impression is not valid. Normalisation is not well understood. It needs a lot of extensions and corrections. It must also be completely revised for the modern DBMS technology. One essential revision is the flexible choice for set-based or multi-set-based semantics. This extension opens the path towards list, pointer, multi-list, etc. semantics that is supported nowadays by systems.

Normalisation is often a performance bottleneck. Repairing this bottleneck is often done on the fly. In practice, skilled consultancy uses here a hands-on, experience-backed approach. The DAMA[10] community and database forums widely discuss in closed groups the experience some people got. We claim that most larger database applications allow coexistence of partial normalisation (in both vertical and horizontal style) and partial denormalisation.

This paper aims now to highlight the path to a coherent theoretical underpinning for this kind of coexistence. We first discussed problems of classical normalisation based on the verticality and locality approach for simple constraints such as functional and multivalued dependencies. The problems discussed can be resolved by pragmatical approaches. Some of them are discussed in the paper. We are not capable to present a full theory which would require a two-volume monograph. So, we restricted only on some parts of this theory.

Normalisation can be understood as a special kind of optimisation. As such it should be treated as "a commandment" [10] unless the database application requires high query performance. Optimisation of schemata is based on some kind of equivalence. We use infomorphisms as one solution for treatment of equivalence. This solution requires deep knowledge of the the given database application. It can be extended to handling of robust constraint sets what is, however, an open issue. It can also be extended to handling by basic-structure normalisation that is neatly supported by interface and exchange tolerance as long as the interfaces and the exchange means provide a support for the other optimisation (or more specifically normalisation) requirements.

4.2 Open Problems

The list of open problems is slowly shrinking and quickly expanding at the same time. We have collected open problems since MFDBS'87[11], have extended this list, and observed whether some of them have been resolved. The latest version in [33] contains 22 open problems which are directly related to normalisation theory.

---

[10] https://www.dama.org.

[11] With 21 open problems from which 13 are not yet solved.

Normalisation theory is currently a theory for system structures in the small. Global normalisation will be theory for system structures in the large. The world is now changing to systems in the web and systems that are based on completely different performance challenges such as big data massives. *Normalisation in the world* is a really big issue for future research. It goes far beyond theories we know for distributed databases.

Revolution Instead of Unworthy Extension. Already research on the OO identifier and the OID pollution has been demonstrating that parts and pieces of database theory must be revised. Many assumptions taken for granted are not valid anymore and will never be valid again for challenging applications such as big data massives. Set semantics was a nice tool in the past. It is not the right one - at least for SQL applications and multi-sets in practice. we might ask why not also to use multi-list semantics. Big data requires a different FD logic.

References

1. Abiteboul, S., Hull, R., Vianu, V.: Foundations of Databases. Addison-Wesley, Reading (1995)
2. Beeri, C., Thalheim, B.: Identification as a primitive of database models. In Proceedings of the FoMLaDO 1998, pp. 19–36. Kluwer, London (1999)
3. Benczúr, A.A., Kiss, A., Markus, T.: On a general class of data dependencies in the relational model and its implication problems. Comput. Math. Appl. **21**(1), 1–11 (1991)
4. Bick, M.: Denormalisierung. Master's thesis, CAU Kiel, Department of Computer Science (2015)
5. Biskup, J.: Foundations of Information Systems. Vieweg, Wiesbaden (1995). (in German)
6. Buxton, S., et al.: Database Design - Know It All. Morgan Kaufmann, Burlington (2008)
7. Celko, J.: Joe Celko's SQL for Smarties - Advanced SQL Programming. Morgan Kaufmann, San Francisco (1995)
8. Celko, J.: Joe Celko's Data and Databases: Concepts in Practice. Morgan Kaufmann, Burlington (1999)
9. Codd, E.F.: The Relational Model for Database Management (Version 2). Addison-Wesley, Reading (1991)
10. Date, C.J.: Database Design and Relational Theory - Normal Forms and All That Jazz. O'Reilly, Sebastopol (2012)
11. Date, C.J.: Go Faster - The TransRelational Approach to DBMS Implementation. C.J. Date & Ventus Publishing ApS, Frederiksberg (2011)
12. Demetrovics, J., Molnar, A., Thalheim, B.: Graphical and spreadsheet reasoning for sets of functional dependencies. In: Proceedings of the ER 2004, LNCS, vol. 3255, pp. 54–66 (2004)
13. Demetrovics, J., Molnar, A., Thalheim, B.: Graphical and spreadsheet reasoning for sets of functional dependencies. Technical Report 0402, Kiel University, Computer Science Institute (2004). http://www.informatik.uni-kiel.de/reports/2004/0402.html
14. Kiss, A., Markus, T.: Functional and inclusion dependencies and their implication problems. In: 10th International Seminar on DBMS, Cedzyna, Poland, pp. 31–38 (1987)
15. Klettke, M., Thalheim, B.: Evolution and migration of information systems. In: Embley, D., Thalheim, B. (eds.) The Handbook of Conceptual Modeling: Its Usage and Its Challenges, pp. 381–420. Springer, Berlin (2011). https://doi.org/10.1007/978-3-642-15865-0_12

16. Koehler, H.: Autonomous sets – a method for hypergraph decomposition with applications in database theory. In: Hartmann, S., Kern-Isberner, G. (eds.) FoIKS 2008. LNCS, vol. 4932, pp. 78–95. Springer, Heidelberg (2008). https://doi.org/10.1007/978-3-540-77684-0_8

17. Köhler, H.: Autonomous sets for the hypergraph of all canonical covers. Ann. Math. Artif. Intell. **63**(3–4), 257–285 (2011)

18. Köhler, H., Link, S.: SQL schema design: foundations, normal forms, and normalization. Inf. Syst. **76**, 88–113 (2018)

19. Leonard, M.: Database Design Theory. MacMillan, Houndsmills (1992)

20. Lightstone, S., Teorey, T., Nadeau, T.: Physical Database Design. Morgan Kaufmann, Burlington (2007)

21. Makowsky, J.A., Ravve, E.V.: Dependency preserving refinements and the fundamental problem of database design. DKE **24**(3), 277–312 (1998). Special Issue: ER 1996 (ed. B. Thalheim)

22. Mannila, H., Räihä, K.-J.: The Design of Relational Databases. Addison-Wesley, Wokingham (1992)

23. Paredaens, J., De Bra, P., Gyssens, M., Van Gucht, D.: The Structure of the Relational Database Model. Springer, Berlin (1989). https://doi.org/10.1007/978-3-642-69956-6

24. Popkov, G.P., Popkov, V.K.: A system of distributed data processing. Vestnik Buryatskogo Gosudarstvennogo Universiteta **9**, 174–181 (2013). (in Russian)

25. Schewe, K.-D., Thalheim, B.: NULL value algebras and logics. In: Information Modelling and Knowledge Bases, vol. XXII, pp. 354–367. IOS Press (2011)

26. Shasha, D.E., Bonnet, P.: Database Tuning - Principles, Experiments, and Troubleshooting Techniques. Elsevier, Amsterdam (2002)

27. Simsion, G., Witt, G.C.: Data Modeling Essentials. Morgan Kaufmann, San Francisco (2005)

28. Sörensen, O., Thalheim, B.: Semantics and pragmatics of integrity constraints. In: Schewe, K.-D., Thalheim, B. (eds.) SDKB 2011. LNCS, vol. 7693, pp. 1–17. Springer, Heidelberg (2013). https://doi.org/10.1007/978-3-642-36008-4_1

29. Steeg, M.: RADD/raddstar - a rule-based database schema compiler, evaluator, and optimizer. Ph.D. thesis, BTU Cottbus, Computer Science Institute, Cottbus, October 2000

30. Thalheim, B.: Dependencies in Relational Databases. Teubner, Leipzig (1991)

31. Thalheim, B.: Entity-Relationship Modeling - Foundations of Database Technology. Springer, Berlin (2000). https://doi.org/10.1007/978-3-662-04058-4

32. Thalheim, B.: Conceptual treatment of multivalued dependencies. In: Song, I.-Y., Liddle, S.W., Ling, T.-W., Scheuermann, P. (eds.) ER 2003. LNCS, vol. 2813, pp. 363–375. Springer, Heidelberg (2003). https://doi.org/10.1007/978-3-540-39648-2_29

33. Thalheim, B.: Open problems of information systems research and technology. In: Kobyliński, A., Sobczak, A. (eds.) BIR 2013. LNBIP, vol. 158, pp. 10–18. Springer, Heidelberg (2013). https://doi.org/10.1007/978-3-642-40823-6_2

34. Thalheim, B.: Conceptual models and their foundations. In: Schewe, K.-D., Singh, N.K. (eds.) MEDI 2019. LNCS, vol. 11815, pp. 123–139. Springer, Cham (2019). https://doi.org/10.1007/978-3-030-32065-2_9

35. Thalheim, B.: Semiotics in databases. In: Schewe, K.-D., Singh, N.K. (eds.) MEDI 2019. LNCS, vol. 11815, pp. 3–19. Springer, Cham (2019). https://doi.org/10.1007/978-3-030-32065-2_1

36. Thalheim, B., Tropmann-Frick, M.: The conception of the conceptual database model. In: ER 2015. LNCS, vol. 9381, pp. 603–611. Springer, Berlin (2015)

37. Tropmann, M., Thalheim, B.: Performance forecasting for performance critical huge databases. In: Proceedings of the EJC 2010, Jyväskylä, pp. 214–233 (2010)

38. Wang, Q., Thalheim, B.: Data migration: a theoretical perspective. DKE **87**, 260–278 (2013)

39. Webster, B.F.: Pitfalls of Object-Oriented Development: A Guide for the Wary and Entusiastic. M&T Books, New York (1995)

40. Wei, Z., Link, S.: Embedded functional dependencies and data-completeness tailored database design. PVLDB **12**(11), 1458–1470 (2019)
41. Yang, C.-C.: Relational Databases. Prentice-Hall, Englewood Cliffs (1986)

Strongly Minimal MapReduce Algorithms: A TeraSort Case Study

Daniel Xia, Michael Simpson$^{(\boxtimes)}$, Venkatesh Srinivasan, and Alex Thomo

University of Victoria, Victoria, Canada
daniel.f.xia@gmail.com, {simpsonm,srinivas,thomo}@uvic.ca

Abstract. MapReduce is a widely used parallel computing paradigm for the big data realm on the scale of terabytes and higher. The introduction of *minimal* MapReduce algorithms promised efficiency in load balancing among participating machines by ensuring that *partition skew* (where some machines end up processing a significantly larger fraction of the input than other machines) is prevented. Despite minimal MapReduce algorithms guarantee of load-balancing within constant multiplicative factors, the constants are relatively large which severely diminishes the theoretical appeal for true efficiency at scale.

We introduce the notion of *strongly minimal* MapReduce algorithms that provide strong guarantees of parallelization up to a small additive factor that diminishes with an increasing number of machines. We show that a strongly minimal MapReduce algorithm exists for sorting; this leads to strongly minimal algorithms for several fundamental database algorithms and operations that crucially rely on sorting as a primitive. Our techniques are general and apply beyond the analysis of strongly minimal MapReduce algorithms; we show that given a sufficiently high, but still realistic, sampling rate, the approximate partitions obtained from a particular sampling strategy are almost as good as the partitions produced by an ideal partitioning.

Keywords: Distributed sorting · Minimal MapReduce algorithms · Sample-Partition problem

1 Introduction

Data is being generated at an increasing pace that leads to an enormous volume being created and stored each year. As a result, there has been a strong push towards *big data analytics* as industry and governments around the world aim to keep pace with the explosion of information. This has led database organizations to build massive parallel computing platforms that rely upon huge numbers of commodity machines. Among these platforms, *MapReduce* has emerged as the popular choice after years of improvement and advancement.

At a high level, MapReduce algorithms instruct how these machines can perform a given task collaboratively. Typically, the input data is distributed

© Springer Nature Switzerland AG 2020
A. Herzig and J. Kontinen (Eds.): FoIKS 2020, LNCS 12012, pp. 301–317, 2020.
https://doi.org/10.1007/978-3-030-39951-1_18

across the machines and the algorithm executes in *rounds* made up of *map* and *reduce* phases. The *map* phase prepares data to be exchanged to other machines and the *reduce* phase has the machines perform isolated computations on its local storage. Rounds proceed until the given task is complete after a *reduce* phase. Ideally, MapReduce algorithms should aim for the minimization of space, CPU, I/O, and network costs for each machine as well as even load balancing. Despite these principles guiding the design of MapReduce algorithms, most previous work has relied upon heuristic approaches or been driven by an experimental performance basis where less emphasis is placed on enforcing rigorous constraints on these performance metrics.

Tao et al. [24] introduced *minimal* MapReduce algorithms that promise efficiency in multiple aspects simultaneously. The notion of *minimal* MapReduce algorithms bounds the storage space and the amount of information sent over the network of each machine to be optimal up to a constant multiplicative factor. In addition, a constraint is placed that the algorithm terminates in a constant number of rounds and that the algorithm achieves a speedup of a factor t when using t machines in parallel. One of the benefits of designing minimal algorithms includes guaranteeing that *partition skew* (where some machines end up processing a significantly larger fraction of the input than other machines) is prevented. Tao et al. [24] conclude that TeraSort[1], the state-of-the-art MapReduce sorting algorithm, is load-balanced within constant multiplicative factors and thus satisfies their minimality definition. However, the multiplicative factor in the proof is relatively large, up to 32; such a factor severely diminishes the theoretical appeal due to a large imbalance in the workload of machines. Our work is motivated by the question of whether the notion of minimal MapReduce algorithms can be strengthened even further?

The main contributions of our work are as follows: (1) We propose a strengthening of the notion of minimality by introducing an additional requirement called *balanced partition* that restricts the number of objects processed by each machine to be evenly balanced up to an additive factor that diminishes with the number of machines. We say such MapReduce algorithms are *strongly* minimal. (2) As a core result, we prove that sorting, which is a backbone primitive for many algorithms and operations in databases and beyond, has a strongly minimal algorithm, i.e. the workload is evenly distributed across machines up to an additive term rather than a multiplicative factor. This is important because in large data centers of similar machines having an overloaded machine is a critical bottleneck (referred to by practitioners as "the curse of the last reducer", see [23]).

More specifically, we aim for a more accurate analysis of the performance of TeraSort. We give a series of bounds which describes the trade-off between the number of machines and the partition skew (hence the worst case maximum workload on a single machine). We conclude that a larger number of available

---

[1] We use the name TeraSort in this paper to refer to the Sample-Partition-Sort *paradigm* for distributed sorting. We would like to emphasize that the version implemented in Hadoop (a popular MapReduce implementation), while following the general paradigm, is not minimal (please see Sect. 3).

machines allows for more even partitions. In particular, we show that as the number of machines grows, the partition skew approaches its optimal value with high probability.

Our techniques are general and apply beyond TeraSort; we show that given a sufficiently high, but still realistic, sampling rate, the approximate partitions obtained from a particular sampling strategy are almost as good as the partitions produced by an ideal partitioning. The sampling strategy we analyze, self-sampling, has appealing probabilistic properties that we are able to leverage through a more extensible probabilistic method that is capable of generating a series of tight bounds for partition evenness. We use a new and refined analysis technique by an interesting bucketing argument that allows *overlapping* buckets in contrast to the non-overlapping scheme of [24].

1.1   Strongly Minimal MapReduce Algorithms

Instead of mappers and reducers, we refer to the workers as *machines*. To accommodate the statelessness of mappers and reducers, we assume that unmentioned data in the algorithm is "carried forward" implicitly to the reducer with the same index in the next round.

Let A be the input for the underlying problem, n be the number of objects in A, and t be the number of machines in the network. Define $m = n/t$ as the number of objects per machine when A is evenly distributed across the machines. Then, as defined in [24], a minimal MapReduce algorithm for a problem on A has the following properties:

- *Minimal footprint:* each machine uses $O(m)$ storage at all times.
- *Bounded net-traffic:* every machine sends and receives at most $O(m)$ words of information over the network in each round.
- *Constant round:* the algorithm terminates after a constant number of rounds.
- *Optimal computation:* the algorithm achieves a speedup of t when using t machines in parallel. Precisely, each machine performs $O(T_{seq}/t)$ work in total over all rounds where T_{seq} is the time required by a *fixed* algorithm $\mathcal{A}$ to solve the problem on a single machine.

Now, we introduce the notion of a *strongly* minimal MapReduce algorithm which strengthens the minimality conditions above by adding an additional condition that we refer to as *balanced partition*. A strongly minimal MapReduce algorithm for a problem on A has the following property:

- *Balanced partition:* each machine processes $m(1 + o(1))$ objects.

Remark: Here, $o(1)$ denotes a lower order term that is independent of the input and goes to 0 as t grows. Note that strong minimality implies that, as t grows, the number of objects processed by each machine approaches the optimal value. As a consequence, the hidden constant in footprint and net-traffic in strongly minimal MapReduce algorithms is substantially smaller than 32, typically close to 1.

In the following sections, we prove that TeraSort is strongly minimal by showing that the *balanced partition* condition is satisfied using an interesting bucketing argument.

2 Related Work

The existing investigation on MapReduce can be broadly classified into two categories: (1) a focus on improving the internal working of the framework, and (2) developing novel MapReduce algorithms to solve interesting problems. On the framework implementation side there has been a variety of work that typically focuses on performing well on a subset of the minimality conditions. These range from specialized methods to rectify skewness [11,14,16] to optimizing the network traffic by keeping relevant data at the same machine [6,12]. On the algorithms side, there has been extensive work dedicated to developing MapReduce algorithms for important database problems [1,4,18,19,26], graph processing [2,13,17,23,25], and statistical analysis [5,7,9,10,21].

Tao et al. [24] justify theoretically the good performance of TeraSort [20] (the state-of-the-art MapReduce sorting algorithm) observed in practice which inspired the new definition of minimal MapReduce algorithms. Their goal is accomplished by specifying how to set a crucial parameter of TeraSort that ensures minimality. Designing minimal algorithms is highly sought after since a minimal algorithm excels on all the minimality conditions simultaneously. Often, it is easy to perform well on certain aspects, while failing on others. Furthermore, Tao et al. [24] point out that even a minimal algorithm can benefit from clever optimization at the system level, and the minimality property may considerably simplify such optimizations. For instance, as the minimality requirements already guarantee good load balancing in storage, computation, and communication, there would be less skewness to deserve specialized optimization.

Studying the minimality of MapReduce algorithms is similar in goal to other models of theoretical parallel computing. We give two such examples now. Karloff et al. [13] put forth the notion of $\mathcal{MRC}$, a class of MapReduce algorithms computable by a MapReduce system characterized by a certain amount of resources. Further, class $\mathcal{MRC}^i$ runs in $O(\log^i n)$ rounds and $\mathcal{MRC}$ is defined by the union of $\mathcal{MRC}^i$ over i. When the algorithm is randomized, it must output the correct answer with probability at least $3/4$. The deterministic subset of $\mathcal{MRC}$ is called $\mathcal{DMRC}$. Note, not all algorithms in $\mathcal{MRC}$ are efficient; rather it only offers to characterize them. One would expect efficient algorithms in $\mathcal{MRC}^0$ and $\mathcal{MRC}^1$ since they consist of constant rounds and logarithmic rounds, respectively. [13] shows that a variety of problems have solutions in $\mathcal{MRC}^0$ and $\mathcal{MRC}^1$, such as finding an MST in dense graphs, frequency moments, and undirected *s-t* connectivity. Recent work following the $\mathcal{MRC}$ model include [3,15,22].

Massive, unordered, distributed (MUD) is a class of MapReduce algorithms proposed by [8] to compute a *distributed* stream. The MUD algorithms consist of three components: a *local function*, an *aggregator*, and *post-processing*. The algorithm designer must ensure that the overall output is independent of the order of application of the aggregator. A connection between MUD algorithms and the MapReduce framework can be established where the local function can be implemented by mappers, and the independence of the post-processing and the order of application implies that we can divide and conquer the output of the local function in a series of rounds. From this construction, it is easy to see that MUD algorithms can be computed very efficiently in a MapReduce system and mostly independent of the underlying computing capability.

3   Sorting with MapReduce

For sorting, the input is a set A of n objects drawn from an ordered domain. Suppose that t machines store A and are indexed from 1 to t, namely $\mathcal{M}_1, ..., \mathcal{M}_t$. A parallel algorithm that solves the sorting problem should terminate with all the objects distributed across the t machines in a (total) sorted fashion. That is, for each machine $\mathcal{M}_i$, the objects that end up in $\mathcal{M}_i$ are in sorted order. Further, this implies that all objects in $\mathcal{M}_i$ precede those in $\mathcal{M}_j$ for all $1 \leq i < j \leq t$.

It is well known that sorting can be solved in $O(n \log n)$ time on a single machine, while there has been a substantial amount of progress on sorting in parallel. *TeraSort* is the state-of-the-art MapReduce algorithm for sorting and the work of [24] proves that it is minimal when a crucial parameter of the algorithm is set appropriately.

3.1   Sampling and Partitioning

Sampling and partitioning form the central idea behind TeraSort. TeraSort conceptually consists of three steps: Sample, Partition, and Sort. First, the algorithm extracts a random sample set from the input and then computes t partition elements from the sample. The partition elements, referred to as *boundary elements*, divide A into t partitions. In the second round, each machine receives all the elements from a distinct partition and sorts them locally using an apriori fixed algorithm $\mathcal{A}$. As the performance of TeraSort is sensitive to the quality of the partition, it is worth examining potential sampling strategies. We describe two such sampling strategies below.

In the current implementation of TeraSort included in Hadoop, the sample is created by reading a elements in total from b locations which are evenly spread across the input dataset. a and b are configurable by users. At each location, $\frac{a}{b}$ elements are read. No guarantee exists that such sampling scheme yields good partitioning. In fact, there are bad cases for every a, b in which the partitions are extremely unbalanced. When the sample comprises elements concentrated in a few small ranges, it may lead to uneven buckets.

Tao et al. [24] discuss the strategy of *self-sampling*, where each element is selected into the sample independently with the same probability. Self-sampling is a good fit for the MapReduce framework as mappers are assumed to have no other knowledge than the input item currently being processed. As we will show later, self-sampling has very appealing probabilistic properties and it achieves asymptotically optimal evenness with high probability. [24] also report the experimental results of another strategy: sampling without replacement. The results are promising and comparable to self-sampling: the unevenness remains low when the sample size is no less than the expected size of self-sampling, regardless of how much it exceeds the latter. However, there is no further investigation on how the evenness is affected by the layout of the input dataset, nor is any bound provided for all input layouts.

In this paper, we focus on analyzing TeraSort with self-sampling. However, our techniques are general and apply beyond TeraSort; we show that given a sufficiently high, but still realistic, sampling rate, the approximate partitions obtained from self-sampling are almost as good as the partitions produced by an ideal partitioning.

3.2   Even Partitions

In TeraSort, we use the notion of an ordered even t-partition which divides a set as evenly as possible.

Definition 1 (ordered t-partition). *An ordered t-partition divides an ordered set of n elements into t partitions. Elements of partition i are smaller than those of partition j for all $i < j$. The first element of every partition except for the first one is called a t-partition element as they describe the partitions in full.*

Definition 2 (ordered even t-partition). *An ordered even t-partition is an ordered t-partition in which the sizes of the partitions differ by at most 1.*

An ordered even partition always exists for any dataset. In fact, we may construct one in the following way. Let $n = ts_1 + s_2$ where $s_1 = \lfloor \frac{n}{t} \rfloor$. The indices of the partition elements are $d_j = d_{j-1} + (s_1 + 1)$ for $1 \leq j \leq s_2$ and $d_j = d_{j-1} + s_1$ for $s_2 \leq j \leq t$ with $d_0 = s_1$.

3.3   TeraSort

Recall, TeraSort consists of three steps: Sample, Partition, and Sort. First, the algorithm extracts a random sample set from the input and then computes t partition elements from the sample. The partition elements, referred to as *boundary elements*, divide A into t partitions. In the second round, each machine receives all the elements from a distinct partition and sorts them locally using a fixed algorithm $\mathcal{A}$. Importantly, the construction of the sample is crucial to efficiency since the partition elements may be insufficiently scattered among the input leading to partition skew in the second round. On the other hand, while it usually implies better partitioning, large samples could incur expensive overheads. We measure the *unevenness* of the partitions in TeraSort as a ratio of the maximum partition size to the optimal size m.

Tao et al. [24] conclude that TeraSort is load-balanced within constant multiplicative factors and thus satisfies their minimality definition. However, the multiplicative factor in the proof is relatively large (16 to 32). Moreover, the proof itself does not extend to substantially smaller bounds. In this work, we seek a more accurate description of the performance of TeraSort. We give a series of bounds which describes the trade-off between the number of machines and the evenness of the partition (hence the worst case maximum workload on a single machine). We conclude that a larger number of available machines allows for more even partitions. In particular, we show that as t grows, the evenness approaches exactly m with high probability.

Initially, the n elements are distributed evenly across the machines, each storing m or $m + 1$ elements. Parameterized by $\rho \in (0, 1]$, TeraSort runs as follows:

Map 1
   Each element is selected into the sample S with probability ρ.
Reduce 1
   S is sent to $\mathcal{M}_1$. $\mathcal{M}_1$ uses $\mathcal{A}$ to compute an ordered even t-partition of S made up of b_i, $i = 1, ..., t - 1$. Each b_i is a *boundary element*.
Map 2
   (Assume that b_i's have been broadcast to all machines.) Element x is sent to $\mathcal{M}_i$ if $b_{i-1} \leq x < b_i$, where $b_0 := -\infty$ and $b_n := +\infty$.
Reduce 2
   On each machine, sort elements locally using $\mathcal{A}$.

In [24], it was shown that TeraSort is minimal when $\rho = \frac{1}{m} \ln nt$ using a detailed analysis of the *minimum footprint* and *bounded net-traffic* conditions. Note that the broadcast assumption in the algorithm may incur a network outflow of size $O(t^2)$ (or $O(t)$ depending on the size of message) at $\mathcal{M}_1$, which would make TeraSort non-minimal when t^2 is no longer $O(m)$. However, in practice the broadcast can be implemented in Hadoop as $\mathcal{M}_1$ writing to a shared file which is then read by all machines. This way, the broadcast cost is evenly distributed among machines. This is an approach that [24] follows as well. Furthermore, [24] shows in Sect. 3.3 that this is not a restrictive constraint because it can be overcome by additional techniques. Finally, the experimental analysis provided in [24] for *pure TeraSort* (their implementation of TeraSort with $\rho = \frac{1}{m} \ln nt$) exhibits very even partitions. Specifically, it can be observed that the load balancing ratio does not exceed a factor of 2 across all the datasets considered. Therefore, our results can be viewed as giving a sound theoretical explanation for the experimental observations in [24].

In the following sections, we prove that TeraSort is *strongly* minimal by showing that the *balanced partition* condition is satisfied, using an interesting bucketing argument.

4   A New Proof of TeraSort's Minimality

In this section we give a new proof of the results in [24] that prove the minimality of TeraSort, but using a different and a more extensible probabilistic method that is capable of proving that TeraSort is strongly minimal.

4.1 Probability Tools

Chernoff bounds restrict from above the tail probability of sums of independent Bernoulli random variables. There are several forms/variants of Chernoff bounds of similar restrictive power. In this work, we use the following form:

$$\Pr\left[\sum_{i=1}^{n} X_i > (1+\delta)\mu\right] \le \exp\left\{-\frac{\delta^2 \mu}{\delta+2}\right\}, \ \delta > 0$$

$$\Pr\left[\sum_{i=1}^{n} X_i < (1-\delta)\mu\right] \le \exp\left\{-\frac{\delta^2 \mu}{2}\right\}, \ 0 < \delta < 1$$

where the X_i's are independent and $X_i = 1, 0$ with probability $p_i, 1 - p_i$ respectively. The mean is $\mu = \sum_{i=1}^{n} p_i$.

4.2 Minimality

The minimality of TeraSort is equivalent to the following claims.

Claim 1. *In Map 1:* $|S| = O(m)$

Claim 2. *In Reduce 2: every machine ends up with $O(m)$ elements*

Claim 1 limits the size of the sample S and thus the amount of traffic that machines send and receive in the first round. Claim 2 limits the machine sizes and the network input in the second round, since the map phase of round 2 never violates minimality as long as every machine holds $O(m)$ elements at the beginning of the algorithm. In the following we show Claims 1 and 2 hold with high probability (at least $1 - O(\frac{1}{n})$). It is straightforward to show Claim 1.

Lemma 1. $\Pr\left[|S| > kn\rho\right] \le \left(\frac{1}{nt}\right)^t$ *when* $\rho \ge \frac{1}{m} \ln nt$ *and* $k \ge 3$.

Proof. $|S|$ is the sum of n Bernoulli random variables of probability ρ; $\mathbb{E}\left[|S|\right] = n\rho$. By Chernoff bounds, we have

$$\Pr\left[|S| > kn\rho\right] \le \exp\left\{-\frac{(k-1)^2}{k+1}n\rho\right\} \tag{1}$$

For the lemma to hold, we require that the exponent on the RHS be bounded above by $\left(\frac{1}{nt}\right)^t$. To show this, we take the minimal values satisfying the inequalities given in the lemma statement: $\rho = \frac{1}{m} \ln nt$ and $k = 3$. It is easy to verify that under these parameter settings the RHS is $\exp\left\{-n \cdot \frac{t}{n} \ln nt\right\} = \left(\frac{1}{nt}\right)^t$.

Theorem 1 (Claim 1). *By setting* $\rho \ge \frac{1}{m} \ln nt$ *and assuming* $m \ge t \ln nt$, *Claim 1 holds with probability* $1 - \left(\frac{1}{nt}\right)^t$.

Proof. By Lemma 1,

$$\Pr\left[|S| > 3m\right] \le \Pr\left[|S| > 3t\ln nt\right] \le \Pr\left[|S| > 3n\rho\right] \le \left(\frac{1}{nt}\right)^t$$

The event that Claim 1 holds is given by complement of the above. Therefore the claim holds with high probability.

In reality, typically $m \gg t$, namely, the memory size of a machine is significantly greater than the number of machines. More specifically, m is at the order of at least 10^6 (this is using only a few megabytes per machine), while t is at the order of 10^4 or lower. Therefore, $m \ge n\rho = t\ln(nt)$ is a (very) reasonable assumption, which explains why TeraSort has excellent efficiency in practice.

Next, we present our approach to the proof of Claim 2 in the form of a few interesting lemmas. First, we formulate a problem closely related to Claim 2.

Problem 1 (Sample-Partition). Let A denote a set of n elements from an ordered universe; a_j denotes the $(j+1)$-th smallest element in A. Construct a sample $S \subseteq A$ by independently picking each element with probability ρ. Let $b_1, b_2, ..., b_{t-1} \in S$ be the ordered even t-partition elements of S. **Question:** how evenly do the b_i's partition A?

Problem 1 captures the probabilistic structure of TeraSort. Clearly, the set of the elements on $\mathcal{M}_i$ in Reduce 2 is exactly $A \cap [b_{i-1}, b_i)$, independent of how the input dataset is spread across machines at the outset. An answer to Problem 1 that the partitions are all $O(m)$ in size proves Claim 2.

The approach of [24] is to suppose we have an ordered partition of A and refer to every partition as a *bucket*. It is easy to observe that if every bucket contains a boundary element, then the distance between any two adjacent boundary elements is less than the sum of the sizes of the buckets in which they exist. This observation will lead to Claim 2 if we additionally ensure that buckets are $O(m)$ in size.

We make a stronger observation: if we allow buckets to overlap with one another, we have a promise of shorter distances between adjacent boundary elements. Formally, consider an ordered even t-partition of A. Let $d(i)$ be the index in A of the i-th smallest partition element; and manually set $d(0) := 1$. Our notion of a bucket is defined by the intervals $I_j := [a_{d(j)}, a_{d(j)+lm})$. The variable $l \ge 0$ controls the length of the interval. The I_j's are well defined for all $j \ge 0$ with $d(j) + lm \le n - 1$. We cover the largest few elements with one additional interval $[a_{n-lm}, a_n]$. Notice that the intervals form a cover of A if $l \ge 1$ and under this condition there can be at most t intervals.

Lemma 2. *If every interval I_j has at least one boundary element, then no two boundary elements are more than $(l+1)m$ away from each other.*

Proof. We prove the contrapositive. Suppose that $|[b_i, b_{i+1}]| > (l+1)m$ for some $i \in \{1, t-1\}$, then there exists an interval $I_j \subseteq [b_i, b_{i+1}]$ which contains no boundary element.

Consider we start at b_i and walk towards larger elements in steps of size $(l+1)m$. Since $|[b_i, b_{i+1}]| > (l+1)m$, we must have not met another boundary elements yet. The interval after the one containing b_i starts at most m away from b_i (due to the spacing of intervals), so its end cannot pass the current element. Therefore, this interval contains no boundary element.

Now, we only need to put a ceiling on the probability that some interval contains no boundary element to conclude Claim 2. This is shown in Lemmas 3 and 4. First, Lemma 3 considers a more generalized notion in which we consider an *arbitrary* subset of A instead of the intervals of sequential elements considered in the definition of buckets.

Lemma 3. *Fix an arbitrary subset $B(x)$ of size x of A. Then $B(x) \cap S$ denotes the set of sampled elements in $B(x)$. With $\rho \geq \frac{1}{m} \ln nt$ and $l \geq 7$,*

$$\Pr\left[|B(lm) \cap S| < \frac{|S|}{t}\right] \leq \frac{1}{nt} + \left(\frac{1}{nt}\right)^t \qquad (2)$$

Proof. Condition on the event $|S| > 3n\rho$ and decompose the probability as follows,

$$\Pr\left[|B(lm) \cap S| < \frac{|S|}{t}\right] \leq \Pr\left[|B(lm) \cap S| < \frac{|S|}{t} \text{ and } |S| \leq 3n\rho\right]$$
$$+ \Pr\left[|B(lm) \cap S| < \frac{|S|}{t} \text{ and } |S| > 3n\rho\right]$$
$$\leq \Pr\left[|B(lm) \cap S| < \frac{3n\rho}{t}\right] + \Pr\left[|S| > 3n\rho\right]$$

By Lemma 1 we can bound the second term in the last line above. Then, we apply Chernoff bounds to the first term of the same line noting that $\mathbb{E}\left[|B(lm) \cap S|\right] = \frac{lm}{n} \cdot n\rho = lm\rho$. We show an upper bound on the RHS by taking the minimal value satisfying the inequality given in the lemma statement: $l = 7$.

$$\Pr\left[|B(lm) \cap S| < \frac{|S|}{t}\right] \leq \exp\left\{-\frac{(l-3)^2}{2l} m\rho\right\} + \left(\frac{1}{nt}\right)^t$$
$$\leq \frac{1}{nt} + \left(\frac{1}{nt}\right)^t$$

This bound proves the lemma.

Now, we are ready to bound the probability that a bucket does not cover a boundary element.

Lemma 4. *For any $0 \leq j \leq t-1$,*

$$\Pr\left[I_j \text{ has no boundary element}\right] \leq O\left(\frac{1}{n}\right) \qquad (3)$$

Proof. In A, if a block of consecutive ordered elements contains no boundary element, then it must contribute no more than $\left\lceil \frac{|S|}{t} \right\rceil$ to the sample S (since a boundary element is taken every $\left\lceil \frac{|S|}{t} \right\rceil$ consecutive samples). Then, Lemma 3 gives us,

$$\Pr\left[I_j \text{ has no boundary element}\right] \leq \Pr\left[|B(lm) \cap S| < \left\lceil \frac{|S|}{t} \right\rceil\right]$$

$$= \Pr\left[|B(lm) \cap S| < \frac{|S|}{t}\right]$$

$$\leq \frac{1}{nt} + \left(\frac{1}{nt}\right)^t$$

By the union bound, with probability $1 - O\left(\frac{1}{n}\right)$ every interval covers at least one boundary element given $l \geq 7$.

As a result, we have that no bucket can fall between two consecutive boundary elements, hence every $A \cap [b_{i-1}, b_i)$ can contain objects in at most 2 buckets. So, by Lemma 2 and setting $l \geq 7$ we have that $\Pr\left[|A \cap [b_i, b_{i+1}]| \geq 8m\right] \leq O\left(\frac{1}{n}\right)$. Finally, we have the following theorem,

Theorem 2 (Claim 2). *By setting $\rho \geq \frac{1}{m} \ln nt$, Claim 2 holds with probability at least $1 - O\left(\frac{1}{n}\right)$.*

5 Proof of TeraSort's Strong Minimality

A natural question is whether a fixed constant factor on m is the farthest we could go with the tools in hand. In other words, can we achieve better guarantees of evenness? Towards this goal, we consider the problem of Sample-Partition as a simple, generic routine potentially used for a variety of problems and so it is worthwhile in pushing the bound further.

Our new analysis is based on a re-examination of the proof of Lemmas 3 and 4. We focus on tighter bounds in Claim 2 (the evenness of partition), though as shown later the choice of parameters also ensures Claim 1.

Please note that in the statement of the following theorem, k and l are free parameters, for which we will later choose appropriate values that will satisfy conditions (4).

Theorem 3. *Given t and $\rho \geq \frac{1}{m} \ln nt$, for any choice of k and l satisfying the constraints below,*

$$\begin{cases} (k-1)^2(t-l) \geq (k+1) \\ ((t-1)l - (t-l)k)^2 \geq 2l \\ (t-1)l > (t-l)k \\ k, t > 1, \ l > 0 \end{cases} \tag{4}$$

we have,

$$\Pr\left[|B(lm) \cap S| < \frac{|S|}{t}\right] \leq \frac{2}{nt} \tag{5}$$

Proof. Let Y_j be the indicator random variable representing whether element $a_j \in A$ is sampled into S. Let $W(x)$ denote the sum of x independent $Bernoulli(\rho)$ random variables. Note, we require that there is no dependency between the random variables underlying two $W(\cdot)$ expressions.

$$\Pr\left[|B(lm) \cap S| < \frac{|S|}{t}\right] = \Pr\left[\sum_{j:a_j \in B(lm)} Y_j < \frac{1}{t}\sum_{j=0}^{n-1} Y_j\right]$$

$$= \Pr\left[(t-1)\sum_{j:a_j \in B(lm)} Y_j < \sum_{j:a_j \notin B(lm)} Y_j\right]$$

$$= \Pr\left[(t-1)W(lm) < W(n-lm)\right] \tag{6}$$

To establish an upper bound on (6), we use the same decoupling technique as in Lemma 3 again. Let,

$$W_1 = W(lm)$$
$$W_2 = W(n-lm)$$

we have,

$$\Pr\left[(t-1)W(lm) < W(n-lm)\right] \leq \Pr\left[W_1 < \frac{W_2}{t-1}, \ W_2 \leq k(n-lm)\rho\right]$$

$$+ \Pr\left[W_1 < \frac{W_2}{t-1}, \ W_2 > k(n-lm)\rho\right]$$

$$\leq \Pr\left[W_1 < \frac{k(n-lm)\rho}{t-1}\right] + \Pr\left[W_2 > k(n-lm)\rho\right]$$

By Chernoff bounds, given $k > 1$ and $(t-1)l > (t-l)k$, we get

$$\Pr\left[W_2 > k(n-lm)\rho\right] \leq \exp\left\{-\frac{(k-1)^2}{k+1}(t-l)m\rho\right\} \tag{7}$$

$$\Pr\left[W_1 < \frac{k(n-lm)\rho}{t-1}\right] \leq \exp\left\{-\frac{lm\rho}{2}\left(1 - \frac{(t-l)k}{(t-1)l}\right)^2\right\} \tag{8}$$

Next, we can bound the probability of each of W_1 and W_2 independently. We make the initial observation that $\rho \geq \frac{1}{m}\ln nt$ implies $m\rho \geq \ln nt$.

First, using the observation above, the RHS of (7)

$$\exp\left\{-\frac{(k-1)^2}{k+1}(t-l)m\rho\right\} \leq \exp\left\{-\frac{(k-1)^2}{k+1}(t-l)\ln nt\right\} \tag{9}$$

Therefore, the probability that $W_2 > k(n-lm)\rho$ is less than $\frac{1}{nt}$ exactly when $\frac{(k-1)^2}{k+1}(t-l) \geq 1$, or $(k-1)^2(t-l) \geq (k+1)$. The resulting inequality yields the second set of constraints from (4).

Second, again using the observation above, the RHS of (8)

$$\exp\left\{-\frac{lm\rho}{2}\left(1-\frac{(t-l)k}{(t-1)l}\right)^2\right\} \le \exp\left\{-\frac{l}{2}\left(1-\frac{(t-l)k}{(t-1)l}\right)^2 \ln nt\right\} \quad (10)$$

Therefore, the probability that $W_1 < \frac{k(n-lm)\rho}{t-1}$ is less than $\frac{1}{nt}$ exactly when $\frac{l}{2}\left(1-\frac{(t-l)k}{(t-1)l}\right)^2 \ge 1$, or $((t-1)l-(t-l)k)^2 \ge 2l$. The resulting inequality yields the third set of constraints from (4). Finally, we enforce the sanity conditions given in the third inequality from (4) to ensure that the RHS of (8) remains positive.

We remark that although Theorem 3 focuses on Claim 2, the system of equations given by (4) also ensures that Claim 1 holds, because $(k-1)^2(t-l) \ge (k+1)$ is a stronger condition than that required for Claim 1 to hold: $(k-1)^2t \ge (k+1)$ (see Eq. 1).

Theorem 3 describes a family of bounds for $(l+1)m$ given admissible values for l, t and k. The choices of l and t dictates the trade-off between the evenness and number of partitions. For a fixed k, as t increases, lower l is accessible and therefore a greater number of partitions implies better evenness. First, we present the following corollary.

Corollary 1. *Let $0 < \epsilon < \frac{1}{2}$ be a parameter and set $k := 1 + \frac{1}{t^\epsilon}$ and $l := 1 + \frac{2}{t^\epsilon}$. Given $\rho \ge \frac{1}{m} \ln nt$, it holds that*

$$\Pr\left[|B(lm) \cap S| < \frac{|S|}{t}\right] \le \frac{2}{nt} \quad (11)$$

It can be verified that the chosen values for k and l satisfy the system of equations given by (4) for $0 < \epsilon < \frac{1}{2}$ and t large enough. Furthermore, the system of equations given by (4) is always satisfied for large enough t given fixed ϵ, k and l. Therefore, we obtain arbitrarily strong evenness as long as t is allowed to be sufficiently large. Given our chosen value of l and Corollary 1 we arrive at the following.

Corollary 2. *With probability at least $1 - O(\frac{1}{n})$, the size of every partition is less than $(2 + \frac{2}{t^\epsilon})m$ where $0 < \epsilon < \frac{1}{2}$.*

In fact, it can be shown that for $\epsilon = \frac{1}{4}$, we can satisfy the system of equations given by (4) when $t > 10$. By combining all of above, we arrive at Corollary 3.

Corollary 3. *Given $\rho \ge \frac{1}{m} \ln nt$ and $t > 10$,*

$$\Pr\left[|A \cap [b_i, b_{i+1}]| > \left(2 + \frac{2}{\sqrt[4]{t}}\right)m\right] \le O\left(\frac{1}{n}\right) \quad (12)$$

for $0 \le i \le t-1$ and where $b_0 := -\infty$ and $b_t := +\infty$.

Finally, we observe that a stronger version of Theorem 1 (Claim 1) exists by noting that plugging $k := 1 + \frac{1}{\sqrt[4]{t}}$ into Lemma 1 yields an analog to Theorem 1 that bounds the probability that $|S| > \left(1 + \frac{1}{\sqrt[4]{t}}\right)m$ to at most $O\left(\frac{1}{n}\right)$.

5.1   Tightening the Bound

Given the memory available in modern machines and typical values for t, we observe that probabilities of failure to produce strongly even partitions on the order of $O(\frac{1}{m})$ are negligible. This is reasonable since current main memories of computation nodes are on the order of gigabytes. If we consider m (the average workload) to be safely within the capacity of main memory (e.g. m is on the order of 2^{20}), then $\frac{1}{m}$ is very small.

Thus, we are able to further restrict the size of the intervals which corresponds to partitions that are more even. The trick is to construct the intervals $\{I_j\}_j$ by placing the left endpoint of the interval $j+1$ a distance $\frac{m}{t}$ away from the left endpoint of interval j (see Fig. 1). As a result,

- Lemma 2 has a stronger form: no two adjacent boundary elements are more than $(l + \frac{1}{t})$ away from each other.
- When we apply union bound with the at most t^2 intervals, we obtain a probability of failure no larger than $O(\frac{1}{nt}) \times t^2 = O(\frac{1}{m})$.

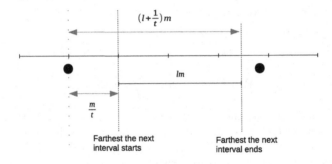

Fig. 1. With $\frac{m}{t}$ spacing, Lemma 2 is strengthened to ensure adjacent boundary elements are within $(l + \frac{1}{t})$ of each other with high probability.

This argument applies to every bound shown in the previous section. We phrase the counterpart of Corollary 3 as an example.

Theorem 4. *Given* $\rho \geq \frac{1}{m} \ln nt$ *and* $t > 10$,

$$\Pr\left[|A \cap [b_i, b_{i+1}]| > \left(1 + \frac{2}{\sqrt[4]{t}} + \frac{1}{t}\right) m \right] \leq O\left(\frac{1}{m}\right) \tag{13}$$

for $0 \leq i \leq t - 1$ *and where* $b_0 := -\infty$ *and* $b_t := +\infty$.

As a result, we see that as the number of machines increases, the evenness of the partitions approach its optimal value m with high probability.

Remark: Strongly Minimal Algorithms for Databases. As mentioned in the introduction, Tao et al. [24] show how a minimal algorithm for sorting leads to minimal algorithms for other database problems by using a single additional round after sorting. The problems considered include *ranking, group-by, semi-join,* and *2D skyline.* As a result of our analysis for TeraSort, these problems have strongly minimal MapReduce algorithms.

6 Conclusions

Despite the great variety of algorithms developed for MapReduce, few are able to achieve the ideal goal of parallelization: balanced workload across the participating machines, network traffic bounded by the total input data size, and a speedup over sequential algorithms linear in the number of machines available.

In this paper we introduce the new notion of *strongly minimal* MapReduce algorithms. Our definition strengthens the minimality criteria of minimal MapReduce algorithms as defined in [24]. Precisely, strongly minimal algorithms have partitions that approach the optimal evenness value of m as the number of machines t grows with high probability. We prove that the popular parallel sorting paradigm, TeraSort, is strongly minimal under the self-sampling strategy with a sampling rate $\rho = \frac{1}{m} \ln nt$. Additionally, this leads to strongly minimal algorithms that settle an array of important database problems.

Finally, our techniques are general and apply beyond the analysis of strongly minimal MapReduce algorithms to any setting that fits the Sample-Partition problem model; we show that given a sufficiently high sampling rate the approximate partitions obtained from self-sampling as the number of partitions increases approach the partition sizes obtained from an ideal partitioning with high probability. We believe that the refined bucketing arguments we use in our analysis are of independent interest and are likely to have other applications.

In future work, our goal is to continue this line of research and identify other fundamental problems that have strongly minimal algorithms.

References

1. Afrati, F.N., Ullman, J.D.: Optimizing multiway joins in a map-reduce environment. IEEE Trans. Knowl. Data Eng. **23**(9), 1282–1298 (2011)
2. Bahmani, B., Kumar, R., Vassilvitskii, S.: Densest subgraph in streaming and mapreduce. Proc. VLDB Endow. **5**(5), 454–465 (2012)
3. Behnezhad, S., Derakhshan, M., Hajiaghayi, M., Karp, R.M.: Massively parallel symmetry breaking on sparse graphs: MIS and maximal matching. arXiv preprint arXiv:1807.06701 (2018)
4. Blanas, S., Patel, J.M., Ercegovac, V., Rao, J., Shekita, E.J., Tian, Y.: A comparison of join algorithms for log processing in MapReduce. In: Proceedings of the 2010 ACM SIGMOD International Conference on Management of data, pp. 975–986. ACM (2010)
5. Das, A.S., Datar, M., Garg, A., Rajaram, S.: Google news personalization: scalable online collaborative filtering. In: Proceedings of the 16th International Conference on World Wide Web, pp. 271–280. ACM (2007)

6. Eltabakh, M.Y., Tian, Y., Özcan, F., Gemulla, R., Krettek, A., McPherson, J.: Cohadoop: flexible data placement and its exploitation in hadoop. Proc. VLDB Endow. **4**(9), 575–585 (2011)
7. Ene, A., Im, S., Moseley, B.: Fast clustering using MapReduce. In: Proceedings of the 17th ACM SIGKDD International Conference on Knowledge Discovery and Data Mining, pp. 681–689. ACM (2011)
8. Feldman, J., Muthukrishnan, S., Sidiropoulos, A., Stein, C., Svitkina, Z.: On distributing symmetric streaming computations. ACM Trans. Algorithms (TALG) **6**(4), 66 (2010)
9. Ferreira Cordeiro, R.L., Traina Junior, C., Machado Traina, A.J., López, J., Kang, U., Faloutsos, C.: Clustering very large multi-dimensional datasets with MapReduce. In: Proceedings of the 17th ACM SIGKDD International Conference on Knowledge Discovery and Data Mining, pp. 690–698. ACM (2011)
10. Ghoting, A., Kambadur, P., Pednault, E., Kannan, R.: NIMBLE: a toolkit for the implementation of parallel data mining and machine learning algorithms on mapreduce. In: Proceedings of the 17th ACM SIGKDD International Conference on Knowledge Discovery and Data Mining, pp. 334–342. ACM (2011)
11. Gufler, B., Augsten, N., Reiser, A., Kemper, A.: Load balancing in mapreduce based on scalable cardinality estimates. In: 2012 IEEE 28th International Conference on Data Engineering, pp. 522–533. IEEE (2012)
12. He, Y., et al.: RCFile: a fast and space-efficient data placement structure in mapreduce-based warehouse systems. In: 2011 IEEE 27th International Conference on Data Engineering (ICDE), pp. 1199–1208. IEEE (2011)
13. Karloff, H., Suri, S., Vassilvitskii, S.: A model of computation for mapreduce. In: Proceedings of the Twenty-First Annual ACM-SIAM Symposium on Discrete Algorithms, pp. 938–948. SIAM (2010)
14. Kolb, L., Thor, A., Rahm, E.: Load balancing for mapreduce-based entity resolution. In: 2012 IEEE 28th International Conference on Data Engineering, pp. 618–629. IEEE (2012)
15. Kumar, R., Moseley, B., Vassilvitskii, S., Vattani, A.: Fast greedy algorithms in mapreduce and streaming. ACM Trans. Parallel Comput. (TOPC) **2**(3), 14 (2015)
16. Kwon, Y., Balazinska, M., Howe, B., Rolia, J.: SkewTune: mitigating skew in mapreduce applications. In: Proceedings of the 2012 ACM SIGMOD International Conference on Management of Data, pp. 25–36. ACM (2012)
17. Lattanzi, S., Moseley, B., Suri, S., Vassilvitskii, S.: Filtering: a method for solving graph problems in mapreduce. In: Proceedings of the Twenty-Third Annual ACM Symposium on Parallelism in Algorithms and Architectures, pp. 85–94. ACM (2011)
18. Lin, Y., Agrawal, D., Chen, C., Ooi, B.C., Wu, S.: Llama: leveraging columnar storage for scalable join processing in the mapreduce framework. In: Proceedings of the 2011 ACM SIGMOD International Conference on Management of data, pp. 961–972. ACM (2011)
19. Okcan, A., Riedewald, M.: Processing theta-joins using mapreduce. In: Proceedings of the 2011 ACM SIGMOD International Conference on Management of Data, pp. 949–960. ACM (2011)
20. O'Malley, O.: Terabyte sort on apache hadoop, pp. 1–3, May 2008. Yahoo. http://sortbenchmark.org/Yahoo-Hadoop.pdf
21. Panda, B., Herbach, J.S., Basu, S., Bayardo, R.J.: Planet: massively parallel learning of tree ensembles with mapreduce. Proc. VLDB Endow. **2**(2), 1426–1437 (2009)
22. Roughgarden, T., Vassilvitskii, S., Wang, J.R.: Shuffles and circuits (on lower bounds for modern parallel computation). J. ACM (JACM) **65**(6), 41 (2018)

23. Suri, S., Vassilvitskii, S.: Counting triangles and the curse of the last reducer. In: Proceedings of the 20th International Conference on World Wide Web, pp. 607–614. ACM (2011)

24. Tao, Y., Lin, W., Xiao, X.: Minimal mapreduce algorithms. In: Proceedings of the 2013 ACM SIGMOD International Conference on Management of Data, pp. 529–540. ACM (2013)

25. Tsourakakis, C.E., Kang, U., Miller, G.L., Faloutsos, C.: DOULION: counting triangles in massive graphs with a coin. In: Proceedings of the 15th ACM SIGKDD International Conference on Knowledge Discovery and Data Mining, pp. 837–846. ACM (2009)

26. Zhang, X., Chen, L., Wang, M.: Efficient multi-way theta-join processing using mapreduce. Proc. VLDB Endow. **5**(11), 1184–1195 (2012)

Event Sequence Interpretation of Structural Geomodels: A Knowledge-Based Approach for Extracting Tectonic Sequences

Xianglin Zhan[1] , Cai Lu[1(✉)], and Guangmin Hu[2]

[1] School of Information and Communication Engineering,
University of Electronic Science and Technology of China, Chengdu 611731,
People's Republic of China
xianglin_zhan@163.com, lucai@uestc.edu.cn
[2] School of Resources and Environment,
University of Electronic Science and Technology of China, Chengdu 611731,
People's Republic of China
hgm@uestc.edu.cn

Abstract. The tasks of obtaining past event occurrences and their temporal order information are important parts of the cognition of the external world. We call this kind of tasks *Event Sequence Interpretations* (ESI). In this work, we focus in the ESI in structural geomodels and propose a knowledge-based approach for extracting tectonic sequences, which is crucial for the cognition of structural geomodels.

As a cognitive task, tectonic sequence interpretation has not been highly automated due to the need to use a large amount of expert knowledge for recognition and reasoning. Meanwhile, artificial ESI may introduce cognitive biases that ultimately lead to subjective uncertainty in the results, which affects the credibility of the interpretations and increases risks in oil and gas production. One potential solution is making personal knowledge better available for computers so that computers can also do ESI. Therefore, we proposed a meta-model for formally representing expert knowledge. The instance of the knowledge representation (KR) meta-model is called an *Event Pattern* (EP), which describes the associations between event occurrences and geometric features in the models. Moreover, we proposed a new pattern matching model called *Joint Prototype Model* (JPM) to find evidences of event occurrences from the raw geological data. The temporal relations of the events can be extracted according to the spatial topology of the geological objects. Our approach can also be extended from structural geomodels to other spatial geometric models. We show the effectiveness of the approach by an application to a real structural geomodel dataset.

Keywords: Spatial cognition · Knowledge representation · Knowledge-based approach · Event sequence interpretation

© Springer Nature Switzerland AG 2020
A. Herzig and J. Kontinen (Eds.): FoIKS 2020, LNCS 12012, pp. 318–333, 2020.
https://doi.org/10.1007/978-3-030-39951-1_19

1   Introduction

Trying to let machines cognize the physical world is a difficult problem of artificial intelligence [14,16,25,28]. The inference of the occurrences of past events and the temporal relations of the events from the current state is an important part of the cognition [20–22]. We call this kind of tasks *Event Sequence Interpretations* (ESI). The ESI aims at understanding how the current state of the system was produced and by which events. It comprises the recognition process that starts with the perception of evidences of past event occurrences, and the reasoning process that infers the chronological order of the past events. In this paper, we focus on the ESI of structural geomodels, and propose a knowledge-based approach to extract tectonic sequences from raw structural geological data. Geological structures refer to the forms left by deformations of rock masses under the actions of tectonic events [9]. A tectonic sequence is a sequence of tectonic events in which each event is sorted by their time order, indicating the evolution of the structural geomodel. The tectonic sequence interpretation is crucial for the cognition of structural geomodels, since the tectonic sequences actually determine the morphology and spatial distributions of geological objects. Geological structural data is a typical kind of spatial geometric data, which describes the geometry of geological structures by point coordinates in 3D space, so our approach of ESI can also be extended to other spatial geometric models.

It can be seen form many examples of ESI that the geological ESI mainly relies on manual solutions [4,7,10,24,26,30]. In traditional ESI processes, geologists first derive geological interfaces form geophysical measurements such as seismic data (Fig. 1(a)). Nowadays, some deep learning algorithms can also perform intelligent geological interface identification, such as the examples shown in Fig. 1(b) and (c) [11,19,27]. After the extraction of geological interfaces, which tectonic events have occurred can be learnt according to the morphology of the interfaces. Then, the spatial topological relations between the geological objects can be determined through geological observations (e.g. outcrop, well logs etc.) and spatial positions showed in seismic images. Geological events can be ordered by time based on the spatial topological relations that contain the time order meanings. However, ESI as a cognitive task is usually not a narrow task with clear protocols, and manual interpretation relies heavily on personal experience, which inevitably bring cognitive bias into the results. Even for the same raw data, there may have multiple interpretations, as shown in [23]. With the reduction of petroleum resources, the structures faced in geological explorations are more complicated. Manual ESI is becoming time-consuming and error-prone.

In an overview, we propose: (1) a workflow of ESI in structural geomodels; (2) an ontology-based meta-model to formally representing event knowledge; (3) an event pattern matching model called *Joint Prototype Model* (JPM) to recognize events and (4) a set of time relationships related to time scales to describe the time order of tectonic events. To our knowledge, our work is the first to make computers do event sequence interpretation in structural geomodels. In particular, our approach works directly on raw geological structural data rather than symbolic data, which is a challenge to recognize evidences of event occurrences.

Proposing a computer method to do ESI provides the following benefits: (1) The interpretations form computers can serve as inspirations to help experts get a higher level of expertise or provide an assistance. (2) Letting computers perform ESI can reduce the possibility of human error. (3) ESI promotes computer cognition of the structural geomodels.

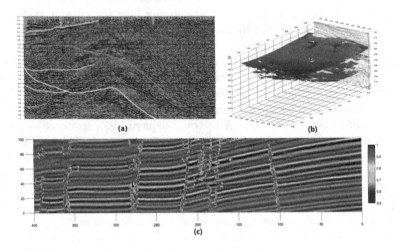

Fig. 1. The examples of (a) geological interfaces identified by manual interpretation in seismic images; (b) interfaces identified by intelligent horizon tracking; (c) faults identified by CNN.

The workflow for EI is shown in Fig. 2. Some explanations for this workflow are as follows: ① Raw geological structural data can be point clouds, meshes, and volumetric data representing geological interfaces or bodies. In this paper, we will use the point clouds; ② Determine geological objects in the raw data, and geological objects refer to geological interfaces in this paper, which are three-dimensional surfaces; ③ The basic features are the "level-0" geometric features that inherent in geological objects. We will describe it in detail in Sect. 2; ④ Event pattern matching is to interpret basic features as the event feature with high-level semantics. The new matching model is call JPM and described in Sect. 3; ⑤ Event patterns are instances of the KR model, representing the event semantics of a group of geometric features of the geological interface; ⑥ The spatio-temporal relation table shows the temporal relation meaning of the spatial topological relations of geological objects; ⑦ Spatial relation refers to the spatial topological relation between interfaces; ⑧ Transform the spatial relations of interfaces to temporal relations of the events related to the interfaces;

Works such as Abel et al. and Carbonera et al. are similar to ours [1,6], since both of them deal with problems of event interpretation and their approaches are also based on knowledge. In [1], it presented the *PetroGrapher* system to support visual analysis tasks. The system was developed based on knowledge inspired by perceptual chunks. The knowledge is used to manage heterogeneous

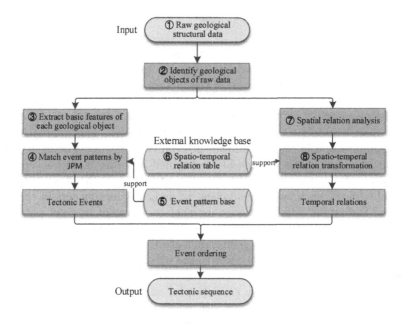

Fig. 2. The workflow of *Event Sequence Interpretation*.

data and perceptual chunk are the knowledge representation primitives. Our approach is also inspired by organizing knowledge as chunks (the event patterns in our work). However, the chunks in our work are knowledge representation structures, instead of knowledge representation primitives. Our structures are explicit and have rich internal contents. In [6], an expert system for automating the task of visual interpretation of depositional processes within the domain of sedimentary stratigraphy is proposed. This approach combines a meta-model for representing inferential knowledge with foundational ontologies such as UFO. However, we consider the parthood relations in UFO are not completely suitable to describe the logic relations of features. Because although the visual chunks describe visual features, the visual features are still event-centered, and the logic describing the visual information should better conform to the event logic. Our approach enhances the representation of logical relations within the chunks by defining a correlation coefficient and refining three logical relations (*subsumption*, *concurrency*, and *composition* instead of *partOf*). Besides, our approach can deal directly with raw data, while Abel and Carbonera emphasize that their methods are only used to process symbolic representations. Besides, the above two works only interpret the events themselves, while we also analyze the time relations of the events. Both of the above two works apply their approaches in sedimentary stratigraphy domain while we develop our approach within structural geology domain. Compared to considering only depositional events in well processes, the geometric information contained in a structural geomodel is significantly more complex, and the cognition of events is more challenging.

2   Knowledge Representation

In this section, we present a reusable meta-model for knowledge representation. The significance of the meta-model is to assign explicit roles to the elements in knowledge. We call each instance of the meta-model an *Event Pattern* (EP). The elements of KR model are defined under the constraints of ontological meta-properties provided by DOLCE, which is a foundational ontology [5]. It ensures that our model conforms to a widely accepted framework. This approach facilitates communication in the field and the possibility of the model being expanded in the future. The content of each EP is specified in terms of the concepts provided by specific domain ontologies. For example, reusing the classes of StructuralGeoOntology, such as the Fold class, to describe the concepts in the event pattern of folding [3]. The knowledge we want to represent in event patterns is the correlation between geometric features of geological objects and the tectonic event occurrences, that is, what geometric features can be considered as the evidences of tectonic events. In other words, the event patterns explicitly show the external manifestations of tectonic events in structural geomodels.

We define the EP as a triple consisting of an *GeoObject* (GO), an *TectonicEvent* (TE), and an *EventFeature* (EF): $EP = \langle GO, TE, EF \rangle$. The Object refers to the conceptual mapping of an observed object that is the participant of the event. The *Event* is the event semantics of this EP. The nature of *EventFeature* is a collection of geometric features that represents the geometric performance of the event. Each instance of *Event Pattern Elements* (EPE) has a mapping to some specific type of concepts or relations of the foundational ontology, respecting constraints of meta-properties. The entities involved in DOLCE can be called *particulars* (P). DOLCE partitions particulars into six basic categories: *objects* (O), *events* (E), *individual qualities* (Q), *regions* (R), *concepts* (C), and *arbitrary sums* (AS) [5]. These three elements of EP can map to the categories of *object*, *event*, and *arbitrary sum* respectively according to DOLCE. AS collects mixed entities that are obtained as sum of elements in different basic categories. The internal elements of *EventFeature* will be described below.

In EP, the internal structure of *EventFeature* is the focus of our research. In [12,13], there is a *determinate-determinable relation* (dD) used to assign partial orders to properties. We consider that the property mentioned here can be considered equal to the individual quality in DOLCE. dD(F,G) means that entities that have the property F also have the (more general) property G and entities that have the property G have at least one of the (more specific) properties that are the *determinates* (D) of G, among which there is F [5]. This relation allows us to organize properties in a hierarchical way (tree-leaves structure). The most specific properties are called basic properties. Except the basic properties, other properties (called general properties) are all composed of basic properties, which express a higher level of semantics. In EP, we call each property a feature (all features can be mapped to individual qualities of *DOLCE*). Then features can also be divided into *basic features* (BF) and *general features* (GF). However, as a whole element, EF should contain another kind of elements to combine those features together. This element is the relation that describe the logical relation between features. There are three kinds of relations defined for entities in mixed entities (aka AS) in DOLCE: P(x,y) stands for "x is part of y", O(x,y) for "x overlaps with y", and SUM(z,x,y) for "z is the mereological sum of x and y". When they are mapped to EP, we can define the corresponding relations as *subsumption relation* (SubR), *concurrency relation* (ConR), and *composition relation* (ComR). The three redefined relations are given event-related semantics. SubR, ConR, and ComR map to part_of, overlap, and sum respectively.

- Subsumption relation: A feature can subsume or be subsumed by other features; a feature subsumed by another is called a sub-feature (*subf*) of the super-feature (*supf*).
- Concurrency relation: If feature f_i is accompanied with feature f_j at above a specified probability threshold, there is a concurrency relation between f_i and f_j.
- Composition relation: If a feature f can be decomposed to several features $f_i(i > 0)$ with smaller granularity, and while all f_i exist means the existence of f, there exists composition relation between f and f_i.

In summary, *EventFeature* can be partitioned into two categories: features, and relations, which is the reason EF maps to the AS category in DOLCE. We have the representation $EF = \langle SF, SR \rangle$, where SF refers to the set of features and SR refers to the set of relations. $SF = \{gf_0, \ldots, gf_n, bf_0, \ldots, bf_m\}$, where gf_i and bf_i represent the instances of general and basic feature. $SR = \{subr_0, \ldots, subr_n, conr_0, \ldots, conr_m, comr_0, \ldots, comr_k\}$, where $subr_i$, $conr_i$, and $comr_i$ represent instances of subsumption relations, concurrency relations, and composition relations respectively. To enhance the representation of logical information in EP, here we provide a numerical measurement to measure the strength of the correlation between two features. The measurement is called *correlation coefficient* (r). The correlation coefficient of two features f_i and f_j is $r(f_i, f_j) \in [0, 1]$. It is obvious that correlation coefficient corresponds to the category *quality* (Q). For the subsumption relation, the correlation coefficient represents the probability of the existence of the super-feature when the sub-feature exists. For concurrency relations, the correlation coefficient represents the probability of one feature existence when another feature exists. For composition relations, the correlation coefficient represents the probability of the feature existence when the smaller feature exists. We have more specific representations of relations: $SubR = \langle supf, subf, r \rangle$, $ConR = \langle f_i, f_j, r \rangle$, and $ComR = \langle f, f_i, r \rangle$.

We mentioned above that all features are specifically described at the most basic level by basic features, and of course with specific values. We consider that a basic feature is universally applicable to a domain and basic features can be extracted from all objects in the domain. When the domain is within geometric objects, the type of basic feature includes *point, edge, plane, location, direction, type, size, angle, vector*. Each basic feature is associated with a *quale structure* (QS), which defines the data type (*quale type*, QT) and possible values the basic feature can assume (*value domain*, VD). The types of data include *point, vector, angle, symbol*. Therefore, we have the representations that the basic feature contains a *feature type* (FT) and a *quale structure* (QS): $BF = \langle FT, QS \rangle$ where

$QS = \langle QT, VD \rangle$. The basic features also bridge the data and the event pattern (provide the way to map data to conceptual domain). General features can be represented as a pair: $GF = \langle SD, SAF \rangle$. SD refers to the set of determinates of the GF while SAF is a set of associated features that have logical relations with the GF. SAF and SD could contain both general features and basic features: $SAF = \{gf_0, \ldots, gf_n, bf_0, \ldots, bf_m\}$, $SD = \{gf_0, \ldots, gf_n, bf_0, \ldots, bf_m\}$.

Synthetically speaking, the overall structure of the KR meta-model is described in UML manner as shown in Fig. 3. The proposed meta-model differs from the traditional ontology based KR model. Traditional ontologies are concept-centered, there exist deficiencies while modeling knowledge: (1) separateness of concepts (the observed entity, event, features and logical relations of features are not organized as a whole knowledge unit); (2) lack of the direct representation of inference knowledge (traditional conceptual relations like constitution, dependence, parthood etc. are not enough to express the event-level associations among features); (3) lack of the constraints from foundational ontology, which leads to the significant differences between some KR models, even in the same domain. These differences can make it difficult to share and manage knowledge. To our knowledge, our approach is the first to propose the pattern of the performances of events (not describing the event itself). Event patterns organize two kinds of semantics (event and geometry) as a knowledge unit. Our meta-model provides an easily reusable and well-founded framework to model event-performance knowledge in spatial geometric models.

Here we take the representation of the folding event knowledge as an example. Folding is a common tectonic event that widely exist in subsurface and one of the major tectonic deformation events. The event pattern of folding is shown in Fig. 4 (at the end of the document). Domain concepts in the event pattern are provided by the StructuralGeoOntology [3]. Composition relations can be directly extracted form the memberships of features, so we did not express them in the UML graph. Event patterns can be easily produced through concepts and concept hierarchies provided by specific domain ontologies.

3   Recognition of Events

In this section, we will present our approach to events recognition by pattern matching techniques. We want to emphasize that it is almost impossible to capture the complete tectonic sequence in some situations. Because the evidences of event occurrences may be completely overprinted by later events and cannot be found in the current state.

3.1   Events and Objects

Before describing the specific method of event recognition, we need to clarify which events and objects are within our scope of discussion. According to the affects to rock masses, tectonic events are inherited by three subclasses: rock generation, rock destruction, and rock deformation. Rock generation represents

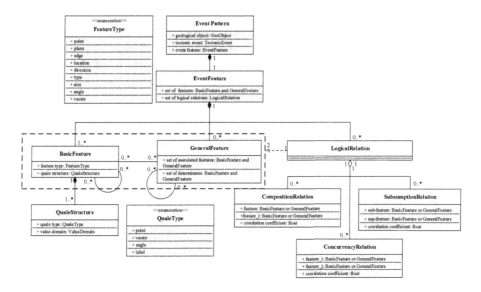

Fig. 3. The knowledge representation meta-model described by UML.

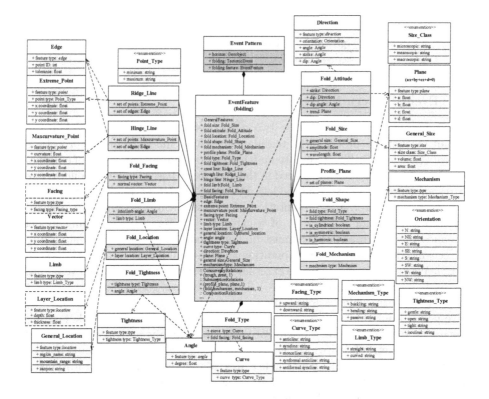

Fig. 4. The event pattern of folding.

processes in which non-rock mass materials (magma and sediment) are converted into rock masses. This subclass includes two events: *crystallization*, and *sedimentation*. Rock destruction refers to processes in which of rock mass is transformed into non-rock mass material, including *dissolution, erosion*. Rock deformation corresponds to processes with only the shape or volume of the rock changes, including *faulting, folding* and *diapirism*. The above six events are the main considerations in this paper. For the sake of simplicity of descriptions, structural geomodels usually use geological interfaces to represent geological structures. These geological interfaces are the objects to be observed. Table 1 shows the correspondence between the events and the observed objects.

Table 1. Events with their corresponding geological objects.

Event	Object (geological interface)
Crystallization	Intrusion interface
Sedimentation	Horizon
Dissolution	Karst cave interface
Erosion	Unconformity
Faulting	Fault
Folding	Fold
Diapirism	Diaper interface

3.2   Event Pattern Matching Model

In cognitive psychology, there are two most used models that can explain the human recognition process: the template matching model, and the prototype matching model [17]. The template matching model states that copies of external models, called templates, are stored in the knowledge base. They have one-to-one correspondences with external models. Prototype matching model states that a prototype is an internal representation of a class of objects, that is, a general representation of all individuals of a category [8]. Template matching model demands that the input object must be highly compliant with the template to be recognized, which is a strong condition and unsuitable for cognitive tasks such as ESI. We also consider that the external manifestations of a type of events has no fixed "form", but it should have a fixed "structure". This fixed structure is the event pattern, which is essentially a complex prototype. At the same time, we have pointed out that in EP, the event feature is composed of multiple features (general and basic features). Each feature is also a highly general description of a class of geometric features, so the features in the event patterns should also be prototypes of the class of geometric features. Matching of multiple prototypes is involved in the matching process of an event pattern. Therefore, we proposed a new pattern matching model called the *joint prototype model* (JPM).

Latecki et al. have mentioned human perception of shape is based on the visual part of the object, making a single important visual part sufficient to identify the entire object [15]. Therefore, as a task similar to human perception of shape, what we need to do is to identify key features in the observed object to identify the whole object. Here comes the problem of determining which features can lead to the matching of the pattern. Obviously, different features should have different degrees of influence on the pattern matching. We divide the features of the pattern into three categories (not on the ontological level): *exclusive features* (EF), *strong features* (SF), and *common features* (CF). The exclusive feature is a feature that exists only in a certain pattern and can directly determine the recognition of this pattern. The strong features appear only in a small number of patterns, and the matching of a strong feature can greatly reduce the search range of the patterns. The common feature widely appears in multiple patterns, and other features need to be matched to recognize patterns. We construct a Bayesian network shown in Fig. 5 to determine whether an event pattern is matched. Each directed edge in the network can be assigned to a conditional probability by users. The given conditional probability meets the following rules: $P(EP \mid EF) = 1, 1 > P(EP \mid SF) \gg P(EP \mid CF) > 0$. If two features have $cc = 1$, then $P(f_a \mid f_b) = 1$. The patterns are also associated with a probability threshold value provided by users that indicates probability requirement for the recognition of the pattern.

As for the matching of individual prototypes, due to the limit of space, we will not dwell on specific methods. The matching of each prototype can follow the method proposed in [18].

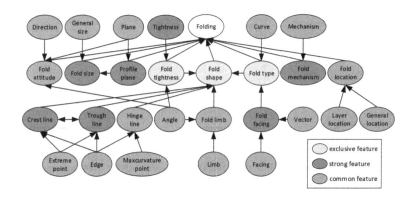

Fig. 5. The Bayesian network example of the folding event pattern.

4   Events Ordering

One of the main steps of ESI is to determine the temporal relations of the events according to the spatial relations of the observed objects. First, the binary temporal relations between events needs to be clearly defined. Although all tectonic events take place in a time interval, the formation of geological models takes a long time. When the occurrence time of events is very short relative to the history of the entire model, they can be considered to occur at a certain moment. For example, people are accustomed to not considering the fault formation process, and the formation time of the stratum is still significant compared to the history of the entire model. So in the definition of time relations, we must not only consider the relations between two time periods, but also the relations between a time period and a moment and two moments to account for the completeness. Therefore, time relations can be divided into three categories: interval to interval, interval to moment, and moment to moment. Allen has already introduced an interval-based temporal logic [2]. The whole temporal relation series are expressed in Fig. 6. The spatial topological relations can be extracted according to the positional status of the point clouds. The GIS software *ArcGIS* provides the *Aggregate Points* tool to create polygons around clusters of proximate points, which can be used to find the boundaries of point clouds. By computing the intersections of these closed boundary polygons, the spatial topological relations of interfaces are obtained.

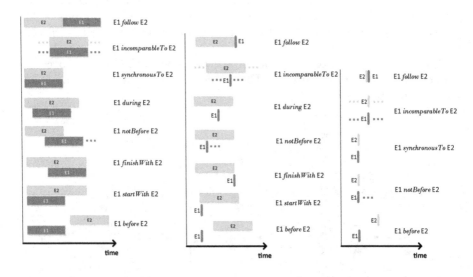

Fig. 6. Binary relations that describe the temporal order between events.

According to the correspondence between temporal relations and spatial topological relations of geological interfaces as shown in Fig. 7 (topological relations are described in [29]), the temporal relations of objects can be derived from the spatial relations. DOLCE states that there are different ways for objects and events of being in time: existing in time vs. occurring in time. So there should be a conceptual transformation from the temporal relations of objects to the temporal relations of events. Previously we have performed event recognition on each object, and each object has been mapped to the events that create it according to Table 1. An object may map to two types of events: the event that generated the object (called the formation event) and the events that deformed the object (called the deformation event). An object must have one and only one formation event, and can have multiple deformation events. Obviously, the object is directly associated with its formation event, so the time order of the formation events is exactly the time order of the objects. For deformation events, they must occur after the formation events (i.e. deformation events *notBefore* formation events). We assume that the deformation events do not affect subsequent formation events.

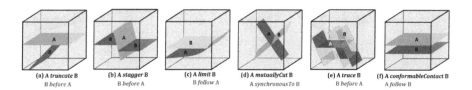

(a) A *truncate* B   (b) A *stagger* B   (c) A *limit* B   (d) A *mutuallyCut* B   (e) A *trace* B   (f) A *conformableContact* B
B *before* A   B *before* A   B *follow* A   A *synchronousTo* B   B *before* A   A *follow* B

Fig. 7. The correspondences of spatial topological relations and temporal relations of geological interfaces.

5 Experiment

In this section, we apply our approach to the raw structural data of Hashan survey (located Zhungeer Basin, Xinjiang province, China) to show the effectiveness of our approach. The recognition of folding event is one of the most challenging tasks in EI, and our approach is the first to consider folding in event interpretation. Due to lack of space, we will show only the recognition of folding events in detail. In the Hashan dataset, the folds are identified as shown in Fig. 8. There are 7 horizons (namely H1–H7) and 4 faults (namely F1–F4) as in Fig. 8(a). The basic properties extracted in Fig. 8(b) are the extreme points on the point clouds of horizons, where the basic properties of H7, H6 and H5 are matched to crest line features and trough line features by prototype matching. According to the Bayesian network in Fig. 5, the prototype matching results lead to the matching of folding event pattern of H7, H6 and H5 (related to the specific conditional probabilities in the Bayesian network).

When combining tectonic events into tectonic sequences, our general idea is first to find the formation sequence of strata. Because the formation of strata (i.e. the formation of horizons) are the most basic tectonic events. Intuitively, other events occur on the basis that the strata have formed. Then find the relationships between other events and the formation events of strata. The horizon sequence is a linear sequence without branches because the relations between strata formations are determinate (not *incomparableTo*), and the process of events does not overlap. Steps are as follows:

1. According to the geologic age of each stratum (usually provided in raw data), the order of the horizons can be obtained.
2. Extract the spatial topological relations among interfaces.

3. According to the spatial topological relations between the interfaces, the pre-order and post-order surfaces of horizons and unconformable surfaces can be determined.
4. Determine whether the surfaces are folded surfaces. If the surface is folded, insert a folding event after the surface formation event.

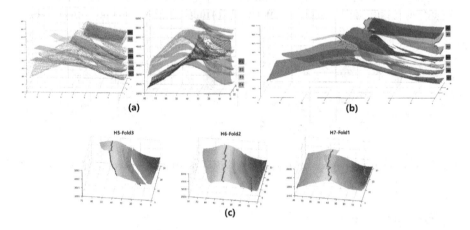

Fig. 8. The recognition of folding events in Hasan survey. (a) The raw point cloud data of geological interfaces. (b) The extracted basic properties of horizons. (c) The identified crest lines (connected red dots) and trough lines (connected green dots). (Color figure online)

We finally obtained the tectonic sequence of Hashan survey as shown in Fig. 9. The time order cannot be determined between parallel events, like F1, F2, F3 and F4. Symbols H1-H7 refer to the sedimentation events that form the corresponding horizons. Symbols F1-F4 refer to the faulting events that form the corresponding faults. Symbols Fold1-Fold3 refer to the folding events that deform H7, H6 and H5.

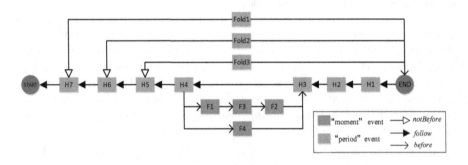

Fig. 9. The tectonic sequence of Hashan survey.

6   Conclusions

In this paper, we present a knowledge-based approach for the event sequence interpretation task in structural geomodels. Its supported by event patterns that instantiated from an ontology-based KR model. When working in event interpretation problems, we face that the interpretation of tectonic events is a kind of cognitive task without clear protocols, which relies heavily on expert knowledge. Therefore, we proposed a KR meta-model to make personal knowledge better available for computers to address issues of subjective uncertainty of manual interpretation. Our KR meta-model is built on the constraints of the foundational ontology DOLCE, and the content is provided by specific domain ontologies. Thus, our work explores the roles that foundational ontologies played in problem solving methods involving visual information. We applied the proposed model to build a well-founded representation of the knowledge about the geometric performances of tectonic events. Instances of the proposed meta-model are called event patterns that are used to recognize the evidences of event occurrences.

References

1. Abel, M., Silva, L.A., De Ros, L.F., Mastella, L.S., Campbell, J.A., Novello, T.: Petrographer: managing petrographic data and knowledge using an intelligent database application. Expert Syst. Appl. **26**(1), 9–18 (2004)
2. Allen, J.F.: Maintaining knowledge about temporal intervals. In: Readings in Qualitative Reasoning About Physical Systems, pp. 361–372. Elsevier (1990)
3. Babaie, H.A., Oldow, J.S., Babaei, A., Lallemant, H.G.A., Watkinson, A.J., Sinha, A.: Designing a modular architecture for the structural geology ontology. In: Special Papers-Geological Society of America, vol. 397, p. 269 (2006)
4. Boote, D.R.: The geological history of the Istria 'depression', Romanian Black Sea shelf: tectonic controls on second-/third-order sequence architecture, vol. 464, no. 1, pp. 169–209. Geological Society, London, Special Publications (2018)
5. Borgo, S., Masolo, C.: Foundational choices in DOLCE. In: Staab, S., Studer, R. (eds.) Handbook on Ontologies. IHIS, pp. 361–381. Springer, Heidelberg (2009). https://doi.org/10.1007/978-3-540-92673-3_16
6. Carbonera, J.L., Abel, M., Scherer, C.M.: Visual interpretation of events in petroleum exploration: an approach supported by well-founded ontologies. Expert Syst. Appl. **42**(5), 2749–2763 (2015)
7. Chen, D., Tucker, M.E., Jiang, M., Zhu, J.: Long-distance correlation between tectonic-controlled, isolated carbonate platforms by cyclostratigraphy and sequence stratigraphy in the devonian of South China. Sedimentology **48**(1), 57–78 (2001)
8. Gero, J.S.: Design prototypes: a knowledge representation schema for design. AI Mag. **11**(4), 26–26 (1990)
9. Groshong Jr., R.H.: 3-D Structural Geology. Springer, Heidelberg (2006). https://doi.org/10.1007/978-3-540-31055-6
10. Haproff, P.J.: Tectonic evolution of the easternmost Himalayan collisional system. Ph.D. thesis, University of California, Los Angeles (2018)

11. Hoyes, J., Cheret, T.: A review of "global" interpretation methods for automated 3D horizon picking. Lead. Edge **30**(1), 38–47 (2011)
12. Johansson, I.: Determinables as universals. Monist **83**(1), 101–121 (2000)
13. Johnson, W.E.: Logic. University Press (1921)
14. Kiefer, P., Giannopoulos, I., Raubal, M., Duchowski, A.: Eye tracking for spatial research: cognition, computation, challenges. Spatial Cogn. Comput. **17**(1–2), 1–19 (2017)
15. Latecki, L.J., Lakämper, R., Wolter, D.: Shape similarity and visual parts. In: Nyström, I., Sanniti di Baja, G., Svensson, S. (eds.) DGCI 2003. LNCS, vol. 2886, pp. 34–51. Springer, Heidelberg (2003). https://doi.org/10.1007/978-3-540-39966-7_3
16. Li, J., Li, Z., Chen, F., Bicchi, A., Sun, Y., Fukuda, T.: Combined sensing, cognition, learning and control to developing future neuro-robotics systems: a survey. IEEE Trans. Cogn. Dev. Syst. **11**, 148–161 (2019)
17. Lin, H.Y.S., Liao, H.Y.M., Li, J.C.: A cognitive psychology-based approach for 3-D shape retrieval. In: 2005 IEEE International Conference on Multimedia and Expo. IEEE (2005). pp. 4-pp
18. Lladós, J., Martí, E., Villanueva, J.J.: Symbol recognition by error-tolerant subgraph matching between region adjacency graphs. IEEE Trans. Pattern Anal. Mach. Intell. **23**(10), 1137–1143 (2001)
19. Lomask, J., Clapp, R.G., Biondi, B.: Application of image segmentation to tracking 3D salt boundaries. Geophysics **72**(4), P47–P56 (2007)
20. Malik, J., Binford, T.O.: Reasoning in time and space. IJCAI **83**, 343–345 (1983)
21. Mastella, L.S., Abel, M., De Ros, L.F., Perrin, M., Rainaud, J.F.: Event ordering reasoning ontology applied to petrology and geological modelling. In: Castillo, O., Melin, P., Ross, O.M., Sepúlveda Cruz, R., Pedrycz, W., Kacprzyk, J. (eds.) Theoretical Advances and Applications of Fuzzy Logic and Soft Computing. Advances in Soft Computing, vol. 42, pp. 465–475. Springer, Heidelberg (2007). https://doi.org/10.1007/978-3-540-72434-6_46
22. Mastella, L.S., Abel, M., Lamb, L.C., De Ros, L.F.: Cognitive modelling of event ordering reasoning in imagistic domains. In: International Joint Conference on Artificial Intelligence, vol. 19, p. 528. Lawrence Erlbaum Associates Ltd. (2005)
23. Perrin, M., Zhu, B., Rainaud, J.F., Schneider, S.: Knowledge-driven applications for geological modeling. J. Petrol. Sci. Eng. **47**(1–2), 89–104 (2005)
24. Rey, J., Somoza, L., Martínez-Frías, J.: Tectonic, volcanic, and hydrothermal event sequence on deception Island (Antarctica). Geo-Mar. Lett. **15**(1), 1–8 (1995)
25. Silvey, P.E., Norman, M.D.: Embodied cognition and multi-agent behavioral emergence. In: Morales, A.J., Gershenson, C., Braha, D., Minai, A.A., Bar-Yam, Y. (eds.) ICCS 2018. SPC, pp. 189–201. Springer, Cham (2018). https://doi.org/10.1007/978-3-319-96661-8_20
26. Smalley, I., O'Hara-Dhand, K., Wint, J., Machalett, B., Jary, Z., Jefferson, I.: Rivers and loess: the significance of long river transportation in the complex event-sequence approach to loess deposit formation. Quatern. Int. **198**(1–2), 7–18 (2009)
27. Wu, X.: Directional structure-tensor-based coherence to detect seismic faults and channels. Geophysics **82**(2), A13–A17 (2017)

28. Xu, Y., et al.: High performance large scale face recognition with multi-cognition softmax and feature retrieval. In: Proceedings of the IEEE International Conference on Computer Vision, pp. 1898–1906 (2017)
29. Zhan, X., Liang, J., Lu, C., Hu, G.: Semantic description and complete computer characterization of structural geological models
30. Zhang, Z., Wang, S., Yang, X., Jiang, F., Shen, J., Li, X.: Evidence of a geological event and environmental change in the catchment area of the Yellow River at 0.15 Ma. Quatern. Int. **117**(1), 35–40 (2004)

Author Index

Aït-Ameur, Yamine   263
Alattar, Munqath   1

Beierle, Christoph   22
Benyagoub, Sarah   263
Brijder, Robert   42

Dvořák, Wolfgang   79

Ferrarotti, Flavio   90

Galliani, Pietro   106
Ghilezan, Silvia   122
González, Senén   90
Grant, John   140
Gyssens, Marc   42

Haldimann, Jonas   22
Hu, Guangmin   318

Ivetić, Jelena   122

Kašterović, Simona   122

Lu, Cai   318

Mahmood, Yasir   157
Medved, Dennis   175
Meier, Arne   157
Mittelmann, Munyque   191
Murai, Ryo   211

Nilsson, Johan   175
Nugues, Pierre   175

Ognjanović, Zoran   122

Pandžić, Stipe   222
Parisi, Francesco   140
Perrussel, Laurent   191

Rapberger, Anna   79
Ravve, Elena V.   242

Sali, Attila   1
Sano, Katsuhiko   211
Santos, Yuri David   58
Savić, Nenad   122
Schewe, Klaus-Dieter   90, 263
Simpson, Michael   301
Srinivasan, Venkatesh   301

Thalheim, Bernhard   281
Thomo, Alex   301
Turull-Torres, José María   90

Väänänen, Jouko   106
Van den Bussche, Jan   42

Wallner, Johannes P.   79
Woltran, Stefan   79

Xia, Daniel   301

Zhan, Xianglin   318

Author Index

Printed in the United States
By Bookmasters